Modern biology
for first examinations

Modern biology
for first examinations

R. Soper B.Sc., M.I.Biol.

S. Tyrell Smith B.Sc., M.A., Ph.D.

M

Macmillan Education

First published 1979

Published by
Macmillan Education Limited
Houndmills Basingstoke Hampshire RG21 2XS
and London
Associated companies in Delhi Dublin
Hong Kong Johannesburg Lagos Melbourne
New York Singapore and Tokyo

Printed in Hong Kong

British Library Cataloguing in Publication Data

Soper, Roland
Modern biology for first examinations
1. Biology
I. Title II. Smith, S. Tyrell
574 QH308.7

ISBN 0-333-26304-9

Contents

	Preface	vi
I	Introductory chapter	1
1	Nutrition	17
2	Transport	62
3	Other types of nutrition	93
4	Storage	101
5	Respiration	111
6	Excretion and regulation	130
7	Response	146
8	Reproduction	164
9	Genetics	184
10	Growth	203
11	The skeleton and locomotion	215
12	Disease	227
13	Soil	245
14	Ecology	254
15	Lower plants and animals	278
16	Higher plants	305
17	Vertebrate animals	315
18	Answers and discussion	329
	Index	355

Preface

In this book the authors have attempted at all times to integrate plant and animal studies, and to give the reader a dynamic approach to the study of biology.

It is essentially designed to give the student a sound training in the science of Biology and is based on a background of factual material combined with a modern inquiry approach. The examination syllabus of every G.C.E. Board has been considered and the book should cover the material required for any one of these boards, its method being suitable for a wide range of teaching techniques. The inquiry method and the factual material would also be suitable for most C.S.E. board syllabuses. Experiments and investigations are suggested through out the book, and questions are asked of the student on virtually every topic involved. The last chapter consists of Answers and Discussion, where the student can assess his or her success in the analysis of data, evaluation of results, observational techniques and other scientific skills.

The Introductory Chapter is most helpful to students, preparing them for biological techniques such as the use of microscopes and the drawing of specimens. The construction of graphs, and the use of S.I. units, pH values and simple chemical reagents are discussed. These are all essential to an understanding of modern Biology.

Chapters on Genetics and Ecology have been included since these topics are now appearing on most examination syllabuses. Besides presenting the basic facts, these chapters give plenty of suggestions for practical investigation. There are three chapters on classification of animals and plants together with examples of organisms that illustrate the wide variety of life.

Each chapter has been completed by the inclusion of typical examination questions requiring an extended essay-type answer. These test the recall and application of knowledge, experimental design and certain other higher abilities. Objective questions of the multiple choice and structured types are available in a separate book of questions based on the text of this book and written by the authors.

Introductory Chapter

I 1.00 S.I. Units

Historical note
The idea of units based on the decimal system was conceived by Simon Stevin (1548–1620). In the early days of the French Academie des Sciences (founded 1666) decimal units were also considered, but the development and adoption of the metric system followed the French Revolution.

Talleyrand, the statesman, was advised by scientists to establish a system of weights and measures based on the metre as the unit of length and the gramme as a unit of mass. The metre was intended to be one ten-millionth part of the distance from the North Pole to the equator at sea level, and the gramme was to be the mass of 1 cubic centimetre of water at 0°C.

In 1873, the British Association for the Advancement of Science selected the centimetre and the gramme as basic units of length and mass for physical purposes. The second was adopted as the base-unit for time and thus gave rise to the *centimetre-gramme-second* (c.g.s.) system. At the beginning of the twentieth century, practical measurements in this system were replaced by larger units, the metre and the kilogramme. These combined with the second, as the unit of time gave the *metre-kilogramme-second* (MKS) system.

Certain other base units of temperature, electric current and light intensity have been added since, and in 1960 the comprehensive system was named the *International System of Units*, abbreviated to S.I. units in all languages.

I 1.10 Important rules for use of S.I. units

There are three categories of S.I. units:

a) Base units
There are six base units.

Quantity	S.I. unit	Symbol
Length	metre	m
Mass	kilogramme	kg
Time	second	s
Electric current	ampere	A
Thermodynamic temperature	kelvin	K
Luminous intensity	candela	cd

Table I.1 Base Units

Quantity	S.I. unit	Symbol	Expressed in terms of S.I. units or derived units
Force	newton	N	$1N=1\ kg\ m/s^2$
Work, energy quantity of heat	joule	J	$1J=1\ Nm$
Power	watt	W	$1W=1\ J/s$
Quantity of electricity	coulomb	C	$1C=1\ As$
Electric potential	volt	V	$1V=1\ W/A$
Electric resistance	ohm	Ω	$1=1\ V/A$

Table I.2 Derived units

b) Derived units

These are stated in terms of base units, and for some of these derived units, special names and symbols exist. The useful ones in biology are listed in table 1.2.

c) Supplementary units

This third type of S.I. unit is not likely to be used in biology at this level. It includes the units for plane and solid angles.

I 1.11 Decimal multiples and sub-multiples are given below:

Multiplication factor for the unit	Prefix	Symbol
10^{12}	tera	T
10^{9}	giga	G
10^{6}	mega	M
10^{3}	kilo	k
10^{2}	hecto	h
10	deca	da
10^{-1}	deci	d
10^{-2}	centi	c
10^{-3}	milli	m
10^{-6}	micro	μ
10^{-9}	nano	n
10^{-12}	pico	p
10^{-15}	femto	f
10^{-18}	atto	a

Table I.3 Decimal multiples and sub-multiples

Thus:

$$1\ cm^3 = (10^{-2}m)^3 = 10^{-6}m^3$$

$$1\ \text{centimetre (cm)} = 1 \times 10^{-2}\ m$$

$$1\ \text{nanometre (nm)} = 1 \times 10^{-9}\ m$$

It should be noted that the kilogramme is somewhat out of place in the above table since it is a basic S.I. unit. It is sometimes more convenient in the school laboratory to work in terms of grammes and cubic centimetres, but where possible the basic units should be used.

I 1.12 Special units with names not included in the S.I. system

Certain specially named units in biology are in common use but do not form part of the S.I. system. These should be progressively abandoned.

The following can continue to be used:
minute (min) = 60 s; hour (h) = 60 min = 3600 s; day (d) = 24 h.

Quality	Unit name and symbol	Conversion factor
Length	Ångström (Å)	10^{-10} metre = 0.1 nm
Length	micron (μm)	10^{-6} metre = 10^{-3} mm
Volume	litre (*l*)	10^{-3} metre^3 = 1 dm^3
Mass	tonne (t)	10^3 kilogramme = 10^3 kg
Heat energy	calorie (Cal)	4.1855 joule = 4.1855 J

Table I.4 Special Units

I 1.13 Conventions agreed for writing S.I. units

a) Units should be written out in full or using the agreed symbols.

b) The letter 's' is never used to indicate a plural form, thus 10 kg and 10 cm^3, **not** 10 kgs or 10 cms^3.

c) A full stop is **not** written after symbols except where the symbol may occur at the end of a sentence.

d) Capital initial letters are never used for units except in the case of certain units named after famous scientists e.g. N (Newton) and W (Watt). Names of units written in full do not have a capital letter, e.g. metres, kilogramme and second. Even those named after scientists do not have a capital when written, thus newton and watt.

e) When symbols are combined as a quotient e.g. metre per second, they can be written as m/s or ms^{-1}. The use of the *solidus* (stroke) must be restricted to once only. In acceleration it must be m/s^2 and not m/s/s.

f) The use of the raised decimal point 2·1 is **not** correct. The internationally accepted decimal sign is placed *level with the feet* of the numerals, e.g. 2.1.
The comma is no longer used in large numbers to divide them into groups of three digits. A space should be left instead so that the figure appears as 345 423 123 and **not** 345,423,123.

g) The mole replaced the gramme-molecule, gramme atom etc. It is not based upon the kilogramme, which is the S.I. unit, but on the gramme. The mole is the amount of substance which contains as many entities as there are atoms in 12 g of carbon 12. The number of entities in a mole is $(6.022\,169) \times 10^{23}$ and is called the *Avogadro constant.* Molar means *per mole.*

I 1.14 Values of imperial units in terms of S.I. units

Length	1 yd = 0.9144 m 1 ft = 304.8 mm 1 mile = 1.609 344 km	Area	1 in² = 645.16 mm² 1 ft² = 0.092 903 m² 1 yd² = 0.836 127 m²
Volume	1 in³ = 16 387.1 mm³ 1 ft³ = 0.028 31 168 m³	Mass	1 lb = 0.453 592 37 kg
Density	1 lb/in³ = 2.767 99 × 10⁴ kg/m³ 1 lb/ft³ = 16.018 5 kg/m³	Power	1 horse power = 745.700 W

Table I.5 Imperial units in terms of S.I. Units

I 2.00 Use of the hand lens and the microscope

The biologist should only accept the evidence of his senses when investigating organisms. His eyes are the most useful, and they can be helped by instruments which produce a magnified image of the object under investigation.

I 2.10 The hand lens

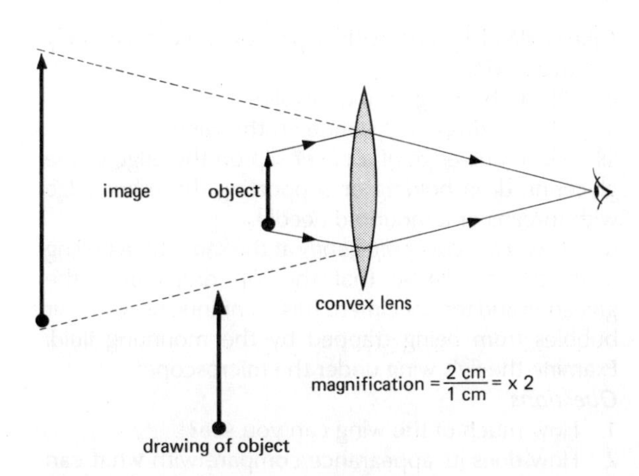

fig I.1 The hand lens and magnification

A hand lens is a convex lens generally mounted in a frame. In order for it to be used as a magnifier, the lens should be placed a short distance from the eye, and the object under investigation brought towards the lens until the enlarged image can be seen. The lens is marked with its magnifying power such as × 8 or × 10 which is an indication of how much larger the image has been made compared with the object. If a drawing is to be made with the help of the lens, then the magnification of the drawing in relation to the size of the object must be calculated. The magnification of the lens bears no relation to the size of the drawing.

Drawing magnification

$$= \frac{\text{linear dimension of the drawing}}{\text{linear dimension of the object}}$$

I 2.20 The microscope

The structure of the microscope is essentially that of an instrument using two convex lenses to obtain a greatly enlarged image of a very small object. Examine a microscope and note its parts, as shown in fig I.2.

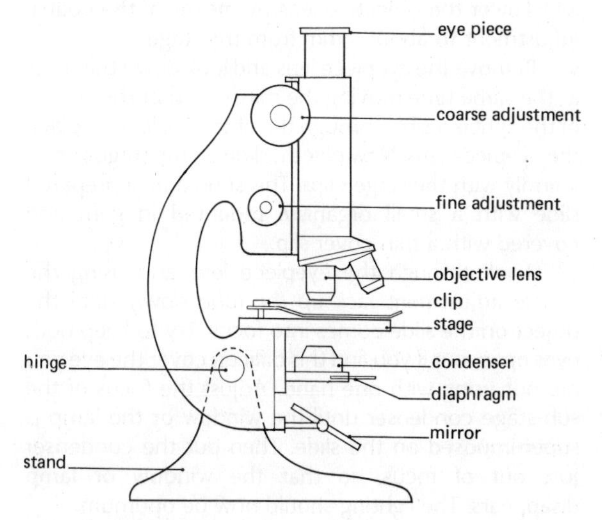

fig I.2 The structure of the microscope

I 2.21 Handling the microscope

A microscope is a very expensive instrument and should be *handled with great care* in order to maintain its precision. Pay particular attention to the following rules when you use the microscope.

i) Always *use two hands* to carry the instrument. Hold the microscope with one hand and place the other under the base.

ii) When putting the microscope on the bench or table, place it *down carefully* in order that the delicate mechanism is not jarred.

iii) *Clean the lenses* by wiping them with lens paper. Never touch the lenses with the fingers or use a coarse cloth for cleaning purposes. The lenses should never be wetted.
iv) Keep the stage of the microscope *dry and clean.* Wipe it immediately if it becomes wet.
v) *Do not tilt* the microscope when using a wet preparation on the slide.
vi) In order to protect the objective lens always cover the object on the slide with a cover *slip.*
vii) Always move the lens *up* when focusing, to avoid breaking the slide.

I 2.22 Using the microscope

i) The instrument should be placed on the bench or table with the arm towards you and the stage away from you. Sit behind the microscope in a comfortable position.
ii) Light must be made to shine through the object: this could be light from a window or an electric lamp on the bench. The mirror under the stage is used to reflect the light. Swing the mirror so that the flat surface is uppermost.
iii) Rack up the sub-stage condenser until it is within 5 mm of the stage.
iv) Lower the objective lens by means of the coarse adjustment to about 5 mm from the stage.
v) Remove the eyepiece lens and look down the tube, at the same time moving the mirror so that the source of the light is visible through the objective lens. Replace the eyepiece lens. Now place a slide on the stage and fix it firmly with the stage clips. This should be a prepared slide with a small organism mounted in gum and covered with a thin cover slip.
vi) Look through the eyepiece lens and using the coarse adjustment *rack up the tube slowly* until the object on the slide comes into focus. Try to keep both eyes open, but if you find this difficult cover the eye you are not using with one hand. Adjust the focus of the sub-stage condenser until the window or the lamp is super-imposed on the slide. Then put the condenser just out of focus, so that the window or lamp disappears. The lighting should now be optimum.
vii) The object can be focused clearly at different levels, according to the thickness of the specimen on the slide, by using the fine adjustment screw.

I 2.23 Investigating the image

Procedure

Cut out a letter 'p' from a newspaper, place it in the middle of a clean slide and cover with a cover slip. Put the slide on the stage of the microscope and focus as described above.

Questions

1 Describe the appearance of the newspaper. How is it different from that seen by the naked eye?
2 Does the letter look larger?
3 Is the letter still the same? Which of the following shapes is similar to the image of the letter 'p'?

p b q d

4 How would you describe the image of the letter 'p' as seen under the microscope?

2.24 Making a temporary slide mount

Procedure

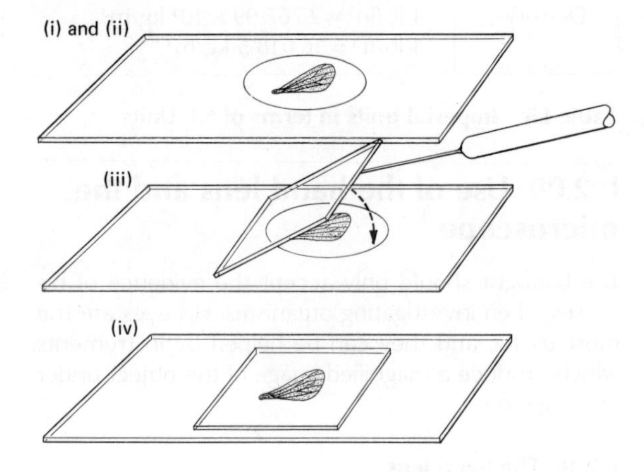

fig I.3 A temporary mount of a fly's wing

Take a dead fly and with a pair of forceps carefully remove a wing.
i) Place the wing on a clean slide.
ii) Place a drop of glycerine on the wing.
iii) Place one edge of a cover slip on the edge of the glycerine drop holding or supporting the other edge with forceps or a mounted needle.
iv) *Lower the cover slip gently* at the same time pulling away the needle so that the slip drops onto the glycerine and the specimen. This technique prevents air bubbles from being trapped by the mounting fluid. Examine the fly's wing under the microscope.

Questions

1 How much of the wing can you see?
2 How does its appearance compare with what can be seen with a hand lens?
3 Focus carefully on the small bars in the wing. Now move the slide to the right whilst observing down the microscope. In which direction does the image move?
4 Now push the slide away from you. Which way does the image move now?

I 2.25 Using the high power of the microscope

Procedure

i) If the microscope is not already adjusted for light and low power work, proceed as in section I 2.22.
ii) Position the object on the slide *in the centre of the field* of view. This is important for tiny structures since the field of view is smaller under high power.

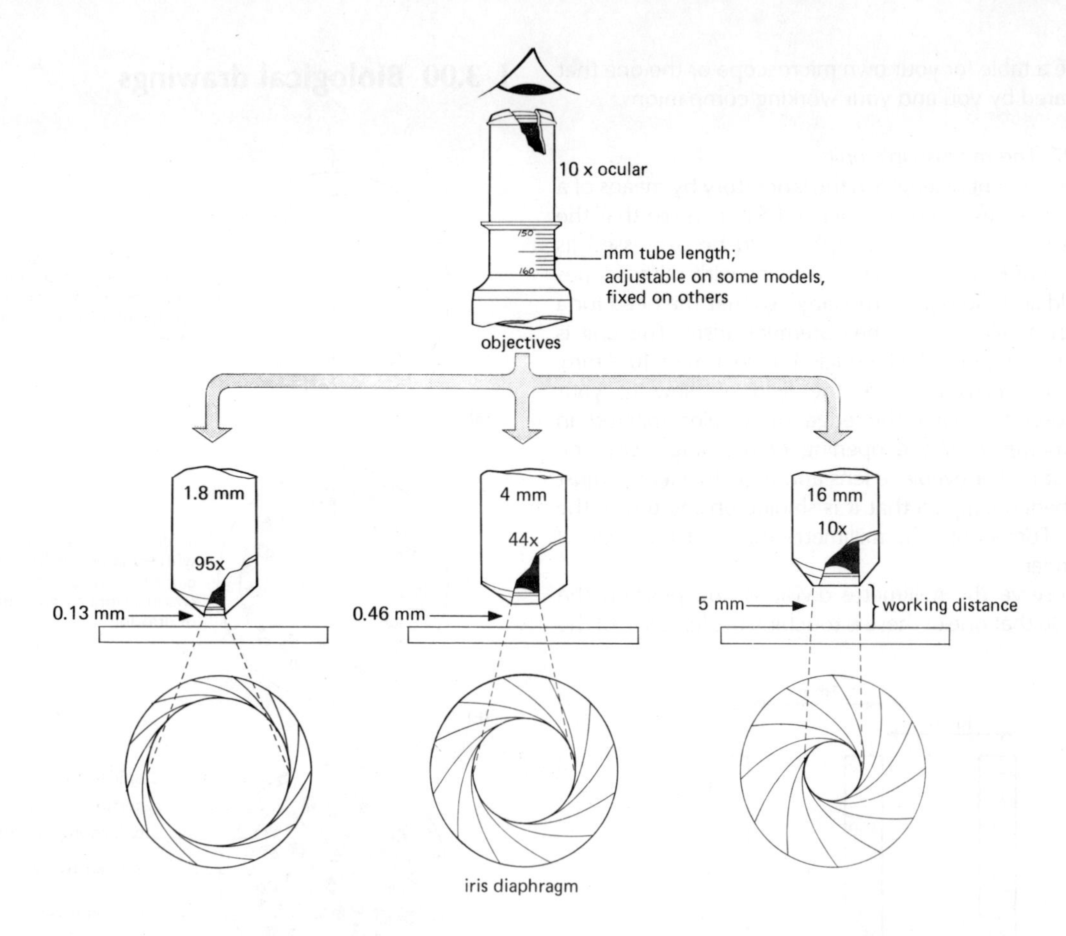

fig I.4 The relationship between the working distance of the objective lens and adjustment of the diaphragm

iii) Bring the high power objective lens into position by swinging the revolving nose piece or screwing the appropriate lens into the tube.
iv) Look at the stage from the side and lower the tube and *objective lens* until it is *almost touching the slide*. It is of help when performing this operation to watch the reflection of the objective lens in the glass slide. Aim to make the image and lens almost meet.
v) Look into the microscope through the eyepiece lens and *slowly* raise the tube by means of the *fine adjustment screws* until the image comes into focus. Move the slide so that the image is in the centre of the field of view (remember your investigation of this aspect in section I 2.24).
vi) The diaphragm can now be adjusted to allow the light to give the best viewing conditions. Fig. 1.4 shows the relationship between objective lens and diaphragm that gives the best results.

I 2.26 Magnification and the microscope
The microscope makes things look much bigger than a hand lens and this magnification can be up to 500 times or more on your microscope. Your microscope may have two or three objective lenses and one or two eyepiece lenses. The top of the eyepiece lens will have its particular magnification written on it, such as ×7 or ×10. Sometimes these figures may be marked on the sides of the lenses. The objective lenses are also marked in this way, and the low power objective that you used in section I 2.22 will probably have ×10 (or 16 mm) as its magnification. There is a high power objective lens which may have ×44 or (4 mm) as its magnification.

The total magnification using any combination of these lenses is obtained simply by multiplying the magnification of the eyepiece lens by that of the objective lens.
Thus:

Objective lens	Eyepiece lens	Magnification of the microscope
×10	×7	×70
×44	×7	×308
×10	×10	×100
×44	×10	×440

Table I.6 Magnification

Make a table for your own microscope or the one that is shared by you and your working companions.

I 2.27 The microscopic unit
Measurement of length in the laboratory by means of a ruler involves the use of derived S.I. units, so that the length of a root for example could be expressed as 6.5 cm or 65 mm. Biologists working with microscopes would find these units too large, so that the need for a much smaller unit of measurement arises. This unit is the *micron* (μm) which equals 1/1000 mm or 10^{-3} mm.

In order to measure the field of view of your microscope, place the edge of a ruler marked in millimetres over the opening of the stage. Use the lowest power eyepiece lens and objective lens. Adjust the bench lamp so that it is shining on the top of the ruler. Focus onto the millimetre marks at the edge of the ruler.

Observe the millimetre divisions and position the ruler so that one of these is touching the left edge of the

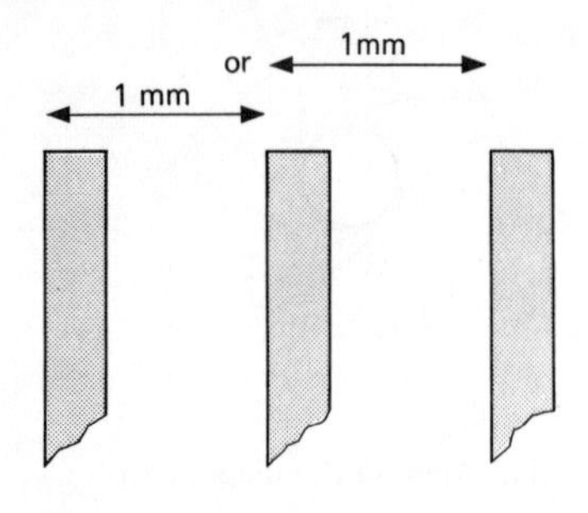

fig I.5 View of marks of a ruler under the microscope

field. Now count the number of millimetres between left and right edges of field at its widest part. Remember 1 mm is the distance from the edge of one mark to the same edge on the next mark.

Questions
1 What is the approximate diameter of the low power field?
2 Convert this figure to microscopic units.
3 Repeat this exercise and complete the following table:

I 3.00 Biological drawings

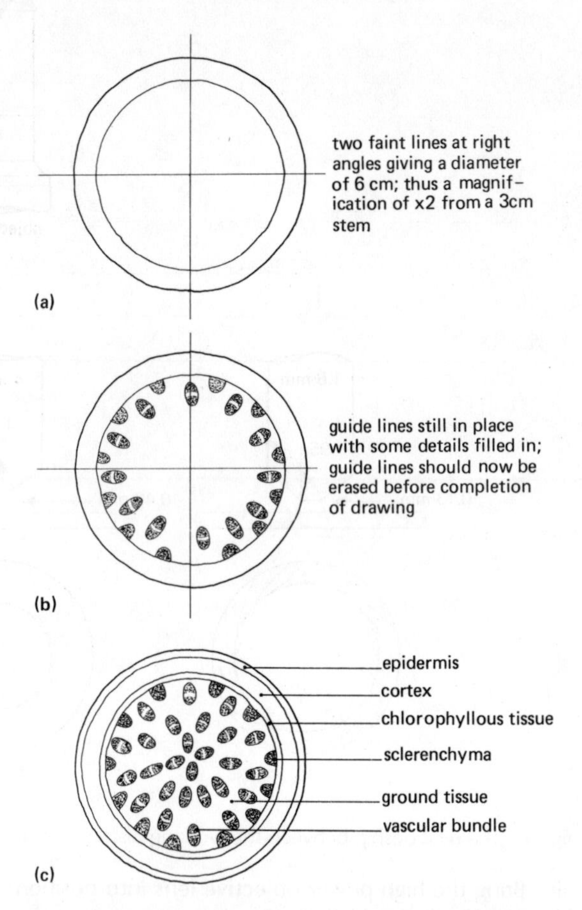

fig I.6 Three stages in drawing a T.S. through the stem of a lily
(a) A preliminary sketch (b) An intermediate stage of drawing
(c) A completed drawing and labels

The biologist must include as part of his work the accurate and careful study of living organisms. His observations are made on the structure and behaviour of animals and plants, and these may include dissections or sections in order to display the internal parts. The observations of structure are improved by the use of a hand lens. These observations are recorded in a *drawing* as a permanent record of what has been seen.

Lens combinations	Eyepiece lens	Objective lens	Magnification	Field of view (μ)
1				
2				
3				
etc				

The drawing made in this way is an important part of the training of a biologist, and there are a number of points to remember.

i) Draw in *lead pencil* only. Do not use coloured pencils, coloured crayons or pens when drawing from specimens during practical work. Coloured crayons or pens may be useful when producing diagrams for theory notes or examination questions.

ii) Keep the *pencil sharp* and use an HB grade. This is hard enough for all thicknesses of line in drawings.

iii) All drawings should be as *large as possible* within the space available. Never take less than half a page, and if possible use a whole page. Leave enough space at either side for labels or annotations, and leave additional space beneath the drawing for a full title.

iv) Observe carefully and *draw first a faint outline* determining the overall magnification. The specimen should be measured in at least two dimensions to ensure that it is drawn accurately in the correct proportions. Never draw less than ×2 for a small specimen such as an insect leg or flower.

v) Fill in the details of your drawing using *firm lines* and *simple outlines. Do not use shading.* Decide where to draw your lines and then do so without lifting the pencil from the paper until the line is completed. A bold single stroke is required, not a succession of half-hearted scratchings. Observe carefully and put all your observations into your drawing.

vi) Take great care with labels and label lines. The latter should *never cross* each other. *Print* the labels *in pencil* then you can rub out mistakes. The label line should be exactly at the centre of the structure labelled. Labels must be correct e.g. never use a plural for a single structure: petal not petals.

vii) Drawings should have a *full title below* to show:

a) the *name of the organism,*

b) the *position* and *type* of section (or dissection) and

c) its *magnification.*

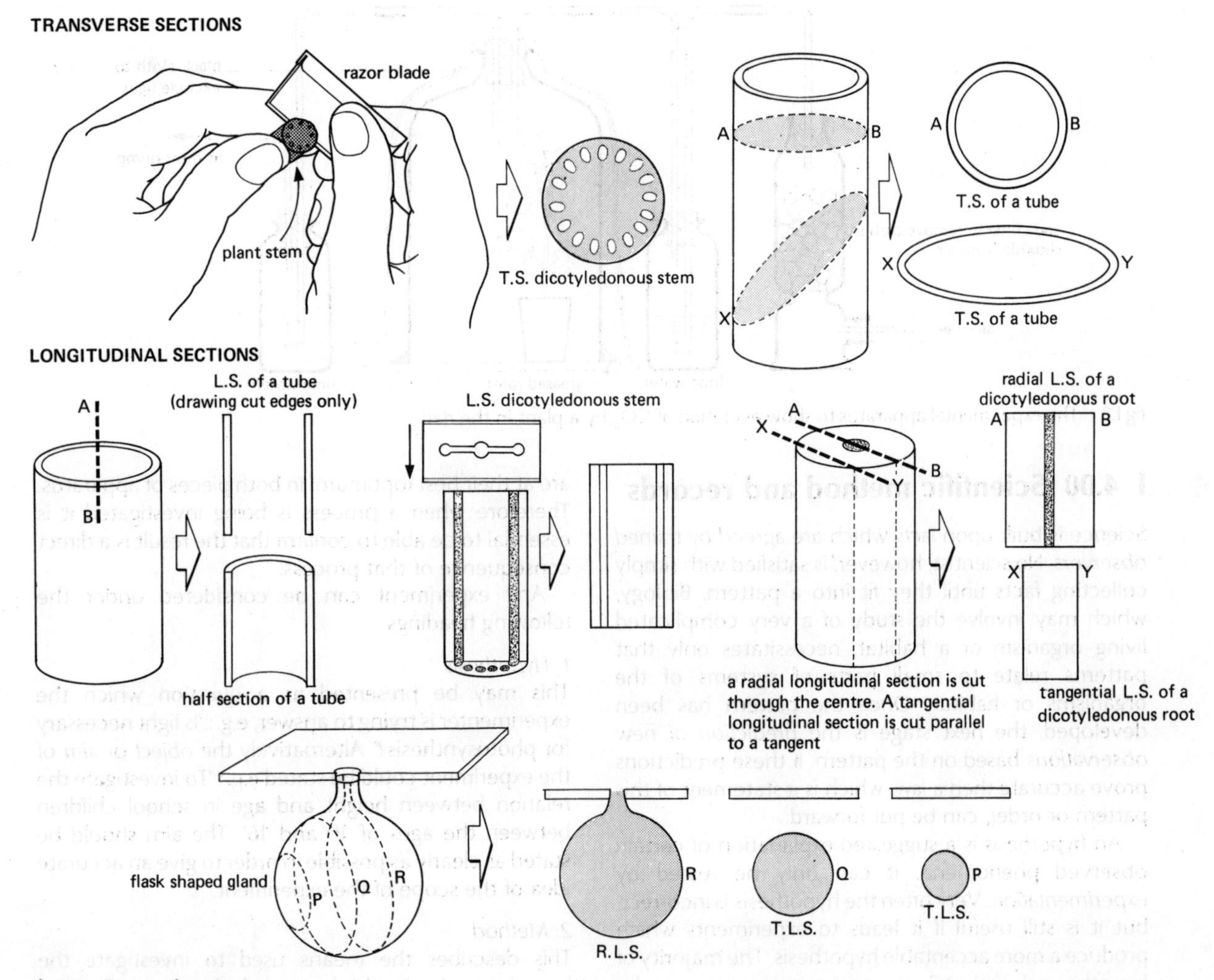

fig I.7 Types of section

viii) Where a large number of small structures are present in a specimen there is no need to repeat all of them in the drawing. This often leads to careless rapid repetition and consequently a poor drawing. Draw with great care from 5 to 10 structures to show their arrangement e.g. fish scales, bristles on the leg of an insect, or filaments on a gill arch.

I 3.10 Types of section

There are a number of different types of section that can be drawn of plant and animal organs, and the most important of these are shown in fig. I.7.

I 3.20 Apparatus

Apparatus should always be *drawn in section* as though the cut edge were being viewed. Thus any tube which permits the flow of liquids or gases will be shown open.

Fig. I.8 shows apparatus designed to investigate the evolution of carbon dioxide by a plant in the dark. Notice that there is a free flow for the air through the soda lime, the lime water, the bell jar containing the plant and finally the second bottle of lime water.

incorrect and so they are discarded.

An experiment is usually performed *to test an hypothesis*, but it may equally produce unexpected facts which give rise to further hypotheses. Thus the scientific enquiry is sent in a new direction. Experiments are *investigations* carried out *under carefully controlled conditions* so that the various factors operating are identifiable. Generally only *one variable factor is observed*. All experiments need to be planned very carefully so that they do in fact investigate the variable with which the hypothesis is concerned.

In order to ensure that conclusions drawn from the experiment are valid, it is essential to carry out at the same time a *control experiment*. The variable factor under investigation is often absent from the actual experiment but present in the control experiment. For example in Chapter 1 section 1.22 the investigation into the uptake of carbon dioxide in photosynthesis involves its elimination in one piece of apparatus, but its presence in the second apparatus. All other factors are at their best (optimum) in both pieces of apparatus. Therefore when a process is being investigated it is essential to be able to confirm that the result is a direct consequence of that process.

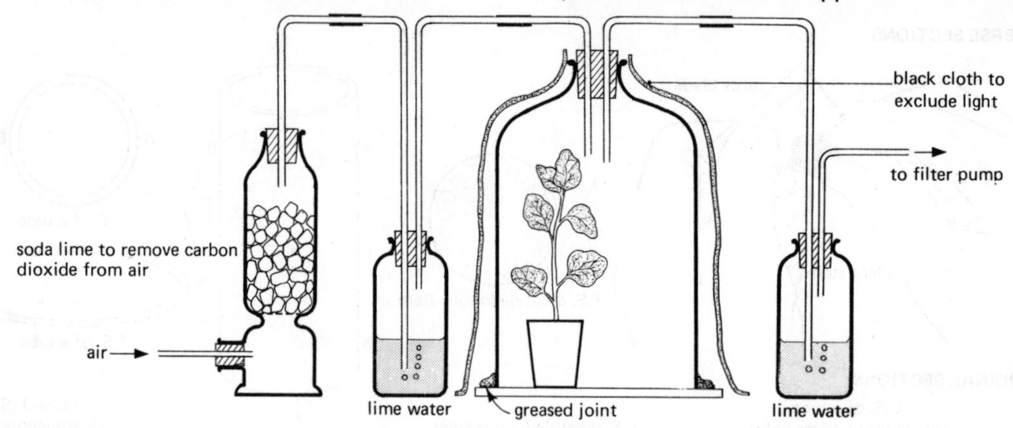

fig I.8 The experimental apparatus to show evolution of CO_2 by a plant in the dark

I 4.00 Scientific method and records

Science is built upon *facts* which are *agreed by trained observers*. No scientist, however, is satisfied with simply collecting facts until they fit into a pattern. Biology, which may involve the study of a very complicated living organism or a habitat, necessitates only that patterns relate to small parts of systems of the organisms or habitat. Once the pattern has been developed, the next stage is the *prediction of new observations* based on the pattern. If these predictions prove accurate then a *law*, which is a statement of this pattern or order, can be put forward.

An *hypothesis* is a suggested explanation of certain observed phenomena. It can only be tested by *experimentation*. Very often the hypothesis is incorrect, but it is still useful if it leads to experiments which produce a more acceptable hypothesis. The majority of hypotheses in scientific experimentation prove to be

Any experiment can be considered under the following headings.

1 Hypothesis

This may be presented as a question which the experimenter is trying to answer, e.g.: 'Is light necessary for photosynthesis?' Alternatively the *object or aim* of the experiment could be stated e.g.: 'To investigate the relation between height and age in school children between the ages of 10 and 16'. The aim should be stated as clearly as possible in order to give an accurate idea of the scope of the experiment.

2 Method

This describes the means used to investigate the hypothesis. It should be a full description of the apparatus, the chemicals used, the living organisms, the measurements made, and any other information that is necessary for any other experimenter to repeat the experiment. **No experiment is valid unless its methods and results can be repeated by another scientist.** The method description should include *drawings* of experimental *apparatus*, together with the *control* apparatus. If apparatus is not used then details of the test group and the control group, measurements and any other relevant facts are required.

3 Results

The results are always expressed as a *series of measurements* or *observations*, and it is important that laboratory notebooks should include full information. It may be that there are minor results that do not fit in the general pattern and such results should not be overlooked. The degree of accuracy of the measurement should always be stated. Results are often best displayed in tabular form. *Visual interpretation* by *histograms or graphs* are a final part of the presentation.

4 Discussion

The final part of the experiment is the discussion of the results and how they affect the hypothesis. Great care must be taken when generalising from them. It may be that *no* final conclusions can be drawn but *further hypotheses suggested* as a result of this particular experiment must be tested. Thus the discussion may form a basis for further experimentation and not until a whole series of experiments have been performed can conclusions be drawn.

In the chapters which follow, all practical experimental work should be written up in this way. Where questions are asked about experiments, the *answers should be included in the discussion*. The method need not be written out completely since the procedure is clearly described in the text of each chapter. It can save time to refer simply to the page and section of the book where the method is set down. The results and discussions should always be complete because this is your own work. Below is an example of the recording of an experiment by a fourteen year old school boy. This is quite a high standard of writing and presentation.

Experiment 8.21

Hypothesis

In this experiment we were trying to find out whether or not plants give out water into the atmosphere after taking it in.

Method

We put a very leafy plant in a large flask of water with a layer of oil on the top of the water. We then put the flask under a bell jar ~~and~~ so that any water which might be given off by the plant would condense on the sides of the bell jar. Also a thin layer of vaseline was put around the base of the bell jar and so the water could not escape (this applied to the water vapour also).

A control experiment was set up using identical apparatus to that described, but no plant was placed in the flask.

Apparatus

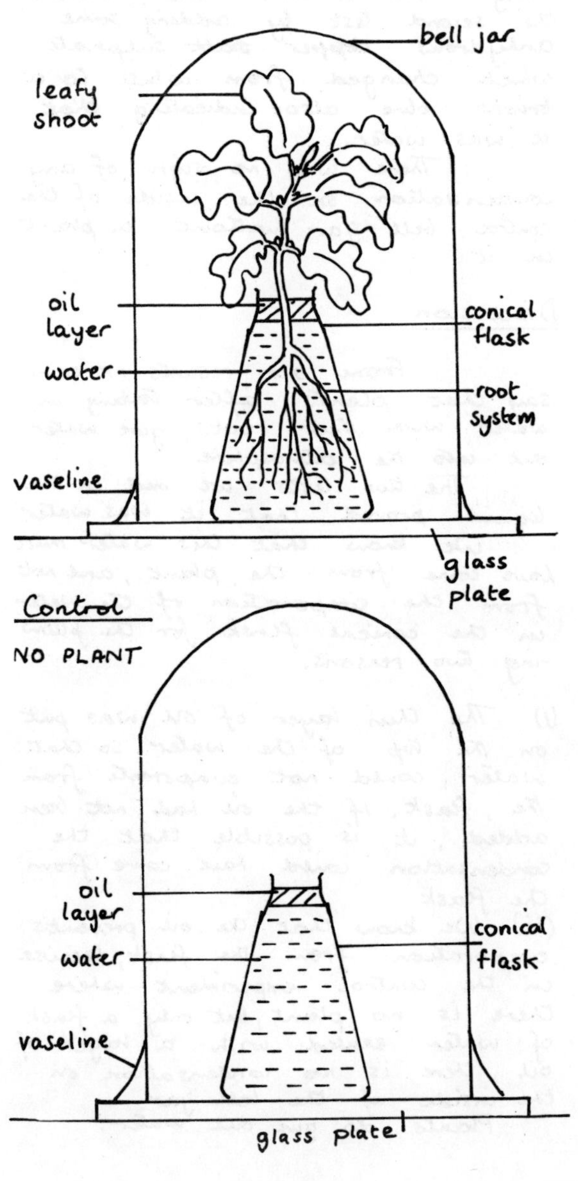

Results

The following week we looked at the bell jars, and condensation had been formed on the inside of the bell jar containing the plant. We thought that the liquid was water, but we could not be certain and so we carried out two tests on it.

The first test was to put the liquid in a small dish and we also put some cobalt chloride paper in with it. The paper changed from blue to pink, indicating that it was water. We did the second test by adding some anhydrous copper ~~salt~~ sulphate which changed from white to a bright blue also indicating that it was water.

There was no sign of any condensation on the inside of the control bell jar without a plant in it.

Discussion

From our results we can say that plants, after taking in water with their roots, give water out into the atmosphere.

The two tests we did on the liquid proved that it was water.

We know that this water must have come from the plant, and not from the evaporation of the water in the conical flask for the following two reasons.

(i) The thin layer of oil was put on the top of the water so that water could not evaporate from the flask. If the oil had not been added, it is possible that the condensation could have come from the flask.

(ii) We know that the oil prevents evaporation from the flask, because in the control experiment where there is no plant, but only a flask of water sealed with a layer of oil, there is no condensation on the inside of the bell jar.

Plants do give out water.

I 4.10 Representing data obtained from experiments

An experiment needs to be *repeated several times* to ensure that results obtained are consistent, but when working in class, there is seldom time for this to be done. The same result can be arrived at when the experiment is performed by *several groups* of pupils in a class. The results of the class can be tabulated on the blackboard, and used for analysis to test the hypothesis.

Suppose that the heights of all the pupils in the class are measured, and they are to be recorded and analysed to find out how they vary. The results would first be set out as follows:

Pupil	Height (cm)	Pupil	Height (cm)
1	132	19	146
2	140	20	128
3	135	21	145
4	129	22	136
5	133	23	135
6	127	24	142
7	134	25	143
8	139	26	121
9	141	27	148
10	130	28	134
11	136	29	127
12	123	30	138
13	139	31	136
14	132	32	130
15	137	33	137
16	131	34	129
17	133	35	144
18	125	36	126

Table I.7 Height range of 36 students

Height (cm)	Frequency f	Total in each group
120–122	1	1
123–125	11	2
126–128	1111	4
129–131	~~1111~~	5
132–134	~~1111~~ 1	6
135–137	~~1111~~ 11	7
138–140	1111	4
141–143	111	3
144–146	111	3
147–149	1	1

Table 1.8 The frequency of occurrence of each class of height

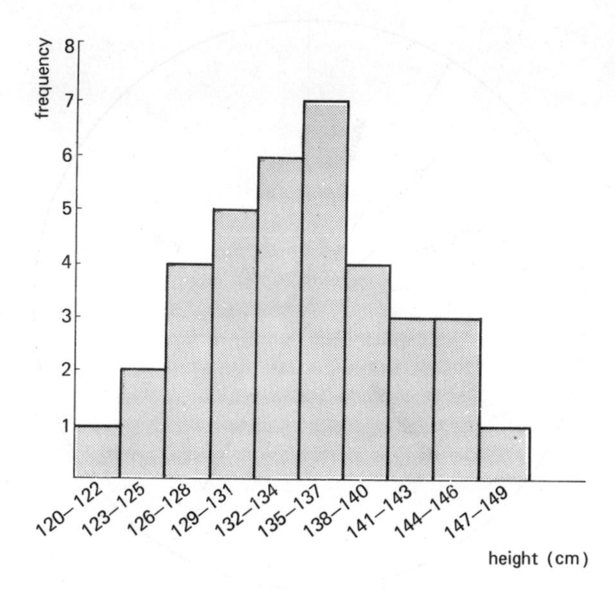

fig I.9 A histogram of heights

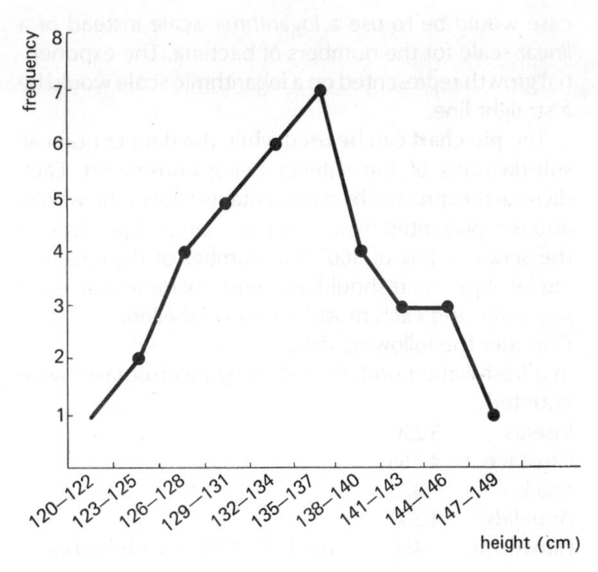

fig I.10 A graph of heights

The range of heights is from 121 cm to 148 cm which gives 28 groups at 1 cm intervals. This is rather a large number and so for simplicity they should be classified into *ten groups* each having *three heights*, e.g. 120 to 122. The number of groups will depend upon the total number of measurements, but as a general rule there should be between 6 and 20 groups.

From the data in table I.7 we can construct the following table I.8 which shows the number of pupils in each height group. Notice that the fifth mark under frequency is used to cross out the previous four, giving a group of five marks.

We now have the *frequency of occurrence* of each height group. One method of displaying this frequency is by a **histogram** or **bar chart.** The groups are shown along the horizontal axis and the frequency on the vertical axis. Rectangles are constructed whose areas are proportional to the frequency in each group. The bases of the rectangles are equal to the group widths.

A second method is by plotting a **graph**. In this case the data in table I.8 is plotted with the median measurement of each group recorded as a point on the graph.

In this particular example for pupil heights it is probably better to display the results as a histogram rather than as a graph. If, however, you were plotting the growth of a colony of bacteria by counting the number of cells at stated intervals, then the graph would be the better method. The resulting curve, shown in fig I.11, is called an **exponential curve.**

This graph also has its limitations, since after about two hours the increase in the number of bacterial cells is so rapid that the scale of the graph is unable to cope with these numbers. A more convenient method in this

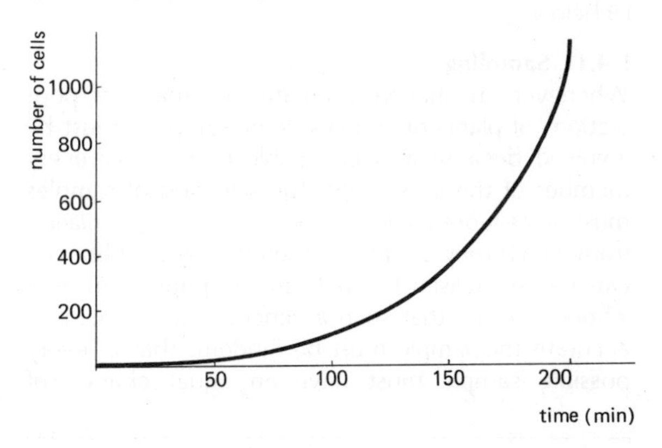

fig I.11 A graph of the increase of bacterial cells

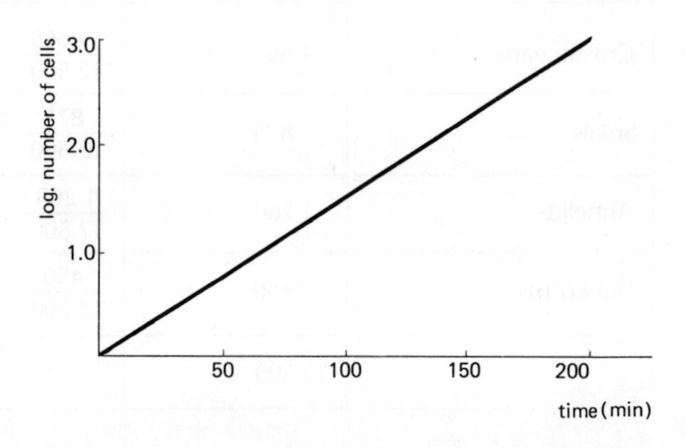

fig I.12 A log graph of the increase of bacterial cells

case would be to use a *logarithmic* scale instead of a *linear* scale for the numbers of bacteria. The exponential growth represented on a logarithmic scale would be a straight line.

The **pie chart** can be used when the data contain all sub-divisions of the subject being considered. Each division is expressed by a percentage (P%) of the whole, and is represented by a sector of a circle. The angle of the sector is P% of 360° (the number of degrees in a circle). A pie chart should *not consist of more than eight segments* and each must be clearly labelled.

Consider the following data:

In a freshwater pond, the following invertebrates were counted.

Insects 5 250
Crustaceans 4 665
Snails 875
Annelids 1 260
Flatworms 450 Total of 12 500 invertebrates

Figure I.13 is a pie chart drawn to illustrate the data. The angles of the sectors were calculated as shown in table I.9 below.

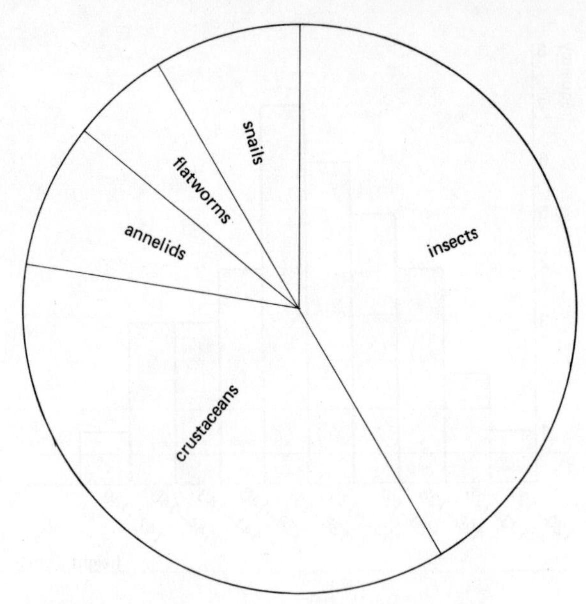

fig I.13 A pie chart based on data in table I.9

I 4.11 Sampling

Whenever quantitative estimates are made of populations of plants or animals, some sampling must be involved because it is impossible to consider every member of the population. The selection of samples must be *as representative as possible of the population* from which they are taken. Then the results obtained can be generalised to apply to the population as a whole. In order that such a generalisation should be accurate the sample must be random; that is, every possible sample must have an equal chance of selection. The choice of each member of the sample must not be influenced by previous choice.

Various factors must be considered before samples are selected, and the most important is the *kind* of information that is required. In this respect the sample must be of one type (*homogeneous*). The class of pupils measured in table I.7 was a *year group* in a school, but there is *no indication* of their *age or sex* in the table. In order to be a homogeneous sample they should be all female (or all male) and all of the same age.

The curve obtained from this sample, shown in fig I.10, is approximately a *bell-shaped* curve, but this

Type of invertebrate	Frequency	Fraction of the total invertebrate frequency	Angle of sector
Insects	5 250	$\frac{5\ 250}{12\ 500}$	$\frac{5\ 250}{12\ 500} \times 360 = 151°$
Crustaceans	4 665	$\frac{4\ 665}{12\ 500}$	$\frac{4\ 665}{12\ 500} \times 360 = 134°$
Snails	875	$\frac{875}{12\ 500}$	$\frac{875}{12\ 500} \times 360 = 25°$
Annelids	1 260	$\frac{1\ 260}{12\ 500}$	$\frac{1\ 260}{12\ 500} \times 360 = 36°$
Flatworms	450	$\frac{450}{12\ 500}$	$\frac{450}{12\ 500} \times 360 = 14°$
	12 500		360°

Table I.9 The method of calculating the sectors of a pie chart

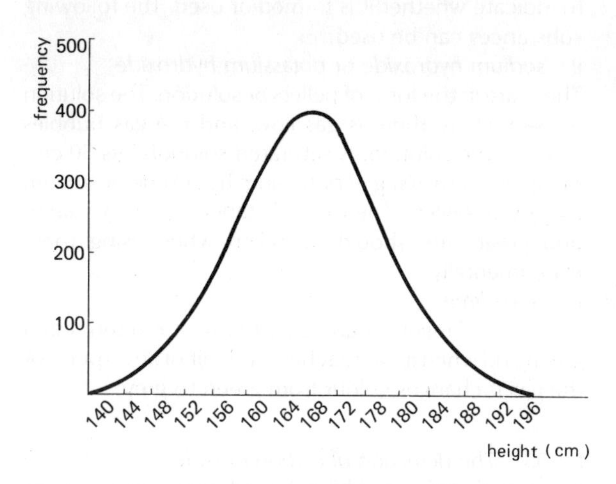

fig I.14 A graph of frequency against height produced as a result of taking a larger sample

would have been a better curve if the sample had been larger, say 100 pupils. Samples of the order of 1 000 or 10 000 are even more valid, and would produce a graph of frequency against height as shown above in fig. I.14. The sample taken in table I.7 can be seen therefore to be too small and not reliable on the basis of age or sex. This raises the further point of the reliability of estimates or generalisations based on a few results. As was stated at the beginning of this section, experiments must always be repeated and plenty of results obtained in order to give a *more reliable answer*.

I 5.00 Elementary chemistry

I 5.10 Elements, mixtures and compounds

An **element** is a substance that cannot be broken down into other, different substances. The smallest particle of an element that can normally exist is called an **atom**. For any one element, the atoms are all similar to each other and different from the atoms of other elements. About 100 different elements have been identified, and each has been given a different symbol: for example, carbon (C), chlorine (Cl), iron (Fe), hydrogen (H), iodine (I), oxygen (O), nitrogen (N), sodium (Na), phosphorus (P) and sulphur (S).

Compounds are substances formed by the combination of two or more elements in a way that causes them to change chemically. The smallest particle of a compound is called a **molecule**, and this consists of atoms of two or more different kinds.

The scientific name for common salt is *sodium chloride*. This is a compound consisting of the two elements sodium (a metal) and chlorine (a gas), which have undergone a considerable change on combining. Thus a molecule of sodium chloride (NaCl) consists of one atom of sodium joined to one atom of chlorine.

Water is a compound of the two gases hydrogen and oxygen. Each water molecule consists of two atoms of hydrogen and one of oxygen, and so the formula for water is H_2O.

The term **mixture** is applied when two substances are mixed together but do not change chemically. Thus, if a spoonful of salt and a spoonful of sugar are placed in a beaker and shaken up, they form a mixture and not a compound. Similarly, **air** is a mixture of different gases, principally nitrogen and oxygen (see section 5.01 – content of air).

I 5.20 Water

Water is the most common and in many ways the most important substance to be found in living systems. The human body consists of about 65% water. From where do we obtain this water, and why is it so important?

The obvious answer to this first question is that we obtain it by drinking. This is certainly true, but in addition much of our food contains a high proportion of water: certain vegetables are 85–90% water and many desert animals never drink but obtain all their water in this way.

The answer to the second question is more complex and is related to a number of very special properties which water possesses.

1 Water is a very effective **solvent**. This means that a large number of substances dissolve in it and thus it is a medium in which all the important chemical reactions of the body can take place.

2 Water has a **high specific heat capacity**. In other words, quite a lot of heat energy is required in order to raise the temperature of a given quantity of water by a small amount. This enables organisms to be relatively independent of fluctuations in climatic conditions and also prevents them from overheating as a result of chemical reactions taking place within their bodies.

3 Water is **liquid** at the temperatures found over much of the earth's surface. It does not freeze at too high a temperature or boil at too low a temperature. Thus, as well as ensuring that enzymatic reactions (which operate most effectively at 35–40°C) can take place in a liquid medium, it is suitable for locomotion by swimming animals, for transport of materials within organisms and for the support of those plants that rely upon turgor pressure for rigidity.

4 Water **expands on freezing**. Like many other substances water contracts on cooling, but contraction stops at about 4°C. Thereafter it expands and thus a given volume of water at just above 0°C weighs less than the same volume at 4°C.

Thus cold water rises to the top of lakes and oceans in winter and forms ice there first, giving some protection to the organisms at greater depths. If this were not the case, arctic lakes would freeze solid and be barren of life.

5 Water has a **high surface tension**. This is the tendency of the surface of a liquid to contract to a minimum area. It is an important factor in the water-retaining properties of soil and in the movement of water upwards through the stems of higher plants.

6 Finally, water has a **neutral pH**: it is neither too acid nor too alkali for many of the reactions taking place within living cells. Linked to this is the fact that it is a major source of hydrogen and oxygen for the body's reactions.

I 5.21 Absorption of water

It may be necessary to keep substances dry or to provide a situation in which there is little or no water in the atmosphere. This is especially necessary in hot, humid climates where delicate instruments may be affected by the growth of fungi under these conditions. The substances that can be used are *calcium chloride* and *silica gel*. A third substance, rather more dangerous but equally effective, is concentrated *sulphuric acid*.

I 5.22 The detection of water

The test reagents used to determine whether or not a colourless liquid is water are *anhydrous copper sulphate* and *cobalt chloride*. Anhydrous copper sulphate is a *white* powder which turns *blue* when wetted. Cobalt chloride is generally used after it has been soaked into a filter paper and dried. When exposed to water or water vapour it changes colour from *blue* to *pink*.

The final tests for the colourless liquid which is suspected of being water are the physical ones to determine whether it boils at 100°C or freezes at 0°C, and whether it has a density of 1 g/cm³.

Usually, there is not time in the school laboratory to complete these tests and so the chemical confirmation, must be sufficient.

I 5.30 Gases in experimental work

I 5.31 Absorption of oxygen

In order to absorb oxygen during experimental work, the liquid used is *potassium pyrogallate* solution. This must be kept under liquid paraffin in order to prevent it absorbing oxygen from the air. It is made by adding *pyrogallol* (125 g) to 65 cm³ of boiled and cooled tap water and 850 cm³ of saturated *potassium hydroxide solution*. This should be transferred to a small jar and covered with liquid paraffin.

This liquid will absorb carbon dioxide as well as oxygen but, since it is normally used in an experiment where carbon dioxide has already been extracted, this dual role is not important.

I 5.32 Absorption of carbon dioxide

It is often necessary during experiments on respiration and photosynthesis to absorb carbon dioxide in order to indicate whether it is formed or used. The following substances can be used:

i) *sodium hydroxide* or *potassium hydroxide*

These are in the form of pellets or solution. The solution is used where there is gas flow, and the gas bubbles through the solution. A saturated solution has 50 cm³ of tap water to 100 g of potassium hydroxide or sodium hydroxide pellets. The two substances are *very caustic* and great care should be taken when using them experimentally.

ii) *soda lime*

This is less dangerous and easy to use. It is often tinted green and when it has reached the limit of absorption of the gas it changes colour from green to grey.

I 5.33 The detection of carbon dioxide

If carbon dioxide is bubbled through *lime water (calcium hydroxide)* it causes very small chalk particles to be set free in suspension and the lime water develops a milky appearance.

A more delicate method of detecting very small changes in carbon dioxide concentration involves using *bicarbonate indicators*. These are indicators of the hydrogen ion (pH) values, which vary according to the acidity of the substance. These indicators are particularly useful in respiration and photosynthesis experiments. A bicarbonate indicator appropriate for these experiments is made up as follows:

Dissolve 0.2 g of *thymol blue* and 0.1 g of *cresol red* in 20 cm³ of *ethanol*. Weigh out 0.84 g of *sodium hydrogen carbonate* and dissolve it in 900 cm³ of de-ionised or distilled water. Add the dye solution to the salt solution in a graduated flask and make up to 1 litre with pure water. Glassware must be clear of all traces of dust or dirt.

A working solution of this stock is made by pipetting 25 cm³ into a 250 cm³ graduated flask, and making up the volume to 250 cm³ with distilled water. Before the reagent is used, air from outside of the laboratory must be bubbled through by means of a filter pump. This will stabilise the colour at deep-red in the bottle or orange red in test tubes. The colour indicates the concentration of carbon dioxide in the air at about 0.03% or 300 parts per million. When this bicarbonate indicator solution is placed in any situation where carbon dioxide is added or withdrawn, it will give a colour change as follows:

Less CO_2 ←	0.03% CO_2	→ More CO_2
purple	orange-red	yellow
pH 9	8	7

This change is caused by the carbon dioxide forming more, or less, *carbonic acid* thus decreasing or increasing the pH value.

I 5.40 Acids, bases and pH

Acids are a large group of chemicals which have the property of giving up *hydrogen ions* (H^+). This capacity is responsible for certain chemical reactions by which acids may be identified. For example, when an acid solution comes into contact with a metal, bubbles of hydrogen gas are formed.

Bases are the chemical opposites of acids. They attract hydrogen ions and react with them. The best known base is the *hydroxyl ion* (OH^-) which reacts with hydrogen ions to form water (H_2O). This particular reaction reduces the number of hydroxyl ions and hydrogen ions in a solution so that neutralisation occurs. A solution is acidic if a surplus of hydrogen ions is present and basic if a surplus of hydroxyl ions is present.

Very often in biology it is necessary to indicate how acidic or basic a solution, a soil or some other substance might be. In order to express this, there has been devised a numerical scale, *the pH scale*. This scale runs from one to fourteen.

Numbers 1 to 7 indicate decreasing acidity; pH7 is neutral; and numbers 7 to 14 show increasing basicity or *alkalinity*.

More acid ← Neutral → more basic
pH 1 2 3 4 5 6 7 8 9 10 11 12 13 14

A number of dyes change colour at different points on the pH scale and they can be used as *pH indicators*. The dyes can be absorbed onto strips of paper and these, when dipped into a solution, will change colour according to the pH. By comparing the colour with a reference chart, the pH can be determined.

The most versatile indicator is Universal Indicator, a mixture of several different indicators with a bigger range of colour changes than simple indicators such as litmus. The latter turns red in the presence of acid and blue with an alkali, but Universal Indicator shows the following range of colours in response to pH change:

pH 1 2 3 4 5 6 7 8 9 10 11 12 13 14
red pink yellow green blue indigo violet
acid neutral alkaline

Universal Indicator is used as a solution with a dropper or as an indicator paper which can be dipped in the liquid being tested.

Experiment

To determine the pH of biological substances.

Procedure

i) Obtain several substances for testing, such as vinegar, milk, urine, raw egg, lemon juice, orange juice and blood.

ii) Use small pieces of pH paper to measure the pH of the materials provided. Make a table of your results.

iii) Obtain the following liquids: hydrochloric acid (dilute, say 0.01 M), sodium hydroxide (dilute, say 0.001 M) and distilled or deionised water.

iv) Use small pieces of pH test paper to measure the pH of the three liquids in step (iii).

Questions

1 What is the pH of the two solutions of hydrochloric acid and sodium hydroxide?

2 How does the pH of the distilled water compare with that of the other two liquids?

3 How does the pH of lemon juice compare with that of the solution of hydrochloric acid?

4 Dilute the acid and the alkali with distilled water and test with pH paper. What happens to the pH value in each case?

I 6.00 What is life?

The term *biology* is derived from two Greek words; *bios*, meaning life, and *logos*, meaning discourse. Thus biologists are scientists who study living things. But how can we tell living things from non-living things? In some cases it is easy: a horse is living, a window is not. In other instances the distinctions are not apparently so clear-cut. A car, for instance, may appear to show more characteristics of life than a lichen found on a rock, although we know that the lichen is living and the car is not.

In general, however, it is possible to draw up a list of seven processes which are carried out by *all* living things in one way or another.

1 *Nutrition* (see Chapter 1)
All living organisms need food. The division of the living world into plant and animal kingdoms is based on the way in which this food is obtained. In most cases, plants make their own food by means of a process called *photosynthesis*. This is known as *autotrophic nutrition*. Animals, on the other hand, usually obtain food by eating other organisms, breaking them down with digestive enzymes and absorbing the breakdown products. This is known as *heterotrophic nutrition*.

2 *Respiration* (see Chapter 5)
Much of the food obtained by either autotrophic or heterotrophic nutrition is used to provide energy for the other vital processes. Energy is released from food when it is broken down, often by combustion or 'burning' in the presence of oxygen. This process is known as *respiration*; it should not be confused with *breathing*, a term which refers specifically in mammals to the means by which the oxygen is brought, via the lungs, into the blood system.

3 *Excretion* (see Chapter 6)
Just as all life processes require energy, they all result in the production of waste materials. The removal from the body of these waste materials is known as *excretion*. Living organisms excrete a variety of substances; for

example, carbon dioxide and water from the breakdown of sugars and, particularly in animals, chemicals rich in nitrogen from the breakdown of proteins. The removal of undigested food material in the faeces of animals does not qualify as excretion, since this material is not *produced* by the living processes of the animals. Urine, however, contains excreted water and nitrogen.

4 *Response* (see Chapter 7)

The environments inhabited by living organisms are constantly changing. Both animals and plants have the ability (sometimes termed *irritability*) to respond to these changes and thus ensure, to a greater or lesser degree, that they are not affected adversely by them. For example, the dermal blood vessels of a man may dilate in response to a rise in the temperature of his surroundings, thus allowing him to lose more heat by radiation and so maintain a constant body temperature: a potted plant which has been grown outdoors may be seen to grow towards a window if it is brought indoors, a response ensuring maximum light for photosynthesis.

5 *Reproduction* (see Chapter 8)

Every living organism has a limited life span. However, all plants and animals have the ability to pass on life, and thus ensure the survival of the species, by producing new individuals with the same general characteristics as themselves. This process is known as reproduction.

6 *Growth* (see Chapter 10)

While a great deal of the food obtained by living organisms through autotrophic and heterotrophic nutrition is respired to produce energy, part is converted into protoplasm. The formation of additional living tissue in this way is termed growth.

7 *Movement* (See Chapter 11)

One of the features that distinguishes animals from plants is the ability of the former to move from place to place: locomotion. This is necessary for animals in order for them to obtain food, whereas plants are able to make their own food by photosynthesis. However, even plants need to spread their species in order to obtain adequate supplies of mineral nutrients and to avoid loss of light through overcrowding. This is achieved in many cases by the development of spores, seeds or fruits with their own dispersal mechanisms. Moreover, plants exhibit a limited range of movements, such as the closing of daisy flowers at night, that do not actually involve locomotion. In addition, it should be noted that the cell contents of all living organisms are in a constant state of motion, termed *cyclosis*.

These seven characteristics of living organisms are investigated in some detail throughout the following chapters. However, they are only the observable manifestations of a single all-important property of living material (protoplasm): the ability to extract, convert and use energy from its environment, and thus to maintain and even increase its own energy content. In contrast, dead protoplasm and non-organic material tend to disintegrate as a result of the physical and chemical forces of the environment: their energy content thus decreases.

The maintenance by living organisms of an ordered, self-regulating system without net energy loss is referred to as *Homeostasis* and is dealt with in Chapter 6.

1 Nutrition

1.00 Chemical compounds made by living organisms

Biology has moved away from being the study of natural history only, that is a purely descriptive science, to an emphasis on experimentation. This change is reflected in biology at all levels from current research interests in universities to first biology courses in secondary schools. Experimental biology has been aided by the development of physical, chemical and mathematical techniques whereby processes can be followed and analysed in a more quantitative fashion than they have been in the past. One relatively new branch of study is biochemistry which is the chemistry of living things. This involves investigating not only the chemical structure of the compounds of life but also such essential processes as energy exchange in *respiration* and *photosynthesis*, synthesis of protein, the transport of compounds and so on. The first priority therefore is to have some knowledge of the basic chemistry of the compounds found in living organisms. Life as we know it on this earth has made use of the element *carbon* as the basic component of all complex materials that circulate throughout living things.

Carbohydrates

The simplest of these carbon compounds in living organisms are the *carbohydrates*, so called because they are characterised by having carbon, hydrogen and oxygen in their molecules, with the ratio of hydrogen and oxygen in the same proportion as in water. Thus:

$$\text{carbo hydrate} = \underset{\text{carbon}}{C} + \underset{\text{water}}{H_2O}$$

$= C_n(H_2O)_m$ where n and m may be the same or different numbers.

They are extremely important compounds because:

i) they are *sources of energy*, and
ii) they have *structural uses* in plant and animal tissues.

Carbohydrates have a basic molecule (building brick) which can be put together to form much more complex molecules (houses). The building brick molecule of carbohydrates is a *simple sugar* or *saccharide*.

The carbohydrate group of substances can be subdivided according to the number of simple sugar molecules:

1 Monosaccharides – one molecule
2 Disaccharides – two molecules
3 Polysaccharides – many molecules

Monosaccharides
These are the simplest sugar molecules. *Glucose* is an example of a monosaccharide. Its chemical formula is $C_6H_{12}O_6$ and it is commonly found in living cells. *Ribose* is another example of a monosaccharide, with the formula $C_5H_{10}O_5$.

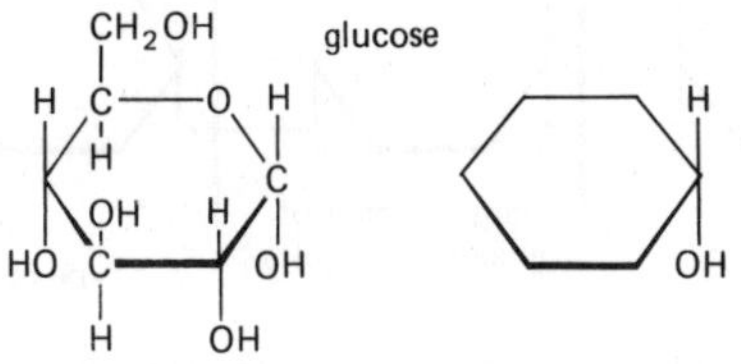

fig 1.1 The formula of glucose and a simplified convention of a glucose molecule

Disaccharides
Sucrose (cane sugar or table sugar), $C_{12}H_{22}O_{11}$ is formed by the joining together of a fructose molecule and a glucose molecule, with the loss of a molecule of water. This method of joining two molecules is called *condensation*.

Maltose $C_{12}H_{22}O_{11}$ is formed by the condensation of two molecules of glucose.

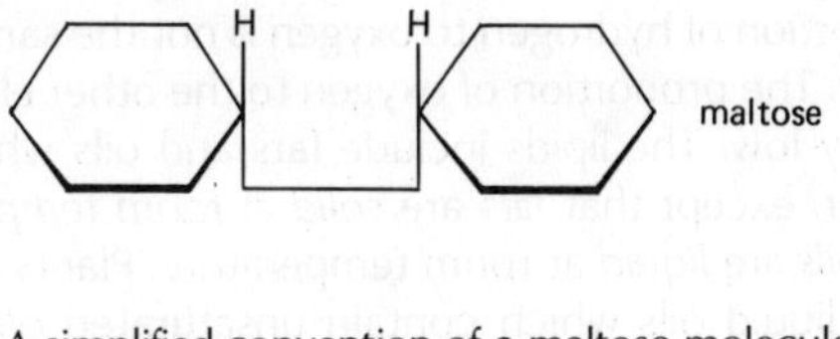

fig 1.2 A simplified convention of a maltose molecule

Lactose (milk sugar) is formed by the condensation of a glucose molecule and another monosaccharide called galactose. Since a molecule of water is lost during condensation the number of atoms in the disaccharide is not exactly twice the number of atoms in a monosaccharide:

$$\underset{\text{glucose}}{C_6H_{12}O_6} + \underset{\text{fructose}}{C_6H_{12}O_6} = \underset{\text{sucrose}}{C_{12}H_{22}O_{11}} + H_2O$$

Maltose and lactose are reducing sugars, but sucrose is a non-reducing sugar.

Polysaccharides
These are the largest carbohydrate molecules and show particularly the two main functions mentioned above:
i) *energy source* – starch and glycogen are used for storage in plants and animals respectively.
ii) *structural molecules* e.g.: *cellulose* forms the cell walls of plants and *chitin* forms the exoskeletons of arthropods.

Starch has the formula $(C_6H_{10}O_5)_n$ where n is a large number. Starch is formed by the condensation of molecules of simple sugar to give chains several thousand sugars in length.

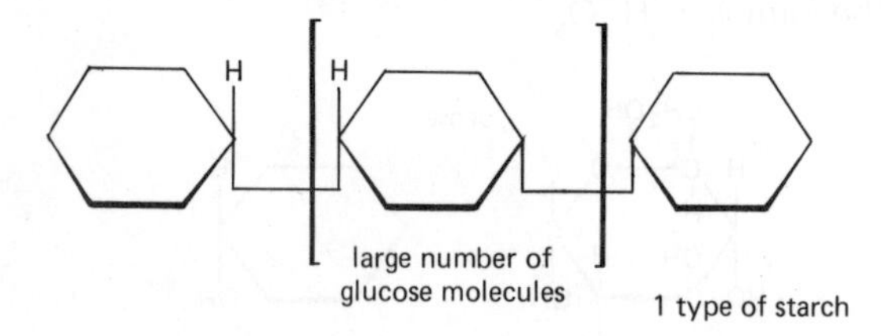

fig 1.3 A simplified convention of a starch molecule

Cellulose, which makes up the cell walls of plants contains many thousands of monosaccharide molecules forming a single long chain molecule.

Chitin is found in the exoskeletons of insects and other animals, as well as in the cell walls of fungi, and again is made up of long chains of monosaccharide units.

Inulin is a storage polysaccharide formed by the condensation of fructose molecules.

Lipids
These are organic compounds which contain carbon, hydrogen and oxygen, as do the carbohydrates, but the proportion of hydrogen to oxygen is not the same as in water. The proportion of oxygen to the other elements is very low. The lipids include fats and oils which are similar, except that *fats* are *solid at room temperature* and *oils are liquid* at room temperature. Plants tend to have liquid oils which contain unsaturated *oleic acid* $C_{18}H_{32}O_2$, whereas animals store solid fats which contain *palmitic acid* $C_{16}H_{32}O_2$ or *stearic acid* $C_{18}H_{36}O_2$.

Fats are made from two types of chemical: *fatty acids* and the alcohol *glycerol*: such a combination is called a *glyceride*. Three fatty acids and glycerol form a *triglyceride* such as *tristearin* $C_{57}H_{110}O_6$. *Waxes*, such as are found on the surface of certain leaves of plants and in beeswax are combinations of fatty acids and other alcohols. Fats, like carbohydrates, are important sources of energy and because of their high energy content and chemically inert nature they are the most economical storage materials for living organisms. Some mammals accumulate food reserves of fat; insects have fat bodies and most plant fruits and seeds have food reserves of oil e.g. oil palm fruits, groundnuts, sunflowers, coconut and castor oil plants. Humans eating more food than their bodies can use, also accumulate fat. Animals that live in cold climates have fat reserves under the skin to provide insulation against heat loss e.g. polar bears, penguins and seals. Tropical animals generally lack fat reserves.

Proteins
The living substance, *protoplasm*, is essentially a solution of *proteins in water*. In addition to carbon, hydrogen and oxygen, all proteins contain *nitrogen*. *Phosphorus* and *sulphur* may also be present. As in polysaccharides, proteins are made up of smaller units forming a long chain molecule. These units are called *amino acids*. *Twenty three* different amino acids are found, and twelve of these are particularly common.

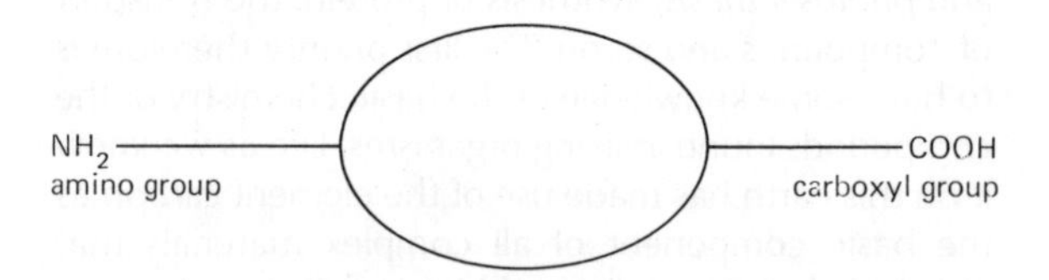

fig 1.4 A simplified convention of an amino acid molecule

Proteins are formed by the condensation of amino acids so that the negatively charged end of one amino acid is attracted to the positively charged end of another amino acid. A molecule of water is eliminated during the process. The bond between two amino acids is called a *peptide bond*. When many amino acids are joined together by this peptide linkage the product is a *polypeptide*. These are joined to form the protein molecule. Below is a list of some amino acids.

Name	Abbreviation
Glycine	Gly
Alanine	Ala
Valine	Val
Leucine	Leu
Aspartic acid	Asp
Glutamine	Glu
Cysteine	Cys
Lysine	Lys

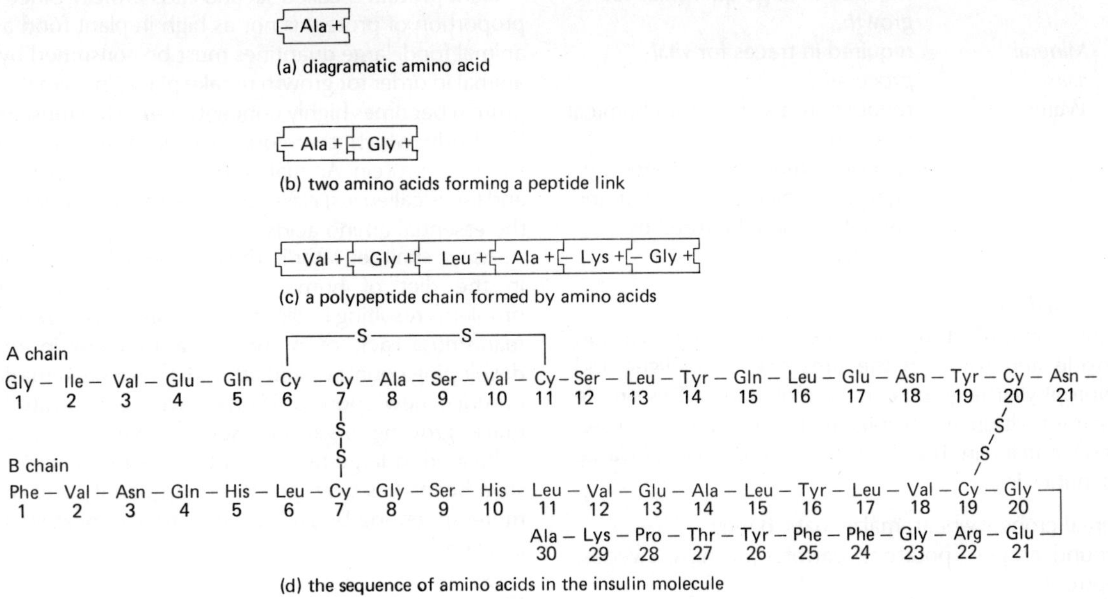

Ala +

(a) diagramatic amino acid

Ala + Gly +

(b) two amino acids forming a peptide link

Val + Gly + Leu + Ala + Lys + Gly +

(c) a polypeptide chain formed by amino acids

A chain

Gly – Ile – Val – Glu – Gln – Cy – Cy – Ala – Ser – Val – Cy – Ser – Leu – Tyr – Gln – Leu – Glu – Asn – Tyr – Cy – Asn –
1 2 3 4 5 6 7 8 9 10 11 12 13 14 15 16 17 18 19 20 21

B chain

Phe – Val – Asn – Gln – His – Leu – Cy – Gly – Ser – His – Leu – Val – Glu – Ala – Leu – Tyr – Leu – Val – Cy – Gly
1 2 3 4 5 6 7 8 9 10 11 12 13 14 15 16 17 18 19 20

Ala – Lys – Pro – Thr – Tyr – Phe – Phe – Gly – Arg – Glu
30 29 28 27 26 25 24 23 22 21

(d) the sequence of amino acids in the insulin molecule

fig 1.5 Polymerisation showing the amino acid sequence in an insulin molecule

Since amino acids can be joined in many different combinations an almost infinite number of proteins can exist. It is fairly certain, however, that living organisms have been economical in the development of new proteins and the number in existence may be much lower than that which is possible theoretically. The amino acids in a protein molecule can form long chains, or the chains can be folded in a complicated fashion to form a *globular molecule*. Recent research has demonstrated the structure of a few of these complex protein molecules, such as *insulin* and *haemoglobin*.

Living processes are controlled by enzymes, and each of these is a protein molecule. The importance of this group of substances in forming the structure and controlling the working of the living organisms is therefore very great. Amino acids and proteins are very sensitive to certain physical changes such as heat. When heated, protein molecules lose their special properties and become *'denatured'*. The coagulated egg-white in a boiled egg is the protein *albumen*, which has become denatured.

1.10 Food

All living organisms require organic compounds for their living processes. The most obvious use of these compounds is for **growth**. Plants are able to manufacture organic compounds from raw materials (carbon dioxide, water and mineral salts containing nitrogen etc.), whereas animals must be supplied with organic compounds in the form of food. Once growth is completed the adult still needs food for replacement and repair of worn out or damaged tissues.

In addition the living organism requires **energy** for its chemical processes, movement and heat production. This energy is produced (released) by burning (combustion) food in the process of **respiration**. Like a machine, every organism must be maintained in good working order. This condition we refer to as **health**.

We can summarise the reasons why plants and animals require food:

1 Growth – the initial increase of cells in organisms to grow to adult size, and the repair of damaged or worn-out tissues.

2 Energy – to provide energy to drive the chemical processes, for mechanical work of muscles, and for the maintenance of body temperature in warm-blooded animals.

3 Health – to afford protection against disease and to provide raw materials for the manufacture of secretions such as hormones and enzymes.

1.11 Classes of food

Animals and plants must be supplied with food or be capable of making their own food from raw materials. The process of making or obtaining food is called *nutrition*.

There are six classes of food:

1 *Carbohydrates* } required in large quantities for
2 *Fats* } *energy* production.

3	*Proteins*	required in large quantities for *growth*.
4	*Mineral salts*	required in traces for *vital processes*.
5	*Water*	required as a *solvent* for chemical reactions.
6	*Vitamins*	specific compounds required by animals for *health*. Most vitamins cannot be manufactured by animals.

Carbohydrates

Plants manufacture glucose by using carbon dioxide and water in the presence of sunlight and chlorophyll (the green colouring of plants). This simple sugar is built up into storage materials such as sucrose, starch and inulin. These substances are present in large quantities in:

Cereal crops – wheat, maize, oats, barley.
Ground crops – potatoes, carrots, parsnips, swedes, beetroot.
Leguminous crops – beans, peas.

Man and animals can make use of these food stores to provide their own carbohydrate requirements. The carbohydrates built up or taken in by living organisms are broken down to release energy. If large quantities of carbohydrates are consumed by animals over and above their energy needs, the carbohydrates are changed into fats and stored in the body. Man throughout the world generally has enough carbohydrate in his diet, except in special circumstances such as drought or flooding, which result in famine.

Fats and oils

Simple carbohydrates manufactured by plants can be changed into oils for storage purposes in fruits and seeds. The food reserves of oil in these structures are used for growth and energy when the seeds germinate and produce the young plants. Animals can feed on oily fruits and seeds and so obtain energy-rich foods.

Fruits and seeds rich in oils include:

peanuts, coconut, maize, cashew nuts, soya bean, sunflower seed.

Proteins

The living material (protoplasm) of plants is manufactured from simple carbohydrates, with the addition of elements obtained from soil salts through the roots. Thus all plant food consumed by animals contains a certain amount of protein, but some plants have a particularly high concentration of protein in certain organs. *Leguminous crops* have a high protein content in their seeds e.g. peas, peanuts, haricot beans and soya beans. In addition sunflower seed, maize, cashew nuts and wheat all contain considerable amounts of protein.

Plant protein is called *second class protein*. Since the proportion of protein is not as high in plant food as in animal food, large quantities must be consumed by an animal in order for growth to take place. In animals, the protein becomes highly concentrated in the muscles of the body which therefore provide a much greater source of protein. Animal protein eaten as food in meat and fish is called *first class protein* because it contains all the essential amino acids.

As we shall see later in the chapter, a lack of protein in the diet of humans causes serious deficiency problems resulting in diseases such as *kwashiorkor* and *marasmus*. Lack of protein is a problem in many developing countries and attempts have been made to produce new sources. Efforts have concentrated on quick growing organisms such as yeast and algae cultivated in large tanks. Plant protein obtained from soya beans has been produced in quantity and made more appetising by giving it a taste like meat such as chicken.

Mineral salts

Chemical analysis of the content of animal tissues produces the following percentages of elements by weight.

Element	%		
Oxygen	65%	93% of the total weight	99.45%
Carbon	18%		
Hydrogen	10%		
Nitrogen	3.0%	6.45% of the total weight	
Calcium	1.5%		
Phosphorus	1.0%		
Potassium	0.35%		
Sulphur	0.25%		
Sodium	0.15%		
Chlorine	0.15%		
Magnesium	0.05%		

Iron, copper, iodine, manganese, zinc, fluorine, molybdenum and others are present only in minute traces – 0.55%.

The diet of an animal must therefore contain these elements. We have seen how carbohydrates, fats and proteins do contain the three major elements constituting 93% of the total weight. The remainder are also present in the diet, particularly in the amino acids, and although they make up only a small percentage they can represent a considerable quantity in a man.

A man weighing 70 kg will contain:

Calcium	1.5%	1 050 g
Phosphorus	1.0%	700 g
Sodium	0.15%	105 g
Magnesium	0.05%	35 g

An analysis of plant structure will give very similar percentages for the same elements: these are absorbed by the plant from the soil.

Element	Source from normal food	Importance to the mammalian body
Nitrogen N	Protein foods, lean meat, fish, eggs, milk.	For synthesis of protein and other complex chemicals, formation of muscle, hair, skin and nails.
Sulphur S	As for nitrogen.	As for nitrogen.
Phosphorus P	As for nitrogen.	For synthesis of protein and other complex chemicals, formation of bones and teeth, formation of ATP. Absence causes rickets.
Iron Fe	Liver, green vegetables, yeast, eggs, kidney.	Forms haemoglobin in red blood cells. Absence causes anaemia.
Calcium Ca	Milk, cheese, green vegetables.	Formation of bones and teeth, necessary for muscle contraction and blood clotting. Absence causes rickets.
Iodine I	Sea fish, and other sea foods, cheese, iodised table salt.	Formation of hormone in the thyroid gland, absence causes goitre and reduced growth.
Sodium Na	Table salt, green vegetables.	Maintenance of tissue fluids, blood and lymph, transmission of nerve impulses.
Potassium K	Vegetables.	Transmission of nerve impulses.
Chlorine Cl	Table salt.	Maintenance of tissue fluids, blood and lymph.

Table 1.1 Mineral elements: their sources and characteristics

Mineral elements – their source and characteristics
These are not taken in as elements but as salts in their ionic form, e.g. sodium chloride (common salt, table salt) as sodium and chlorine ions.

The formation of proteins and other complex compounds in the plant body incorporates the elements into the structure of the plant. Animals acquire their elements by feeding on plants. The facts about mineral salts are given in table 1.1.

Water
The chief constituent of living matter is water which is so familiar that its importance in structure and functioning is often overlooked. Below is the approximate composition of a man weighing 70 kg:

Water	70%	49 kg	85% fluid
Fat	15%	10.5 kg	
Protein	12%	8.4 kg	15% solid parts of cells and support structures
Carbohydrate	0.5%	0.35 kg	
Minerals	2.5%	1.75 kg	

Water is most important as a solvent in living organisms and thus plays a fundamental part in cellular reactions. It has a *high heat capacity*, so that in an animal or a plant, chemical reactions producing considerable

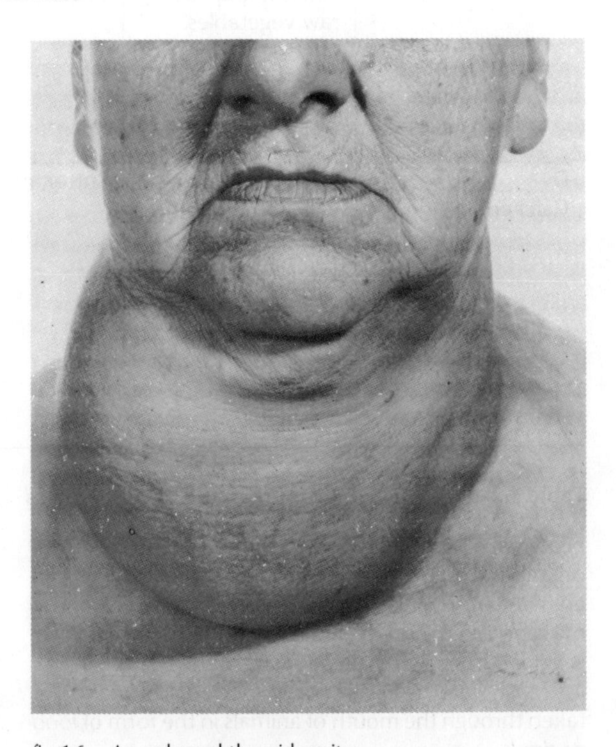

fig 1.6 An enlarged thyroid: goitre

Vitamin	Source from normal food	Special source	Symptom of deficiency	Special notes
A Retinol	Liver, egg-yolk, green vegetables, cocoa, carrots.	Butter, margarine, cod-liver oil.	Sore eyes, reduced night vision, colds and bronchitis, unhealthy skin.	Carotene from plant pigment converted to vit A in intestine walls.
B_1 Thiamine	Unpolished cereals, beans, lean meat, egg yolk.	Bread, milk, kidney.	Retarded growth, lack of appetite in children, nervous inflammation and weakness, paralysis, the disease called beri-beri.	Likely in rice-eating peoples of Asia.
B_2 Riboflavin	As for B_1 plus green vegetables.	As above for B_1 plus Marmite and liver.	Skin disorders, eye and mouth membrane sores, the disease called dermatitis.	
Nicotinic acid	As above for B_1 plus green vegetables.	As above for B_1.	Digestive disorders, mental disorders and skin disease, called pellagra.	Likely in maize-eating peoples of Africa.
C Ascorbic acid	Fresh fruit, citrus fruit (e.g. oranges, lemons, grapefruit), raw vegetables.	Prepared concentrated juices.	Bleeding from gums and other membranes, teeth disorders, reduced resistance to infection, the disease called scurvy.	Common fatal disease on old sailing ships where sailors had no fresh fruit.
D Calciferol	Liver, fat, fish, egg yolk, formed in the skin by sunlight.	Butter, margarine.	Weak bones, particularly leg bones, poor teeth, the disease called rickets.	Young mammals susceptible to disease.
K	Liver, green vegetables, egg yolk, unpolished cereals.		Prolonged bleeding, essential for blood clotting.	Made by bacteria in the gut.

Table 1.2 Vitamins: their sources and characteristics

heat make little alteration in the temperature of the organism. Large amounts of water are lost daily from animals and plants and so a corresponding amount must be taken in to maintain the *water balance*. Water is gained from two main sources:

i) Most water is *absorbed by the roots* of plants, or taken through the mouth of animals in the form of *food or drink*.

ii) Small amounts are formed in the tissues by *oxidation of the hydrogen* in food.

Vitamins

It was not until the beginning of the twentieth century that it became clear that humans and other animals could not be kept healthy on a diet of pure carbohydrates, fats, proteins, mineral salts and water. Small amounts of a number of other substances were found to be necessary. Gowland Hopkins, an English chemist, first gave the clear proof of these facts by his experiments on rats. He showed that the addition of milk to food consisting of pure protein, starch, sucrose, lard, inorganic salts and water brought about greatly increased growth in a group of rats compared with a control group. He deduced that certain 'accessory factors' which are essential for health were present in milk. These factors are today called *vitamins* and the facts about them are given in table 1.2.

Their absence from the diet causes certain *deficiency diseases*. This was a strange new idea at the time – that disease could be caused by absence of food, coming so shortly after Pasteur had found that most disease is caused by microbial organisms in the body.

The vitamins are only required as traces in the diet, but once inside the body their importance is considerable. They act as *co-enzymes* in certain reactions. These co-enzymes cannot be synthesised in the body cells as are all true enzymes, hence they must be obtained from foods. If they are not present in the diet then the body functions are disrupted.

Each vitamin was first named with a letter before its chemical nature had been established. Once this had occurred then the vitamin was given a chemical name. In some cases the vitamin has been synthesised in the laboratory.

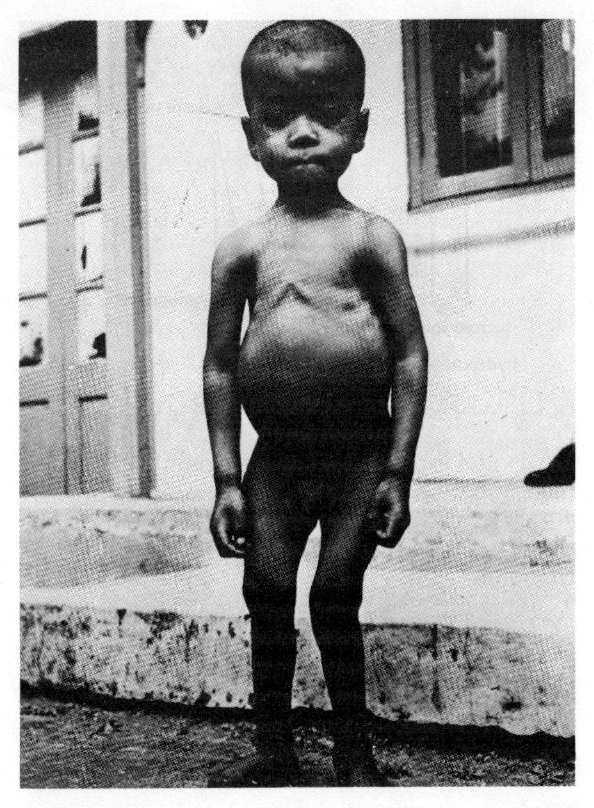

fig 1.7 Rickets

Experiment

Each class of food substance can be detected by a chemical test. The following investigations establish each test on a pure food substance:

Carbohydrates

Test for starch

Procedure

Take 1 cm^3 of starch suspension in water and add two drops of *iodine solution* (aqueous solution of iodine in potassium iodide), in a test-tube. This can also be performed using a small quantity of powdered starch.

Question

1 What colour change do you see?

Test for reducing sugar

Procedure

1 Take 1 cm^3 of glucose solution and place in a test-tube. Add 2 cm^3 of *Benedict's* or *Fehling's* solution.

2 Place the test-tube in a water-bath (beaker or tin) of boiling water for five minutes.

3 Take 2 cm^3 of Benedict's or Fehling's solution and place in a second test-tube with 1 cm^3 distilled water. Place the tube in the water-bath as in 2 above. This is

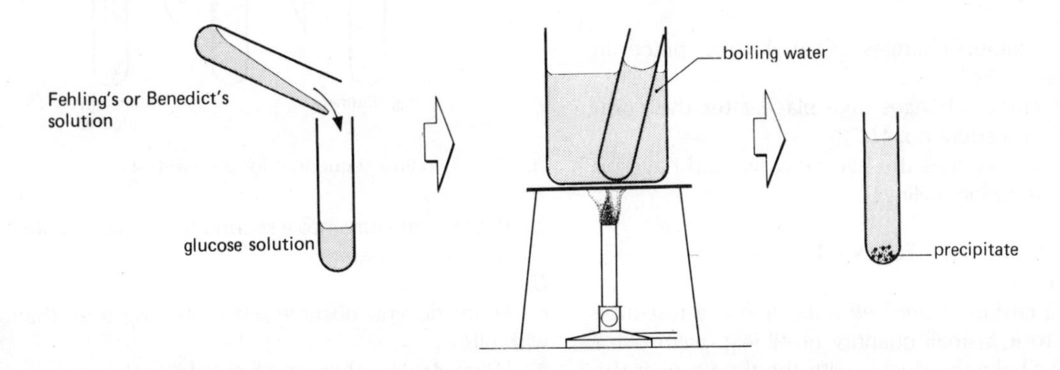

fig 1.8 A flow sequence for the reducing sugar test

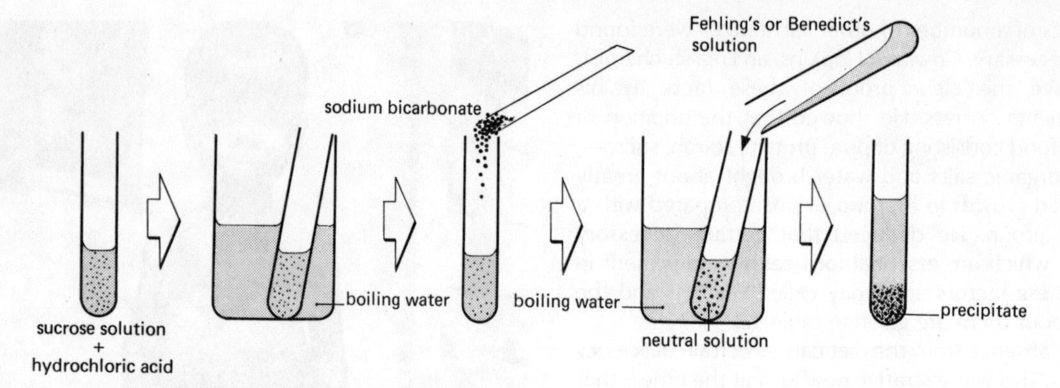

fig 1.9 A flow sequence for a non-reducing sugar test

the control experiment, for comparison with the first tube.

4 Shake each tube at intervals and examine the colour of the liquid.

Questions

2 What is the final colour seen in the first test-tube, containing glucose?

3 Did you notice any intermediate colour changes before five minutes elapsed?

4 What colour changes are shown by the second test-tube?

Test for non-reducing sugar

Procedure

1 Repeat the Benedict's or Fehling's test with a sucrose (table sugar) solution. Set up a control tube. Compare the two test-tubes.

2 Take 1 cm³ of sucrose solution in a test-tube and add three drops of dilute hydrochloric acid. Place in a water-bath and boil for 2–3 minutes.

3 Cool the test-tube under a cold tap, or in beaker of cold water. Add solid sodium bicarbonate slowly until the fizzing stops. This indicates that the solution has been neutralised.

4 Add 2 cm³ of Benedict's or Fehling's solution to the *neutral* (*or slightly alkaline*) solution. Place in the boiling water of the water-bath for 5 minutes.

Questions

5 What colour changes (if any) take place in procedure no. 1?

6 What colour changes take place after the completion of procedure no. 4?

7 What action does the hydrochloric acid have on the sucrose during boiling?

Test for fats and oils Test No. 1

Procedure

1 Take 2 cm³ of *ethanol* (*ethyl alcohol*) in a test-tube and add to it a small quantity of oil (e.g. *palm oil* or *castor oil*). Shake thoroughly with the thumb over the end of the test-tube.

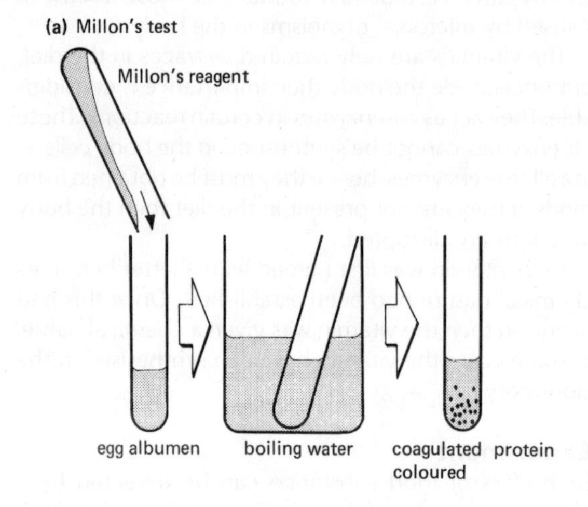

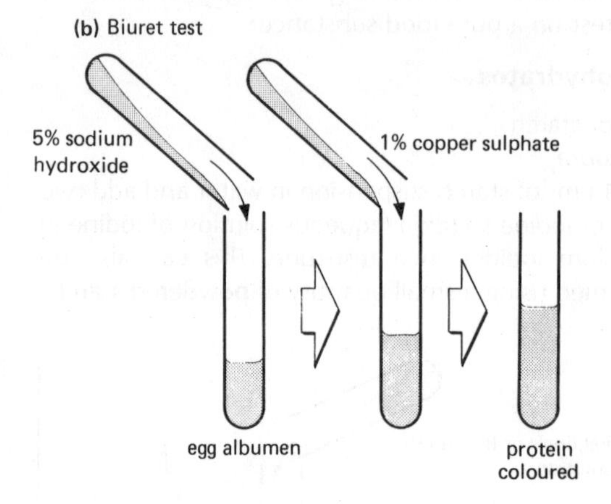

fig 1.10 Two flow sequences for protein tests

2 Pour the mixture into a second test-tube containing about 2 cm³ of water.

Questions

8 What do you observe when shaking the ethanol with oil?

9 What do you observe when adding the mixture to water?

Test No. 2
Procedure
1 Place a spot of oil on a clean sheet of absorbent paper.
2 Alongside of the oil spot place a spot of water.
3 Leave the sheet of paper for ten minutes, then hold the paper to the light.
Question
10 What differences do you observe in the appearances of the oil and water spots?

Test for proteins Test No. 1 – Millon's test
Procedure
1 Place 2 cm³ of egg *albumen* (fresh egg white or dried egg albumen in solution) into a test-tube and add 2 cm³ of *Millon's reagent*.
2 Set up two control tubes, one with glucose solution and one with oil, and add 2 cm³ of Millon's reagent to each tube.
3 Place the test-tubes in a water-bath of boiling water for five minutes.
Questions
11 What colour changes occur with the egg albumen and the Millon's reagent?
12 What colour changes occur in the control test-tubes?

Test No. 2 – Biuret test
Procedure
1 Place 2 cm³ of egg albumen into a test-tube and add 1 cm³ of *sodium hydroxide* solution.
2 Add 1% *copper sulphate* solution drop by drop, shaking at each drop (do not add the copper sulphate too rapidly).
Question
13 What colour changes occur when the copper sulphate solution is added?

Test for vitamin C (ascorbic acid)
Procedure
1 *Ascorbic acid* is a powerful reducing agent. It can be detected by using the dye *Dichlorophenolindophenol, DCPIP*. The dye is blue in colour and is decolourised on *reduction* by ascorbic acid.
2 Place 2 cm³ of 0.1% DCPIP solution into a clean test-tube. This must be an exact amount. Use a syringe or pipette to measure the quantity required.
3 Make up a 0.1% solution of ascorbic acid (1 mg per cm³). Using a syringe, add it to the DCPIP drop by drop until the blue colour disappears. This happens quite suddenly. Record the number of drops of ascorbic acid required to decolourise the DCPIP.
4 Repeat the test with the same amount of DCPIP solution using the juices of raw fruit and vegetables. For example, squeeze an orange so that the juice is collected in a beaker and add it drop by drop until the DCPIP in a test-tube is decolourised. For each juice count the number of drops.
5 Record your results in the form of a table with the name of each juice tested, together with the number of drops required to change the dye.

Juice	Ascorbic acid 0.1%	Orange juice	Lemon juice
Number of drops			

Questions
14 Work out the actual amount of ascorbic acid in one mg of each of the various juices.
15 Leave some of the juices exposed to air for several days and test again for ascorbic acid. What do you find?

Now that you have established the various tests for the different classes of food substance, carry out the same tests on different common food materials in order to find out what classes of food substance they contain. Draw the following table in your practical book and then carry out tests on any common foodstuffs such as cabbage, carrot, broad bean, fish, milk, peanuts, etc.

Each food should be *cut or ground into small pieces* in order to break up the cell structure so that the organic compounds can be released for the tests. Grinding the food in a pestle and mortar is the best method.

Note that a non-reducing sugar test should only be carried out if the reducing sugar test gives a negative result.

Check your results with table 1.6 which shows the actual quantities of each class of food per 100 g. Carbohydrates are not subdivided into sugars and starch.

Food tested	Reducing sugar	Non-reducing sugar	Starch	Fats or oils	Proteins
Soya bean	nil	nil	present	present	present

1.20 Nutrition in green plants

All animals obtain food either directly or indirectly from green plants. (The green colour of plants is caused by a pigment called *chlorophyll*.) Large meat-eating animals feed on smaller animals but these smaller animals feed on vegetation. For example, lions eat antelopes and antelopes graze on grass. This kind of relationship is called a *food chain*. Every food chain includes an animal which takes in plant matter for food. These animals are called *herbivores* and must be able to obtain all their requirements for energy production, growth, repair and health from plant material. The herbivores are then consumed by *carnivores*.

Let us consider therefore what types of food material are present in plants.

1.21 What foods are present in the green leaf?

Experiment

Testing leaves for starch

Procedure

1 Take a green broad-bladed *dicotyledonous*, leaf, e.g. geranium, balsam, lilac, that has been exposed to sunlight. (This experiment can be done on a small scale by using discs cut from the leaf with a cork borer.

2 Dip the leaf or the discs into boiling water. This kills the living substance of the leaf.

3 The green colouring matter must now be removed. Half fill a test-tube with *ethanol* (ethyl alcohol) and place in it the leaf (rolled up) or the leaf discs. Prepare a water-bath and heat the water to boiling point. **Extinguish the lighted bunsen**. Place the test-tube containing ethanol and leaf into the boiling water.

4 The ethanol will boil and dissolve out the green matter. When the leaf or discs are *colourless*, remove them from the ethanol with forceps, and wash them in cold water. This *softens* the leaf structure by returning water removed by the ethanol. A brittle leaf will disintegrate.

5 Place the softened leaf into a petri dish and cover with *iodine solution* for several minutes. Wash away the iodine solution with tap water.

6 Repeat the experiment with a leaf from a shoot of a similar plant which has been kept in the dark for 24 hours.

Questions

1 What do you observe regarding the ethanol that remains in the test-tube? Explain your observation.

2 What do you observe regarding the first leaf after testing with iodine solution? What do you conclude from this result?

3 What do you observe regarding the second leaf after testing with iodine solution? What do you conclude?

Testing leaves for reducing sugars

Procedure

1 Take a *monocotyledonous* leaf such as onion or lily and grind up the leaf with a little sand using a pestle and mortar. Add a small quantity of water.

2 Pour off the water and test for reducing sugar with Benedict's or Fehling's solution.

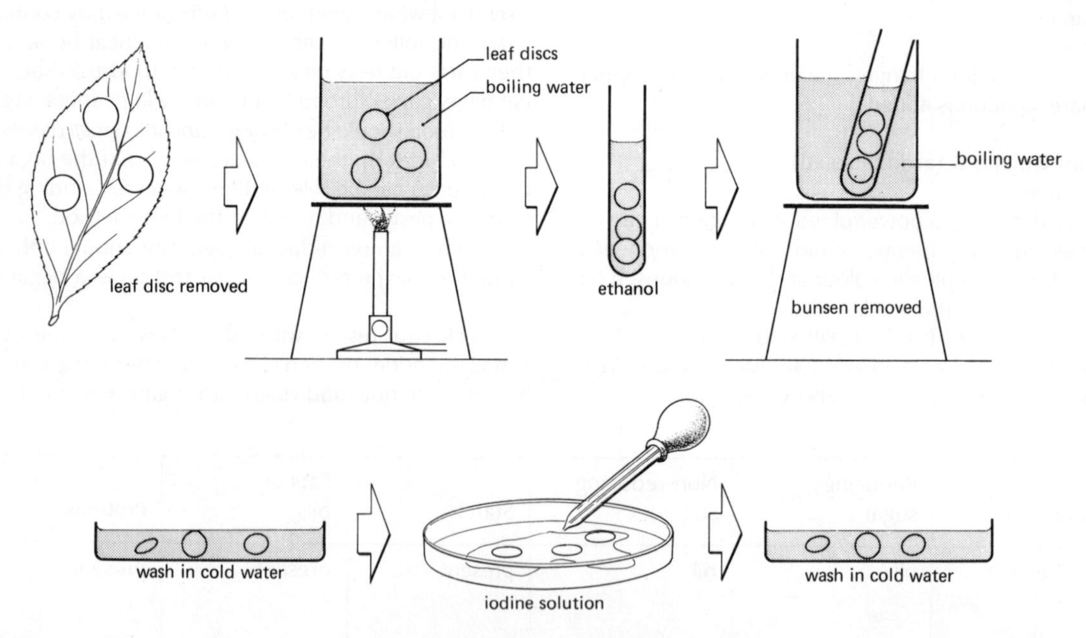

fig 1.11 A flow sequence for decolourising a leaf followed by the starch test

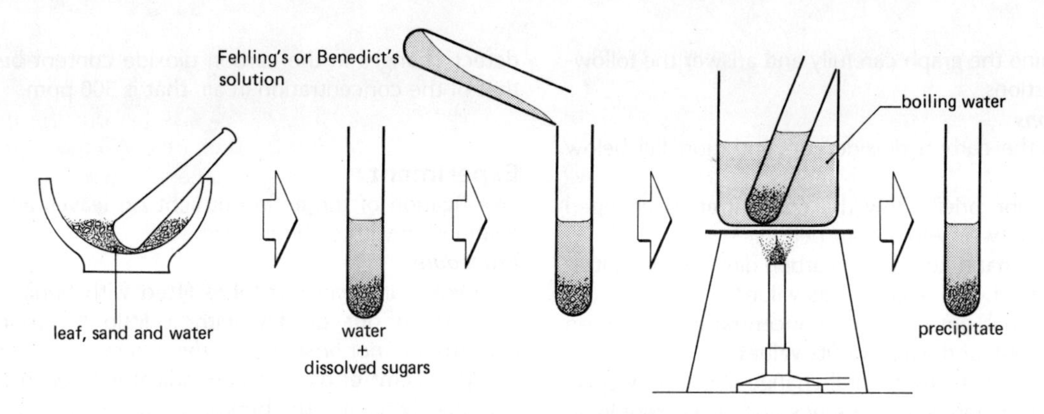

fig 1.12 A flow sequence for the sugar test on a leaf

Question

4 What coloured precipitate was obtained as a result of the reducing sugar test? What do you conclude?

Testing a variegated leaf for starch

Procedure

1 Take a *variegated* leaf (e.g. *Tradescantia, Coleus,* variegated maple or ivy) which has been exposed to sunlight and make a *sketch* of the distribution of green and white areas of the leaf.

2 Carry out a starch test on the complete leaf.

3 Compare the distribution of starch in the leaf with your original sketch of the green and white areas.

Questions

5 Where, in terms of green and white areas, is starch produced in the leaf?

6 What conclusions can you draw from these observations?

You have shown that a carbohydrate (starch) is present in the leaf, and this has been produced by the plant for its own use. It is of course available to animals when the leaves are eaten as part of the diet. It is also clear from the experiments that light is necessary for starch formation and also that chlorophyll is necessary for starch formation.

1.22 Carbon dioxide

The average *carbon dioxide* content of the atmosphere has been found to be fairly constant at about 0.03% or *300 parts per million* (ppm). It is possible to measure the carbon dioxide content of air accurately, and fig 1.13 shows the measurements for the air in a forest over a period of twenty-four hours. The results of this experiment show that in this particular forest the carbon dioxide concentration was always greater than 0.03%. Perhaps a little later you will be able to explain why the atmosphere of the forest investigated in this experiment contained an average carbon dioxide concentration higher than is normally found in other vegetation.

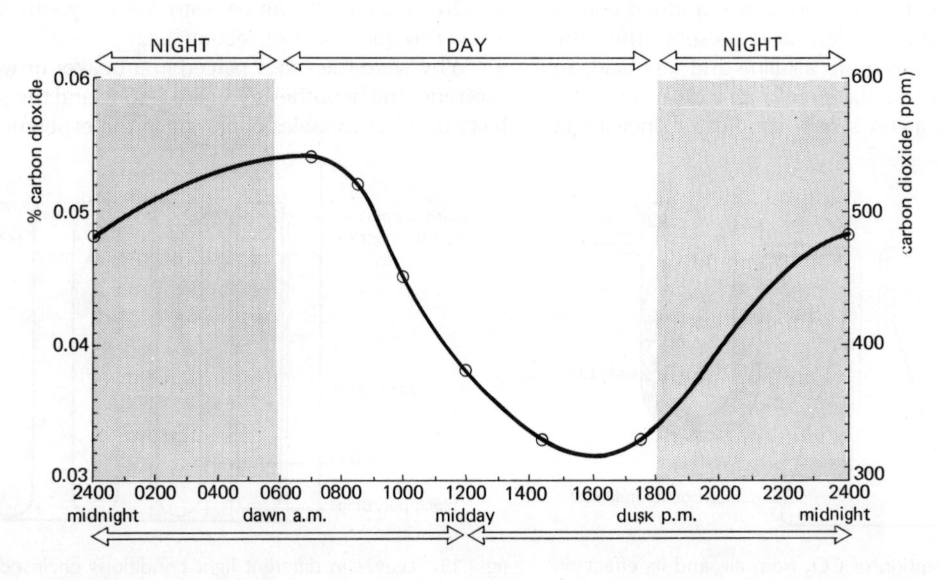

fig 1.13 A graph of CO_2 concentration in the air in a forest, over a 24-hour period

Examine the graph carefully and answer the following questions.

Questions

1 Did the carbon dioxide concentration fall below 0.03%?

2 Describe briefly how the concentration changed during the twenty-four hour period.

3(a) At what time was the carbon dioxide concentration greatest, and what was its value?

(b) At what time was the concentration of carbon dioxide least, and what was its value?

4 The main environmental change over the twenty-four hour period is the presence and absence of light. Can you produce an hypothesis to suggest an explanation for your observations in terms of the forest vegetation, the carbon dioxide and the light changes?

In order to test this hypothesis we can use an indicator which is extremely sensitive to concentrations of carbon dioxide. It is called a bicarbonate indicator (see Introductory Chapter) and it has a dull red colour established by drawing air through it from outside the laboratory. We can draw through this indicator air which has had the carbon dioxide removed by an absorbent substance called soda lime (see Introductory Chapter).

Procedure

1 Set up the apparatus as shown in fig 1.14.

2 Turn on the tap in order to extract air from the flask. Fresh air will thus bubble into the flask, having been drawn through the soda lime. Keep the tap running until the colour of the indicator changes.

You will see that the indicator becomes *purple* in colour. The air with no carbon dioxide in it must have drawn carbon dioxide from the indicator. Carbon dioxide in solution forms *carbonic acid* and if it comes out of solution there is less acid present. Thus the solution has become more alkaline and less acid, so that the change in colour is really an indication of the pH. This is clearly a very sensitive indicator since it can detect changes in the carbon dioxide content below that of the concentration in air, that is 300 ppm.

Experiment

Investigation of the action of light on leaves and its relationship with carbon dioxide.

Procedure

1 Take four clean test-tubes fitted with bungs and wash them out quickly with a little bicarbonate indicator (*do not breathe near the tubes*).

2 Add 2 cm^3 of bicarbonate indicator to each tube and quickly replace the bungs.

3 Label the tubes A, B, C and D. Into the first three tubes place a broad-bladed leaf above the indicator solution. Ensure that the bungs are quickly and tightly replaced. Leave the tube D empty.

4 Completely cover tube A with black *polythene* fixed by elastic bands. Completely cover tube B with some thin muslin or mosquito netting, also secured with rubber bands. Leave tubes C and D uncovered.

5 Put the tubes in a clean beaker of tap-water, and place a powerful lamp next to the beaker. Alternatively, place them in strong sunlight.

6 Record the time at which the experiment was set up. Shake the tubes at frequent intervals. Observe the tubes for any colour change in the bicarbonate indicator (this experiment will take between one and three hours).

Questions

5 What colour changes do you observe in the indicator solution in tubes A, B, C and D?

6 What explanation can you give in terms of uptake or release of carbon dioxide?

7 What is the function of tube D?

8 Do your results agree with your hypothesis in answer to question 4 of section 1.22?

9 Why were the tubes placed in a beaker of water? (Consider the hypothesis you are testing and the possibility of other variables operating in the experiment.)

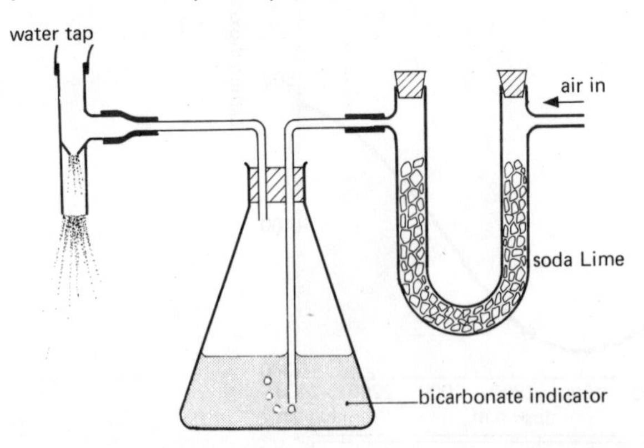

fig 1.14 The absorption of CO_2 from air, and its effect on bicarbonate indicator solution

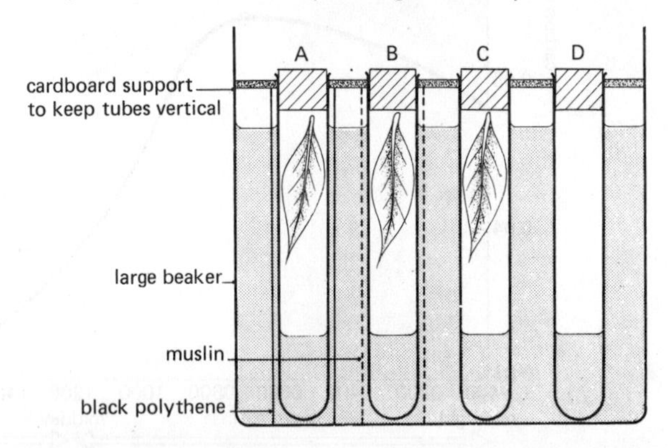

fig 1.15 Leaves in different light conditions enclosed with bicarbonate indicator solution

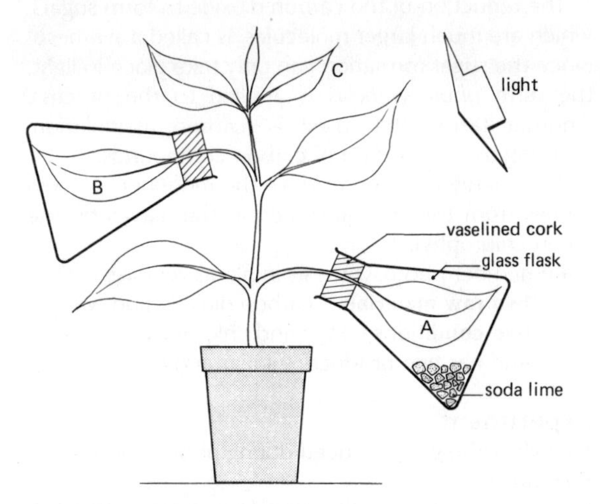

fig 1.16 The experimental apparatus to investigate the use of CO_2 by the leaves of a green plant

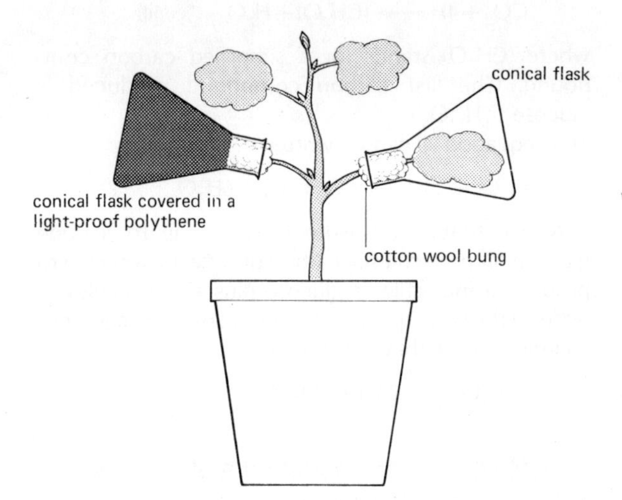

fig 1.17 The experimental apparatus to investigate the use of light by the leaves of a green plant

Experiment

What use is made of carbon dioxide by a plant in the light?

Procedure

1 Remove food reserves (starch) from the leaves of a potted plant (e.g. *Balsam* or *Pelargonium*) by keeping it in the dark (e.g. in a cupboard) for twenty-four hours. This is called *destarching*.

2 Take the potted plant and enclose one leaf in a glass flask with some soda lime to absorb carbon dioxide. The leaf stalk should pass through a split cork which has been *vaselined* to make it airtight. Support the flask with a clamp on a retort stand.

3 Enclose a second leaf in a similar flask *without* the soda lime.

4 Put the potted plant and flasks in sunlight for three hours (or expose to bright artificial light in the laboratory).

5 Test the two enclosed leaves and another leaf of the plant for the presence of starch (see section 1.21).

Questions

10 What were your results with the three leaves after testing with iodine solution for the presence of starch?

11 What do you think happens to the carbon dioxide absorbed by the plant in daylight?

12 Since we are looking at the uptake of a gas from air, how could you criticise the apparatus used, and suggest a method of improving the apparatus to overcome your criticism?

Experiment

Is light necessary for starch formation?

Procedure

1 Remove the food reserves from (destarch) a potted plant (*Balsam* or *Pelargonium*) as in the previous experiment.

2 Take the potted plant and enclose one leaf in a glass flask. Lightly surround the leaf stalk with a small cotton wool bung (this allows gas to circulate). Cover the flask *with light-proof material* e.g. a black polythene bag.

3 Enclose a second leaf in a similar flask *without the light-proof cover*, and again surround the leaf stalk with a small cotton wool bung.

4 Put the potted plant and flasks in sunlight for three hours (or expose to bright artificial light in the laboratory).

5 Test the two enclosed leaves and another leaf of the plant for the presence of starch (see section 1.21).

Questions

13 What were your results on testing the three leaves with iodine solution for the presence of starch?

14 What do you conclude about the relationship between light and starch formation?

15 In the light of your experimental work in sections 1.21 and 1.22, criticise the apparatus used, and suggest a method of improving the apparatus to overcome your criticism.

1.23 Formation of carbohydrates in plants

Your experiments have shown that, in order to form starch in the leaves of a plant, *light, carbon dioxide* and the green pigment *chlorophyll* are necessary. Carbohydrates consist of the elements C, H and O so that clearly the third element H must be supplied in addition to the C and O provided by carbon dioxide. Scientists have discovered that the *function of light* absorbed by the leaf is to *split water*.

$$2H_2O \longrightarrow 4H + O_2 \; \ldots\ldots \text{(i)}$$

The carbon dioxide molecules are *reduced* by the *hydrogen* produced as a result of the breakdown of water.

$$CO_2 + 4H \longrightarrow (CH_2O) + H_2O \ldots\ldots \text{(ii)}$$

where (CH_2O) stands for a simplified carbon compound. (The first carbon compound produced is glucose $C_6H_{12}O_6$.)
Thus equation (ii) can be written:

$$6CO_2 + 24H \longrightarrow C_6H_{12}O_6 + 6H_2O$$

Notice that if we write equation (ii) to include glucose, six molecules of carbon dioxide are reduced to produce a molecule of glucose plus six molecules of water. If the two equations (i) and (ii) are put together to include glucose they are written:

$$12H_2O \longrightarrow 24H + 6O_2 \uparrow$$

$$6CO_2 + 24H \longrightarrow C_6H_{12}O_6 + 6H_2O$$

This can be simplified as:

$$12H_2O + 6CO_2 \longrightarrow C_6H_{12}O_6 + 6H_2O + 6O_2$$

The glucose is rapidly converted to starch in most plants but in others such as *sugar cane* the storage carbohydrate is *sucrose*. Starch is not found in the leaves of other monocotyledonous plants such as onion and cereals. You will note that oxygen is given off during this process. This oxygen *comes from the water molecules*, whereas the oxygen in the carbon dioxide molecules goes partly into the carbohydrate molecule and partly into the new water molecules.

The reduction of the carbon dioxide to form sugars, which are much larger molecules, is called a *synthesis*. Since the sugar formation can only take place in light, the term *photosynthesis* is applied to the process. Another term often used is 'carbon assimilation' referring to the build up of carbon compounds.

The energy required to drive the reduction process comes from the light absorbed in the leaves by the green chlorophyll.

Remember photosynthesis in plants requires:

two raw materials – carbon dioxide and water,
two conditions – light and chlorophyll,
and has **two products** – sugar and oxygen.

Experiment

To collect the gas produced during photosynthesis.

Procedure

1 Take a beaker of rain water (or tap-water that has been standing to eliminate chlorine) and place in it some pondweed such as *Elodea, Hydrilla* or *Ceratophyllum*. Cover the weed with an inverted funnel (see fig 1.18(a)). Make sure that the top of the funnel stem is about 2–3 cm below the surface of the water.

2 Fill a test-tube completely with water, place your thumb over the end and invert the tube. Place the covered end of the test-tube under the water surface in the beaker, and remove your thumb. The test-tube should then be moved over the funnel stem and lowered.

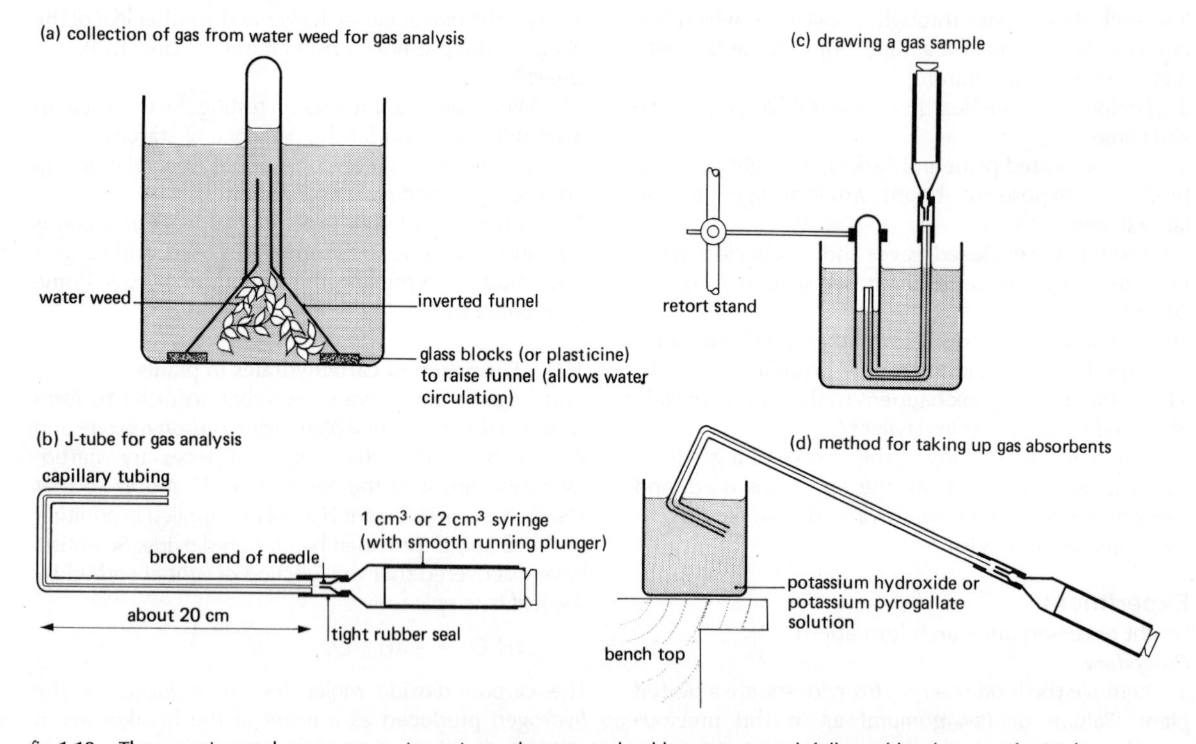

fig 1.18 The experimental apparatus to investigate the gas evolved by water weed, followed by the use of a J-tube

3 A small amount of *sodium bicarbonate* can be placed in the water of the beaker to ensure that sufficient carbon dioxide is released to enable the water weed to photosynthesise.

4 The apparatus can now be placed on a bench near the window of the laboratory so that it receives strong sunlight. At night it should be kept illuminated with a powerful electric light bulb.

5 Gas will start to collect in the top of the test-tube. Leave the apparatus until the test-tube is about half full of gas.

The gas which collects in the test-tube must be *analysed* in order to determine its constituents, particularly with regard to carbon dioxide and oxygen. In order to do this you will need a J-tube (made of capillary tubing with a fine bore) attached to a syringe of 1 or 2 cm^3 capacity. The syringe should have a needle attached which has been cut off near its base. This gives a finer adjustment to the syringe (see fig 1.18(b)).

Experiment

To confirm that oxygen is given off by pondweed in the light.

Procedure

1 Lift the test-tube from the funnel and place a thumb over the open end. Transfer the tube to another large beaker of water and hold the tube in the clamp of a retort stand (see fig 1.18(c)).

2 Push the plunger of the syringe in fully, and place the open end of the capillary tube into the water of the beaker. Gently pull out the plunger and draw water into the capillary tube. The length of the water enclosed in the J-tube should be about 5 cm. Throughout the experiment, try to handle the J-tube as little as possible, since the heat of the body will cause expansion changes in the gas sample.

3 Without removing the J-tube from the water, place the open end up into the test-tube until it is in the gas at the top of the tube (see fig 1.18(c)). Gently pull out the plunger in order to draw a sample of the gas into the tube. The length of this gas bubble should be about 10 cm, twice the length of the first sample of water. Now lower the J-tube and draw in another 5 cm of water to seal off the gas bubble. These measurements of the water and gas bubble need not be accurate.

4 Remove the J-tube from the beaker and place it in a trough of water that has been standing for a while at laboratory temperature. This will bring the J-tube and its contents to a constant temperature. Leave for 5 minutes and then measure the length of the gas bubble to the nearest millimetre.

5 Gently push down the plunger to eliminate most of the water from the seal at the end of the capillary tube. **Ensure that no gas escapes**. Now dip the end of the capillary tube into a 10% solution of *potassium hydroxide* solution in a beaker. The beaker must be at the edge of the bench since the J-tube must be inverted to perform this operation (see fig 1.18(d)). Draw in about 2–3 cm of potassium hydroxide solution by pulling back gently on the plunger.

6 Take the J-tube out of the potassium hydroxide and by pushing and pulling the plunger move the contents of the tube backwards and forwards several times in order to mix the gas bubble with the chemical. Leave the apparatus in the water bath for 5 minutes to stabilise the temperature.

7 Measure the length of the gas bubble again. Any shortening of its length will be due to the carbon dioxide absorbed.

8 Repeat steps 5–7 using *potassium pyrogallate* solution. This is kept in a beaker with a layer of liquid paraffin over its surface to prevent the absorption of oxygen from the air. Make sure that the open end of the J-tube is **pushed through** the paraffin into the pyrogallate solution. Push out nearly all of the potassium hydroxide solution and pull in about 2–3 cm of pyrogallate solution. Move the contents backwards and forwards as in step 6 above. Place the J-tube in the water trough and leave for 5 minutes.

9 Measure the length of the gas bubble in the J-tube. This time any shortening of its length will be due to the oxygen absorbed.

10 Write down your results as follows:

Initial length of gas bubble	= A cm
Length of bubble after absorption by potassium hydroxide	= B cm
Length of bubble after absorption by potassium pyrogallate	= C cm
Therefore length of carbon dioxide absorbed	$= A - B$
Therefore percentage of carbon dioxide present	$= \frac{A - B}{A} \times 100\%$
Therefore length of oxygen absorbed	$= B - C$
Therefore percentage of oxygen present	$= \frac{B - C}{A} \times 100\%$

Note Since the bore of the capillary tube is uniform we can assume that the length of the gas bubble is proportional to its volume. The percentage needs to be expressed to the nearest whole number.

11 In a class situation *tabulate* the class results and obtain the *mean value* for the percentages of carbon dioxide and oxygen respectively.

Questions

1 Why was the J-tube placed in the water trough after each operation?

2 Why was the gas bubble moved backwards and forwards several times with each absorbent?

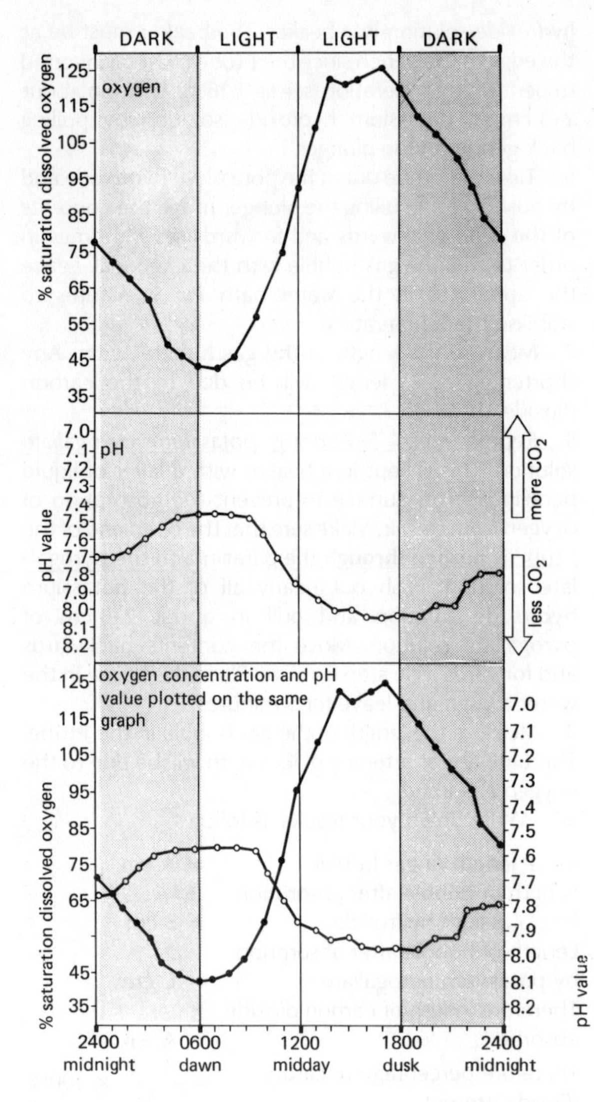

fig 1.19 Graphs of changes in river water over a 24-hour period

3 What results were obtained after analysis of the gas bubble? What is their significance? (Remember that air in the laboratory has about 21% oxygen and 0.03% carbon dioxide.)
4 Do you think that variation in light intensity would have any effect on this experiment?

Further evidence of the exchange of gases by plants can be obtained by examination of data obtained from water weed growing naturally in a habitat such as a small river.

Examine fig 1.19 and answer the following questions.
5 Look at the curve for *oxygen saturation* against time. At what times was the oxygen present in (a) its greatest concentration and (b) its lowest concentration?
6 During what period of the twenty four hours was the oxygen concentration increasing?
7 The *hydrogen ion concentration* (pH) is related to carbon dioxide concentration because dissolved carbon dioxide forms carbonic acid. In the middle graph pH value and carbon dioxide concentration are shown. What is the pH value at the highest concentration of oxygen? Can you suggest why this value coincides with the lower CO_2 level as shown on the graph?
8 What is the pH value at the lowest concentration of oxygen? Why does this value coincide with the greater CO_2 level as shown on the graph?
9 Which of the figures so far shown in this chapter corresponds with the middle graph of fig 1.19?

The bottom of fig 1.19 shows the other two graphs put together, indicating a *clear relationship* between oxygen and carbon dioxide concentrations. At night oxygen is absorbed and carbon dioxide released, while during the hours of daylight, when photosynthesis occurs, carbon dioxide is absorbed and oxygen released. The significance of the absorption of oxygen and release of carbon dioxide at night will be investigated later, but it should be noted now that *gas exchanges* during night and day *compensate* partially for each other.

1.24 Is all light important?

If you look at a detached leaf against the light, you will notice some light is passing through it. This light is a pale green colour. The surface of a leaf on which the sun is shining has a bright green colour, and this light must have been reflected. Thus the leaves can only be absorbing part of the available sunlight for use as energy during photosynthesis (see fig 1.20(a)). A leaf extract can be produced by grinding green leaves with ethanol, or with a mixture of 80% acetone and 20% water. If white light from a projector is passed through a *60° glass prism*, the white light is split into its *constituent colours*. When these are projected onto a screen they are seen as the colours of a rainbow. Their order can be distinguished as red, orange, yellow, green, blue, indigo and violet. If the green leaf extract is poured into a flat-sided container and placed between the prism and the

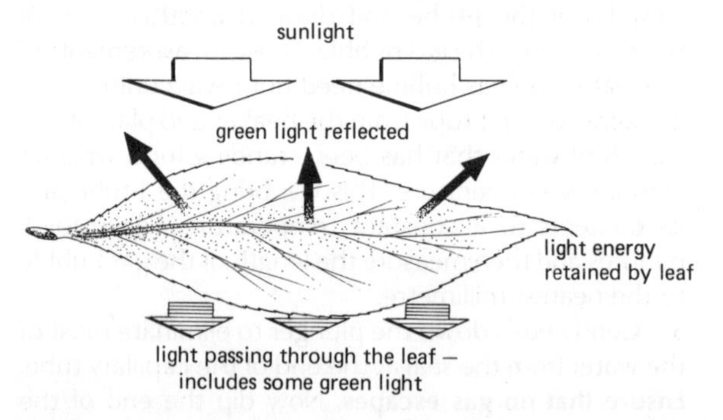

fig 1.20 (a) The relationship between a leaf and light

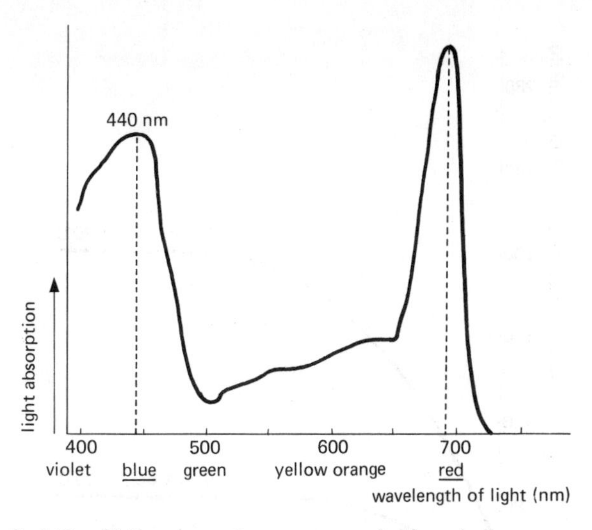

fig 1.20 (b) The absorption spectrum of chlorophyll

screen, the change in the colours on the screen is quite striking, for some of the colours are *absorbed by* the *green extract.* The greatest absorption occurs in the blue and red parts of the spectrum (see fig 1.20(b)). The colours have been extracted by the green pigments (chlorophyll) in the leaf extract. It is probable that the plant uses the *light energy* from these particular colours and converts it to *chemical energy*. Thus it is enabled to split the water molecules at the beginning of the photosynthetic process. It takes a great deal of energy to split a molecule of water so that our first equation in section 1.23 can now be written to take account of this fact.

$$2H_2O + \text{energy} \longrightarrow 4H + O_2 \uparrow$$

Further energy is incorporated into the glucose molecules during their formation and when they are built up into starch molecules. It is this energy which is released in animals when the food is utilised in the body (see section 1.11). The chlorophyll which absorbs the light is located inside the leaf cells in the *green disc-shaped chloroplasts* which line the internal cell walls. They are very numerous in the *palisade cells* of the upper parts of the leaf and for this reason the upper surfaces of most leaves are darker than the lower surfaces. In this position the chloroplasts receive most light, and so the major part of photosynthesis takes place in the cells of the upper surface. An interesting confirmation of the relation between chloroplasts and light absorption is the fact that the chloroplasts can adjust their position to absorb maximum light. In *bright* light they are *end-on* to the incident light, but in *dim* light they move *side-on* at right angles to the light rays. Furthermore they move closer to the surface of the leaf and towards the light during dull conditions (see fig 1.21), enabling more light to be absorbed.

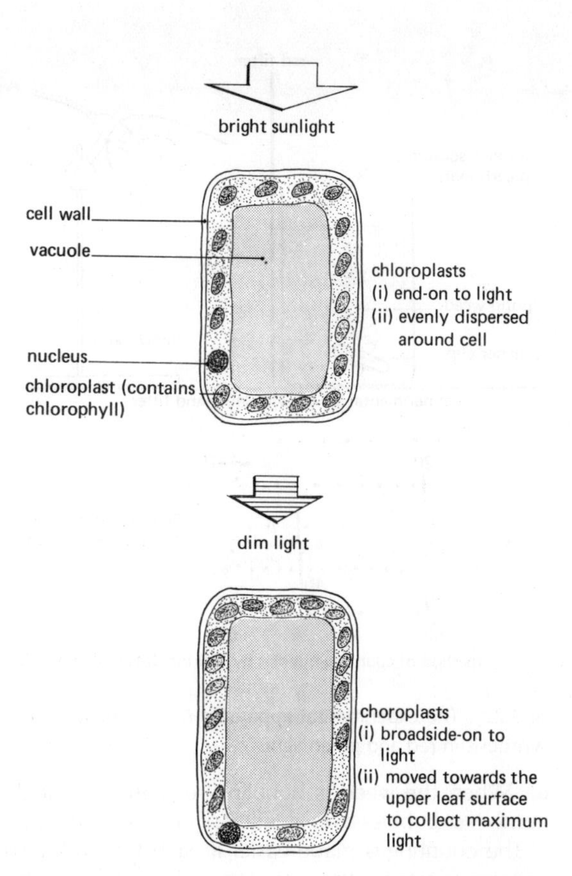

fig 1.21 The distribution of chloroplasts in the upper cells of a leaf

Experiment

Do differences in the wave length of light have an effect on photosynthesis?

Procedure

1 Take a beaker or glass jar filled with water and add a small amount of sodium bicarbonate. Stir to dissolve.

2 Place a bunch of water weed (*Elodea* or *Ceratophyllum*) in another jar near to an electric light, so that photosynthesis and gas production are stimulated.

3 Take a small piece of water weed about 5 cm in length and weight the uncut end (near the growing point) with a small paper clip. Place the pond weed in the beaker prepared in no. 1 above with the cut end upwards, but well beneath the water level. The paper clip will cause the weed to float vertically.

4 Place a *red light filter* between a bench lamp (100 watt) and the beaker containing the pond weed (see fig 1.22). Switch on the lamp so that the light shines through the red filter. The apparatus should be away from bright sunlight.

5 Bubbles will soon appear through the cut end of the stem. If they do not, then cut off a small portion of the stem, and if this fails then try again with a fresh piece of water weed.

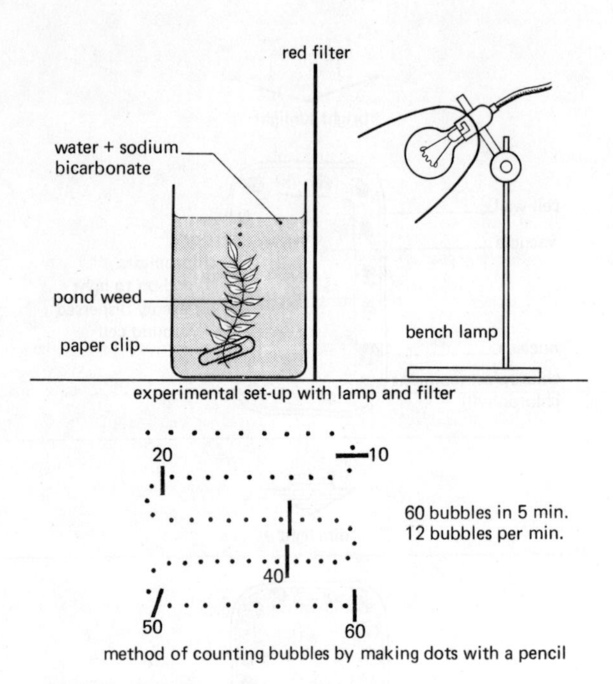

fig 1.22 The experimental apparatus to investigate photosynthesis in red and green light

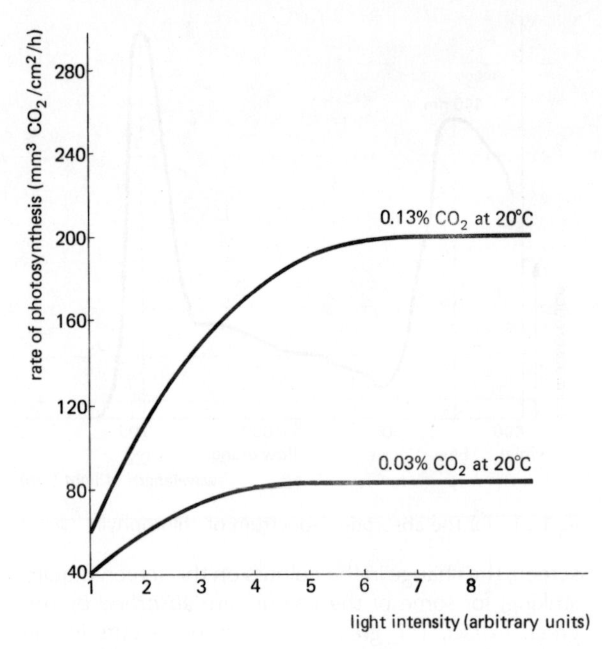

fig 1.23 A graph showing the effect of increase in CO_2 concentration on the rate of photosynthesis

6 When the weed is bubbling regularly, count the number of bubbles over a period of time, say 5 minutes.

The counting is made easier if each time a bubble appears, you tap with a pencil on a piece of paper, thus obtaining a series of dots which can be counted later (see fig 1.22).

When several readings have been made, replace the shoot and change the red filter to a *green filter*. Allow the bubble rate to adjust, and then count again.

7 Record your results in the form of a table.

Number of dots (therefore number of bubbles) in 5 min.		Rate of bubbling per min.	
red light	green light	red light	green light

mean rate of bubbling in red light =
mean rate of bubbling in green light =

Questions

1 In what part of the shoot is the gas produced? Why is it bubbling from the cut end?

2 Which light filter produced gas at the faster rate?

3 What explanation can you give for this difference in the rate of bubbling?

1.25 Limiting factors

When a plant is exposed to *increasing light intensity*, the *rate* of photosynthesis steadily *increases*. If the temperature is kept constant at 20°C and the carbon dioxide concentration at 0.03% (average for the atmosphere), the rate of photosynthesis can be plotted against increasing light intensity (see fig 1.23, graph A). It can be seen from the graph that, at a certain light intensity, the rate of photosynthesis levels off. What causes this levelling off in the rate of photosynthesis? There are two possible causes.

i) The *temperature* is *too low* for the chemical reactions to increase their activity further.

ii) The *carbon dioxide concentration* is *too low* to allow any further increase in the rate of photosynthesis.

If the temperature is raised to 30°C it makes little difference to the rate of photosynthesis. This rules out cause (i). However, if the experiment is repeated at 20°C with an additional 0.1% of carbon dioxide, there is a startling increase in the rate of photosynthesis, as shown in graph B in fig 1.23. Carbon dioxide is therefore a *limiting factor* in this process. Any chemical reaction dependent on a number of factors will always proceed at a rate controlled by the factor that is operating at full capacity. In the graphs in fig 1.23 as long as the *rate of photosynthesis is rising*, the *light intensity* is a limiting factor, but directly the graph levels out *some other factor* must be limiting the rate of photosynthesis.

In the natural environment, *light and temperature* can also be *limiting factors* for photosynthesis, especially early in the morning and in the evening. Plants living in shady places in a forest have only a short period of the day when light intensity is not a limiting

fig 1.24 A methane burner for CO_2 production in glasshouses

factor. In such places, there is intense competition between plants for the available light. Since it is advantageous for a plant to be taller than its neighbour, most shoot systems tend to grow upwards, towards the light. In tropical rain forests the trees with their great height and foliage dominate the vegetation below. Epiphytes are able to grow high up in the trees, thus capturing some of the light essential for photosynthesis.

Carbon dioxide remains constant in the atmosphere at an average of 0.03%, and for this reason it is a considerable *limiting factor under natural conditions*. As seen from the graph, fig 1.23, the rise of 0.1% in the concentration of carbon dioxide increases the rate of photosynthesis and therefore the productivity of the plant. In the search for increased food production, scientists have devised a method of increasing carbon dioxide concentration in glass houses in temperate climates. By the use of *methane burners* as in fig 1.24 additional carbon dioxide is produced and this, combined with a slight increase in temperature, increases considerably the productivity of the plants in the glass houses (see fig 1.25 and fig 1.26). In temperate climates, light and temperature are more likely to be limiting factors than they are in tropical countries, and therefore glass houses are in common use. In such enclosed conditions, increasing carbon dioxide concentration to 1.0% can mean *greater production*. It is of course an expensive procedure and furthermore great care has to be taken to ensure that no dangerous fumes are produced by the burners.

fig 1.25 Lettuce grown in air (CO_2 at 0.03%)

fig 1.26 Lettuce grown in enriched air (CO_2 at 1.00%)

1.26 Protein

In many plants the sugars formed by photosynthesis are immediately changed into starch grains in the leaf cells. At night this starch is changed back into sugar and transported around the plant to where it is needed for growth and energy production. The movement of the products of photosynthesis is called **translocation**. The sugar may be used immediately to produce energy or it may be stored for long periods in swollen plant organs (e.g. sweet potato) where it is changed back into starch. In other organs such as fruits and seeds, it may also be changed to oils or carbohydrates for storage purposes. At growing points in roots and shoots new protein is required and the sugars contribute towards its production.

Prior to the formation of *6-carbon sugars* (e.g. glucose $C_6H_{12}O_6$) in photosynthesis, a *3-carbon* compound *phosphoglyceraldehyde* is formed in the leaf. It is from this basic compound that a whole series of rapid reactions occurs as part of photosynthesis.

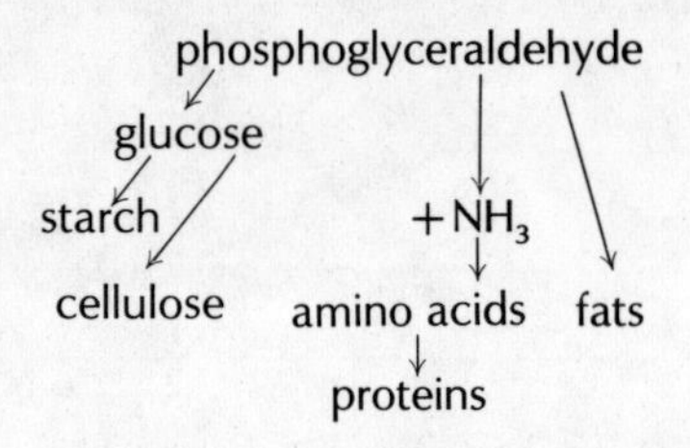

Similar syntheses of complex compounds occur in other parts of the plant to where sugar has been translocated. The sugar is changed to phosphoglyceraldehyde and from this the other substances can be built up. Phosphoglyceraldehyde already contains the C, H and O required for carbohydrate and fat synthesis, but proteins also contain *nitrogen*. This element enters the root system most commonly as the *nitrate ion* NO_3 and goes through a series of reductions until it becomes *ammonia*.

NO_3 (nitrate) $\longrightarrow$ NO_2 (nitrite) $\longrightarrow$ NH_2OH
NH_2OH (hydroxylamine) $\longrightarrow$ NH_3 (ammonia)

Ammonia is used in the formation of *amino acids*, the building blocks of proteins. The first amino acid formed is *glutamic acid* and from this the other essential amino acids can be produced. The nitrate ion is the form in which most plants absorb their nitrogen. The extensive root system in the soil takes up the nitrogen salts through the root hairs. Complex plant protein is built up, but ultimately the plant dies and decays, or is eaten by animals and turned into animal protein. Animals likewise will die and decay, but before this they produce nitrogenous excretory matter (urine) and faeces. *Putrefying bacteria* break down the protein of dead organisms, and also their excretory matter into *ammonium compounds*. *Nitrifying bacteria* in the soil change the *ammonium compounds* to *nitrites* and then to *nitrates*, so that once again these ions are available to the plant (see fig 1.27). *Nitrosomonas sp.* act on ammonium compounds, and *Nitrobacter sp.* on nitrites.

Some of the nitrates are broken down by *denitrifying bacteria* and the nitrogen is released into the atmosphere, but *nitrogen fixing bacteria* in the soil can carry out the reverse process and restore nitrates to the soil by using nitrogen taken from the air. One particular group of plants, the *Leguminosae* (including peas, beans and clover), can utilise *atmospheric nitrogen* by means of nitrogen-fixing bacteria which live in swollen *nodules* on their roots. The great majority of plants, however, cannot make direct use of the vast store of nitrogen in the air. Scientists are trying to introduce nitrogen-fixing bacteria into plants such as cereals.

During thunderstorms some nitrogen is added to the soil by the action of lightning which forms *nitric acid* and *nitrous acid* in the rain. These acids enter the soil and combine with the metallic parts of salts to form

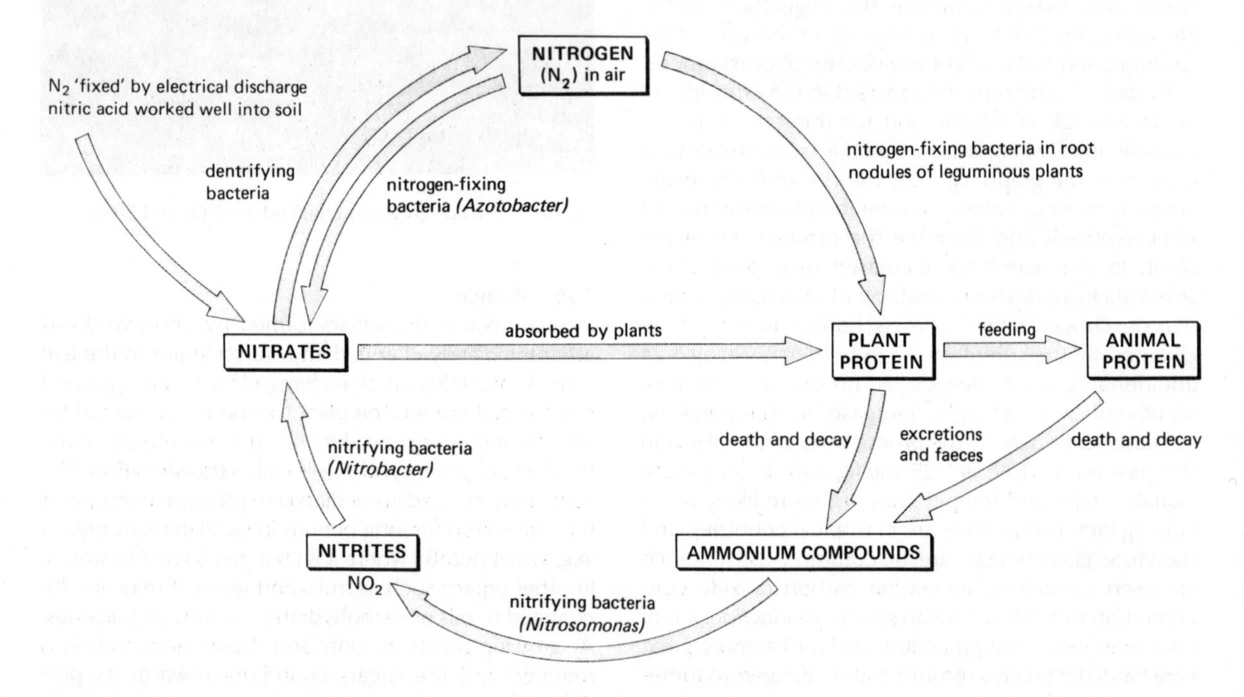

fig 1.27 The nitrogen cycle

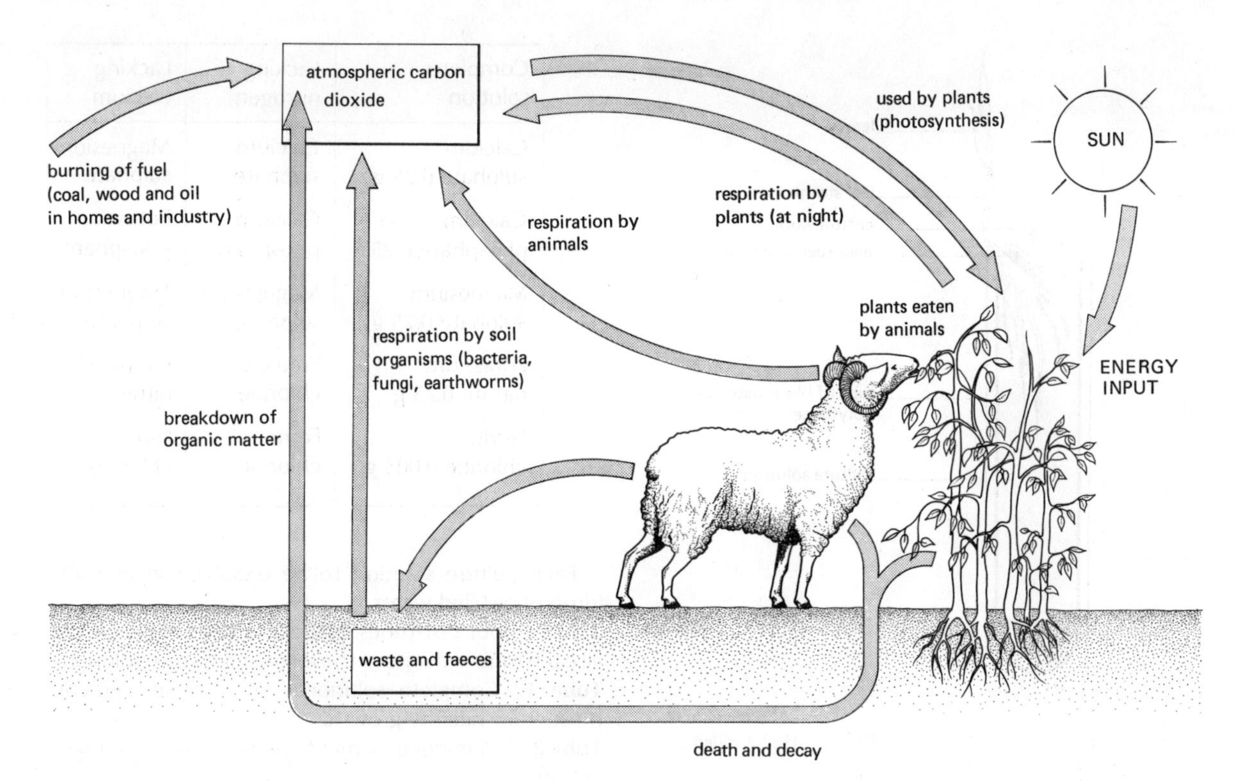

fig 1.28 The carbon cycle

nitrates. This is an important source of nitrogen for tropical soils. This build up of compounds of nitrogen and their eventual breakdown and return to the soil or air can be seen as a continuous process which is called a *cycle*. Nature provides continual renewal of this element, which is so important for the manufacture of protein. Unfortunately, this cycle can be broken where natural habitats are replaced by man's agriculture, which in many cases does not return plant and animal material to the soil.

There are a number of elements in nature in addition to nitrogen that travel through a cycle. Fig 1.28 shows the *carbon cycle* which involves photosynthesis and respiration (see Chapter 5). Both of these processes exchange carbon dioxide with the air. Combustion of plant tissue releases carbon dioxide into the air.

The manufacture of plant protein always requires nitrogen, but other elements, such as *sulphur* and *phosphorus*, may also be necessary. These, and other elements are obtained from the soil in the form of mineral salts which are absorbed by the roots. In addition to C, H, and O there are seven 'major' elements required in plant nutrition: nitrogen, sulphur, phosphorus, potassium, calcium, iron and magnesium. Also needed but in extremely small quantities are certain 'minor' elements, such as: zinc, manganese, cobalt, boron, copper, chlorine and silica. Analysis of a plant such as maize shows that the following elements are present:

Elements as a percentage of dry weight

Carbon	43.5	Nitrogen	1.5	Silica	1.2
Oxygen	44.4	Sulphur	0.2	Chlorine	0.1
Hydrogen	6.2	Phosphorus	0.2	Aluminium	0.1
	94.1%	Calcium	0.2		1.4%
		Iron	0.1		
		Magnesium	0.2		
		Potassium	0.9		
			3.3%		

The three basic elements and the seven major elements constitute 97.4% of the total dry weight. The major elements are needed by the plant for:

i) synthesis of proteins (including enzymes),
ii) formation of the middle lamella (*calcium pectate*) between the cell walls,
iii) maintenance of a suitable medium for the functioning of enzymes.

Where a particular element is missing in the nutrition of a plant, the plant may develop a mineral deficiency disease recognisable by its symptoms as being specific for the missing element. Similarly, *deficiency diseases* in animals may occur through lack of vitamins or minerals (see tables 1.1 and 1.2).

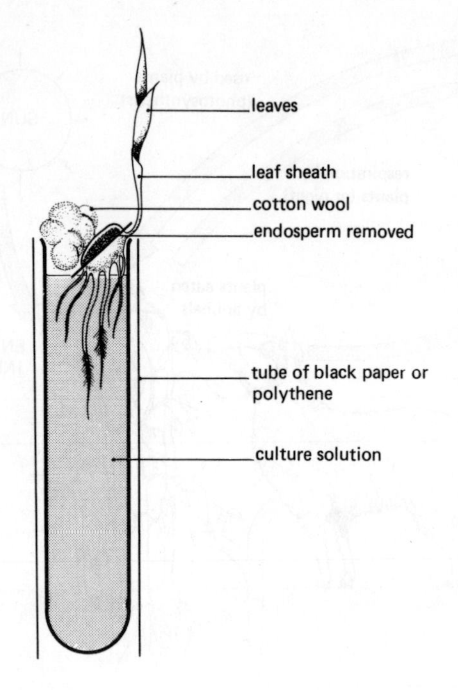

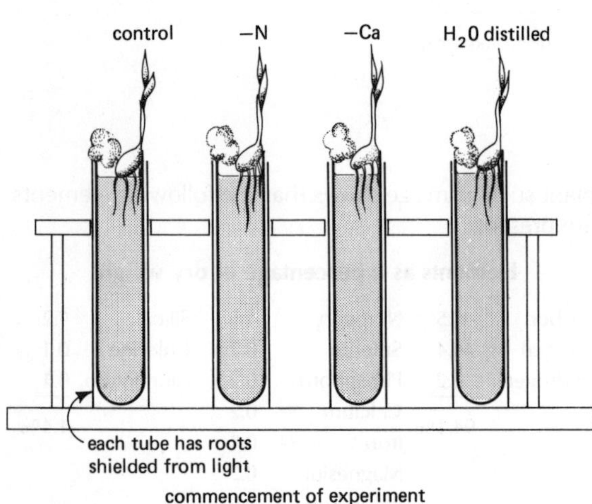

fig 1.29 The experimental apparatus for the culture of seedlings

Experiment

What mineral salts do plants need?

Procedure

1 Make up the following solutions devised by Sachs, a German botanist of the nineteenth century. The *complete culture solution* has a balanced amount of salts considered necessary for plant growth. If one of the salts is left out of the solution for experimental purposes then the solution must be balanced. This is done by replacing the missing element with an equal amount of one already present. The resulting solution is deficient in one element.

2

Complete solution	Lacking nitrogen	Lacking calcium
Calcium sulphate 0.25 g	Calcium sulphate	Magnesium sulphate
Calcium phosphate 0.25 g	Calcium phosphate	Potassium phosphate
Magnesium sulphate 0.25 g	Magnesium sulphate	Magnesium sulphate
Potassium nitrate 0.75 g	Potassium chloride	Potassium nitrate
Ferric chloride 0.005 g	Ferric chloride	Ferric chloride

Each culture solution to be dissolved in one litre (dm^3) of distilled water.

3 Take four test-tubes, place in a test-tube rack and fill each tube with one of the following culture solutions.

Tube 1 Complete solution – all mineral elements present for healthy growth.

Tube 2 All mineral elements present except nitrogen (nitrate).

Tube 3 All mineral elements present except calcium.

Tube 4 Distilled water (no dissolved salts).

4 Select four maize seedlings at about the same stage of development (these should have been germinated about seven days previously). Root systems should be of equal length. By means of cotton wool, wedge the seedling in the mouth of each tube so that the roots are well covered. Do not wet the cotton wool. Cut off the endosperm from each grain (see fig 1.29). This removes any alternate food supply.

5 Note the date and time, and set aside the tubes where the shoots will receive adequate light. The roots should be shielded from light by a black paper or black polythene tube.

6 Inspect the seedlings every day. Top up falling levels in each tube with *distilled water*, taking care each time not to wet the cotton wool.

7 At the end of two to three weeks examine the seedlings carefully and draw up a table of comparison showing:

A leaf colour—greenness, yellowness, dead-patches etc.

B leaf length—cut off each leaf at the base, measure its length and record the total lengths of all the leaves measured.

C root length—cut off the root system below the grain, separate each major root and cut off at the base. Measure the lengths of these roots and record the total root length. Construct the table below for each culture tube:

Major element	Function in the plant	Effect of deficiency	Notes
Nitrogen	Amino acid, protein, nucleotide synthesis.	Small sized plants – yellow underdeveloped leaves.	Frequently deficient in heavily used soils – nitrogenous fertiliser required.
Potassium	Amino acids and protein synthesis and particularly cell membrane formation.	Leaves have yellow edges – premature death.	Soils need potassium after heavy nitrogen manuring.
Calcium	Cell wall (middle lamella) development at root and stem apex.	Very poor root growth.	Little present in acid soils – helps aeration of clay soils.
Phosphorus	Formation of high energy phosphate compounds.	Small plants – leaves dull and dark green.	Frequently deficient – little present in soils over pH 7.
Magnesium	Part of the chlorophyll molecule – activator of enzymes.	Leaves turn yellow – veins remain green.	Deficient in acid soils.
Sulphur	Protein formation.	Leaves turn yellow.	
Iron	Chlorophyll synthesis.	Leaves turn yellow – veins remain green.	Not available in clay soils.
Minor element			
Manganese	Activator of some enzymes.	Shoots die back.	
Zinc	Activator of some enzymes.	Leaves not formed properly	Often deficient in acid soils.
Boron	Influences uptake of calcium ions.	Brown colouration appears in the shoot.	Easily washed out of soil by heavy rain.
Silicon	Cell wall formation in grasses, not essential in most plants.	Decrease in weight of cereal straw.	
Aluminium	Not essential but absence can cause cell division to be upset.		

Table 1.3 The major elements and their importance

	Tube 1 Control All minerals present.	**Tube 2** Less nitrogen	**Tube 3** Less calcium	**Tube 4** Distilled water
A Leaf colour				
B Leaf length				
C Root length				

8 Present the results shown in the table as a series of histograms for leaf length and root length in tubes 1, 2, 3 and 4.

Questions

1 Why should the cotton wool be kept dry?

2 What disadvantage have the roots of the seedlings in the culture solutions compared with roots in soil?

3 Which solution provided:

a) the most growth judged on leaf length;

b) the least growth judged on leaf length?

4 Which seedling had the least growth of root system? Check table 1.3 and explain this result.

fig 1.30 Bean plants in culture solutions

5 Why is it difficult to determine whether the absence of a particular element did more harm to root growth or to shoot growth?
6 Why are the solutions topped up with distilled water rather than with the mineral salt solutions?

1.30 Nutrition in animals

We have seen in section 1.20 that plants can manufacture food to provide for their energy requirements, growth and health. In order to take carbohydrates, fats and proteins into their bodies, animals must break down these complex molecules into their original building blocks. This is partly because food substances must be *transported* by the blood to where they are needed. In order to be *absorbed* by the blood, they need to be *soluble* and *small* enough to pass through a blood vessel wall. This process of food breakdown is called *digestion* and occurs in the alimentary canal (gut) of an animal. Any undigested food is eliminated from the end of the alimentary canal as *faeces*.

Experiment

A model gut and the movement of molecules
Procedure
1 Take a beaker (or tin) of tap-water and warm to about 37°C (body temperature of a mammal).
2 Take about 15 cm of *visking (cellulose) tubing*, wet it under a tap and tie a knot very tightly in one end. Open up the other end and pour in a mixture of 5% starch and 10% glucose solution. Close the open end with a paper clip.
3 Wash the outside of the tubing with tap-water to remove any external traces of starch and glucose.
4 Place the tubing inside a boiling tube containing a little tap water. Place the boiling tube in the water bath.
5 Test some of the water from the boiling tube. Extract a small sample with a pipette, place in a small test-tube and add iodine solution. Test another sample with Benedict's solution. Boil this in a test-tube.
6 Repeat the two tests after twenty minutes. Record your results in the form of a table.

Time of test	Result with iodine solution	Deduction

Time of test	Result with Benedict's solution	Deduction

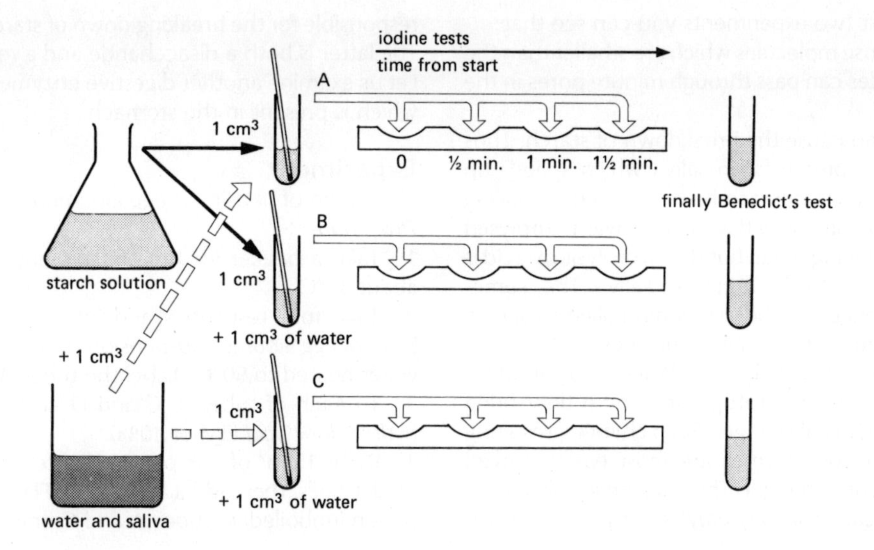

fig 1.31 A flow sequence for the saliva experiment

Questions

1 Was any starch or glucose present in the water in the first tests?

2 Was any starch or glucose present in the water after the second tests, after twenty minutes?

3 If one of the substances has passed through the tubing, what explanation can you give for this result?

4 What control should be set up in this experiment?

The visking tubing was considered to be a model for the gut and we can suppose that the gut might behave in the same way, permitting small molecules to pass, but not large ones. Starch, however, could also be broken down into smaller maltose or glucose molecules. We must look for evidence of this breakdown.

Experiment

What action does saliva have upon starch?

Procedure

1 Heat a beaker or tin of tap water to 37°C.

2 Wash out your *buccal cavity* (mouth) with distilled water and then *suck* a small smooth stone or a rubber band. This results in the collection of *saliva* which should be discharged into a small beaker. Pour the saliva into a test-tube and place the test-tube in the water-bath.

3 Take three test-tubes and label them A, B and C. Add 1 cm^3 of 2% starch solution to tubes A and B. Add 1 cm^3 of distilled water to tubes A, B and C. Measure the quantities carefully.

4 Prepare a spotting tray or white tile with three rows of iodine solution drops as in fig 1.31.

5 Record the time, then add 1 cm^3 of warmed saliva to tube A. Stir in the contents with a glass rod. Immediately remove one drop of the mixture, by means of the glass rod, and add it to the first drop of iodine solution. Record the colour of the iodine solution. Wash the end of the glass rod.

6 After 30 seconds, remove a second drop and put it into the second drop of iodine solution. Record the colour. Rinse the glass rod. Repeat the test every half minute until the colour of the iodine solution no longer changes.

7 Repeat the process for tube B over the same length of time, omitting saliva. Record your results.

8 Add 1 cm^3 of saliva to tube C and repeat the process for tube C. Record your results.

9 Finally test the contents of tubes A, B and C for reducing sugar using Benedict's solution.

Record your results in table form.

	Colour of iodine solution in test-tubes		
Time	A	B	C
0 30 secs. 1 min. etc.			

Questions

5 What do you conclude about the fact that, after a while, the iodine solution no longer changed colour?

6 What could have caused this? Give your reasons.

7 What happened to the starch in tube A? State which test supports your answer.

8 a) How does this experiment help to explain digestion?

b) What control would confirm your conclusion?

From the last two experiments you can see that:

i) the glucose molecules which are smaller than the starch molecules can pass through minute pores in the visking tubing,

ii) saliva can cause the breakdown of starch. Thus there must be 'something' in saliva which speeds up chemical change. Starch can be hydrolysed by boiling with acid in the same way that sucrose was hydrolysed to test it for reducing sugar, but this hydrolysis by acid is slow compared with the action of saliva. The 'something' which speeds up the reaction is called a *catalyst*, and organic catalysts are called *enzymes*.

Enzymes occur widely in plants and animals, particularly within the protoplasm, where they bring about many chemical changes. Such changes, resulting from the action of enzymes are most easily studied when they occur in the gut. The enzyme in saliva is an example. It is called *salivary amylase* or *ptyalin* for it is responsible for the breaking down of starch to maltose. The latter is both a disaccharide and a reducing sugar. Let us examine another digestive enzyme called *pepsin* which is present in the stomach.

Experiment

The action of pepsin on egg albumen

Procedure

1 Take a beaker (or tin) of tap-water and warm to about 37°C.

2 Take four test-tubes and into each place about 5cm³ of *egg albumen suspension* (1% dried albumen in water heated to 90°C). Label the tubes A, B, C and D.

3 To each of tubes B, C and D add three drops of *hydrochloric acid* (2M or 10%).

4 Place 1 cm³ of 1% pepsin in a test-tube and heat until it boils, then add it to tube D. Place 1 cm³ of 1% pepsin (unboiled) in tubes A and C only.

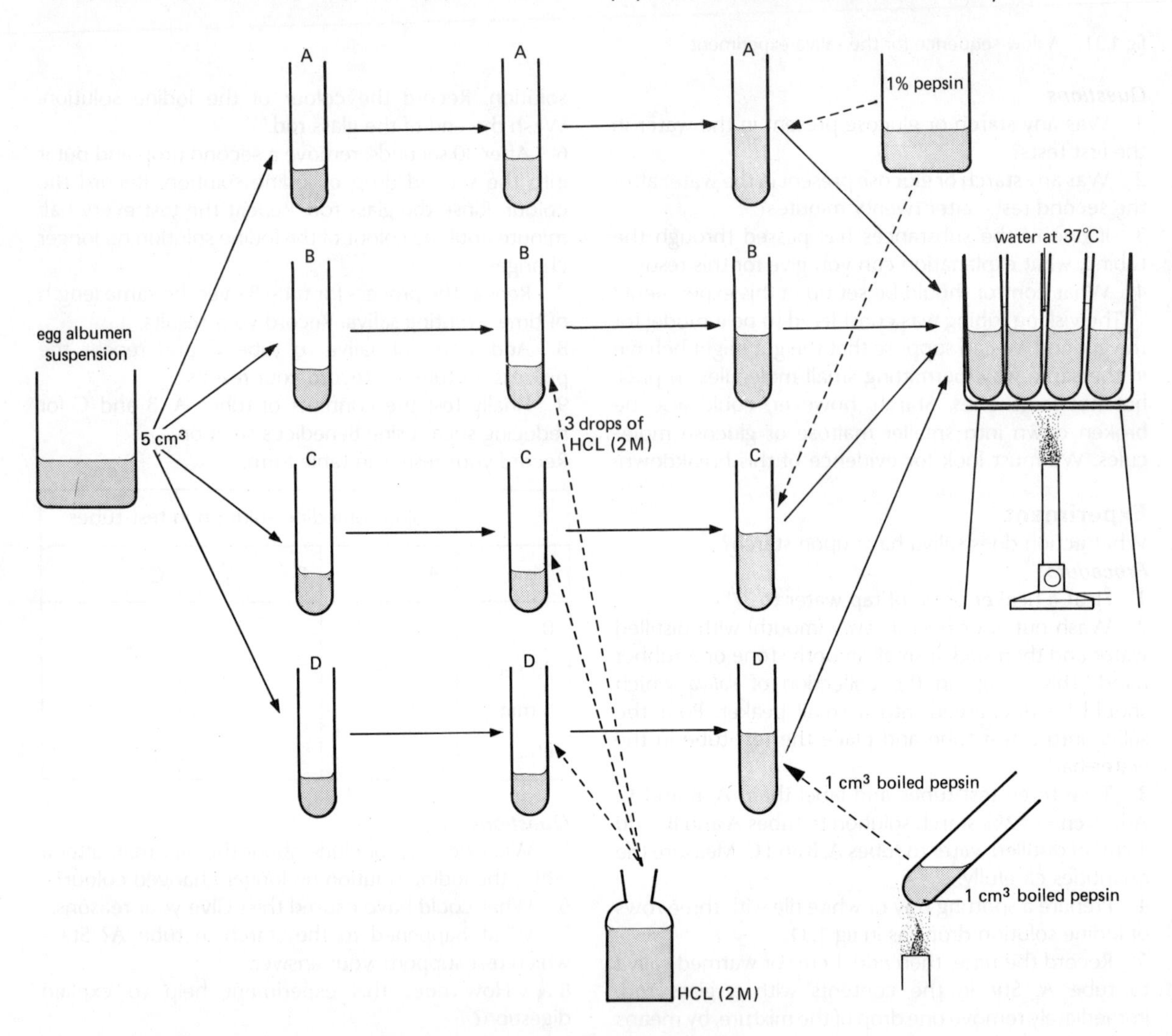

fig 1.32 A flow sequence for the pepsin experiment

5 Place all four tubes in the water bath.
6 Examine the tubes at 2 minute intervals, and after 6–8 minutes remove the four tubes from the water bath and place them in a test-tube rack. Examine the contents of each tube and record your observations in the form of a table.

Tube	Contents	Appearance at beginning of experiment.	Appearance at end of experiment.
A	albumen + pepsin		
B	albumen + HCl		
C	albumen + pepsin + HCl		
D	albumen + boiled pepsin + HCl		

Questions
9 In which tube have the contents a different appearance after six minutes or more? What has happened to the egg albumen suspension?
10 How is the enzyme affected by boiling? What evidence leads you to this conclusion?
11 What other hypothesis could you advance to account for the results in tube C?
12 What are the best pH conditions in which the enzyme pepsin can act?

Pepsin is a *protein-digesting enzyme* which occurs in the *stomach*. The experiment shows that it can digest egg albumen which is mainly protein. It is not possible to generalise about all enzymes from this one experiment. Nevertheless we have now examined two enzymes occurring in two different parts of the alimentary canal, the buccal cavity (mouth) and the stomach. One enzyme digests starch (carbohydrate) and the other digests egg albumen (protein). Digestion is essentially the *breaking down of food* into smaller and smaller parts. This may begin, even before we eat food, by cutting it with a knife into a size we can get into the mouth. Once in the buccal cavity the teeth and the tongue help to cut the food into even smaller pieces, and finally the process is completed by enzymes. We can therefore describe the two types of digestion as (i) mechanical (ii) chemical.

1.31 Ingestion

Many animals take in their food in the form of small organisms and pieces of organic debris which are strained out of the water in which the animals live by filtering mechanisms. Such animals range in size from ciliate protozoans and small bivalve (lamellibranch) molluscs to the baleen whale. In the bivalve mollusc, the organs used are called gills since they have a respiratory function. The gills are hollow and their sieve-like walls are covered with minute hair-like projections known as cilia. The waving motion of these cilia creates a current, which draws water into the gill cavity. Gland cells in the gills produce mucus which entangles the food particles suspended in the water. These are then swept towards the mouth by the action of the cilia and eventually enter the alimentary canal.

In the Crustacea, some groups such as water fleas have their limbs hidden under the shell (carapace). The limbs are fringed with bristles, the movement of which causes a current of water to move towards the head end. This current carries with it small suspended particles of food which are strained out by the bristles before being swallowed.

Detritus feeders include the earthworm and the sturgeon (a large fish), both of which swallow large quantities of earth or mud from which they are able to extract nutrients.

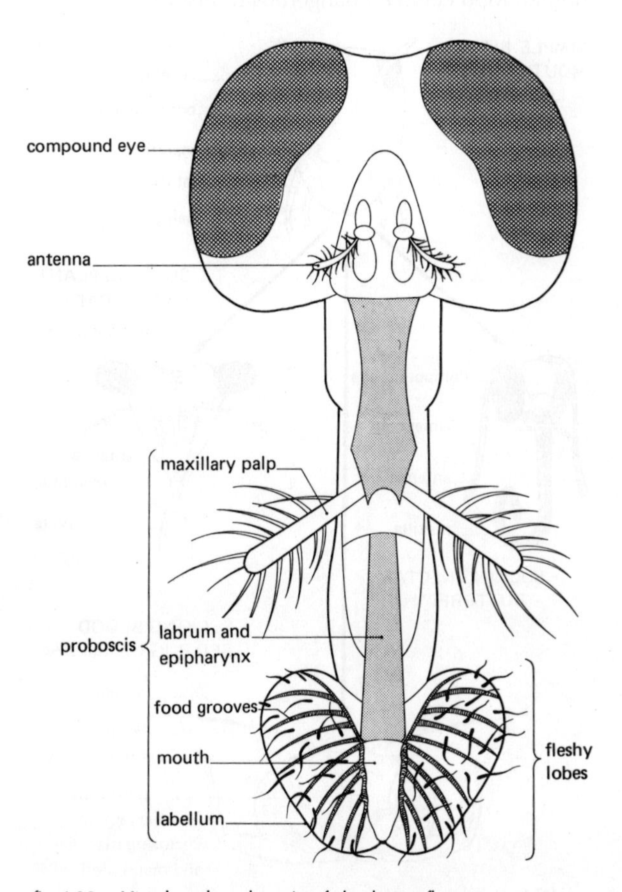

fig 1.33 Head and proboscis of the housefly

Certain fluid feeding animals have mouth-parts specially modified to form minute tubes capable of penetrating animal or plant bodies and sucking up the organic fluids. Many of these organisms are insects such as the mosquito, which sucks blood, and the greenfly, which sucks plant sap.

Investigation 1

How does a housefly feed?

Procedure

1 Capture some houseflies and place them in a glass jar containing a lump of sugar. Cover the container with muslin. Observe the way in which they feed on sugar.

2 Examine the head of a dead housefly. The mouth-parts form a proboscis containing two tubes (see fig 1.33). One tube conducts down saliva to moisten the food and partially digest it before it is drawn through the fleshy lobes of the proboscis and up the second tube.

Questions

1 What organic materials do houseflies normally feed upon? How can this habit be dangerous to humans?

2 List other ways in which houseflies alighting on human food could be dangerous to health.

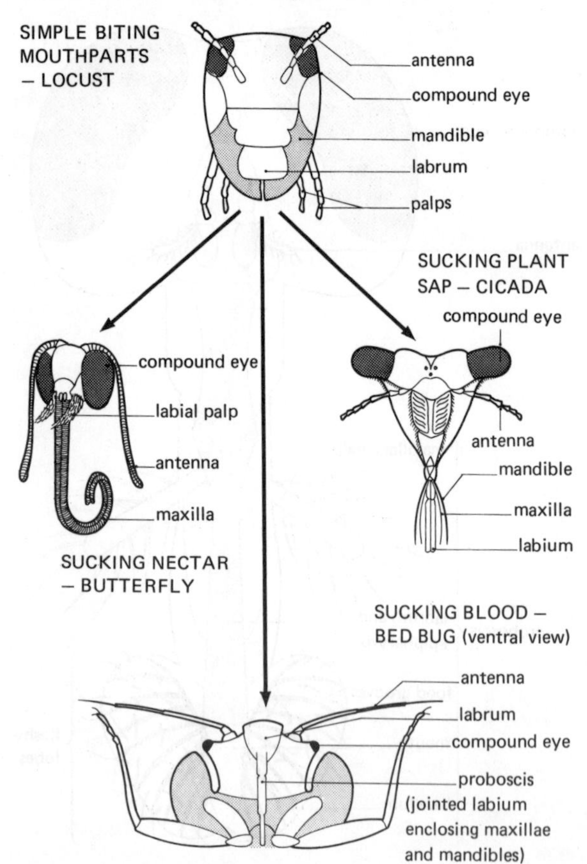

fig 1.34 Insect mouth parts

Notice that when the housefly feeds, the proboscis is extended under the head by the combined actions of muscles and blood pressure. The salivary glands secrete saliva down the duct onto the food and then the food is sucked up by the powerful muscles of the pharynx.

In the aphids (greenfly of roses and blackfly of beans) and cicadas the mandibles and maxillae (see fig 1.34) of the mouthparts are produced into slender pointed stylets, while the labium forms a sucking tube. The stylets pierce the epidermis of the plant shoot so that the labium can enter the wound. The juices of the phloem are sucked into the gut. The Lepidoptera (butterflies and moths) do not possess mandibles, but the maxillae form a sucking tube made of two halves fastened together by spines and hooks. The tube, which is very flexible, is coiled into a spiral beneath the thorax, and when extended it can be used for sucking up nectar from flowers.

Most animals that eat solid food swallow it in comparatively large lumps.

Investigation 2

To examine the mouthparts of a dead locust

Procedure

1 If locusts are kept in the laboratory, place some blades of grass in the cage and watch the insects feeding. Alternatively watch a film loop showing the feeding mechanism.

2 Kill a locust by placing it in a jar with a cotton wool pad soaked in chloroform.

3 Pin the locust onto a cork mat or wax with its ventral side uppermost.

4 Separate the upper lip or labrum from the other mouth parts by grasping the base firmly with forceps and pulling gently (see fig 1.35)

5 A pair of black biting jaws (mandibles) are now visible. Grasp one of these with forceps near its base and remove it, whilst holding the head firmly with your fingers.

6 A second pair of appendages (maxillae) lie beneath the mandibles. Remove one of these in a similar way.

7 The lower lip or labium can now be removed as a single structure.

8 Place the mouthparts on a sheet of white paper and examine them with a hand lens or binocular microscope.

Questions

3 Examine the inner surface of the mandible. Can you see how its structure could be related to its function?

4 What is the function of the maxillae?

5 How do the mandibles move in a living locust? Does the locust feed on the edge of the grass or along its flat surface?

An alternative animal for study is the Cockroach (*Periplaneta americana*). This insect is easily cultured in the laboratory and is unfortunately very common in

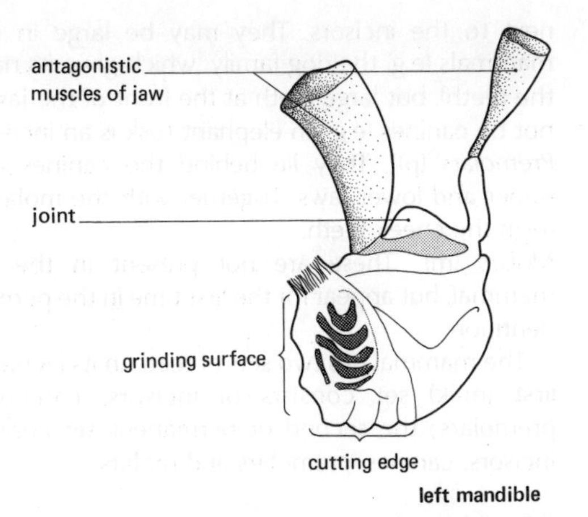

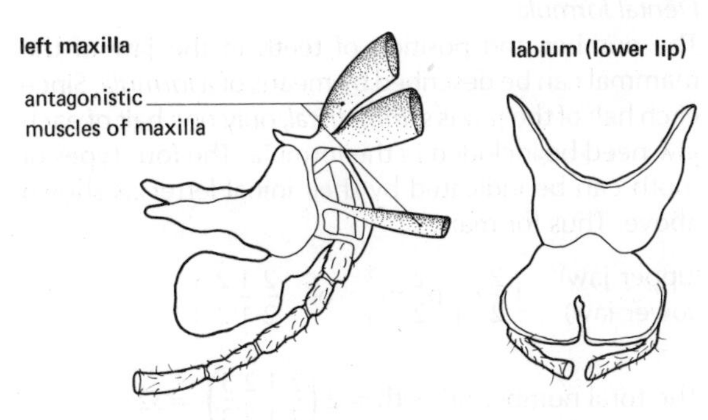

fig 1.35 Mouth parts of a locust

fig 1.36 Desert locust

the kitchens of some houses, hotels and restaurants. It is omnivorous, although it lives mainly on sugary and starchy substances. The food particles are held by the maxillae and labium so that the mandibles can chew the food with a lateral action. The salivary glands on either side of the gut secrete saliva onto the food as it is chewed. The saliva moistens the food, thus assisting swallowing, but at the same time the amylase begins the digestion of starch to form the sugar maltose.

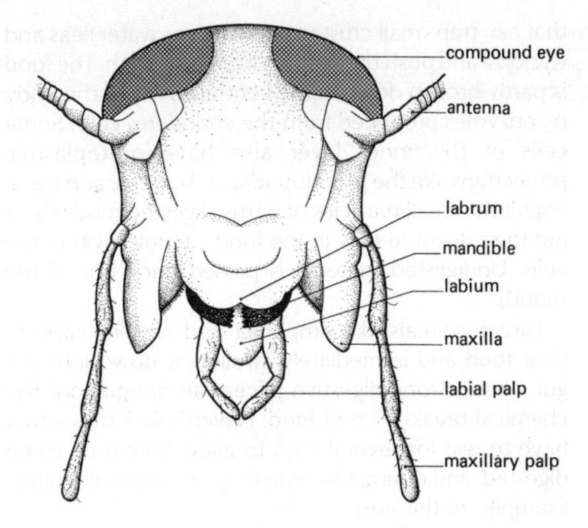

fig 1.37 The head of a cockroach – front (dorsal) view

Smaller animals also take in their food in relatively large lumps. *Amoeba* is a minute protozoan living in freshwater (size 1×10^{-4} m). It produces projections of its protoplasm (pseudopodia) that can surround a piece of food (diatom or bacterium) and enclose it in a food vacuole. The food is digested by enzymes secreted into the vacuole. The final digested products can be absorbed and assimilated into the protoplasm.

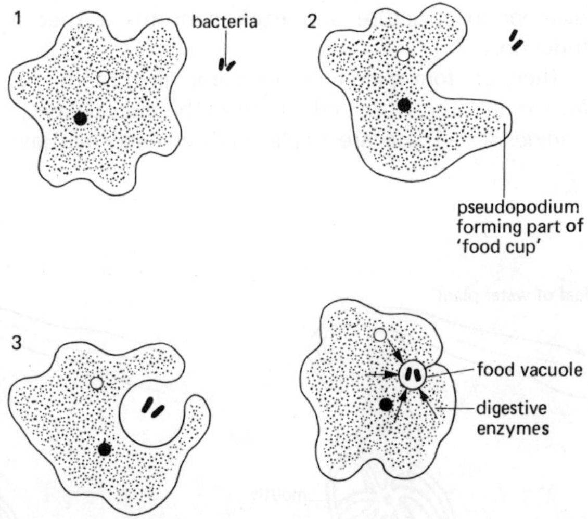

fig 1.38 *Amoeba* feeding

Hydra is a small fresh-water animal of the phylum Coelenterata, which also contains jellyfish and sea anemones. It is just visible to the naked eye, varying in length from 10–20 $\times 10^{-3}$m. The hollow body has two layers of cells forming an outer ectoderm and an inner endoderm. The mouth in the upper part of the body is surrounded by eight tentacles. These have stinging cells that can trap small crustaceans (such as water fleas and *Cyclops*) and push them down into the mouth. The food is partly broken down in the central cavity of the body by enzymes produced from the endoderm cells. Some cells of this inner layer also have protoplasmic projections like the pseudopodia of *Amoeba*, and these engulf the small particles of partly digested food which are then digested fully in the food vacuoles within the cells. Undigested material is passed back out of the mouth.

Larger animals like amphibia and reptiles capture their food and immediately swallow it down into the gut where strong digestive juices can bring about the chemical breakdown of food. Nevertheless they often have to rest for several days to allow their food to be digested, and during this time they are very vulnerable. Examples of this are:

a large python swallowing a pig,
a grass snake swallowing a frog,
a toad swallowing a worm.

The teeth of these organisms are all alike and are simply sharp backward-pointing cones. Mammals are unique in that they cut and grind their food into small pieces before it is swallowed. They are able to do this because of their complex *dentition* (i.e. several types of teeth) and the arrangement of their *jaw bones*, jaw muscles and cheek muscles. All of these, together with the *lips and tongue*, help in their different ways. However, the most important are the *teeth*. Each type is different in structure and each performs a special function.

There are four main types of mammalian tooth.

Incisors (i) These are in the front of the buccal cavity.

Canines (c) These are single teeth in each half jaw next to the incisors. They may be large in certain mammals (e.g. the dog family, which gave its name to the teeth), but large teeth at the front of the jaw need not be canines (e.g. an elephant tusk is an incisor).

Premolars (p) They lie behind the canines on the upper and lower jaws. Together with the molars they form the cheek teeth.

Molars (m) These are not present in the young mammal, but appear for the first time in the permanent dentition.

The mammal has two sets of teeth in its lifetime. The first (milk) set consists of incisors, canines and premolars; the second or permanent set consists of incisors, canines, premolars and molars.

Dental formula

The number and position of teeth in the jaw of the mammal can be described by means of a *formula*. Since each half of the jaw is *symmetrical*, only one half of each jaw need be included in the formula. The four types of tooth can be indicated by their initial letter as shown above. Thus for man:

(upper jaw)
(lower jaw) $i\frac{2}{2}\ c\frac{1}{1}\ p\frac{2}{2}\ m\frac{3}{3}$ or $\frac{2\,1\,2\,3}{2\,1\,2\,3}$

The total number of teeth $= 2\left(\frac{2\,1\,2\,3}{2\,1\,2\,3}\right) = 32$

Investigation 3

How are human teeth distributed?

Procedure

1 Examine a fellow pupil's teeth or those of your brother or sister. How many teeth can you see?

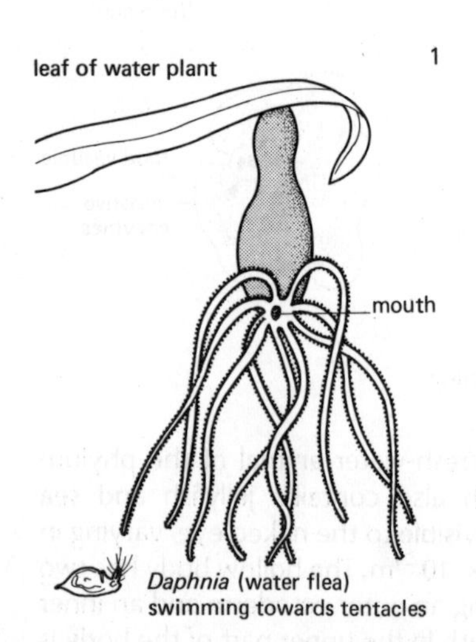

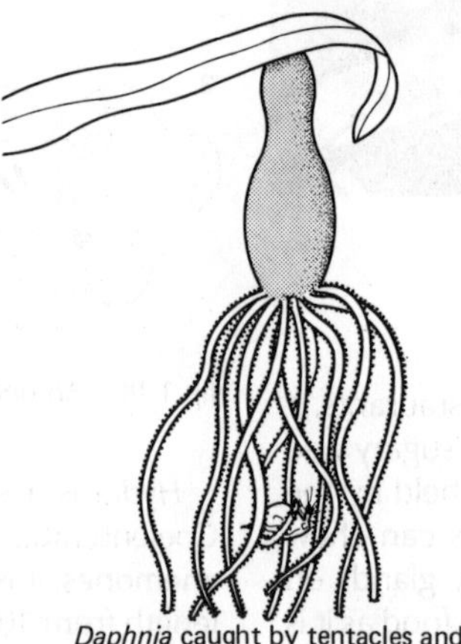

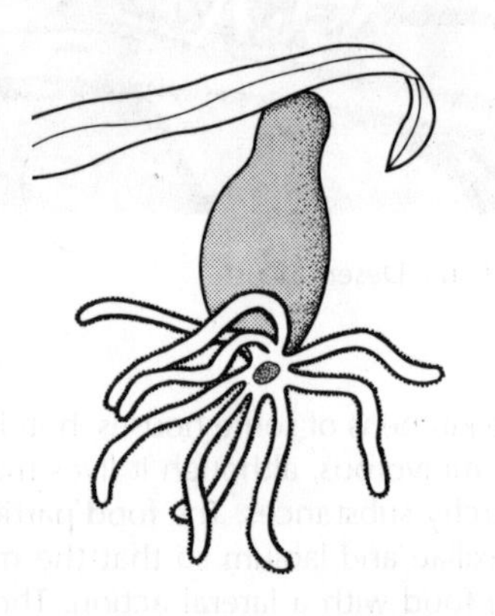

fig 1.39 *Hydra* and the capture of its food – sequence 1 to 3

	Number in upper jaw	Number in lower jaw	Teeth lost or extracted	General shape of tooth	Probable function
Incisor					
Canine					
Premolar					
Molar					

Record the following: Age of subject and total number of teeth.
Complete the above table.

2 Ask your subject to eat a fruit, such as an apple. Observe how the lips, tongue and teeth are used.

3 Look at fig 1.40 and compare your information in 1 above with the number of teeth shown in the figure.

4 Try to find the answer to the following questions by examining your young brothers and sisters and by discussion with your fellow students.

i) At what age do teeth first appear?

ii) How many teeth are present in upper and lower jaws by the age of three years?

iii) Which teeth appear first?

iv) Which are the last teeth to appear?

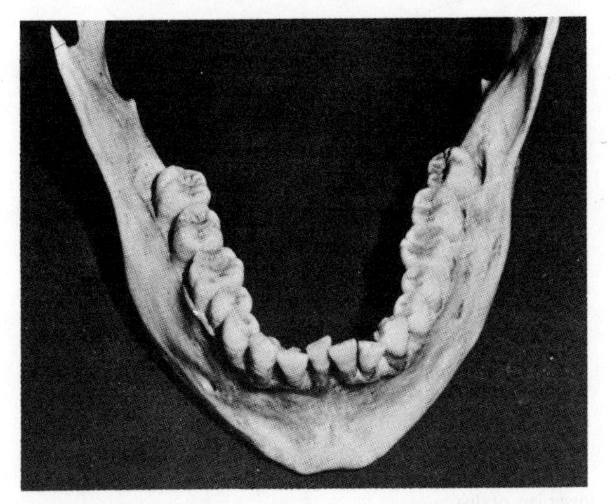

fig 1.40 An adult human lower jaw with teeth

Internal structure of the tooth

If a tooth is sawn in half the internal structure can be seen as shown in fig 1.41. The outer white layer of the *crown* is the *enamel*, which is the hardest organic substance known. Within this lies the *dentine* which is also hard but unlike the enamel contains living tissue. However, it is soft enough to be carved, for the ivory of the elephant's tusk is dentine. Dentine occurs in the crown and in the *root* below the surface of the gum. In the centre of the tooth is the *pulp cavity* containing nerves and blood vessels. The root is buried firmly in the *socket* of the jaw, fixed by *cement* and surrounded by *fibres*. These fibres allow some freedom of movement

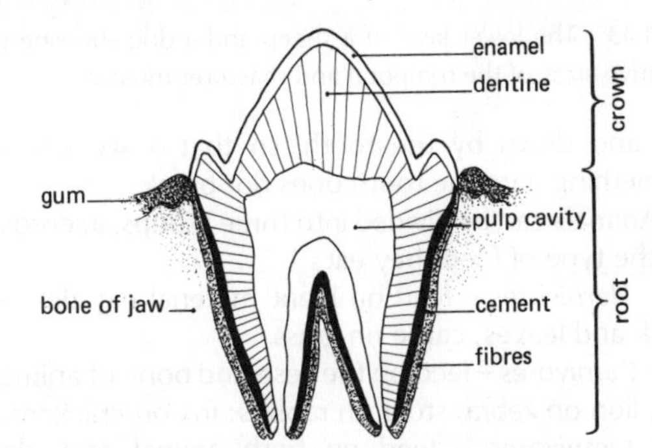

fig 1.41 A vertical section of a premolar tooth of a carnivore

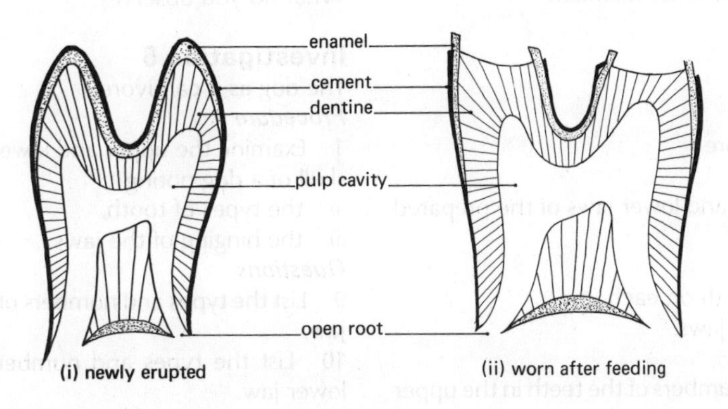

fig 1.42 A vertical section of a premolar tooth of a herbivore

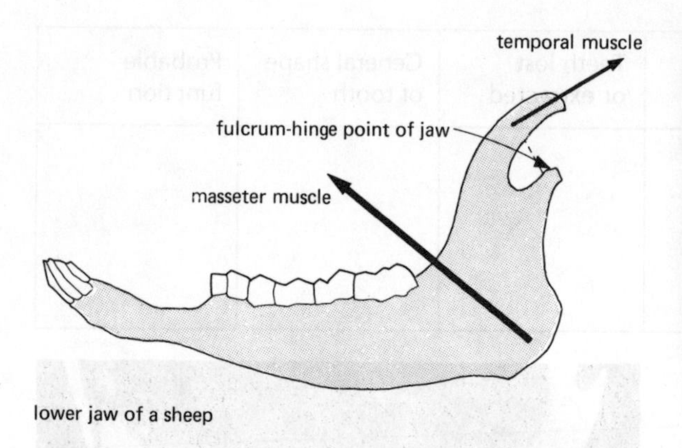

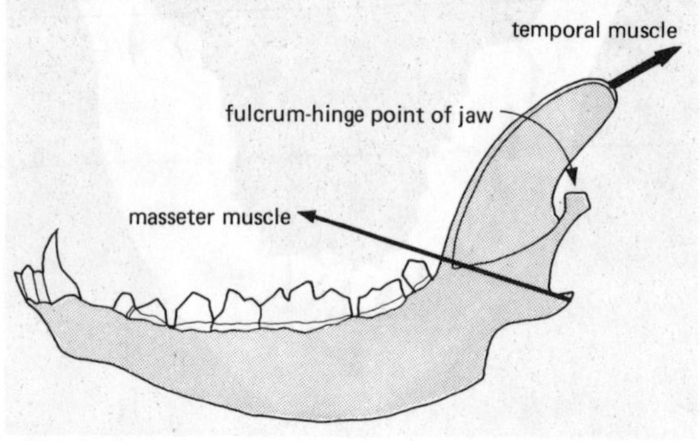

fig 1.43 The lower jaws of a sheep and a dog showing the relative sizes of the temporal and masseter muscles

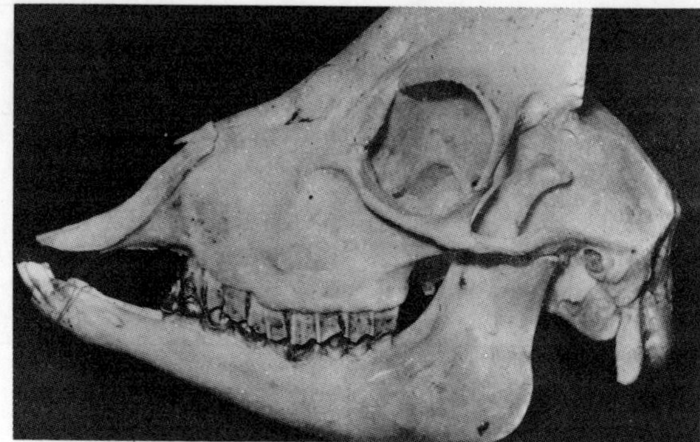

fig 1.44 (a) The skull of a sheep

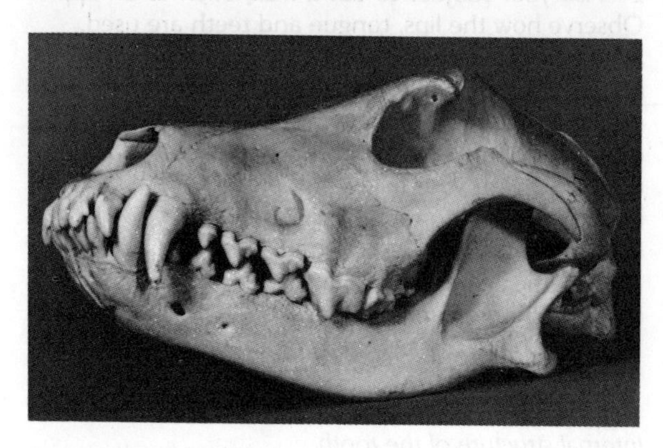

fig 1.44 (b) The skull of a dog

up and down by the tooth, so that if we bite on something hard the tooth does not break.

Animals can be divided into three groups, according to the type of food they eat:

i) *Herbivores* – feed on plant material e.g. deer on bark and leaves; cattle on grass.
ii) *Carnivores* – feed on the flesh and bone of animals, e.g. lion on zebra; stoat on rabbits; fox on chickens.
iii) *Omnivores* – feed on both animal and plant material; e.g. man on meat and vegetables; wild boar on grass, shoots, leaves and small animals.

Investigation 4

The sheep as a herbivore

Procedure

1 Examine the upper and lower jaws of the prepared skull of a sheep noting:
i) the types of tooth,
ii) the gaps in the teeth on each jaw,
iii) the hinging of the jaws.

Questions

1 List the types and numbers of the teeth in the upper jaw.
2 List the types and numbers of the teeth in the lower jaw.
3 How does the sheep eat its food?
4 What is the function of the large gap between the teeth of the lower jaw?
5 What is the dental formula for the sheep?
6 How does the sheep chew?
7 Describe the shape and probable functions of each type of tooth.
8 Extract a molar and look at the end of the root. What do you observe?

Investigation 5

The dog as a carnivore

Procedure

1 Examine the upper and lower jaws of the prepared skull of a dog noting:
i) the types of tooth,
ii) the hinging of the jaws.

Questions

9 List the types and numbers of the teeth in the upper jaw.
10 List the types and numbers of the teeth in the lower jaw.
11 How do such teeth help to obtain food?

12 Describe the action of the hinge and the movement of the lower jaw.
13 What is the dental formula for the dog?
14 Describe the shape and the probable functions of each type of tooth.

The jaw movements of these two animals can be seen if you watch them feeding. The sheep moves the lower jaw from side to central so that the premolar and molar teeth grind the grass. The dog jaw moves only in an up and down plane so that the animal cannot grind, but cuts or crushes flesh and bone. These different actions are possible because of the different ways in which the jaws are hinged. In the sheep the hinge is loose so that the jaw can move in a circular fashion, whereas in the dog it is tightly hinged to permit only up and down movement. The jaws are moved by two muscles, the *temporal* and *masseter*, and their differences in the two animals are shown below.

	Temporal muscle	Masseter muscle
Sheep	A small muscle.	A large muscle, the high jaw joint gives this muscle greater leverage.
Dog	A large muscle, applies greater leverage for cutting and crushing, and also for the stabbing movement of the large canines.	A small muscle.

Table 1.4 Jaw muscles of sheep and dog

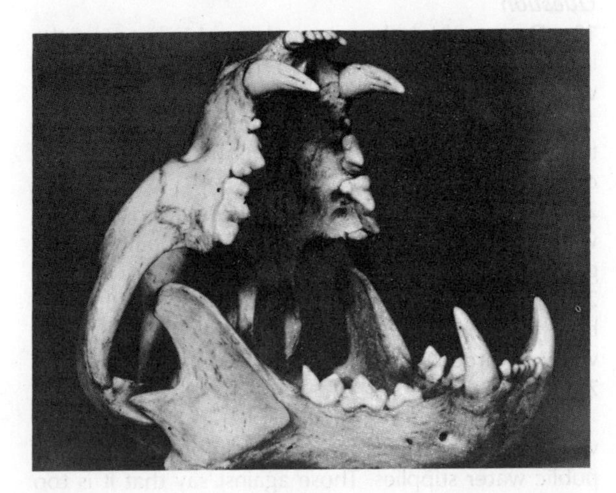

fig 1.45 The skull of a leopard

Herbivores can be divided into
i) those that *cut* roots, bark etc., using very sharp incisors e.g. rabbits, rats, guinea pigs,
ii) those that *pull up* grass using the lower front teeth and a pad on the front of the upper jaw, (or upper and lower incisors) e.g. cows and sheep,
iii) those that use their lips to pull up grass or leaves off trees, e.g. hippopotamus,
iv) those that use a specialised organ to pull leaves, bark and branches off trees, e.g. the trunk of the elephant.

Carnivores do not have the variety of feeding methods of the herbivores, but use their prominent canines and powerful jaws to kill their prey, and then tear off flesh.

Experiment

Tooth decay and its cause

Procedure

1 Take a tooth extracted from a mammal (e.g. sheep) and poke it in the crown region with a sharp needle or forceps.
2 Place the tooth in *dilute hydrochloric acid* for two or three days.
3 Wash the tooth and examine it. Using a needle or forceps again, poke the tooth in the crown region.

Questions

15 What differences can you see in the tooth? Compare it with a similar tooth still in the jaw of the sheep.
16 What happens when you poke the tooth with the sharp instrument, after it has been in the acid?
17 What part of the tooth has become exposed?

Dental care in man

In order to keep teeth healthy, there are three points to consider.
i) They must be *used* in the *correct* way.

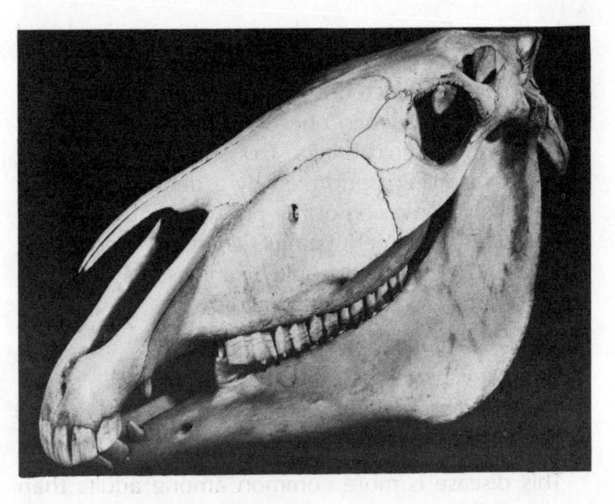

fig 1.46 The skull of a horse

ii) They must be kept *clean*.
iii) They must receive *regular* expert attention.

i) Teeth should never be used for improper purposes such as cracking nuts, opening bottle tops or even breaking cotton. They must be given the right kind of work by chewing hard food, for modern diets are often too refined and prepared to need chewing. Small babies should be provided with a bone ring or some hard object on which to chew. Fibrous tough foods, e.g. apples, raw carrots, nuts, oranges and other fruits, should be included in the diets of children and adults.
ii) Teeth must be kept clean to prevent food residues causing decay. *Bacteria* growing on these pieces of food *produce acids* which can eat away the enamel and the dentine. A toothpick can remove food particles from between the teeth, but they should also be brushed after each meal with a toothbrush or a tooth stick. If no toothbrush is available, a small piece of rag and some salt can be equally effective. The salt acts as an antiseptic and an abrasive.
iii) In spite of care, decay and other diseases may attack the teeth. Treatment will be most effective if the disease can be discovered and controlled in the early stages. If possible, the teeth should be examined at least twice a year by a dentist. Small cavities in the teeth can be filled, while badly decayed teeth can be extracted.

Dental diseases

The two most important diseases are *dental caries* and *periodontal disease*.

i) Dental caries is caused by:

1 lack of hard food;
2 too much sweet food;
3 lack of calcium in the diet;
4 lack of vitamin D;
5 lack of cleaning;
6 general ill-health.

In the earliest stages only *enamel* is involved and since the process is painless, it escapes notice. If this early stage is neglected it spreads *to the dentine* and this is noticeable when one is eating sweet foods or drinking hot or cold drinks. This stage can be easily halted by a 'filling' performed by a dentist. The third stage involves invasion of the pulp cavity by bacteria and can be extremely painful when the nerves are attacked. The final result could be an abscess, and then the tooth may have to be extracted.

ii) Periodontal disease is caused by:

1 lack of vitamins A and C;
2 lack of massage of the gums;
3 imperfect cleaning.

This disease is more common among adults than among children. It causes the *gums* to become *soft and flabby* so that they do not properly support the teeth. The reddening of the gums, bleeding and the presence of *pus* are symptoms of the disease described as *pyorrhoea*. It does not occur if the diet is correct, if the gums are brushed regularly to encourage blood circulation and if particles of food are not allowed to accumulate. When food accumulates, it leaves a film on the teeth which builds up to form a hard layer called 'plaque'. Plaque is the main cause of periodontal disease.

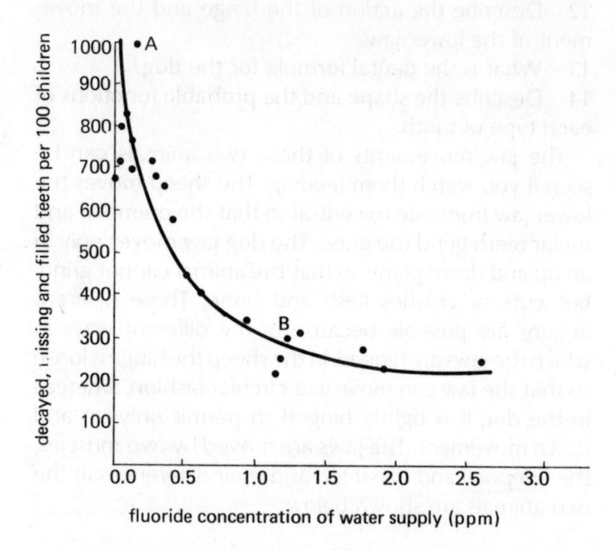

fig 1.47 A graph showing the effect of fluoride concentration on tooth decay in children

Fluoridation

Recent research has shown that *fluoride salts* in the drinking water may have some effect on reducing dental caries. Examine the graph shown in fig 1.47 and answer the following questions.

Question

18 Do you think there is a relationship between the incidence of dental caries and the fluoride content of water? If so, in what way?
19 What is the number of cavities per child where the water contains about 0.2 ppm of fluoride (point A on the graph)?
20 What is the number of cavities per child where water contains about 1.3 ppm of fluoride (point B on the graph)?
21 If your drinking water contains 0.1 ppm of fluoride, how much more fluoride would you need to add to the water in order to reduce the incidence of caries by 75%?

Scientific opinion is divided on the question of whether or not this chemical should be added to the public water supplies. Those against say that it is too imprecise a way of ensuring that safe limits will not be exceeded. The amount of water consumed by people varies considerably and it is now known that in polluted industrial areas the amount of fluoride contained in food alone may already be too high. Certainly it does involve a risk to add 1 ppm of fluoride to a water supply system, and many countries have now had second thoughts about the safety of such a scheme. In Sweden fluoridation is banned for medical reasons, in France scientific and medical opinion has always been against it; Holland, which has had fluoride in its water supply for a long time, has now decided against it.

Caution would seem to be necessary, particularly in highly industrialised countries, in order to avoid overdosage.

Muscular action on food

The stomach has a muscular coat of four layers arranged in different directions. The *churning action* brought about by the contraction of these muscles, and the secretion of *gastric juice* by the walls of the stomach, results in the food becoming a liquid called *chyme*.

The movement of food through the *oesophagus* towards the *stomach* and onwards through the small intestine after leaving the stomach, is also brought about by the muscular action of the gut wall. When a portion of food is swallowed it stretches the wall of the oesophagus. This contains circular and longitudinal muscles which, by their contraction, force the food on down the oesophagus. Similar waves of contraction and relaxation flow along the small intestine, pushing food in front of the contracting muscles. This process of moving food is called *peristalsis*.

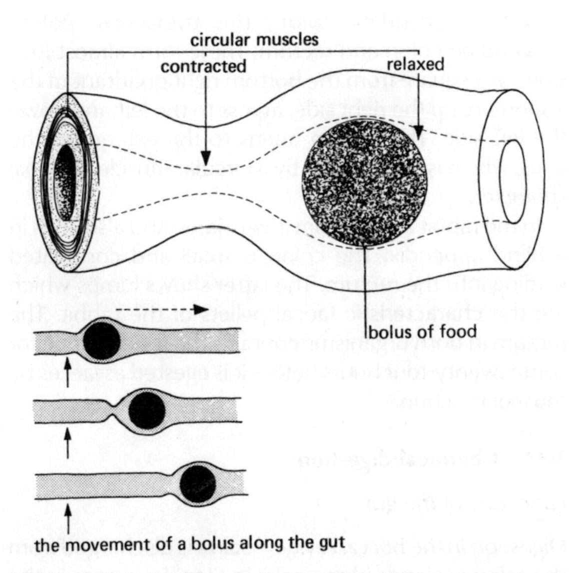

fig 1.48 A diagram showing peristalsis and how food is moved along the oesophagus and intestines

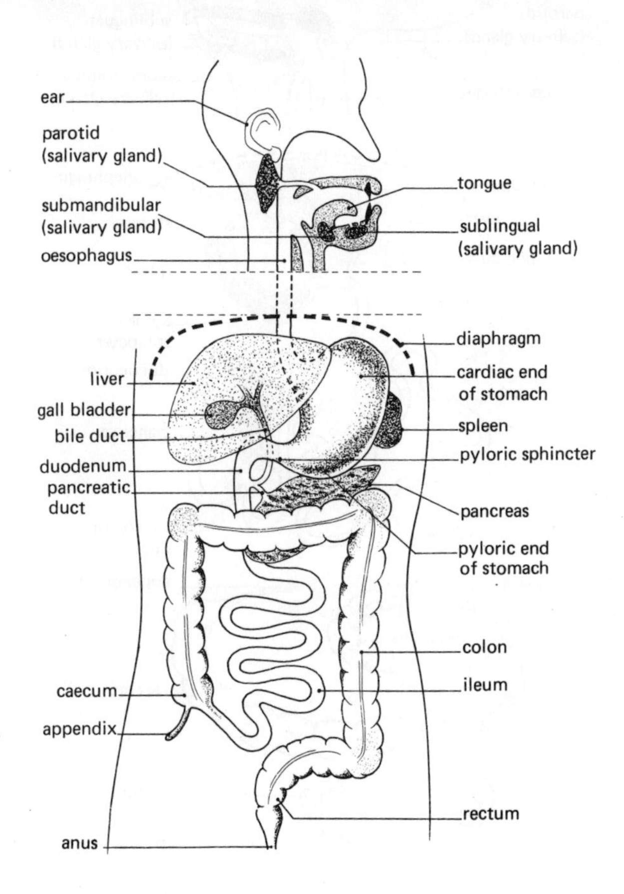

fig 1.49 A diagram of the alimentary canal of Man

1.32 Structure of the alimentary canal (gut)

Examine fig 1.49 and fig 1.50 which shows the comparable structures of the gut in Man and the rabbit. In all mammals there are corresponding sections of the gut, but in different species there may be different emphasis on certain sections. This is particularly noticeable in the case of the *caecum* and *appendix* of the rabbit, compared with those of Man.

Food is taken in through the mouth and its digestion begins in the buccal cavity where it is chewed and moved around. As the food is pushed to the back of the buccal cavity an automatic swallowing action (a reflex) causes it to be pushed into the oesophagus. This is a straight tube passing through the chest region into the stomach, which lies just under the *diaphragm*. Food does not drop down the oesophagus, like a stone down a well, but is pushed down by the contraction of the muscular wall. This can even occur when you are standing on your head and drinking water!

The stomach retains the food for three to four hours, churning it into *chyme*. Digestion, begun in the buccal cavity, is continued by mechanical and chemical

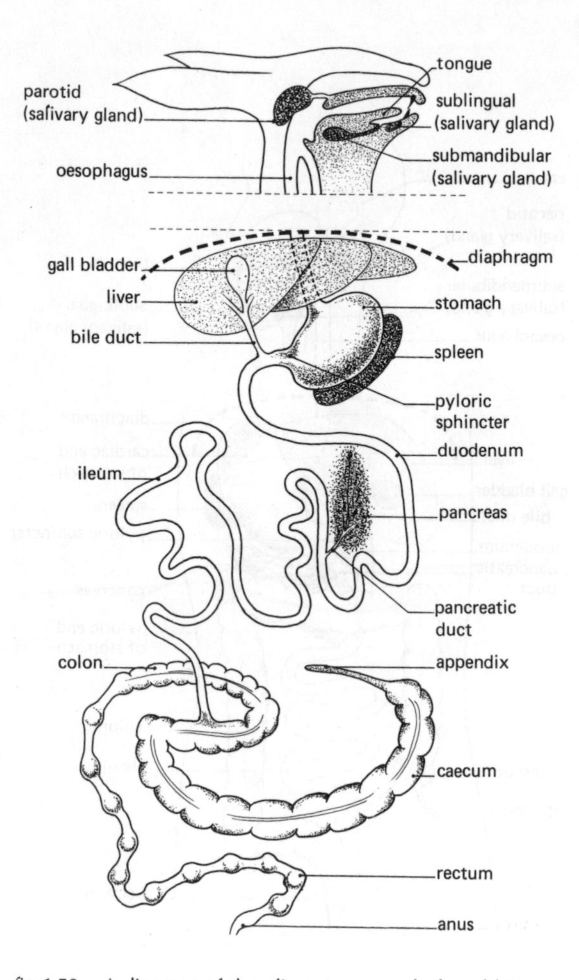

fig 1.50 A diagram of the alimentary canal of a rabbit

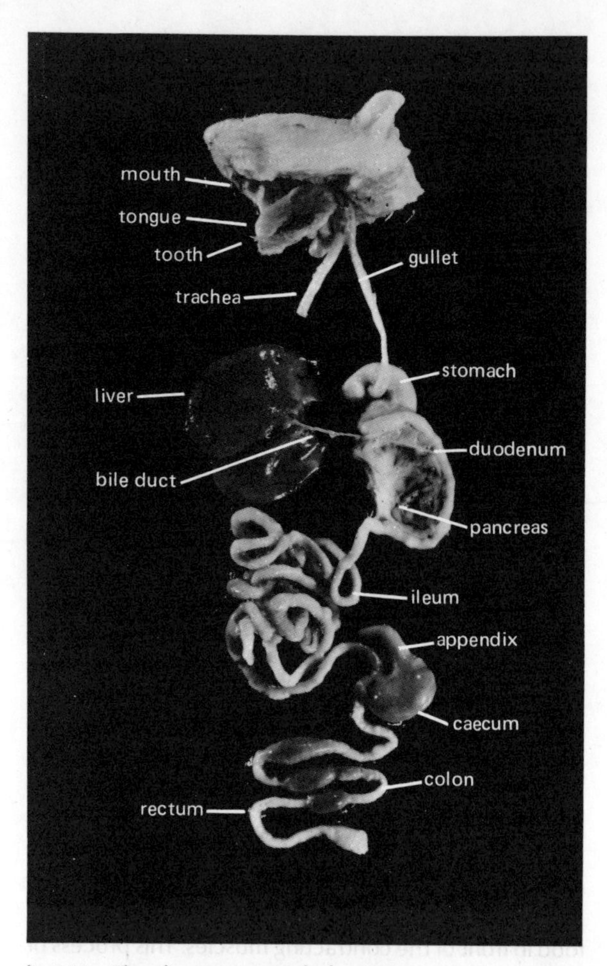

fig 1.51 The alimentary canal of a rat

means. The exit from the stomach is surrounded by a circular muscle (sphincter) called the *pylorus*. This opens at intervals to allow chyme to pass into the small intestine. Some mammals have more complicated stomachs than Man or the rabbit. The sheep, for example, has a stomach with four compartments.

The first part of the small intestine is the *duodenum*. Two ducts enter this section, the *bile duct* from the liver and the *pancreatic duct* from the *pancreas*. In Man these two ducts join to form a common duct, whereas in the rabbit they open quite separately in different parts of the duodenum. The rest of the small intestine is divided into the *jejunum* and the *ileum*. In both Man and the rabbit these two sections form the longest section of the gut. In Man it is about 7 m in length and in the rabbit about 1 m. In order to compress these lengths into the abdominal cavity, the small intestine forms a series of loops and folds held in position by a tough membrane, the *mesentery*. The next section of the gut is the *large intestine*. At its junction with the small intestine is a side branch, the *caecum,* which narrows into the blind ending *appendix*. In Man, the large intestine is divided into the ascending *colon*, the transverse colon, descending colon and rectum. These form almost four sides of a square from the bottom right quadrant of the abdomen, up the right side, across to the left and down the left side. The *rectum* opens to the exterior at the *anus*, which is surrounded by a circular muscle, the *anal sphincter*.

In the rabbit the caecum is very large and also ends in a blind appendix. The colon is small and corrugated leading into the rectum. The latter shows lumps which are the characteristic faecal pellets of the rabbit. The rectum in both organisms contains the food residue for some twenty-four hours before it is egested as *faeces* by muscular action.

1.33 Chemical digestion

Functions of the gut

Digestion in the buccal cavity. *Saliva* is discharged from the salivary glands (three pairs in Man, four pairs in the rabbit). The saliva of Man contains *amylase,* which begins the *digestion of starch*. The rabbit does not possess this enzyme. The saliva also contain *mucus* to soften the food and to lubricate its passage. In addition, salts provide a neutral to alkaline medium suitable for the action of the enzyme *ptyalin* (amylase). The food is formed into a ball or *bolus* by the tongue and then swallowed.

Digestion in the stomach. The stomach wall secretes *gastric juice* containing *rennin* which clots the soluble protein of *milk*, (particularly important in juveniles), and *pepsin*, which begins the breakdown of protein molecules into smaller units called *polypeptides*. The acidity of the juice is provided by *hydrochloric acid*, which kills off any bacteria and also provides an acid medium (pH 1.5 to 2.0) suitable for the action of pepsin. The acid stops the activity of ptyalin. The food becomes very liquid in the stomach, owing to mucus, the fourth constituent of gastric juice. The chyme begins to leave the stomach through the pylorus after two to three hours. Chyme is only allowed through the pylorus in small, easily digestible amounts.

Digestion in the small intestine. *Bile* is secreted from the *liver* through the bile duct. This juice contains *salts* which break up the liquid fats into an *emulsion*, (emulsification) such that the tiny globules provide a larger surface area on which the fat-digesting enzymes can act. The alkaline bile also *neutralises* the acid chyme.

Pancreatic juice is alkaline and contains *three enzymes* acting on peptides, fats and starch. The intestinal wall secretes a group of enzymes called the *succus entericus,* which help the pancreatic juice to finally complete digestion of:

proteins to form amino-acids,
carbohydrates to form glucose,
and fats to form fatty acids and glycerol.

These are the final breakdown products which are soluble and small enough to be absorbed through the intestinal wall into the bloodstream. The vegetable diet of the rabbit contains a great deal of *cellulose*, and although Man cannot digest this carbohydrate, the rabbit must do so or its food intake would be very limited. The digestion of cellulose occurs in the caecum with the aid of large numbers of bacteria which produce *cellulase*, the required enzyme. The faecal pellets, however, still contain undigested food material and the rabbit will eat these pellets during the night hours. This dung-eating is called *coprophagy*. It is the only means by which rabbits obtain protein.

Enzymes have the following important properties.

i) They act as *organic catalysts, speeding up* the rate of food breakdown.

ii) Each enzyme causes the breakdown of the *one substance* (that is they are *specific*) to form a particular end-product.

iii) They are affected by *temperature*. They work best at 37°C to 40°C which is about body temperature of birds and mammals.

iv) They are affected by the *pH of their surroundings*. Each enzyme works most effectively at its own specific pH.

Gland	Secretion	Enzymes	Action substrate	Products
Salivary glands	Saliva	Amylase	Starch	Maltose
Stomach wall – Gastric glands	Gastric juice	Pepsin + hydrochloric acid + rennin	Protein, soluble milk protein	Polypeptides insoluble milk protein
Pancreas	Pancreatic juice	Trypsin	Protein	Peptides and *amino-acids*
		Lipase	Fats	*Fatty acids* and *glycerol*
		Amylase	Starch	Maltose
Wall of small intestine	Succus entericus	Sucrase	Sucrose	*Glucose* and *Fructose*
		Maltase	Maltose	*Glucose*
		Peptidase	Polypeptides	*Amino acids*
		Lipase	Fats	*Fatty acids* and *glycerol*

The substances in italics are the final soluble breakdown products of digestion.

Table 1.5 The major enzymes concerned with digestion

Enzyme	Source	Optimum pH
amylase	saliva	6.8
pepsin	gastric juice	2.0
rennin	gastric·juice	2.0
trypsin	pancreatic juice	7.0 to 8.0

v) They are required only in *small quantities*.
vi) The *rate* of an enzyme reaction will *increase* with increasing *substrate concentration*, up to a certain limit.

1.34 Absorption of digested foods

It can be seen from table 1.5 that food is broken down by digestion into the end-products of amino-acids, glucose, fatty acids and glycerol. These together with mineral salts, vitamins and water must now be absorbed through the walls of the alimentary canal. This occurs in the region of the small intestine, where a large surface area for absorption is available because of:

i) the great length of the small intestine,
ii) the folded inner surface,
iii) the small projections, called *villi*, which cover the inner folds. Each square millimetre of surface has about twenty to forty villi.

It has been shown that even the surface of each villus has tiny projections (*microvilli*) of the cell surfaces which must increase greatly the total internal surface area of the small intestine.

What happens to the digested food materials when they have passed through the wall of the gut? This important question will be dealt with in Chapter 2 where we consider the problem of transporting digested food around the body. Once in the bloodstream, however, the food is available to perform the functions which were discussed in section 1.10. The *glucose* and *amino-acids* pass into the *blood capillaries* of the villus, and the *fatty acids* and *glycerol* pass into the *lacteal*. Both of these vessels carry the digested food away from the intestinal wall. Water is absorbed through the walls of the small intestine and large intestine, so that the watery contents of the gut gradually become solidified to form the faeces. Many substances pass along the gut and remain undigested.

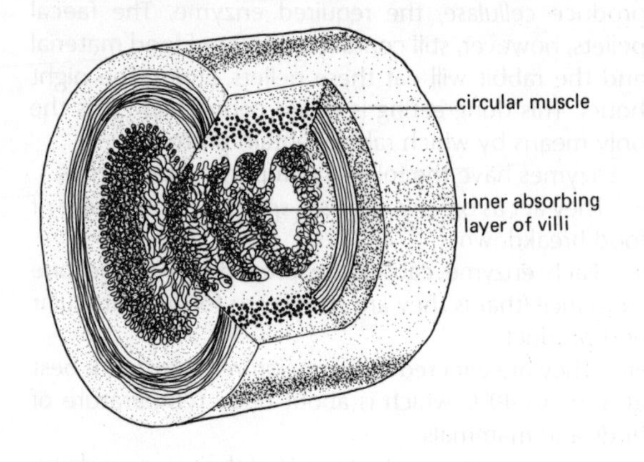

fig 1.52 A diagram of a portion of the small intestine showing the internal folds and the villi

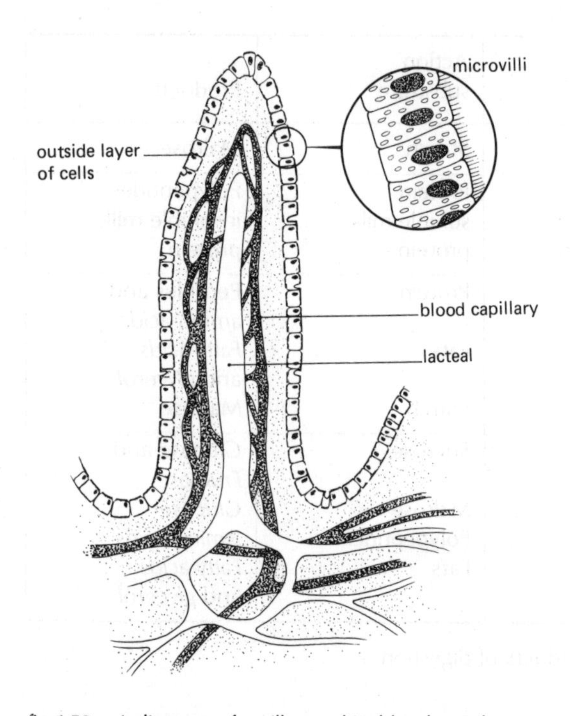

fig 1.53 A diagram of a villus and its blood supply

fig 1.54 The internal wall of the small intestine

Cellulose is the principle example in Man, and this carbohydrate is largely present in the fruits and vegetables of the diet. Cellulose provides the *roughage* or bulk on which the muscles of the rectum can act. The faeces passed from the anus contain, in addition to cellulose, a good deal of mucus, broken down cells and bacteria.

1.40 Energy value of food

One of the functions of food is to produce energy for all our conscious activities and, in addition, for all of the activities of the body over which we have no control, for example, the beat of the heart, muscular contractions of the gut etc. Some of this energy is used to keep the body warm, some for chemical reactions and some for the muscular contractions that produce body movement. We can classify the *energy* we need as follows:

i) To maintain the functioning of the ordinary living processes e.g. heart beat, maintenance of body temperature, breathing, circulation of blood etc. All of these functions can be called the *Basal Metabolism*. The amount of energy required for this is known as the *Basal Metabolic Rate* (BMR).

ii) All other activities including: (a) our everyday activity such as moving, standing, washing, combing hair etc., and (b) the work which we perform. This can vary from sitting at a desk and writing all day (bank clerk), to cutting down trees for eight hours a day (forester). The amount of energy needed for these two very different activities will vary considerably. Further factors will affect the two divisions above.

The BMR will depend upon:

Age In actively growing children the basal metabolic rate is high and decreases rapidly in the first twelve years of life. Thereafter it falls more slowly, and at the age of twenty years decreases very gradually. With increasing age, also, people become less physically active.

Size The body size is important because of the relation of surface area to volume. The heat loss from a tall thin person is much greater than from a short fat one. Men generally are larger than women and thus require more energy. A child, though smaller, can be more active and in proportion needs a greater amount of energy than an adult.

Climate In a warm climate the individual body demands less energy to maintain body temperature. Taking a reference temperature of 10°C, it is estimated that energy requirements are decreased by 5% for every 10°C rise in temperature. Because Man uses up energy to protect himself against cold, it is estimated that there is an increase of 3% for every 10°C fall below the reference temperature. This will help him to maintain his body temperature.

All energy is transferable so that, for whatever purpose it is required, the energy value of foods and energy requirements of Man can be calculated in terms of heat energy. Heat energy for many years has been measured in calories which are defined as follows:

1 *calorie* is the heat required to raise the temperature of 1 g of water through 1 degree C.

For practical purposes in nutrition this unit was too small so that the *kilocalorie* was used.

1 *kilocalorie* is the heat required to raise the temperature of 1 000 g of water through 1 degree C (= 1 000 calories), (or 100 g of water through 10 degrees C, and so on).

With the introduction of S.I. units, the unit of heat used is now the *joule* which is defined in terms of individual work done. James Joule showed that *mechanical work* can be converted into an *equivalent amount of heat*. Therefore both can be measured in the same units. For our purposes all we need to consider is that:

1 calorie (cal) = 4.2 joules (J)
1 kilocalorie (kcal) = 4 200 joules
= 4.2 kilojoules (kJ)

You will still find in many books on nutrition the energy value of foods described in terms of kilocalories, but kilojoules and megajoules are the current units.

joule	J	1 000 J = 1 kJ	1 MJ	= 239 kcal
kilojoule	kJ	1 000 kJ = 1 MJ	1 kcal	= 4.2 kJ
megajoule	MJ	1 MJ = 1 000 000 J		

Remember that 1 g of water has a volume of 1 cm³, thus 10 g of water has a volume of 10 cm³.

Questions

1 How many kilojoules are required to heat 50 cm³ of water through 1 degree C?

2 How many kilojoules are required to heat 50 cm³ of water from 20°C to 25°C?

You can use this knowledge to find out how much heat energy is released from food when it is burned. Carbohydrates and fats in the food are the principal sources of energy for the body, but protein can also be used for this purpose. If these substances are burned under controlled conditions in the laboratory the energy produced can be calculated. Typical results:

1 g of carbohydrate (as glucose) produces in the body about 16 kJ.
1 g of fat produces in the body about 38 kJ.
1 g of protein produces in the body about 17 kJ.

Notice that *fats* have about 2½ times the energy per unit mass of *carbohydrates*.

Using these values the energy content of any food can be calculated from the proportions of nutrients that it contains. For example 100 g of maize (whole grain) contains 10 g of protein, 4.5 g of fat and 70 g of carbohydrate. Thus, the total energy value can be worked out as shown:

70 × 16 kJ	=	1120 kJ from carbohydrate
4.5 × 38 kJ	=	171 kJ from fat
10 × 17 kJ	=	170 kJ from protein
		1461 kJ

Experiment

To measure the heat energy produced by burning food

Procedure

1 Take a small test-tube, place in it 20 cm³ of water and fix the test-tube in the clamp of a retort stand.

2 Weigh a piece of cashew nut or groundnut (about 0.5 g) and attach the nut to a pin held upright by a piece of plasticine (see fig 1.55).

3 Position the nut immediately beneath the test-tube. Place a thermometer in the test-tube together with a small piece of bent wire to act as a stirrer. Plug the mouth of the test-tube with cotton wool.

4 Record the temperature of the water.

5 Mark the position of the plasticine on the bench with a piece of chalk, then remove the plasticine, pin and nut. Hold the nut in a bunsen flame until it starts to burn and then place it beneath the test-tube in its original position.

6 Move the stirrer up and down as the nut is burning. Adjust the height of the tube on the retort stand so that the flame is beneath it, and no flame flows around the side of the tube.

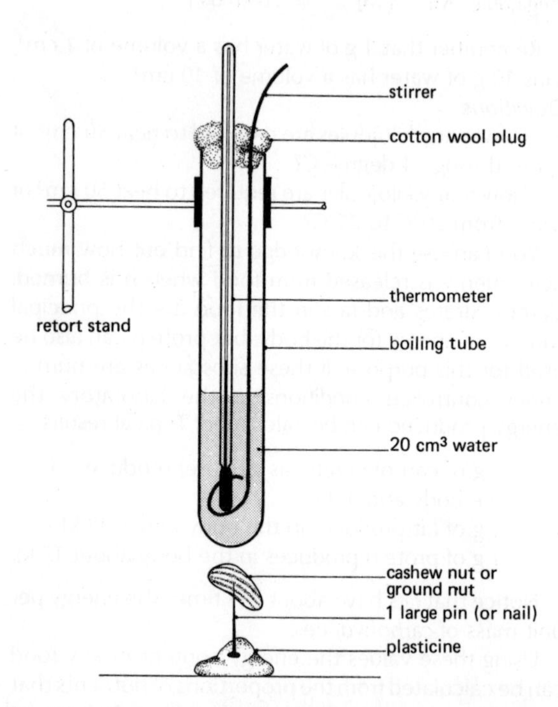

fig 1.55 The experimental apparatus for measuring heat energy produced by the burning of food

7 When the nut has burned out continue stirring the water until the temperature reaches its highest point, then *record this temperature.*

8 Work out your results as follows:

temperature of the water at the start = A°C
highest temperature reached by the water = B°C
mass of the cashew (or ground) nut = C g
volume of water = 20 cm³
therefore mass of water = 20 g
heat gained by water = 20 g × (B – A) deg. C × 4.2
= 20 × (B – A) × 4.2 J

$$\text{Heat produced by 1 gm nut} = \frac{20 \times (B - A) \times 4.2}{1000 \times C} \text{ kJ/g}$$

9 *Tabulate* your class results and work out the *mean value.*

The principle behind this experiment is the transfer of heat energy from the food to the water in the test-tube.

Questions

3 This is a very crude experiment. Can you give any criticisms of the method? How could you improve the experiment to make it more accurate?

4 Will the food produce more, the same, or less energy when it is eaten and used in the body?

Occupation

The occupation followed by an individual man or woman has the greatest effect on the rate at which energy from food is utilised. *Occupations* can be grouped broadly into the following classes.

i) *Sedentary* Men involved in these occupations need about 3 400 kJ for an 8-hour day. This class includes office workers, journalists, teachers, doctors, lawyers, clergy and secretaries.

ii) *Moderately active* Men involved in these occupations need about 4 600 kJ for an 8-hour day. Often their work is not very strenuous, but involves a lot of walking or weight lifting. This class includes railway workers, shop assistants and fishermen.

iii) *Very active* Men involved in these occupations need about 6 800 kJ for an 8-hour day. This class includes steel workers, forestry workers, army recruits, farm labourers and builders labourers.

Most women working at home or doing light work would require 3 400 kJ per day, while those involved in great physical activity would require about 4 600 to 5 000 kJ per day.

The energy expenditure, for these three groups (including BMR and everyday activities) is analysed overleaf. Total energy requirement for one day will be equal to total energy expenditure. Thus a sedentary man will require 10 300 kj (10.3 MJ) in his daily diet.

	Sedentary	Moderately active	Very active
BMR	1 900	1 900	1 900
Everyday activity	5 000	5 000	5 000
Work (occupation)	3 400	4 600	6 800
	10 300 kJ	11 500 kJ	13 700 kJ

Women need to expend *comparatively less energy* because of their *smaller size*. If more food is eaten than an individual needs for his activities then some of the *excess food* will be converted into *fat*. All three classes of nutrient can be converted into fat. The most fattening foods are those containing most energy.

1.41 Balanced diet

The individual cells of the body tend to have a short life compared with the length of life of the organism as a whole. In Man, the total body protein is replaced about every six months. *A balanced diet* must take this into account together with energy requirements and health requirements. Daily food intake should be a mixed diet of:

1 meat, eggs, fish or milk
2 fresh vegetable foods such as fruit, citrus fruit juice, tomato
3 grain cereals, potatoes, bread.

Each of these foods provide certain essentials, but may be deficient in others. For example vitamin C, lacking in 1 and 3 above, is present in 2. Vitamin D is lacking in 2 and 3, but present in 1.

A diet consisting mainly of carbohydrate-rich food such as in 3 would be considered *unbalanced*, because it does not contain enough protein, mineral salts and vitamins present in 1 and 2.

To obtain a balanced diet it is necessary to eat a *wide variety of foods*, since a shortage of any of the main classes will result in deficiency diseases.

In the developed countries today, these diseases are rare, as food of sufficient quality and quantity is available. Most people are educated to understand which foods are necessary to keep themselves and their children healthy. In certain developing countries deficiency diseases are common for *three reasons* which are listed below.

i) Shortage of the right kind of food. In many areas foods rich in protein and vitamins are not available. The normal diet consists mainly of cereals, starchy foods and sugar. Animal protein especially is very scarce.

ii) *Low incomes.* Families are often too poor to buy the right kinds of food even if they were available. Foods rich in proteins and vitamins, such as milk, meat, fish, eggs, and many of the fruits are usually the most expensive.

Examine fig 1.56 that shows a child suffering from kwashiorkor.

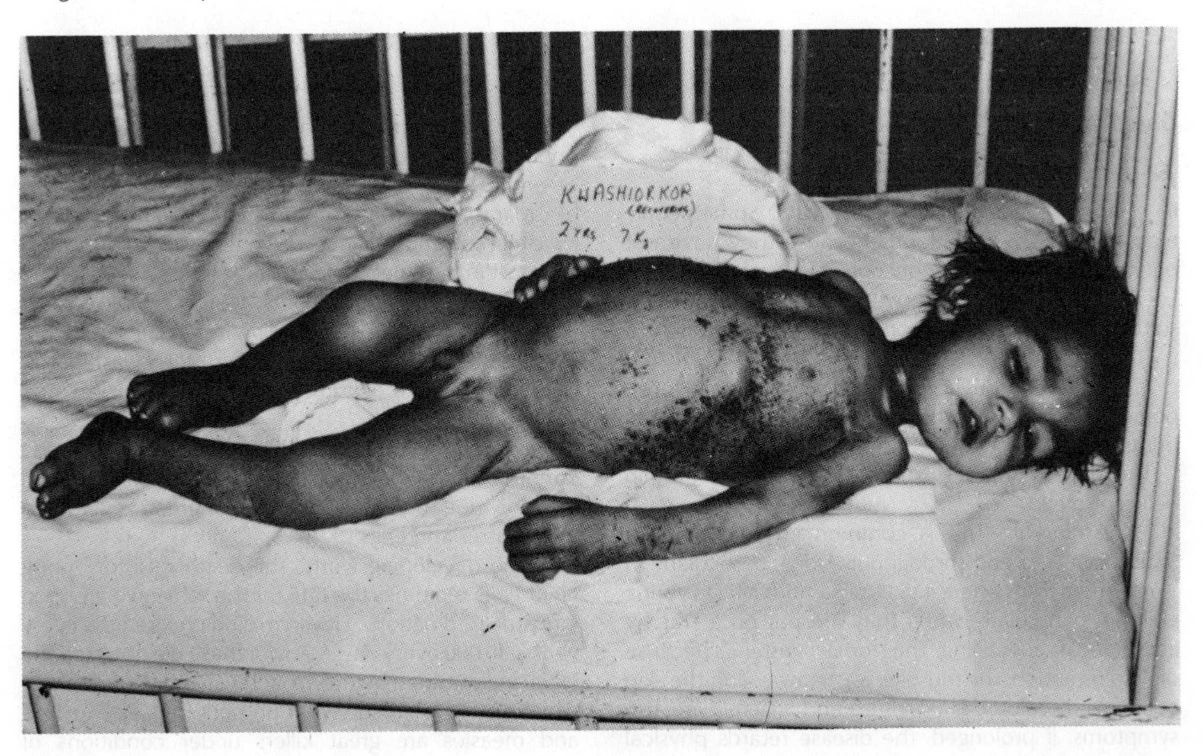

fig 1.56 Kwashiorkor

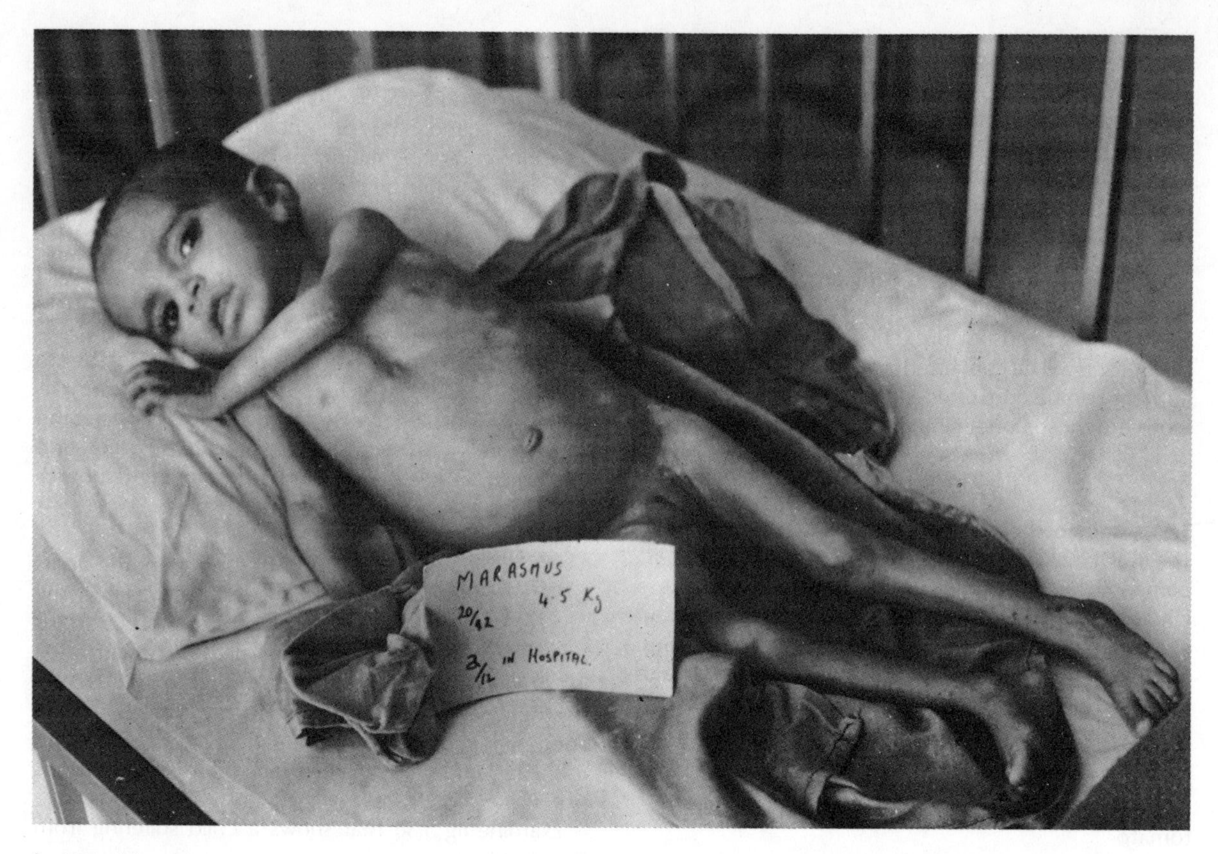

fig 1.57 Marasmus

Questions

1 What symptoms can you see?

2 Why has the child developed this disease?

iii) *Superstition and lack of education.* Some people may not eat the right foods because they have not been educated in the concepts of a balanced diet. Local custom can dictate what is eaten and, even if suitable foods are available, people may be unwilling to try them.

In some countries it is thought that certain foods possess magical properties and that to eat them would bring bad luck.

There are a number of diseases caused by protein deficient diets.

a) *Kwashiorkor* This is common in large parts of South East Asia and throughout Africa, particularly in West Africa from where it gets its name. It affects young children particularly when they are put on a starchy adult diet after weaning. The *stomach distends* because of fluid retention, the muscles waste away and the skin becomes discoloured. *Diarrhoea* and *anaemia* are also symptoms. If prolonged, the disease retards physical and mental growth.

Early treatment, with a special *high-protein* fluid containing skim-milk powder and vegetable protein is essential.

b) *Marasmus* The symptoms are similar to those of kwashiorkor and are due to *general starvation* rather than just protein deficiencies. There is no swelling of the abdomen or skin rash. All body tissues exhibit *general wasting* and a severe *loss of fluid* often results in death. The treatment is the same as that for kwashiorkor, though more carbohydrate may be needed in some cases.

c) *Vitamin deficiency* is still widespread in many countries of the Far East, Latin America and Africa, and is mainly due to insufficient intake of green vegetables, milk, butter and eggs.

In the developing world, the death rate for young children is ten times the rate for the same age group in industrialised nations. Eleven million children die every year—30 000 every day. Many of these die not from the initial malnutrition but from bacterial and virus diseases which they would survive if properly fed. Pneumonia and measles are great killers under conditions of malnutrition.

Table 1.6 Common sources of Food Classes

Food (100 g)	Protein g	Fat g	Carbohydrate g	Calcium mg	Iron mg	Vitamin A μg	Vitamin D μg	Thiamine mg	Riboflavine mg	Nicotinic acid mg	Vitamin C mg	Energy value kJ
Vegetables												
Carrots, old	0.7	0	5.4	48	0.6	2000	0	0.06	0.05	0.6	6	96
Potatoes, raw	2.1	0	18.0	8	0.7	0	0	0.11	0.04	1.2	8-30(h)	318
Potatoes, boiled	1.4	0	19.7	4	0.5	0	0	0.08	0.03	0.8	4-15(h)	331
Potato chips, fried	3.8	9.0	37.3	14	1.4	0	0	0.10	0.04	1.2	6-20(h)	989
Sweet corn, canned	2.6	0.8	20.5	5	0.5	35	0	0.03	0.05	0.9	4	398
Beans, haricot	21.4	0	45.5	180	6.7	0	0	0.45	0.13	2.5	0	1073
Beans, runner	1.1	0	2.9	33	0.7	50	0	0.05	0.10	0.9	20	63
Peas, fresh raw or quick frozen	5.8	0	10.6	15	1.9	50	0	0.32	0.15	2.5	25	264
Peas, canned, processed	7.2	0	18.0	29	1.1	67	0	0.06	0.04	0.5	2	402
Tomatoes, fresh	0.9	0	2.8	13	0.4	117	0	0.06	0.04	0.6	20	59
Brussels sprouts, boiled	2.4	0	1.7	27	0.6	67	0	0.06	0.10	0.4	35	67
Cabbage, boiled	0.8	0	1.3	58	0.5	50	0	0.03	0.03	0.2	20	34
Lettuce	1.1	0	1.8	26	0.7	167	0	0.07	0.08	0.3	15	46
Onions	0.9	0	5.2	31	0.3	0	0	0.03	0.05	0.2	10	96
Parsnips	1.7	0	11.3	55	0.6	0	0	0.10	0.09	1.0	15	205
Fruit												
Apple	0.3	0	12.0	4	0.3	5	0	0.04	0.02	0.1	5	193
Bananas	1.1	0	19.2	7	0.4	33	0	0.04	0.07	0.6	10	318
Black currants	0.9	0	6.6	60	1.3	33	0	0.03	0.06	0.3	200	117
Grapefruit	0.6	0	5.3	17	0.3	0	0	0.05	0.02	0.2	40	92
Oranges	0.8	0	8.5	41	0.3	8	0	0.10	0.03	0.2	50	147
Orange juice, canned unconcentrated	0.8	0	11.7	10	0.4	8	0	0.07	0.02	0.2	40	197
Peaches, canned	0.4	0	22.9	3.5	1.9	41	0	0.01	0.02	0.6	4	369
Pineapple, canned	0.3	0	20.0	13	1.7	7	0	0.05	0.02	0.2	8	318
Raspberries	0.9	0	5.6	41	1.2	13	0	0.02	0.03	0.4	25	105
Rhubarb	0.6	0	1.0	103	0.4	10	0	0.01	0.07	0.3	10	25
Strawberries	0.6	0	6.2	22	0.7	5	0	0.02	0.03	0.4	60	109
Nuts and cereals												
Peanuts, roasted	28.1	49.0	8.6	61	2.0	0	0	0.23	0.10	16.0	0	2455
Biscuits, chocolate	7.1	24.9	65.3	131	1.5	0	0	0.11	0.04	1.1	0	2082
Biscuits, plain, semi-sweet	7.4	13.2	75.3	126	1.8	0	0	0.17	0.06	1.3	0	1806
Bread, brown	9.2	1.8	49.0	92	2.5	0	0	0.28	0.07	3.2	0	993
Bread, white	8.3	1.7	54.6	100	1.8	0	0	0.18	0.02	1.4	0	1060
Crispbread, Ryvita	10.0	2.1	69	86	3.3	0	0	0.37	0.24	1.4	0	1332
Rice	6.2	1.0	86.8	4	0.4	0	0	0.08	0.03	1.5	0	1504
Spaghetti	9.9	1.0	84.0	23	1.2	0	0	0.09	0.06	1.7	0	1525
Meat												
Bacon, average	11.0	48.0	0	10	1.0	0	0	0.40	0.15	1.5	0	1994

(h) vitamin C falls during storage

Food (100 g)	Protein g	Fat g	Carbohydrate g	Calcium mg	Iron mg	Vitamin A µg	Vitamin D µg	Thiamine mg	Riboflavine mg	Nicotinic acid mg	Vitamin C mg	Energy value kj
Beef, average	14.8	28.2	0	10	4.0	0	0	0.07	0.20	5.0	0	1311
Beef, corned	22.3	15.0	0	13	9.8	0	0	0	0.20	3.5	0	939
Chicken, roast	29.6	7.3	0	15	2.6	0	0	0.04	0.14	4.9	0	771
Ham, cooked	16.3	39.6	0	13	2.5	0	0	0.50	0.20	3.5	0	1768
Lamb, roast	25.0	20.4	0	4	4.3	0	0	0.10	0.25	4.5	0	1190
Liver, fried	29.5	15.9	4.0	9	20.7	6000	0.75	0.30	3.50	15.0	20	1156
Pork, average	12.0	40.0	0	10	1.0	0	0	1.00	0.20	5.0	0	1710
Sausage, pork	10.4	30.9	13.3	15	2.5	0	0	0.17	0.07	1.6	0	1546
Steak and kidney pie, cooked	13.3	21.1	16.2	37	5.1	126	0.55	0.11	0.47	4.1	0	1274
Fish												
Cod, haddock, white fish	16.0	0.5	0	25	1.0	0	0	0.06	0.10	3.0	0	289
Fish fingers	13.4	6.8	20.7	50	1.4	0	0	0.12	0.16	1.8	0	804
Herring	16.0	14.1	0	100	1.5	45	22.25	0.03	0.30	3.5	0	796
Kipper	19.0	16.0	0	120	2.0	45	22.25	0	0.30	3.5	0	922
Sardines, canned in oil	20.4	22.6	0	409	4.0	30	7.50	0	0.20	5.0	0	1194
Milk, eggs, fat												
Cream, single	2.8	18.0	4.2	100	0.1	155	0.10	0.03	0.13	0.1	1	792
Milk, liquid, whole	3.3	3.8	4.8	120	0.1	44(a) 37(b)	0.05(a) 0.01(b)	0.04	0.15	0.1	1	272
Milk, whole, evaporated	8.5	9.2	12.8	290	0.2	112	0.12	0.06	0.37	0.2	2	696
Cheese, Cheddar	25.4	34.5	0	810	0.6	420	0.35	0.04	0.50	0.1	0	1726
Eggs, fresh	11.9	12.3	0	56	2.5	300	1.50	0.10	0.35	0.1	0	662
Butter	0.5	82.5	0	15	0.2	995	1.25	0	0	0	0	3122
Margarine	0.2	85.3	0	4	0.3	900(e)	8.00	0	0	0	0	3222
Honey	0.4	0	76.4	5	0.4	0	0	0	0.05	0.2	0	1207
Jam	0.5	0	69.2	18	1.2	2	0	0	0	0	10	1098

(a) Summer value (b) Winter value (e) some margarines contain carotene

Questions requiring an extended essay-type answer

1 a) How would you test maize meal for
 i) reducing sugar,
 ii) starch,
 iii) fat, and
 iv) protein?
b) Describe a comparative test for the amount of vitamin C in lemon juice and tomatoes.

2 a) Describe carefully how you would investigate experimentally that
 i) photosynthesis does *not* go on in darkness, and
 ii) a green potted plant produces carbon dioxide.
b) Construct a *table* to show the main differences between photosynthesis and respiration.

3 a) Why is nitrogen an important element for living organisms?
b) In what forms, and from what sources, is this element obtained by a mammal and a green plant?
c) Describe an experiment to investigate the hypothesis that nitrogen is necessary for healthy growth in a green plant.

4 a) What is the importance of a *'cycle'* in nature?
b) Give a brief account of **two** cycles in nature, and show how they contribute to plant growth.

5 a) Why do organisms require food?
b) What are the **six** main classes of foodstuff? Give examples of food in which *each one* of the main classes is present in large amounts.

c) Many animals do not drink. Explain how the water content of their bodies is maintained.

6 a) Describe the passage and digestion of a meal of meat and potatoes through the gut of man.
b) What use is made of the digested products of this meal after absorption into the blood stream from the intestine?

7 a) State **six** properties of enzymes.
b) *Name* a plant enzyme, state where it acts and its importance to the plant.
c) *Name* an animal enzyme, say where it occurs, and describe how you could investigate its action outside the animal's body.

8 a) What do you understand by the term, '*a balanced diet*'?
b) What are the common reasons for deficiency diseases present in human communities?
c) Describe the symptoms of **two** common deficiency diseases and suggest how the diseases could be cured.

9 Describe experimental investigations to determine whether
a) the leaves of a green plant produce starch in sunlight, and
b) oxygen is released by a green plant in light.

2 Transport

2.00 Why is transport necessary?

All animals need to take in oxygen and foodstuffs from the environment and to eliminate carbon dioxide and other waste products from their bodies. Small animals such as protozoa, jellyfish and flatworms carry out these exchanges by simple *diffusion* through their body surfaces, but with *larger organisms* some cells become *too far away* from the external surface of the body or the internal surface of the gut for simple diffusion to be effective. Thus oxygen would never reach the innermost cells while digested food moving through the gut wall would never supply the outermost cells of the body. There is a *limiting factor* present; this is the rate at which gases and dissolved substances can diffuse through living material.

As animals evolved and increased in size a *transport system* became essential, in order that dissolved materials could be moved rapidly around the organism. In many cases this function is fulfilled by a blood system.

2.10 Surface area to volume ratios

We can examine more closely the need for a transport system if we look at the vital factor of increasing size and the relationship between *surface area* and *volume* in the living organism. This factor affects the rates of gaseous exchange, heat loss, and loss of excretory matter in every animal. In order to understand this problem it is easier to think of the body shape, not as an irregular structure, but as a symmetrical box shape. Consider a simple animal as though it were a cube of size 1 cm: its surface area (S.A.) is $6 \times 1\ cm^2$ and this will be the total area through which oxygen and carbon dioxide passes in and out. The volume (V) of such a cube is $3\ cm^3$.

The surface area: volume ratio is thus

$$\frac{S.A.}{V.} = \frac{6}{1} = 6$$

If the length of the cube is now doubled to 2 cm, then:

$$\frac{S.A.}{V.} = \frac{6 \times 2^2}{2^3} = \frac{24}{8} = \frac{3}{1} = 3$$

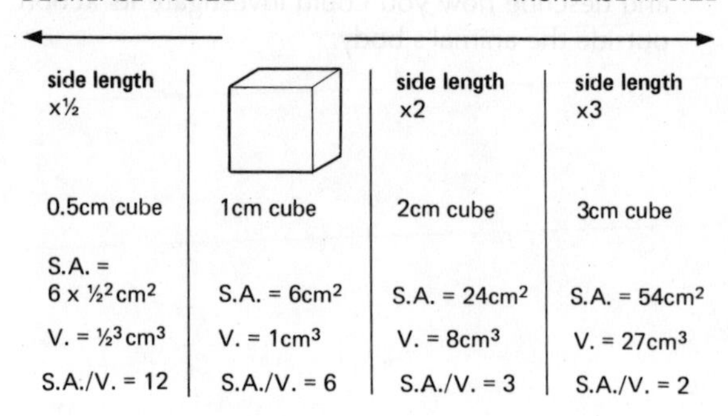

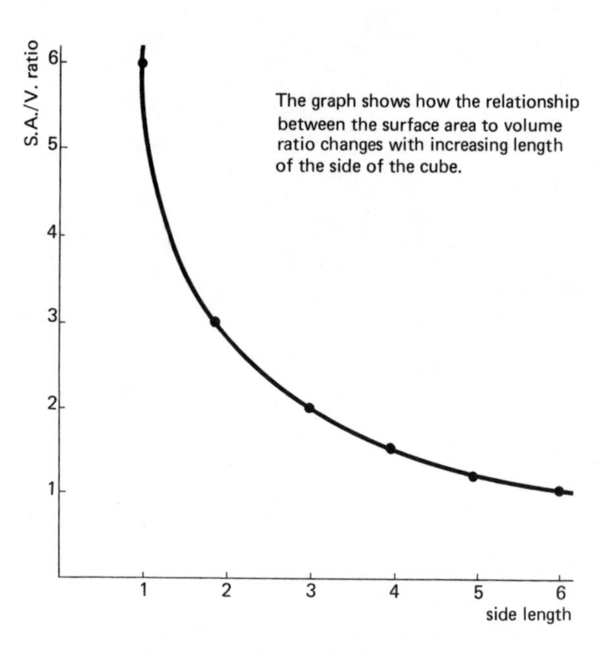

fig 2.1 A graph showing the relationship between the surface area/volume ratio and length

Now examine fig 2.1 and you will see this illustrated in two ways. Taking the 1 cm cube as the reference size it is clear that, as the cube size increases, the surface area: volume ratio decreases, whereas a decrease in cube size brings about an increase in this ratio. This principle is shown in a different fashion in the graph.

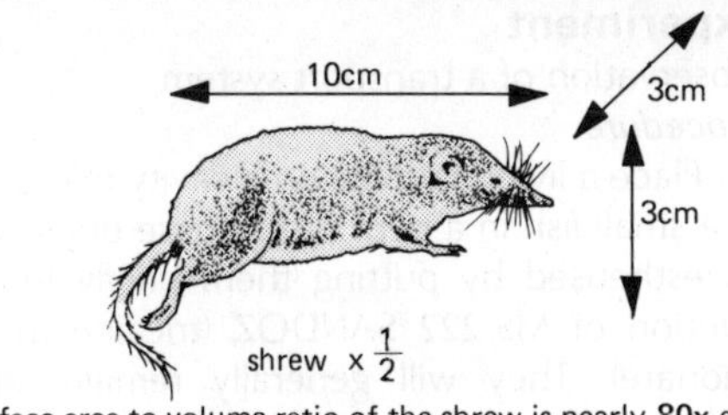

The surface area to volume ratio of the shrew is nearly **80x** greater than that of the elephant

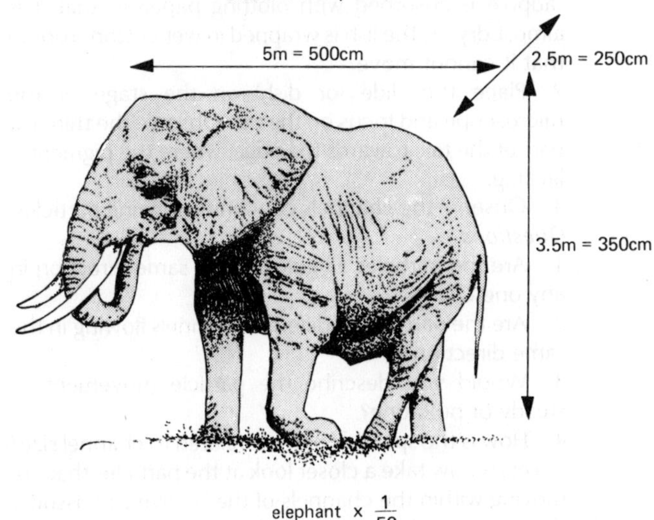

consider the body of each animal as a box

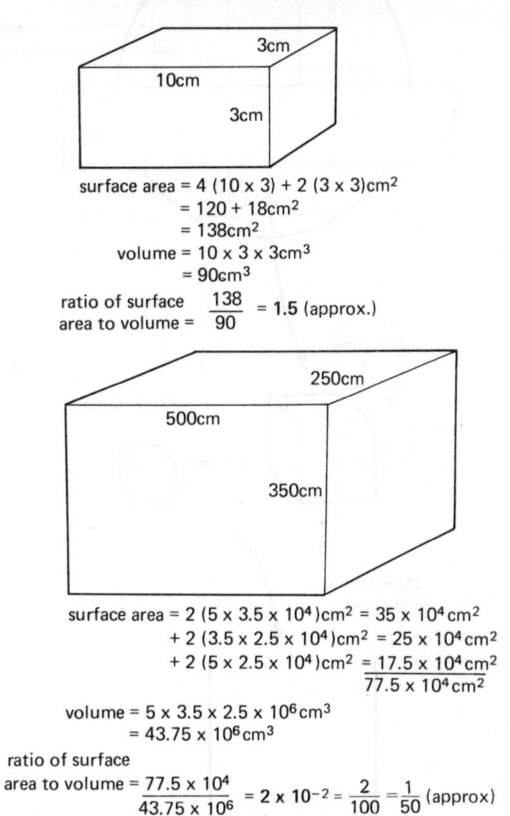

fig 2.2 The surface area/volume ratios of an elephant and elephant shrew

You will now appreciate that, for a *very small* animal body, the surface area to volume ratio is *very great*. Conversely, for a *very large* animal of *the same general shape*, the surface area to volume ratio is *very small*.

Examine fig 2.2. Here we have two animals of greatly differing size, the elephant and the elephant shrew. The bodies can be considered as boxes of similar dimensions to those of the organisms. The calculations shown in the figure indicate that the S.A./V. ratio of the elephant is 80 times smaller than that of the shrew. This difference causes a number of problems for both of these animals. We shall consider these later.

Both the elephant and the shrew are mammals and possess transport systems. However, if we compare the elephant shrew with a single-celled amoeba in the same way, then the latter will have an enormous surface area to volume ratio compared with that of the elephant shrew.

Experiment

Investigation of S.A./V. ratio in agar blocks

Procedure

1 Make up an agar solution in hot water in a beaker, and then pour it into the base of a petri dish to a depth of 1 cm. Allow it to solidify.

2 Place a ruler over the agar and cut parallel lines 1 cm apart. Repeat at right angles, so that the agar is now in 1 cm cubes (see fig 2.3).

3 Take some 1 cm blocks and cut them into 0.5 cm blocks by making three cuts at right angles to each other across the 1 cm blocks.

4 Put three blocks of each size in a beaker and cover with a potassium permanganate solution (0.5%).

5 Remove a small and a large block every three minutes.

6 Cut vertically through each block and observe the penetration of colour.

7 Repeat the experiment using different strengths of solution and shorter time intervals.

Questions

1 What do you observe, concerning the penetration of colour into the blocks?

2 What difference will it make if the experiment is repeated (i) using the same solution but with a shorter time interval, or (ii) using a stronger solution with the same time interval?

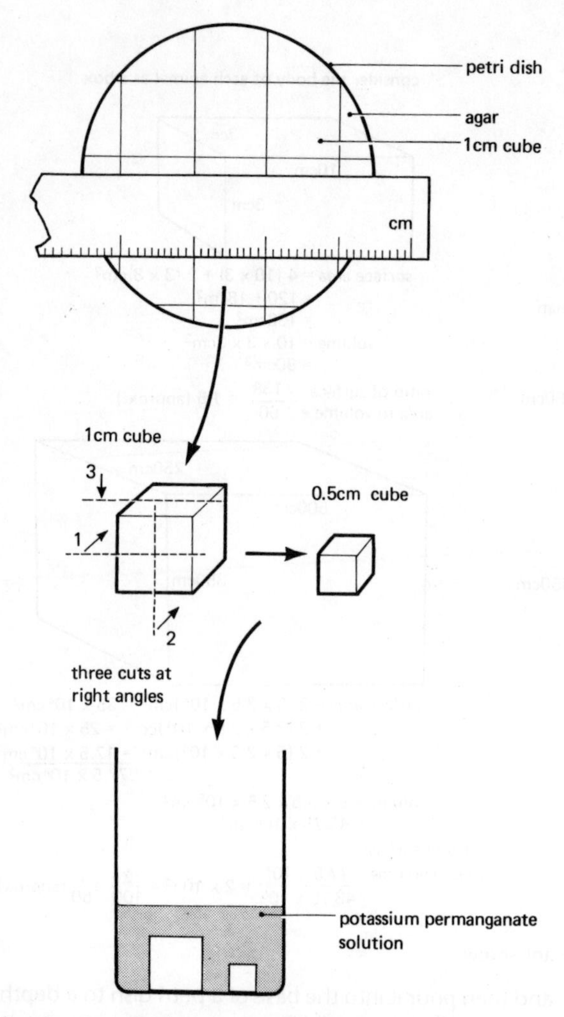

fig 2.3 Agar blocks in potassium permanganate

From the experiment it is clear that, with increasing size of organisms, the ability to absorb through the *body surface* (related to surface area) will decrease in relation to the *needs of the organism* (related to volume). The process of diffusion will need to move dissolved substances over greater and greater distances. Remember the potassium permanganate in the experiment. Thus the S.A./V. ratio and diffusion will become *limiting factors* for activities which depend upon supplies of oxygen and food, and upon the removal of carbon dioxide and waste products.

Large organisms need three changes in their structure:

1 i) Additional surface area externally for the absorption of oxygen, e.g. lungs and gills.

ii) Additional surface area internally for the absorption of food e.g. a long gut, folds and villi.

2 A system whereby a *circulating fluid* carries the absorbed substances at a faster rate than can be effected by diffusion. This is a *mass-flow system.*

2.20 The transport system in action

Experiment

Observation of a transport system

Procedure

1 Place a living tadpole on a cavity microscope slide, or a small fish in a petri dish. (These organisms can be anaesthetised by putting them briefly into a 0.03% solution of MS-222 SANDOZ (tricaine methane sulphonate). They will generally remain still enough without anaesthetic, if (i) the water surrounding the tadpole is absorbed with blotting paper so that it is almost dry; (ii) the fish is wrapped in wet cotton wool so that it cannot move.

2 Place the slide (or dish) on the stage of the microscope and focus on the tail. Observe the thinnest part of the tail, towards the edge, where the pigment is lacking.

3 Observe the channels containing moving particles.

Questions

1 Are the particles all going in the same direction in any one channel?

2 Are the particles in different channels flowing in the same direction?

3 Would you describe the particle movement as steady or pulsating?

4 How is the speed of particle related to channel size?

Let us now take a closer look at the particles that are moving within the channels of the fin. We can assume that there is a liquid carrying the objects along just as a stream carries mud particles after heavy rain. The channels are small *blood vessels* enclosing the blood and forming a *mass-flow system transporting material* around the body of the tadpole or the fish. All vertebrates have blood systems and the most readily observable blood system is our own.

Experiment

What does our blood contain?

Procedure

1 Sterilise the end of the little finger of the left hand by using cotton wool moistened with alcohol.

2 Use a sterile lancet or a needle sterilised in a bunsen flame and, having shaken the left hand downwards several times (to force blood into the finger), jab the point into the skin at the side of the sterile little finger. This is the least sensitive part.

3 Now squeeze along the little finger with the thumb of the left hand, and, as a drop of blood comes from the puncture in the skin, collect it near the end of a clean dry slide.

4 Take a second slide, place its edge on the margin of the drop and draw the blood along the first slide. The blood will be drawn out into a thin film or smear.

5 Place the slide with the blood smear on the bench to dry. Sterilise the puncture on the little finger again with the alcohol-soaked cotton wool.

Precaution: Do not use the lancet or needle if it has been placed on the bench top. You must sterilise it in a flame before using it again.
6 When the blood smear is dry, place it on a petri dish and add one or two drops of Leishmann's stain; then add one or two drops of distilled water. Rock the slide gently to mix the two liquids. Leave for five minutes.
7 Wash off the stain with distilled water. After shaking off the remaining distilled water, leave for five minutes to dry.
8 Examine under a microscope.

Question

5 Draw the different types of cell visible. One type will be very numerous but the others will be hard to find. Look carefully. What colour are the majority of cells? What colour are the cells of the less numerous type?

2.30 Structure of blood

Blood consists of a fluid matrix, the plasma, in which float two types of cell. In mammals it constitutes about 10% of the body weight. Man has about 6 litres of blood.

1 Red blood cells (erythrocytes)

These are biconcave discs which contain the red pigment *haemoglobin*. A nucleus is *not* present in mammalian red blood cells but the erythrocytes of amphibians, reptiles and birds are nucleated. There are 5 000 000 of these cells in every cubic millimetre of blood. They appear yellow when viewed as an unstained smear on a microscope slide, but in bulk their red colour is obvious. They are stained red by Leishmann's stain. They develop in the marrow of bones of the adult, but in the embryo they are formed in the liver and spleen.

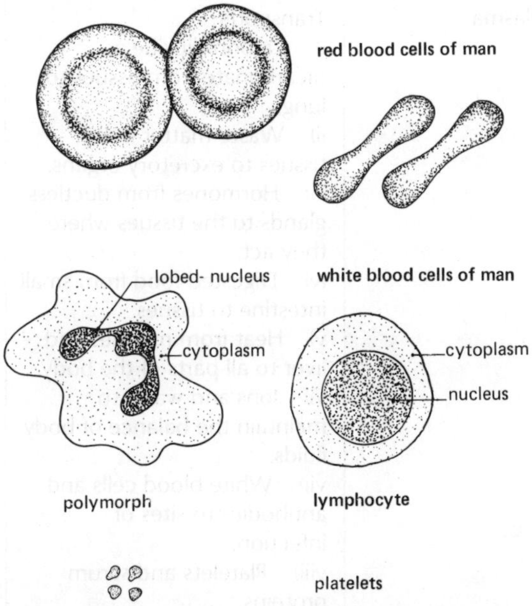

red blood cells of man

lobed- nucleus

white blood cells of man

cytoplasm

cytoplasm

nucleus

polymorph

lymphocyte

platelets

fig 2.4 The different types of mammalian blood cell

2 White blood cells (leucocytes)

You will have noticed amongst the red blood cells on your blood smear, certain blood cells which are stained blue. These are the white blood cells or leucocytes. They do not have a fixed shape and are capable of a type of amoeboid movement brought about by changing shape. The nucleus is present, often with a lobed appearance. It is the nucleus which takes up the blue stain. They are larger than erythrocytes and about 6 000 are present in one cubic millimetre of blood, giving a ratio of approximately 800 red to 1 white blood cells. The number of leucocytes will be greatly in excess of 6 000 in one cubic millimetre during times of infection. As many as 200 000 in one cubic millimetre may be produced to combat micro-organisms. Development takes place in the *bone* and the *lymphatic tissue*. There are five different types of leucocytes each with its characteristic appearance and functions, but two forms (*polymorphs* and *lymphocytes*) constitute 95% of the white cell population.

3 Plasma

This is the fluid matrix of blood which is almost colourless with a slight yellow tinge. 97% of the plasma

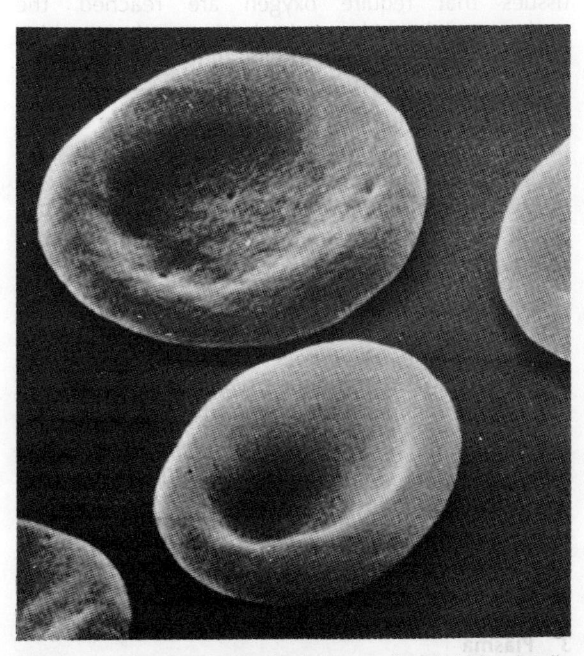

fig 2.5 A photomicrograph of mammalian red blood cells, taken using a scanning electron microscope

is water, the remaining 3% consisting of dissolved substances. Plasma is essentially a means of transport for these dissolved substances, supplying them to the tissues that it bathes when it passes out of the minute blood vessels. It then collects waste products.

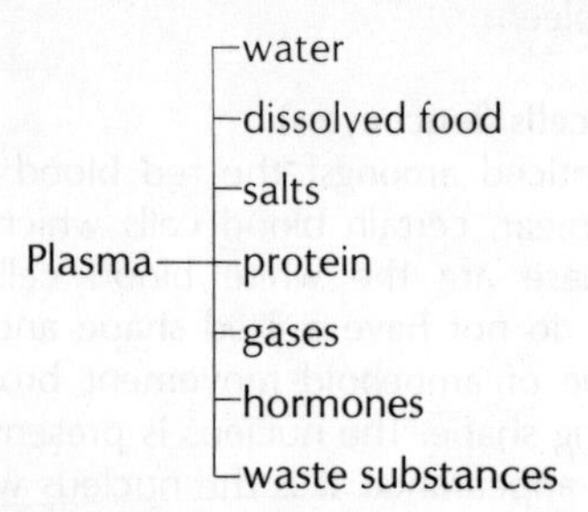

4 Blood platelets (thrombocytes)
These are very small particles which are discoid in shape, but they assume a star-shaped appearance in extracted blood. They are especially concerned with blood clotting and the size of the clot or *thrombus* depends upon the numbers of platelets present. Platelets have no nucleus, and they are believed to originate from detached portions of the cells lining the blood vessels.

Functions of blood

1 Red blood cells
Oxygen diffuses into the blood in the lungs and *combines with* haemoglobin to form *oxyhaemoglobin*. The blood cells are transported around the body. When tissues that require oxygen are reached, the oxyhaemoglobin releases its load and the resulting haemoglobin can now pick up small amounts of carbon dioxide to return to the lungs. The vast numbers of red blood cells provide a sufficiently large surface area to pick up all the oxygen that the body needs. The structure of the red cells enables them to increase their surface area in the blood capillaries by forming a bell shape. The central thinner part of the cell becomes pushed out by the plasma current (see fig 2.6).

2 White blood cells
The polymorphs can pass through the walls of the capillaries into the surrounding tissues and there *engulf* disease-causing bacteria. The bacteria are digested inside the blood cells. In response to the foreign proteins (*antigens*) of bacteria the lymphocytes produce other proteins, called *antibodies*, which immobilise disease organisms by causing them to clump together. These *agglutinated bacteria* cannot reproduce and eventually die.

3 Plasma
This fluid carries the proteins *fibrinogen* and *prothrombin* which together with the thrombocytes help

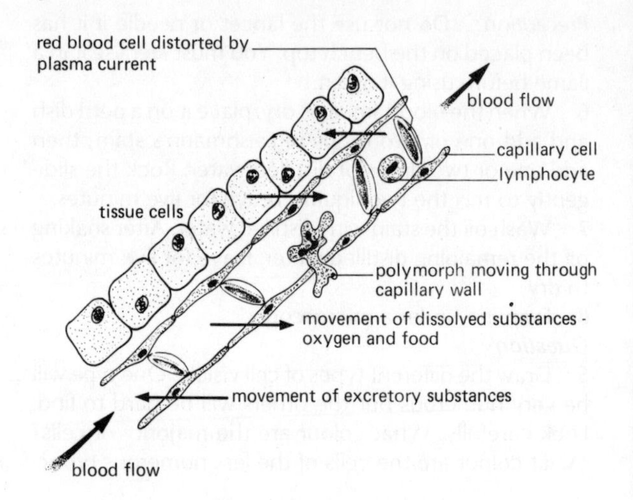

fig 2.6 The movement of blood cells through a capillary

Blood component	Functions
Red blood cells	Transport oxygen as oxyhaemoglobin. Transport carbon dioxide (very small amount).
White blood cells	Attack and engulf bacteria (polymorphs). Produce antibodies (lymphocytes).
Blood platelets	Aid clotting (together with plasma components).
Plasma	Transports: i) Carbon dioxide as bicarbonate from tissues to lungs. ii) Waste matter from tissues to excretory organs. iii) Hormones from ductless glands to the tissues where they act. iv) Digested food from small intestine to tissues. v) Heat from muscles and liver to all parts of the body. vi) Ions and water to maintain the balance of body fluids. vii) White blood cells and antibodies to sites of infection. viii) Platelets and serum proteins.

Table 2.1 Summary of functions of the blood

in blood clotting. An interlacing network of fibres is formed which entangles blood cells and forms a clot. The excretory materials such as *urea* and *uric acid* formed by the liver are carried to the kidneys by the plasma. *Hormones* are transported in minute quantities from the ductless *endocrine glands* to their sites of action. Gases such as *oxygen* and *carbon dioxide* are carried in chemical combination or in solution. This is in addition to gases carried in the red blood corpuscles.

2.31 Blood groups

In the past many attempts were made to save lives after severe injury by transfusing blood from one human to another. Some of these efforts were successful while others resulted in death. Any attempts with non-human blood, such as that of sheep, were always fatal to the recipient. It was not until 1900 that an Austrian, Karl Landsteiner, discovered the reason for these apparent inconsistencies.

He discovered that death was caused by the transfused red blood cells clumping together (*agglutinating*) and blocking the small blood vessels. His investigations showed that two separate factors were operating, one involving the red blood cells and the other involving the plasma. By mixing together red blood cells from certain people with the plasma from other people, he developed the idea that there are four types of blood based on the cell factor (antigen).

Type O – no cell factor
Type A – A type cells
Type B – B type cells
Type AB – both A and B type cells

These cells are contained in plasma, which contains anti-cell factors (antibodies) according to the following pattern:

Type O blood has anti-A and anti-B factors,
Type A blood has anti-B factors,
Type B blood has anti-A factors, and
Type AB blood has no plasma factors.

The facts can be summarised thus:

Blood group	Antigen on the red blood cell	Antibody in the serum of the same individual
O	no antigen	antibody a and b
A	antigen A	antibody b
B	antigen B	antibody a
AB	antigen AB	no antibodies

Thus, for transfusion purposes we need consider **only the effect** that the **recipient's plasma** will have on the **donor's red blood cells**. The plasma of the donor will be so diluted that it will not affect the red blood cells of the recipient.

The proportion of these blood groups varies in different parts of the world, but groups O and A are always the two largest groups. Group AB is the smallest (about 10% of most populations), so that comparatively few people belong to this group, which can receive blood from any other person.

Other factors present in blood have been discovered since the time of Landsteiner which affect only small numbers of people. A particularly interesting example is the *Rhesus factor*, so named because it was first isolated in Rhesus monkeys. About 85% of the population possess this antigen and are described as *rhesus positive* (Rh+), while the remaining 15% lack the antigen and are referred to as *rhesus negative* (Rh−). No antibodies to the Rhesus factor are normally present in Rh− blood; but if Rh+ blood is introduced by transfusion or by 'leakage' from foetal blood during birth, then they *are* formed. This latter fact aroused considerable interest because it provided an explanation of why certain newborn babies suffered a potentially fatal jaundice known as *haemolytic disease of the newborn*. The problem arises only when the baby's mother is Rh− and his or her father is Rh+. The children's blood will usually be Rh+ and some of it leaks into the maternal circulation. This causes the production of *rhesus antibodies* in the blood of the

Patient's (recipients) blood group	Antibody present in serum	Donor's blood group			
		O	A	B	AB
O	ab	√	x	x	x
A	b	√	√	x	x
B	a	√	x	√	x
AB	o	√	√	√	√
Group AB can receive all other groups. Group AB called **Universal recipient**		Group O can be given to all other groups. Group O called **Universal donor**			

√ = compatible with recipient
x = incompatible with recipient i.e. agglutinated

Table 2.2 Blood group compatibilities

mother, and because of their small size, these antibodies pass into the blood of the foetus during the next pregnancy. The red blood cells of the foetus are progressively destroyed because of the antigen–antibody reaction on their surface. With successive pregnancies, the antibody level builds up in the bloodstream of the mother and, consequently, it becomes more likely that the child will be born suffering from haemolytic disease. Providing that the mother and father have been diagnosed for their rhesus factors then the child can be saved by a complete blood transfusion shortly after birth. In recent years a vaccine has been developed to prevent destruction of foetal red blood cells by the mother's antibodies. This vaccine contains a small amount of the Rhesus+ factor. The mother produces a small number of antibodies, and these destroy any foetal cells which cross the placenta. This stops the production of further antibodies by the mother.

Question

1 A woman of child-bearing age requires a blood transfusion before she has had any children. Her blood is group A and Rh−. What blood group or groups can she safely be given, bearing in mind that she is married to a man who is Rh+?

2.40 Essentials of a transport system

From the time of the ancient Greeks to the seventeenth century the blood system was thought to contain the basic principle of life – a spirit, drawn in by the act of breathing. It was thought to enter the body through the wind pipe and then pass to the heart by way of the lungs, encountering other spirits during its passage from the heart to the rest of the blood system. It was not until the seventeenth century, that William Harvey (1578–1657), an Englishman, finally disproved this theory. By 1615, Harvey had developed the concept of the **circulation of blood**. The story of that discovery illustrates two important aspects of scientific investigation, namely **observation** and **measurement**. From these two deductions can be made.

The original theory of the Greeks implied that blood was being *continually formed* in the body. Harvey destroyed this idea with brilliant simplicity by pointing out that the *weight of blood required* to be manufactured by the body *was far too great*. He showed that the blood can only leave the ventricle of the heart in one direction, and then he measured the capacity of the heart. He found that the heart contained about 56 g of blood. It beats 72 times in one minute, so that **in one hour** it throws out:

$(72 \times 60) \times 56$ g
$= 241\,920$ g
or 242 kg

This is about **three times** the average weight for a man!!

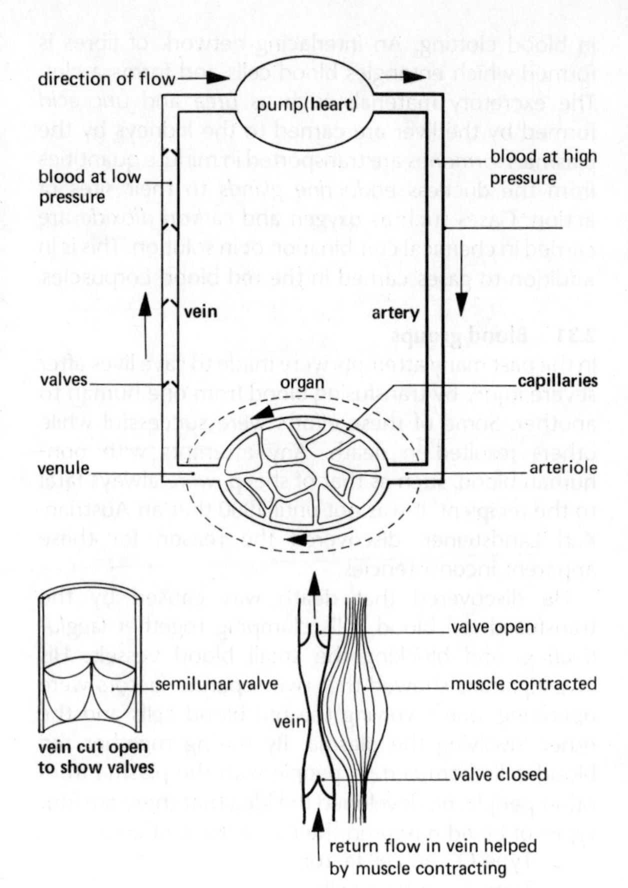

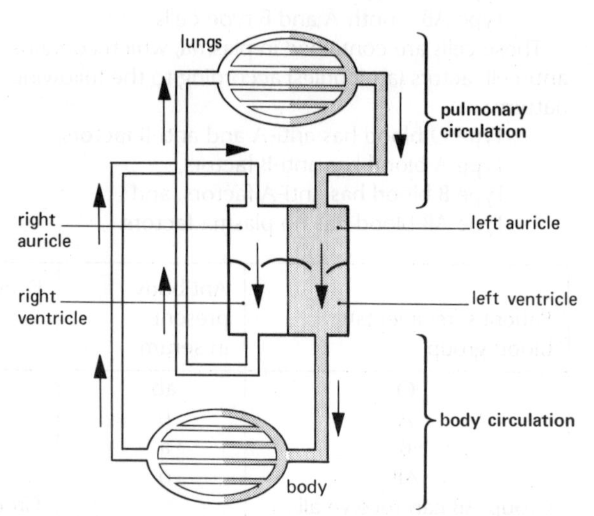

a simple plan of the double circulation of a mammal

fig 2.7 Simple diagrams of circulatory systems

Where does all this blood come from? Where does it go to? The answer can only be that it is the **same blood** returning to the heart, time and time again, to be recirculated around the body. This knowledge that blood circulates formed the foundation on which has been built a great deal of the interpretation of the physical activities of living organisms.

You will note that the application of **observation** and **deduction**, together with **simple measurement**, helped to disprove the misguided theories which had been handed down from the Greeks. Harvey confirmed these observations by many dissections of the heart and blood vessels of men and animals. He was not however able to see the finer blood vessels (capillaries), and it was left to Malpighi to discover these forty years later.

Thus in all vertebrate animals, the transport system consists essentially of a pump (the heart) and a connecting series of vessels through which the blood is pumped. The fluid keeps circulating in one direction aided by valves in the pump and in the vessels. (See fig 2.7.)

2.41 The heart

Experiment

The mammalian heart – external examination

Procedure

1 Take the heart of a sheep with the main blood vessels intact (the hearts of rabbit or rat can be used, but these are on a much smaller scale). Note the general shape and use fig 2.8 to help you identify the various features.

2 Distinguish the two auricles from the ventricles. How would you describe their external appearance?

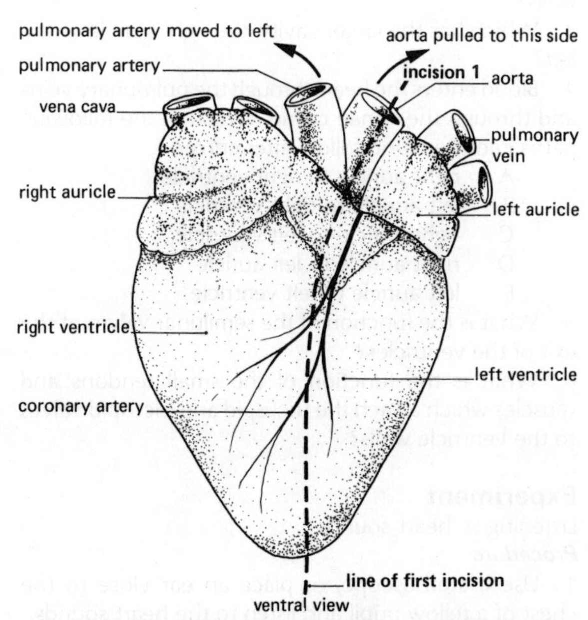

fig 2.8 The external structure of the heart shown in ventral view

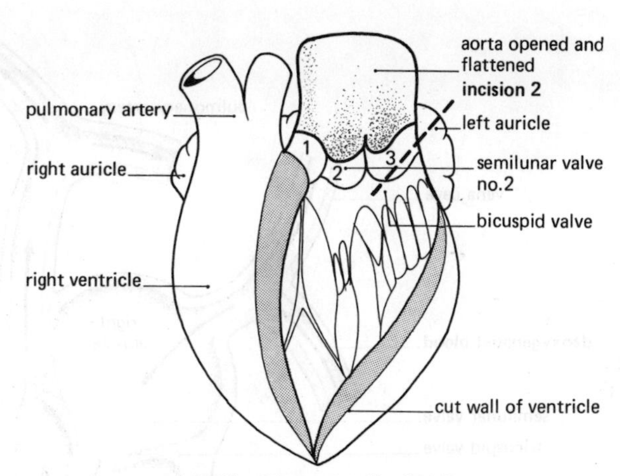

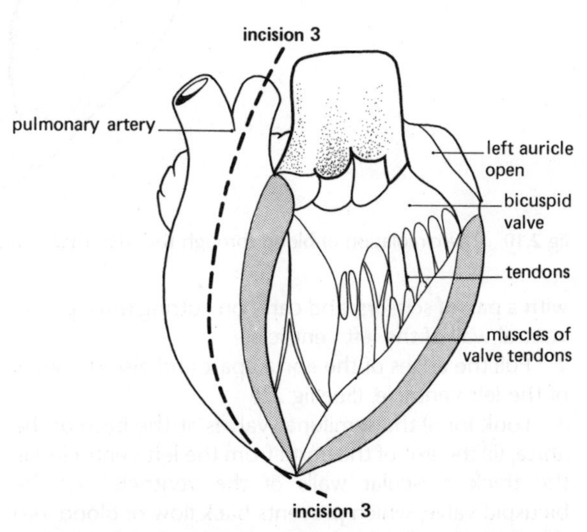

fig 2.9 A series of drawings of a dissection of the heart displaying its internal structure

3 Look for the remains of the arteries and veins, and examine the thickness of their walls in each case. The arteries have the thicker walls.

4 Look to see if there are any valves inside the vessels. Which have valves at the base, arteries or veins?

5 Look for the coronary artery on the wall of the ventricle. What do you think is the function of this vessel?

Experiment

Internal structure of the heart

Procedure

1 *Incision 1* (see fig 2.8). Cut down through the aorta

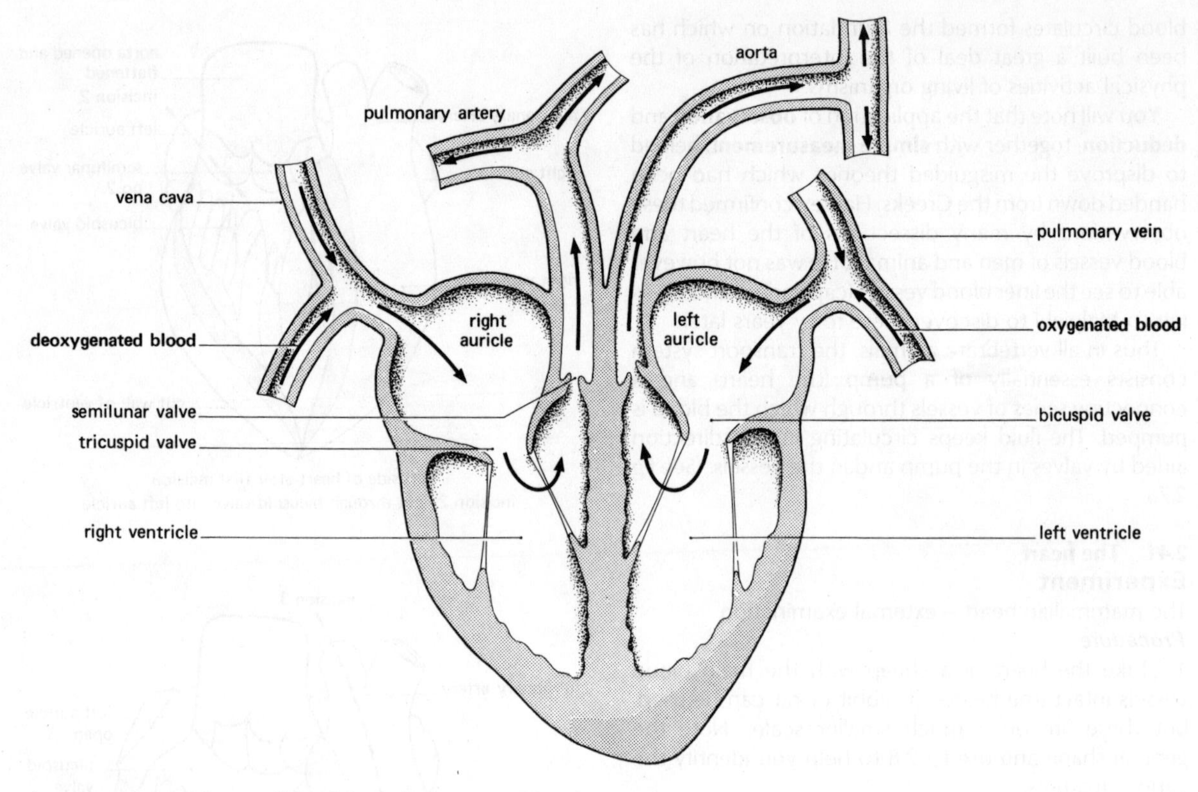

fig 2.10 The circulation of blood through the mammalian heart

with a pair of scissors, and carry on cutting through the ventral wall of the left ventricle.

2 Pull the edges of the aorta apart and also the walls of the left ventricle. (See fig 2.9)

3 Look for (i) the semilunar valves at the base of the aorta, (ii) the exit of the aorta from the left ventricle, (iii) the thick muscular walls of the ventricle, (iv) the bicuspid valve, which prevents back flow of blood into the auricle on contraction of the ventricle, (v) the strings (tendons) and small muscles attached to the bicuspid valve to prevent it from being turned inside out.

4 *Incision 2* (see fig 2.9). Cut through the bicuspid valve with scissors, and continue cutting along the ventral wall of the left auricle.

5 Look for (i) the thin walls of the auricle (compare them with the walls of the ventricle), and (ii) the veins which open wide into the left auricle (compare the thickness of their walls with those of the aorta).

6 *Incision 3* (see fig 2.9). Cut through the pulmonary artery and along the ventral wall of the right ventricle (*incision 3*). Cut through the tricuspid valve and along the ventral wall of the right auricle (*incision 4*).

Questions

1 Is there any opening or passage between right and left sides of the heart?

2 Which has the larger cavity, the right ventricle or the left ventricle? Which has the thicker muscular walls?

3 Which has the larger cavity, the right auricle or the left?

4 Blood enters the heart through the pulmonary veins and through the venae cavae. Which of the following paths can the blood follow after entry?

A right auricle to right ventricle
B left auricle to right ventricle
C right auricle to left ventricle
D right auricle to left auricle
E left auricle to left ventricle

5 What is the function of the semilunar valves at the exit of the ventricles?

6 What is the function of the small tendons and muscles which attach the tricuspid and bicuspid valves to the ventricle walls?

Experiment

Listening to heart sounds

Procedure

1 Use a stethoscope, or place an ear close to the chest of a fellow pupil and listen to the heart sounds.

You will hear two sounds at each beat, often described as 'lub-dupp'. The first sound is a dull thud caused by the closure of the bicuspid and tricuspid valves as the ventricles contract. The second sound is

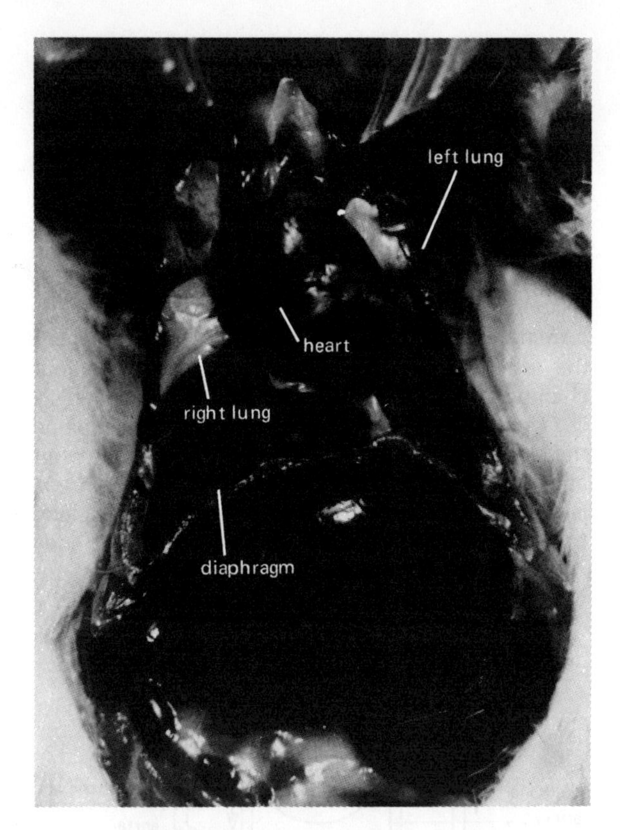

fig 2.11 A dissection of the rat thorax

more abrupt and of a higher pitch, caused by the closing of the semilunar valves and the vibrations of the walls of the arteries.

The heart beats an average of 72–75 times per minute. At higher temperatures and during exercise this can increase up to 200 beats per minute. The rate is generally faster during fevers and slower during sleep. The heart of a child beats faster than that of an adult. Emotion can cause an increase in heart rate. A person reading an exciting book, watching a thrilling film or even looking at pictures of the opposite sex can experience a faster rate of heart beat.

The use of the term 'normal' as applied to rate of heart beat can mean one of two things, either (i) that it has the value found in the majority of people, i.e. the mean or average, or (ii) that the heart is functioning properly, without fault.

Confusion between these two meanings can sometimes cause alarm, especially when people find out that their pulse rate is different from what is said to be 'normal'. Normal in the case (ii) above can cover a wide range. The American Heart Association *accepts as normal*, values between 50 and 110 beats per minute.

The heart is supplied with nerves from the brain to control its beating, but because of its own inherent muscular activity the heart will continue to beat after the nerve supply is cut.

The sequence of the beat is as follows:

auricles contract simultaneously (*auricular systole*);
blood in the auricles is expelled into the ventricles;
ventricles contract simultaneously (*ventricular systole*).
blood in the ventricles expelled into the pulmonary arteries and the aorta;
blood is forced against the bicuspid and tricuspid valves which close preventing back flow of blood; they are prevented from being blown inside out by the tendons;
pause (*diastole*) during which all parts of the heart relax;
cycle recommences.

The muscles of the heart must be supplied with oxygen and food materials. This function is performed by the coronary artery branching from the aorta and spreading through the heart muscle.

2.42 The blood vessels

The channels through which the mass flow of blood occurs are of two types; *arteries* and *veins*. Arteries carry blood away from the heart and veins carry blood towards the heart. The force of contraction of the ventricles squeezes the blood from the heart into the arteries. The blood is therefore under considerable pressure in the arteries. The mass of blood pushed into the arteries presses outwards against the artery walls, the elastic tissue stretches and then contracts inwards to force the blood forwards through the arteries. These two processes produce the characteristic pulsating of blood in the arteries. The blood cannot return to the heart, because of the presence of *semilunar valves* at the bases of the main arteries.

The blood returns from the tissues to the heart in a smooth stream at low pressure through the veins. The pressure in the arteries has been lost as the blood flows through the tissues in narrower and narrower channels. Since the pressure has been lost, how does the blood get back to the heart? The veins have semilunar or pocket valves along their length which prevent the back flow, but onward movement is largely maintained by the incidental action of muscles against the large veins. The continual movement of the limbs, the breathing movements of the thorax and the contraction of the abdominal muscles all maintain this flow towards the heart. Regular exercise is therefore essential. Gravity can also play a small part in those parts of the body above the heart.

Capillaries are the smallest of the blood vessels. They connect the end of the arterial system with the beginning of the venous system. Their non-muscular walls consist of a *single layer of flattened cells* which allows the plasma to leak out very easily. Thus water,

Arteries	Veins
1 Carry blood away from the heart.	Carry blood towards the heart.
2 Carry blood with a high oxygen content (except the pulmonary artery).	Carry blood with a low oxygen content (except the pulmonary vein).
3 Walls are thick, muscular and elastic.	Walls are thin (little muscle).
4 Valves are absent (except at the base of large arteries leaving the heart).	Valves are present throughout their length.
5 Blood flows rapidly under high pressure.	Blood flows slowly under low pressure
6 Blood flows in pulses.	Blood flows smoothly.
7 Tend to lie deeper in the body.	Tend to lie near the body surface.

Table 2.3 Differences between arteries and veins

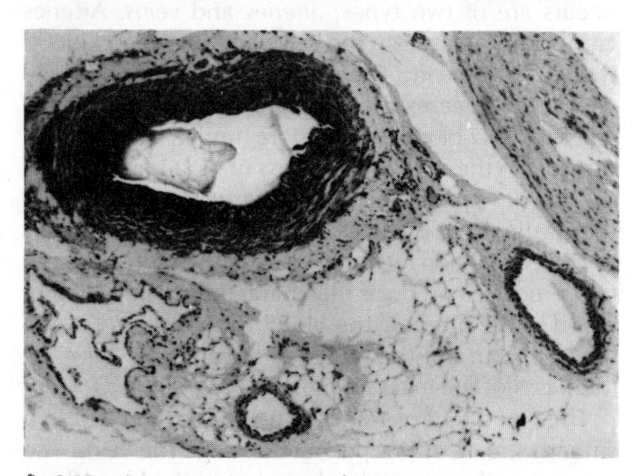

fig 2.12 . A transverse section of an artery and a vein

oxygen, carbon dioxide, dissolved foods and other substances move freely between the cells of the body tissues. Some white blood cells (but not red cells) also find it possible to squeeze out between the cells of the capillary wall.

The capillary network is extremely dense throughout the body. Therefore no cell of the body is far from a supply of food materials and oxygen. The capillaries branch from the *arterioles* (small arteries) which have muscular walls, enabling them to contract and shut off blood from certain sets of capillaries. The capillaries join together to form the first *small veins or venules.* Most of the small vessels seen in the tail of the fish or tadpole (section 2.20) are capillaries. It can be seen that they are so narrow as to allow red blood cells to pass in *single file only.*

2.50 Circulation

Fish have a *single* circulation of blood similar to that shown in fig 2.7. Incorporated into the system is a set of capillaries in the gills where the blood receives oxygen and gives up carbon dioxide. The heart in the fish is a much simpler structure that in the mammal, consisting of one auricle and one ventricle. As we saw in section 2.40 it was Harvey who first demonstrated the relationship between the heart and its blood vessels in forming a continuously circulating flow of blood. Unlike the fish heart, the heart of the mammal has four chambers and the circulation is *double* instead of single.
i) *The pulmonary circulation,* in which the blood travels from the heart to the lungs and back to the heart again.
ii) *The body (systemic) circulation,* in which the blood travels from the heart to all the organs of the body

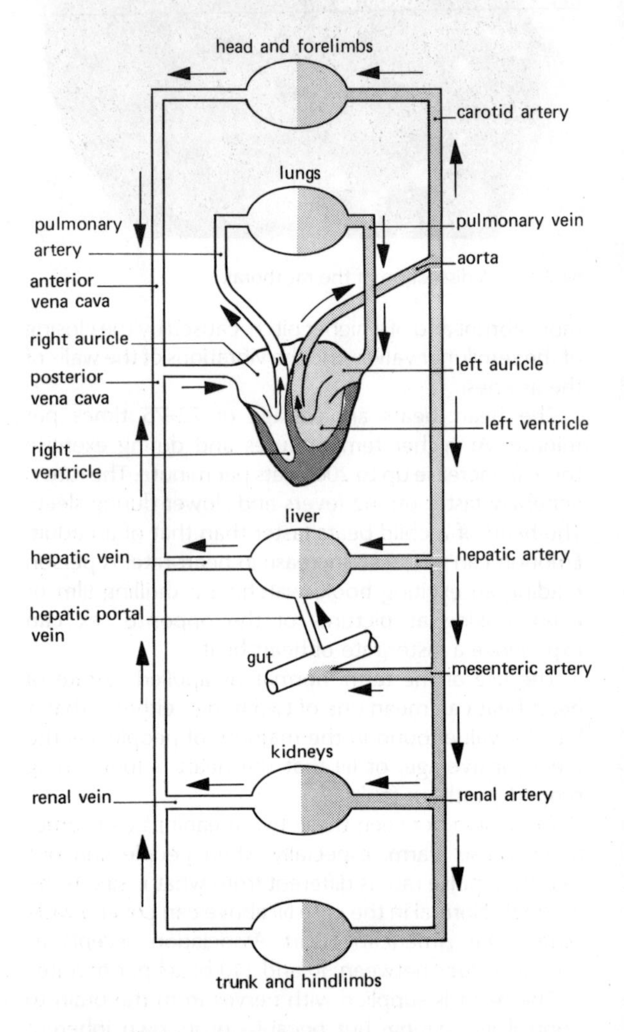

fig 2.13 A diagram of the circulatory system in a mammal

(except the lungs) and back to the heart again.

A simple diagram of this circulation is shown in fig 2.13. If we consider any one blood cell commencing its circulation at the left ventricle, then before that same blood cell returns to the left ventricle it must have passed through at least two sets of capillaries. One set is in the lungs and the second set could be in any part of the body. From fig 2.13 it can be seen that the mammalian blood circulation includes a *portal system*. There is only one of these in the mammal, where the blood from the gut capillaries is carried by the *hepatic portal vein* to the liver capillaries, and finally from these back to the heart by way of the *hepatic vein*. The gut capillaries give up oxygen and take up digested food so that the blood flowing to the liver delivers this food to the liver capillaries. Thus the blood flowing through this portal system goes through *three sets of capillaries* before reaching its starting point at the left ventricle again. The blood in the hepatic portal vein is deoxygenated, and so the liver is supplied with oxygen by way of the *hepatic artery*.

Let us now follow the course of the blood through the double circulation. The oxygenated blood is pumped from the left ventricle into the largest artery, the *aorta*. This distributes the blood to the whole body, from head to toe. Branch arteries carry blood to the gut, kidneys, liver, sex organs, limbs, skin etc. The blood passes through capillaries in all of these organs where exchange of materials (including oxygen and carbon dioxide) takes place. From the capillaries the deoxygenated blood is collected into venules and then into veins which pour the blood back into the right auricle of the heart. The main veins connecting with this side of the heart are called *venae cavae*. The blood in the veins has been deoxygenated, and is a dull red colour easily distinguished from the bright red arterial blood.

The right auricle then contracts, forcing the deoxygenated blood down through the tricuspid valve into the right ventricle, whence it is pumped to the lungs by way of the *pulmonary artery*. In the lung capillaries the blood picks up oxygen and gives up carbon dioxide, returning by the *pulmonary vein* to the left auricle of the heart. Finally contraction of the left auricle forces blood through the bicuspid valve into the left ventricle once more.

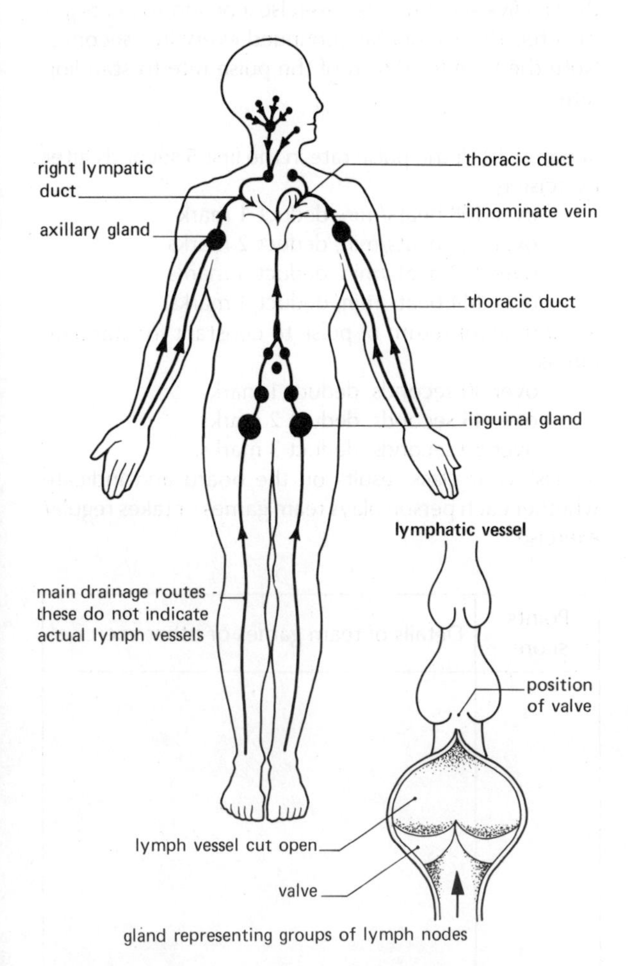

fig 2.14 The position of the main lymph glands and vessels

2.51 Lymph

The functions of the plasma can only be performed if it can be brought into contact with cells in order to exchange dissolved substances. To this end the plasma must pass through the walls of the capillaries and bathe the cells of the tissues. In this phase, plasma becomes known as *tissue fluid*. The pressure of the liquid in the blood vessels prevents the return of the tissue fluid and so it must return by another route. The tissue fluid around the cells drains into the series of vessels making up the *lymphatic system* (see fig 2.14). The tissue fluid, now called *lymph*, is caused to flow along the system by:

i) muscular movement around the lymph vessels which compresses them;

ii) valves preventing back flow of lymph whilst the vessels are compressed;

iii) greater pressure in the lymph capillaries than in the lymphatics (main lymph vessels);

iv) inspiratory breathing movements of the thorax which suck lymph into the thoracic duct, and the expiratory movements which force it into the innominate veins.

Thus the lymph is eventually discharged back into the blood circulation by way of the *innominate veins* near the heart. The lymph is now plasma and contains much of the waste materials to be dealt with by the excretory organs.

Lymph vessels have swellings called *lymph nodes* at certain points. These are glandular structures, which produce white blood cells. In humans they are perhaps most noticeable in the armpit (*axillary glands*) the angle of the jaw, and the groin (*inguinal glands*). Note that the term 'gland' here refers to a group of lymph nodes, and not to an endocrine gland. Any infected wound can cause these glands to swell, as the infection spreads along the lymph vessels. This condition known as *septicaemia*.

2.52 The pulse

The pulse wave in the arteries is dependent on three factors:

i) the *inflow* of blood from the heart,

ii) the *resistance* of the arterioles to outflow of blood into the capillaries, and

iii) the *elasticity* of the arterial walls.

At each beat of the heart (systole), a mass of blood passes down the arteries and at the same time, the elastic tissue is distended. Between the beats of the heart (diastole), the arteries contract because of their elasticity, and thus a steady stream of blood flows along the arteries. The pulse wave is non-existent in the capillaries unless the arterioles are widely dilated. In the veins there is no detectable pulse and the rate of blood flow is uniform.

Experiment

Can the pulse be detected?

Procedure

1 In certain parts of the body, large arteries are situated near the surface. Place the three middle fingers of your right hand over the radial artery of the left wrist. This can be your own wrist or that of your fellow pupil. Move your fingers until you can feel a regular pulse.

2 Count the number of beats in 30 seconds. Repeat this several times and obtain the average rate over a half minute period. The subject must be seated.

3 Calculate the pulse rate per minute for yourself and your partner.

4 Record the pulse rate for a group such as a class of pupils or a group of adults. Draw a histogram to demonstrate the variation of the pulse rate. Find the mean (see Introductory Chapter section 4.10).

Question

1 What is the spread of your histogram?

2 What is the mean?

Experiment

Does exercise have any effect on the pulse rate?

Procedure

1 Use a scoring technique which can give an indication of physical fitness. Start with 12 points and deduct points as indicated in the following tests.

2 Sitting pulse. The subject under investigation should be sitting quietly for at least three minutes. Take the sitting pulse at the wrist.

Scoring: if pulse is over 78/min, deduct 1 mark
if pulse is over 84/min, deduct 2 marks
if pulse is over 96/min, deduct 3 marks.

3 Standing pulse. The person should stand up slowly. Count the number of beats from the fifth to the tenth second after standing, and for every five second period until the rate is constant for three consecutive periods. Convert to beats per minute. This is the standing pulse rate.

Scoring: Increase of pulse rate after standing for fifth to tenth second by:

24 or more – deduct 1 mark
36 or more – deduct 2 marks.

4 Response to exercise. The subject should stand in front of a chair, and then step up on the chair and down again five times in fifteen seconds. Then he should stand still on the floor, and count the number of pulse beats in the first five seconds after exercise (convert to beats per minute). This should be continued every five seconds. Note the time for return of the pulse rate to standing rate.

Scoring: i) If the pulse rate in the first 5 seconds after exercise is:

over 108 beats/min, deduct 1 mark
over 120 beats/min, deduct 2 marks
over 132 beats/min, deduct 3 marks
over 144 beats/min, deduct 4 marks.

ii) If time for return of pulse to constant for standing rate is:

over 30 seconds, deduct 1 mark
over 45 seconds, deduct 2 marks
over 60 seconds, deduct 3 marks.

5 List your class results on the board and indicate whether each person plays team games or takes regular exercise.

Points score	Details of team games or other exercise

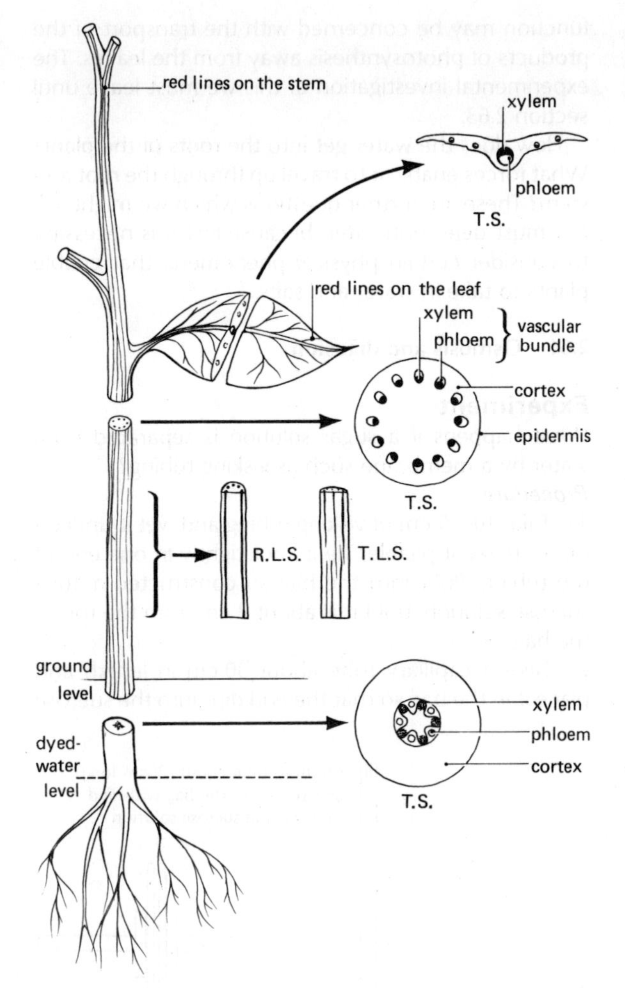

fig 2.15 A dicotyledonous plant sectioned at different levels

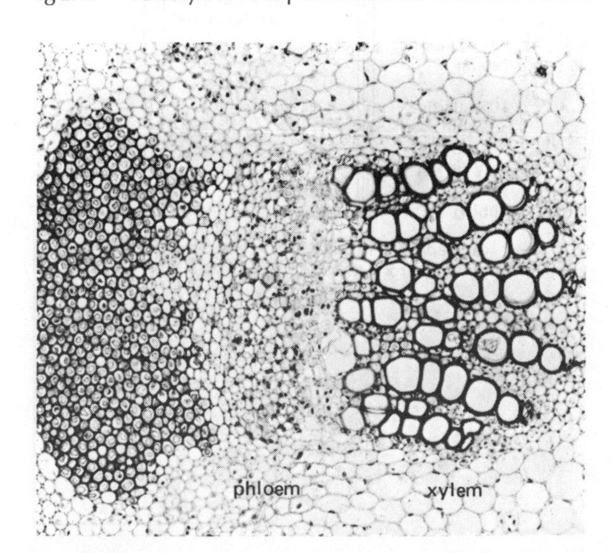

fig 2.16 A transverse section through a dicotyledonous stem showing a vascular bundle

Questions

3 Why does the pulse rate increase when a person changes from a sitting to a standing position?

4 Why is there a marked increase in a person's pulse rate after he has been stepping on and off a chair?

5 Is there any correlation between the test score and the standard of fitness of the person performing the test?

2.60 Transport in the plant

Most higher plants grow in soil and, in order that they should live, must be supplied continually with water. If a well-established plant is cut off at the base of its stem, a fluid will flow from the cut end of the stem. Simple observation tells us, therefore, that water is taken up by the plant, and we can ask the question: 'is there a mass flow system causing movement of fluid in plants as well as in animals?'

Experiment

Is there a movement of fluid in plants?

Procedure

1 Obtain some plants with roots intact and also some cut leafy shoots. Small, quick-growing annual plants and woody shoots, such as *Hydrangea* or privet, are suitable.

2 Place the roots of the small plants and the bases of the leafy shoots (not more than 2–3 cm submerged) in two jars of water,

Jar 1 is coloured with red ink or eosin.

Jar 2 is coloured with carmine which is insoluble in water, but remains as suspended free particles.

3 Place the jars in a well-lit and well-ventilated place such as a side bench near a window. Use a lens to observe the stem and leaves at 30 minute intervals. Alternatively the jars can be set aside for twenty-four hours and examined the next day. (Eosin is toxic and quickly kills the plant material so that they cannot be left for more than twenty-four hours).

4 Use a razor or razor blade to cut transverse and longitudinal sections, as thin as possible, from the root and stem (see fig 2.15).

5 Use a brush for transferring the sections to a slide. Examine them under a microscope or with a strong hand lens.

Questions

1 What are the differences between the plants in the two jars?

2 What evidence is there to show that the red colour is carried up the plant?

3 Examine the sections and decide whether the red colour is scattered over the whole section or confined to particular parts.

4 Can you determine which portion or portions (named in fig 2.15) of the stem and leaves carry the red colour upwards?

Another way in which the movement of materials through a plant can be investigated is by using *radioactive isotopes*. These are forms of elements which emit radioactive particles. The particles can be detected and counted by means of a sensitive instrument called a *Geiger counter*. The number of particles detected is counted by a decade counter.

The roots of a balsam plant were immersed in water containing *radioactive phosphorus* (^{32}P) and during the next twelve hours a count of radioactivity was made on the second-lowest leaf and on the apex of the plant. The results are shown in table 2.4.

Time (hours)	Background Count	Count at second leaf	Count at apex
09 45		23.6	
09 45		26.4	
10 20	20.5	20.2	
10 30		24.2	
11 15	26.3	26.5	
11 30		29.0	
11 45		24.0	20.0
12 00		25.7	23.0
12 30		28.3	25.6
13 45		33.4	36.2
14 00	23.5	38.6	39.9
14 20		41.2	52.3
14 30		46.2	55.4
15 10		52.3	58.2
22 30		89.4	91.4

Table 2.4 ^{32}P count on a small rooted plant placed in water containing radioactive phosphorus

Questions

5 What is meant by the term background count?

6 At what time did the count at the second leaf begin to show a definite rise above the background count?

7 At what time did the count at the apex of the plant begin to show a definite rise above the background count?

8 Can you suggest a use for the phosphorus particles at the apex?

If we examine the data in table 2.4, and recall the first two experiments, it is clear that the plant is moving water up through the root and stem, and into the leaves and growing point.

Dissolved particles such as the radioactive phosphorus are also moving upwards through the conduction channels. Experiments with the dyed water show that it is the *xylem tubes* that deliver water and dissolved substances to the leaves and the apex of the plant. The phloem tubes do not take part in the upward transfer of water. If they are conducting tubes their function may be concerned with the transport of the products of photosynthesis away from the leaves. The experimental investigation of this we must leave until section 2.63.

How does the water get into the roots of the plant? What forces enable it to travel up through the root and stem? These are further questions which we might ask but must defer until later, because first it is necessary to consider certain physical phenomena that enable plants to take in water and salts.

2.61 Osmosis and diffusion

Experiment

What happens if a sugar solution is separated from water by a membrane such as visking tubing?

Procedure

1 Take 10–15 cm of visking tubing and wet it under a tap to make it pliable. Tie a knot tightly in one end of the tubing. Pour into the bag so constructed a 10% sucrose solution, until it is about 3 cm from the top of the bag.

2 Take a capillary tube about 30 cm in length and place it in the bag so that the end dips into the sucrose

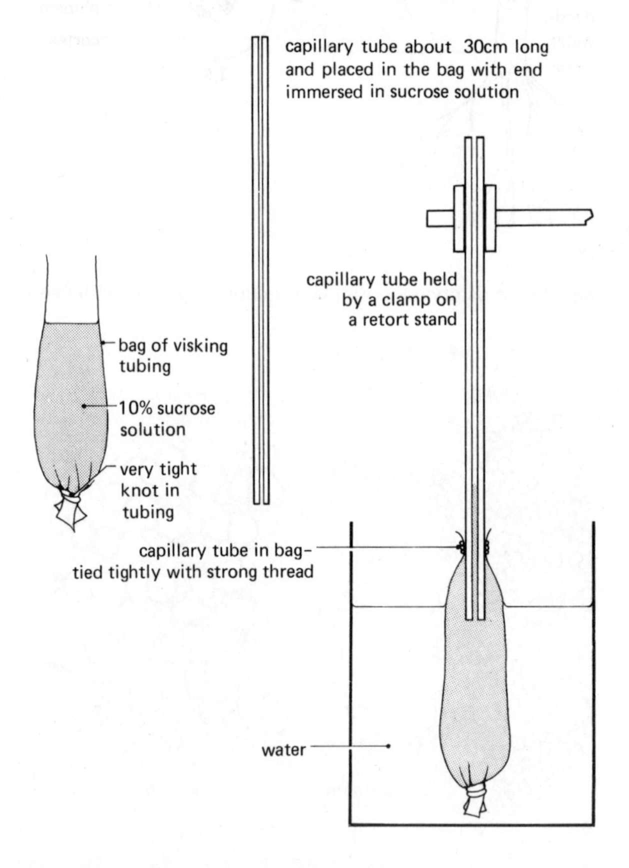

fig 2.17 Visking tubing containing a sugar solution, with capillary tube attached, placed in water

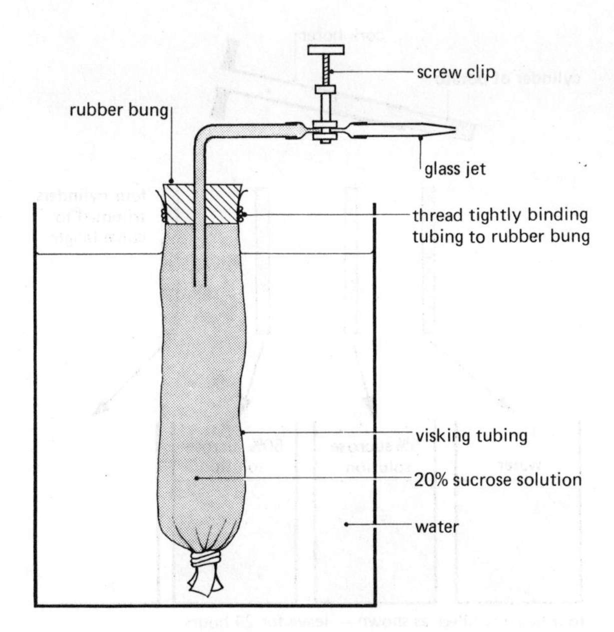

fig 2.18 Visking tubing containing a sugar solution; a closed system with a screw clip, placed in water

solution. Tie the open end of the bag tightly around the capillary tube, using a strong thread (take care not to cut the tubing).

3 Fix the capillary tube to a clamp and suspend the bag of sucrose solution in a beaker of water (see fig 2.17).

4 Mark the initial level of the sucrose solution in the capillary tube and leave the apparatus to stand. Mark the level of the sucrose solution every ten minutes.

Questions

1 What height did the sucrose solution reach in the capillary tube after 30 minutes?

2 What explanation can you give for the change in height of the liquid in the capillary tube?

3 Is there any evidence to indicate that some form of pressure is acting in the apparatus?

Experiment

Is pressure exerted when solutions of different strengths are separated by visking tubing?

Procedure

1 Set up the apparatus as shown in fig 2.18. Use visking tubing of wide diameter (3 cm) and having wet it first, tie it tightly into a knot at one end. Tie the other end tightly to a rubber bung by means of strong thread.

2 The rubber bung should have a piece of right-angled glass tubing pushed through it and into the bag. Attached to the other end of the glass tubing is a short length of rubber tubing into which is pushed another piece of glass tubing drawn into a fine jet. The rubber tubing has a screw clip attached.

3 Remove the rubber tubing from the right-angled bend, and, using a pipette, fill the bag with a 20% sucrose solution. Replace the bung and tubing and then place the bag in water in a beaker, leaving it to stand for twenty-four hours.

Questions

4 How does the appearance of the bag after twenty four hours compare with its state at the beginning of the experiment? How does the bag feel?

5 What explanation can you give for the change in the visking tubing?

6 What happens when you unscrew the clip? Can you give an explanation of the result?

What you have discovered is the phenomenon of *osmosis*. The water surrounding the visking tubing has moved through the tiny pores in the cellulose and increased the quantity of water inside the bag. As a result the liquid has moved up the tube in the first experiment, and filled the bag in the second experiment. If a tightly fitting plunger had been present in the bore of the capillary tube, a pressure could have been exerted preventing the rise of the liquid. This pressure would have to be equal to the *osmotic pressure* of the sucrose solution in the bag. The second experiment demonstrates this pressure, which, having built up, causes a jet of liquid to rush out on release of the screw clip. We can define osmosis as the *movement of solvent* (in this case water) through a *partially permeable membrane* from a *dilute solution* (or solvent only) to a *more concentrated solution*. In nature, the solvent is always water, and the passage of water through the membrane continues until the concentration on both sides of the membrane is equal. The simple explanation of this process is shown in fig 2.19 which illustrates the partially permeable nature of the visking tubing. The tubing acts as a *molecular sieve* whose pores are so

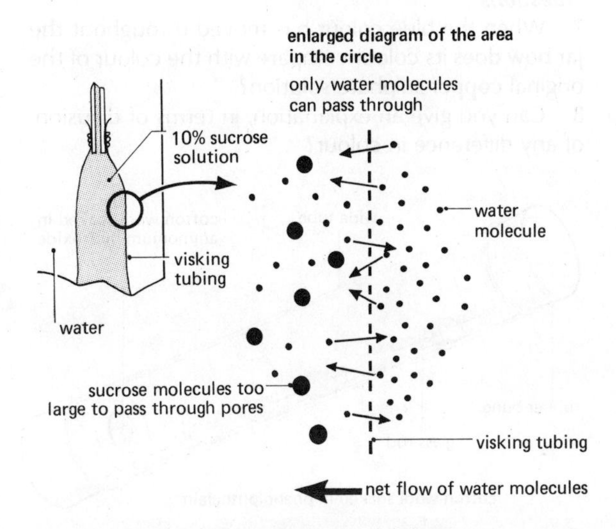

fig 2.19 A diagrammatic representation of the pores present in a membrane, and the movement of molecules

minute that they allow through molecules of water (H_2O), but not the larger molecules of sucrose ($C_{12}H_{22}O_{11}$).

It can be seen in the figure that the molecules of water can pass equally well in either direction through the pores. Since there are more molecules of water outside the tubing than there are inside, there is a net movement of water molecules from the water (or weak solution) to the strong solution.

This flow of water can be regarded as a particular example of the phenomenon of *diffusion*, defined as the process by which *molecules of a substance* present in a *region of high concentration* in a *fluid* (liquid or gas) tend to *move into a region of low concentration* until they are evenly *distributed*.

In the two previous experiments, the water molecules surrounding the visking tubing are present in high concentration, while those in the solution of sucrose are present in low concentration, so that the water molecules flow from where they are in a high concentration to the lower concentration.

Experiment

To demonstrate diffusion in a liquid

Procedure

1 Make up a strong solution (20–30%) of copper sulphate – it must be very blue in colour.

2 Introduce 50 cm³ of the copper sulphate solution into the bottom of a tall jar of distilled water. Use a pipette and let the copper sulphate flow gently so that it forms a distinct layer at the bottom of the jar. The division between water and copper sulphate should show as a clear line.

3 Leave the jar to stand for several days and mark the level reached by the blue colour each day.

Questions

7 When the blue colour has moved throughout the jar how does its colour compare with the colour of the original copper sulphate solution?

8 Can you give an explanation, in terms of diffusion, of any difference in colour?

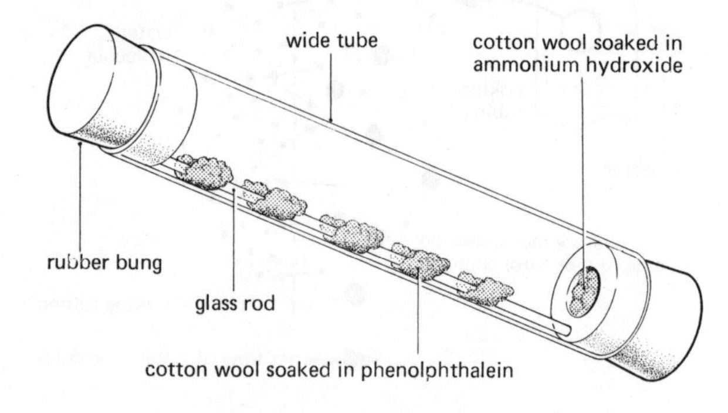

fig 2.20 Diffusion in a gas

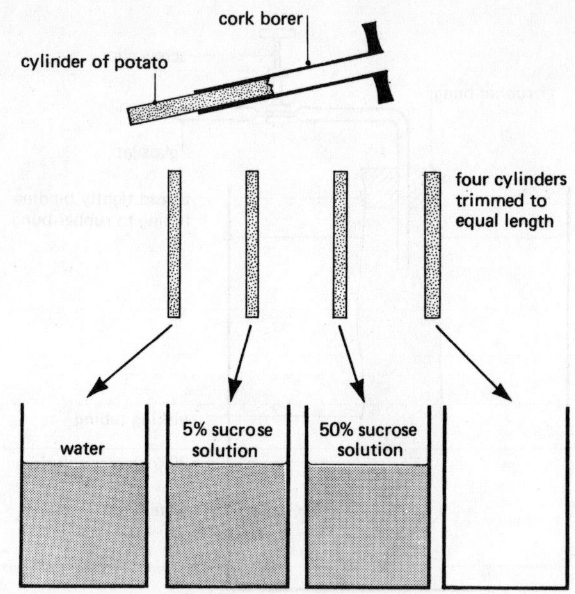

fig 2.21 Potato cylinders placed in different liquids

Experiment

To demonstrate diffusion in a gas

Procedure

1 Take a wide glass tube (50 cm in length; 2.5 cm in diameter), with a rubber bung fitted in each end.

2 Take a 40 cm glass rod, wrap small pieces of cotton wool at each end, and at 10 cm intervals along the rod (5 in all). Each portion of cotton wool must be soaked in phenolphthalein. Alternately, strips of red litmus paper could be used.

3 Place the rod in the wide glass tube, and stopper one end. In the other end place cotton wool soaked in ammonium hydroxide. Immediately stopper this end with the second rubber bung (see fig. 2.20).

4 Observe the cotton wool on the glass rod.

Questions

9 What do you observe?

10 How do you know that this is diffusion?

11 What can you say about the comparative speeds of diffusion in liquids and gases?

The process of diffusion is important to all living organisms. Let us now consider the two processes of osmosis and diffusion with reference to living tissues.

Experiment

What effect do different solutions have on potato tissues?

Procedure

1 Use a cork borer to cut four uniform cylinders of potato tissue. Trim all four to the same length. They should be as long as possible, and not shorter than

Initial length of cylinders (to nearest mm)	Length after 24 hours (to nearest mm)	Difference in length + or −	Percentage change	Texture firm or flaccid

The length of cylinders of potato tissue

5 cm. Record the length of the cylinders of tissue in a table like the one above.

2 Treat the four pieces as follows:
i) immerse in water in a beaker,
ii) immerse in 5% sucrose solution in a beaker,
iii) immerse in 50% sucrose solution in a beaker, and
iv) leave exposed to air in a beaker. (See fig 2.21).

3 Leave the cylinders of potato for twenty-four hours. Remove them from the beakers and dry each cylinder on blotting paper.

4 Measure the length of each cylinder. Record the lengths.

5 Feel each cylinder and describe whether its texture is firm or flaccid.

6 Record your results in the table above. Calculate the percentage change in length for each cylinder.

Questions

12 Piece number (iv) will have decreased in length because of drying out and will feel flabby. Which of the other pieces feels most like piece number (iv)?

13 What evidence is there that substances have entered or left the cells of the potato?

14 Which of the two processes (osmosis and diffusion) investigated is demonstrated by this experiment?

15 How could the experiment be improved to give more data? How could you use these data?

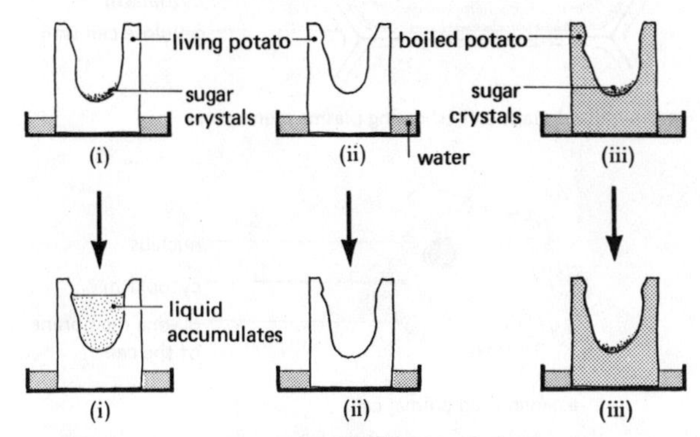

fig 2.22 The movement of water through the cells of a potato

Experiment

Does water flow through a series of plant cells?

Procedure

1 Cut three cubes (about 5–10 cm sides) from potatoes and scrape out a cavity in each tube (see fig 2.22).

2 Immerse one cube in boiling water for about three minutes to kill the living protoplasm.

3 Place each of the three cubes with its base in water in a petri dish. Treat as follows:
i) living potato – place some sugar crystals at the bottom of the cavity,
ii) living potato – leave the cavity empty,
iii) boiled potato (dead cells) – place some sugar crystals at the bottom of the cavity.

4 Leave the three cubes for about two or three hours, and then examine the cavities in each case.

Questions

16 What do you observe? Account for these observations.

17 Why did you set up numbers (ii) and (iii) (in procedure 3 above)? What function do they perform in the experiment?

Experiment

Do the osmotic pressures of the cell contents cause cells to interact with each other?

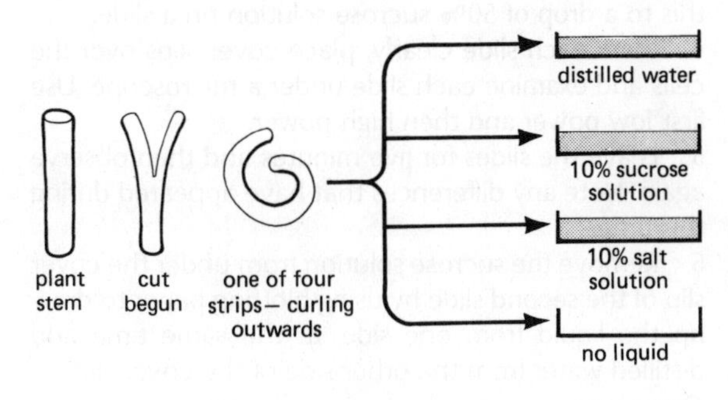

fig 2.23 The effects of different liquids on the cells of dandelion

Procedure

1 Take a 3 cm length of dandelion stem and cut it into four longitudinal strips. The strips will curl outwards with the epidermis or outer layer on the inside of the curve. Observe the structure of one strip with a lens.

2 Place the four strips in petri dishes as follows:

i) in distilled water,

ii) in 10% sucrose solution,

iii) in 10% sodium chloride solution,

iv) in air – no liquid. (see fig 2.23).

3 Examine the strips after 15 minutes and note, in each case, the amount of curling and the position of the epidermis.

Questions

18 Why does the strip curl outwards when it is first cut? (Look at fig 2.23 and consider the types of cell on the outside and the inside of the stem).

19 What effects have the two 10% solutions on the strips? What explanation can you offer?

20 What effect has the air on number (iv)? Why has this occurred?

In view of the evidence we have gathered so far from the behaviour of the potato cylinders in different solutions and from the bending of the strips of plant material, it is obvious that plant cells can change their size. In order to understand how this change occurs, we need a further experiment.

Experiment

How does the plant cell change shape under different osmotic conditions?

Procedure

1 Obtain an onion bulb and separate a fleshy bulb scale. From the inside of the scale peel off a thin layer of epidermal cells (the best type of onion bulb to use is the red-skinned variety). (See fig 2.24).

2 Transfer the thin layer of cells to a drop of distilled water on a slide. In order to prevent damage, move the layer of cells with a thin brush.

3 Peel off another strip of epidermal cells and transfer this to a drop of 50% sucrose solution on a slide.

4 Mark each slide clearly, place cover slips over the cells and examine each slide under a microscope. Use first low power and then high power.

5 Leave the slides for five minutes and then observe again. Note any differences that have appeared during this time.

6 Remove the sucrose solution from under the cover slip of the second slide by using blotting paper to draw up the liquid from one side, at the same time add distilled water from the other side of the cover slip.

Questions

21 Make sketches of the appearance of the cells in each liquid. How would you describe the difference in appearance of the cells?

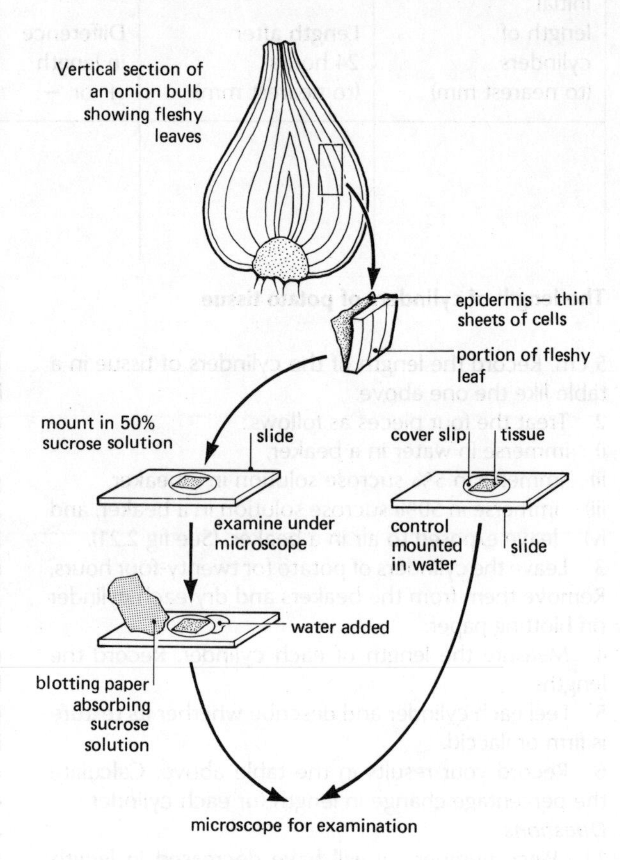

fig 2.24 Epidermal cells in an onion bulb in sucrose solution

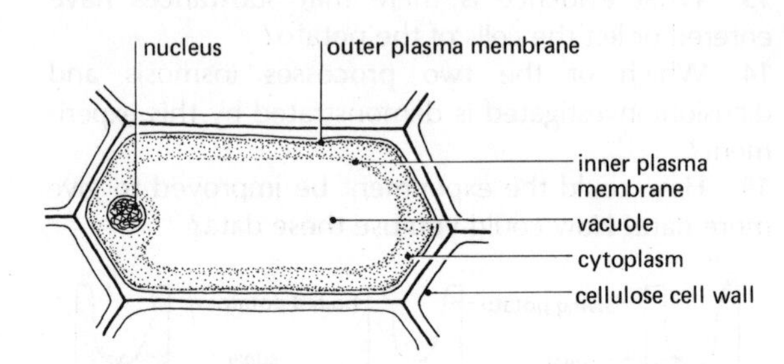

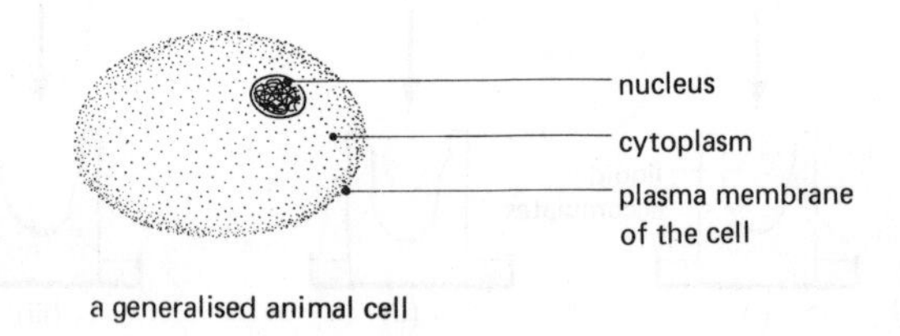

fig 2.25 Generalised plant and animal cells

22 Can you account for the differences with reference to the two liquids in which the cells have been immersed?

The changes that take place in the onion cells are easier to understand if one considers the general structure of plant cells. Each cell has a rigid cell wall of *cellulose*, an inner layer of *living cytoplasm* and within this a *space (or vacuole)* filled with water and various dissolved substances. (Known as cell sap). Compare this with a generalised animal cell shown alongside the plant cell in fig 2.25.

Your observations of the onion cells showed that when they were in sucrose solution the water passed out of the cell vacuole by osmosis (from the weak solution in the vacuole to the strong solution outside), and thus the cytoplasm retracted from the cell wall. The cell wall did not collapse, but the contents became a rounded ball of cytoplasm. Conversely when the onion epidermis was placed in water, the water molecules passed into the cell vacuole, causing the cytoplasm to be pressed outwards against the cell wall.

The experiments with visking tubing were explained on the basis of a molecular sieve through which large molecules cannot pass, but small ones can. The visking tubing is made of cellulose which takes the form of long fibres made up of long chain molecules or polymers. Between the molecules are gaps about 0.4–0.5 nm (4–5 Å) wide. Water molecules are about 0.3 nm wide, whereas a sucrose molecule is about 1.0 nm in width. It can be understood therefore why water molecules and not sucrose molecules can pass through the cellulose membrane of visking tubing. The cell walls of plants are also formed of long fibres of cellulose. In this case the gaps between the fibres are about 10 nm wide. The cell wall does not therefore form a barrier to water or sucrose molecules.

Since the cellulose cell wall is permeable to most solutions, it must be the cytoplasm with its inner and outer membranes which is only partially permeable and can thus produce the osmotic effects we have seen on pages 79–80. This function is probably that of the very thin plasma membranes at the surface of the cytoplasm.

We can now summarise our knowledge about the performance of plant cells and cell models when external conditions are changed:

1 The cells of the plant (onion epidermis) when supplied externally with sucrose solution are in a state of lowered internal pressure (cytoplasm contracts from the cell wall).

2 The vacuole surrounded by cytoplasm becomes smaller and the bag of cytoplasm collapses when in strong sugar solution, but it recovers when placed in water.

3 The cells of the plant (onion epidermis) when supplied externally with water are in a state of internal pressure (cytoplasm pressing against the cell wall).

4 Water passes into cells and causes them to swell up and press against each other (in the cylinders of potato).

5 Water passes from cell to cell from those in contact with the water to those inside the plant structure. This is increased if the innermost cells are in contact with a strong sugar solution (in the cavities of the cubes of potato).

We have not shown clearly that there is two-way traffic of water across the artificial membranes or the cell membranes. This can be done by using a radioactive isotope in a tracer. The tracer in this case is 'heavy water' which contains an isotope of hydrogen called deuterium (^{2}H). Compared with water this tracer is much denser and can be detected by using a sensitive apparatus for measuring the masses of liquids. We could use an apparatus identical to that used earlier in this section. (See fig 2.17).

The visking tubing is permeable to water of both types, but not to sucrose.

Questions

23 How would you know that water is passing into the bag of cellulose by osmosis?

24 How would you know that some water molecules are passing out of the visking tubing into the beaker?

All of our experimental work so far has been concerned with plant tissue. Animal tissue is different in that the cells are not surrounded by cellulose cell walls.

Experiment

Does osmosis take place in animal cells?

Procedure

1 Take three test-tubes containing the following:

Tube (i) 1 cm^3 of distilled water,

Tube (ii) 1 cm^3 of salt solution at the same concentration as blood plasma (0.85% sodium chloride solution),

Tube (iii) 1 cm^3 of salt solution at twice the concentration of blood plasma (1.70% sodium chloride solution).

2 Obtain some drops of blood using sterile techniques exactly similar to those used in section 2.20.

3 Allow some drops of blood to fall into each tube. If they fall on the inside wall of the tube, tilt the tube so that the liquid removes the blood. Shake the tubes gently. See fig 2.26.

4 Resterilise the little finger.

5 Leave the tubes for 2 to 3 minutes.

6 From each tube, remove a drop of liquid on a glass rod and place it on a glass slide. Cover the drop with a cover slip.

7 Examine under low power and then under high power of a microscope. Note the appearance of any blood cells that you can see.

8 Allow the test-tubes to stand for about half an hour and then observe the appearance of the liquids.

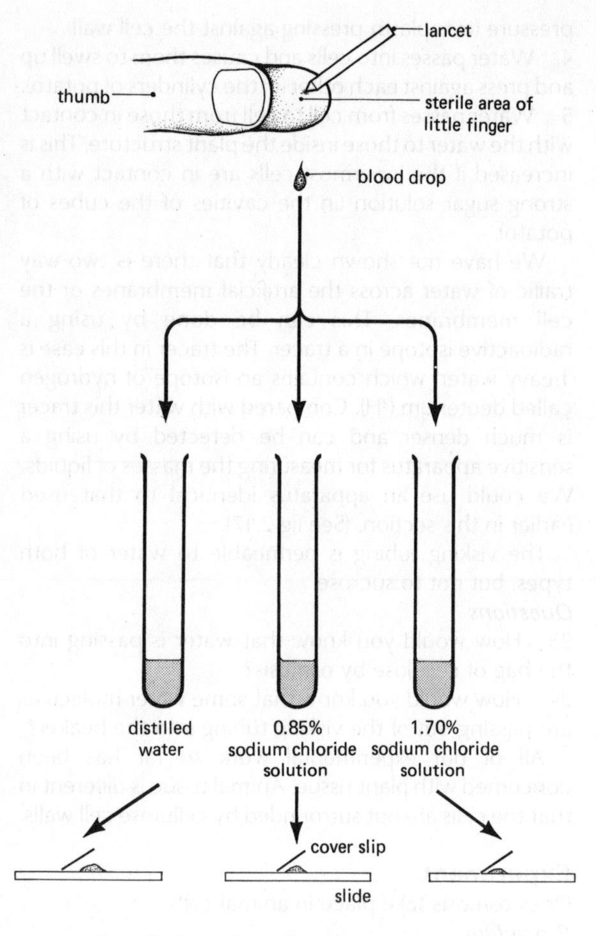

fig 2.26 Blood drops in different salt solutions

Questions

25 Were blood cells present in tubes (i), (ii) or (iii)?

26 If so describe the appearance of the blood cells, indicating in which tube they were found.

27 Can you give an explanation of your observations?

28 What was the appearance of the liquid in each tube after half an hour?

29 Linking all of these observations together, can you give an explanation of your observations in terms of osmosis?

The earlier experimental work on plants in this section 2.61 shows that plant cells can adjust to their surroundings without suffering permanent damage. This is reflected in the behaviour of complete plants when subjected to varying amounts of water: they can wilt, recover, and wilt again and still recover. From the last experiment, it is seen that animal cells cannot adjust in this way, for when the blood cells are exposed to water they absorb so much that they burst. This is because they have *no restraining cellulose cell wall* to enable them to resist the further entry of water once the elastic limit of the cell membrane has been reached.

All animals that live in fresh water have this problem of osmotic uptake of water by the cells of their tissues. The body fluids are more concentrated than the surrounding water so that they tend to absorb water through the skin or the body surface. It is essential that this water is continuously removed otherwise the body fluids such as blood will be diluted, and the cells will swell and rupture. In freshwater vertebrates (fish and amphibia) the kidneys act as *osmoregulators* and excrete a *watery urine* after extracting excess water from the blood.

Sea water animals have the reverse problem, in that their body fluids are less concentrated than sea water so that the body water tends to pass out through the skin into the surrounding medium by osmosis. In order to compensate for the loss of water, many sea fish swallow salt water and absorb it through the wall of the gut. The extra salts taken in have to be eliminated and this is effected mainly through the gills. The urine tends to be much less watery than in fresh water fish.

The kidneys also act as osmoregulators in land animals, where water is lost by evaporation from the body surface and through the excretion of fluids. Water is gained in food and drink but it is the kidneys that are mainly responsible for maintaining a fine balance.

2.62 Turgor and plasmolysis in plant cells

All organisms are absolutely dependent on an adequate supply of water from their environment. In plants it is essential as:

i) the major constituent of protoplasm;
ii) a solvent for organic compounds;
iii) a transport fluid;
iv) a basic raw material for photosynthesis;
v) for supporting herbaceous plants by keeping the shoot systems turgid;
vi) for cooling higher plants by evaporation from the leaves.

The last two features we shall consider in the remainder of this chapter. What do we mean by the term *turgidity*? Our experimental work has shown that plant cells are distended by uptake of water: they are then said to be in a state of *turgor*, while the tissue as a whole is *turgid*. Water no longer enters the cells since the resistance of the cell wall exactly balances any osmotic pressure forcing water in. When plant cells are surrounded by a solution of higher osmotic pressure, water passes out, the cytoplasm collapses from the cell wall and the cell is now said to be *plasmolysed*. The tissue as a whole becomes *flaccid*.

A plant whose roots are in water (or a well watered soil) has most of its cells in a state of turgor. If water is withheld, cells become plasmolysed, the tissue becomes flaccid, and the plant droops or wilts. It is the turgidity of the cells that maintains herbaceous plants, which have no woody supporting tissue, in an erect position.

Wilting due to plasmolysis of the cells can be brought about by a number of conditions e.g. (i) hot, dry, windy weather causing the plant to lose water rapidly, (ii) flooding of land by sea water so that roots can no longer take up water by osmosis.

In fact any condition in which water loss exceeds water gain must result in cells losing their turgor and plants wilting. The importance of water uptake in maintaining turgidity can be seen as vital for the well-being of the plant, for it needs to remain erect in order to perform such vital functions as photosynthesis and transport.

Water is absorbed from the soil by the root system. The part of the root concerned with the absorption of water is the *root hair zone* which lies a few millimetres behind the root tip. Here, the root surface over 3–5 millimetres is covered by minute projections of the surface cells. These projections number several hundred per cm^2, and are known as root hairs. They tend to live for a few days only and then collapse and die. Those farthest from the root tip die first, since they are the oldest, and all the time new ones are being formed on

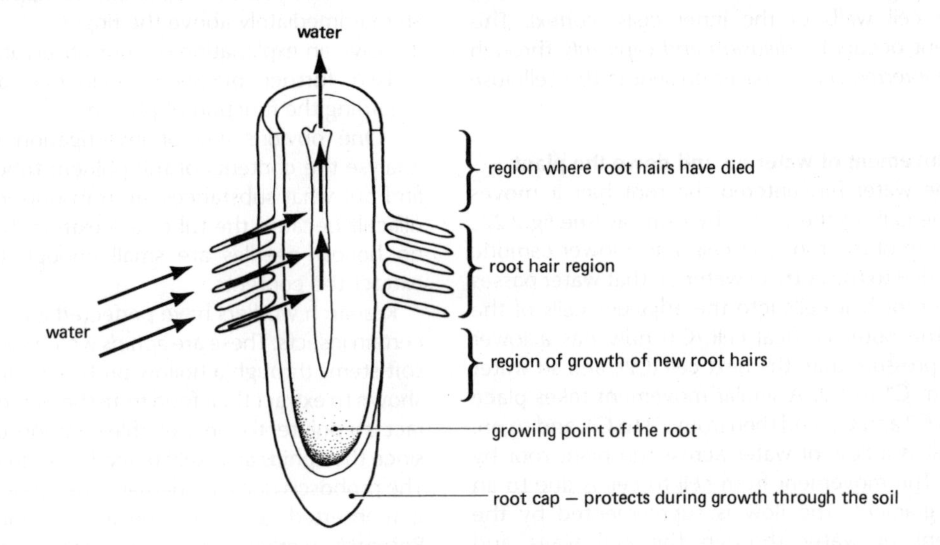

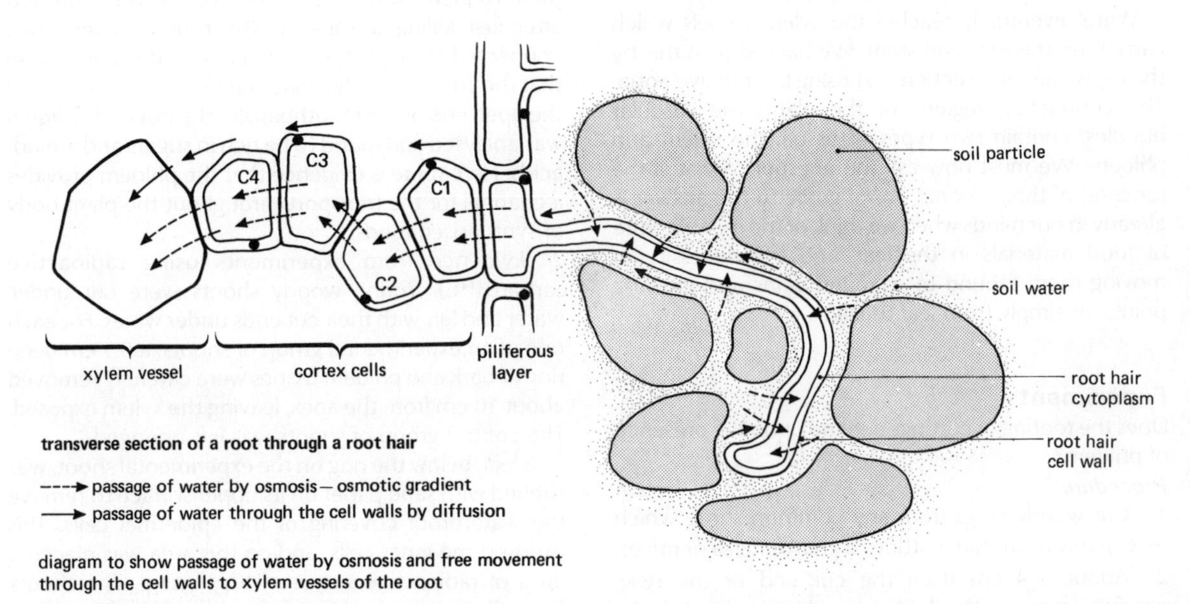

fig 2.27 A root and its associated root hairs, and a root hair, showing the passage of water from soil to xylem

the root tip side of the zone. Thus as growth of the root proceeds the root hair zone always keeps a constant position behind the advancing tip.

Each root hair is a thin-walled outgrowth from a surface cell of the *piliferous layer*. It grows out between the soil particles and through the soil water which adheres as a fine film around each tiny piece of rock. The soil water has mineral salts dissolved in it, so that it is in effect a weak solution. Water is drawn into the root hair by the higher osmotic pressure of the cell sap. The cytoplasm and its membranes act as the partially permeable membrane. Water also enters the plant by moving along the cellulose cell wall of the root hair and into the cell walls of the inner cells (*cortex*). This movement occurs by *diffusion and capillarity* through the *large intermolecular spaces* present in the cellulose fibres.

2.63 Movement of water up and down the plant

Once the water has entered the root hair it moves across the cells of the cortex by osmosis (see fig 2.27). The cell sap of the root hair cells has a lower osmotic pressure due to the entry of water, so that water passes from the root hair cells into the adjacent cells of the cortex. The outer cortical cell (C1) now has a lower osmotic pressure than the next cell (C2), and so water flows from C1 to C2. A similar movement takes place between C2 and C3, and then from C3 to C4, and so on. The result is a flow of water across the plant root by osmosis. The movement from cell to cell is due to an *osmotic gradient*. The flow is supplemented by the movement of water through the cell walls and intercellular spaces by diffusion and capillarity.

Water eventually reaches the xylem vessels which carry it up the root and stem. We have shown this by the experiments in section 2.61 using the red dye eosin. The conducting regions of the plant, the vascular bundles, contain two types of tissue, the xylem and phloem. We must now ask the question, 'what is the function of the *phloem*?' Some guide to the answer is already in our minds when we think of the manufacture of food materials in the leaf, and the necessity of moving them around to storage organs and growing points, or simply from leaf to leaf.

Experiment

Does the rooting of cuttings depend upon the presence of phloem?

Procedure

1 Cut woody twigs from any common shrub which roots easily from stem cuttings e.g. *Hydrangea* or privet.

2 About 3–4 cm from the cut end of the twig, carefully remove the bark and phloem, leaving the xylem (wood) exposed. Use a sharp knife to cut out a ring about 0.5 cm wide.

3 Cut gaps in several twigs and leave other twigs untouched as controls.

4 Place all of the twigs in pots of damp sawdust or sand, and water daily.

5 After three to four weeks wash away the material around the base of the twigs with water. Separate the experimental twigs from the controls.

6 Examine the twigs carefully and note where roots have arisen.

Questions

1 Where have roots grown on the ringed and unringed stems?

2 What do you observe about the appearance of the stem immediately above the rings?

3 Give an explanation of your observations.

Two further pieces of evidence are available regarding the function of phloem.

1 One obvious line of investigation would be to analyse the contents of the phloem tubes in order to find out what substances are transported. This is very difficult because the tubes are extremely narrow, and no hollow needles are small enough to enter and extract the contents.

Research workers have perfected a technique using certain insects. These are *aphids* which suck juices from soft stems through a hollow proboscis, and have been shown to extract their food from the phloem tubes. This fact is a pointer to the probable contents of the phloem since the aphids are likely to feed on high energy foods. The proboscis does not damage the phloem tube when it is inserted, and thus the flow is not interrupted. Research workers have used large aphids, allowed them to pierce a plant, and then cut off the proboscis after first killing the insect. The proboscis remained connected through the plant cells to the phloem, so that the contents of the conducting tissue flowed out of the open end of the mouthparts of the insect. The liquid was analysed and found to be rich in sugars and amino-acids. Here there is evidence that the phloem provides a channel for the transport throughout the plant body of synthesised foods.

2 Evidence from experiments using radioactive carbon (^{14}C). Young woody shoots were cut under water and left with their cut ends under water. For each one of an experimental group of shoots, a 0.5 cm deep ring of bark and phloem tissues were carefully removed about 10 cm from the apex, leaving the xylem exposed. The control group of shoots was left unringed.

A leaf, below the ring on the experimental shoot, was rubbed with sand paper on its upper surface to remove the waterproof covering of the epidermal cells. This exposed the inner cells, and on this area was placed a drop of radioactive sucrose ($^{14}C_{12}H_{22}O_{11}$). The shoots were allowed to stand for twenty four hours. They were washed very carefully to remove the excess sucrose and then pressed between sheets of absorbent paper

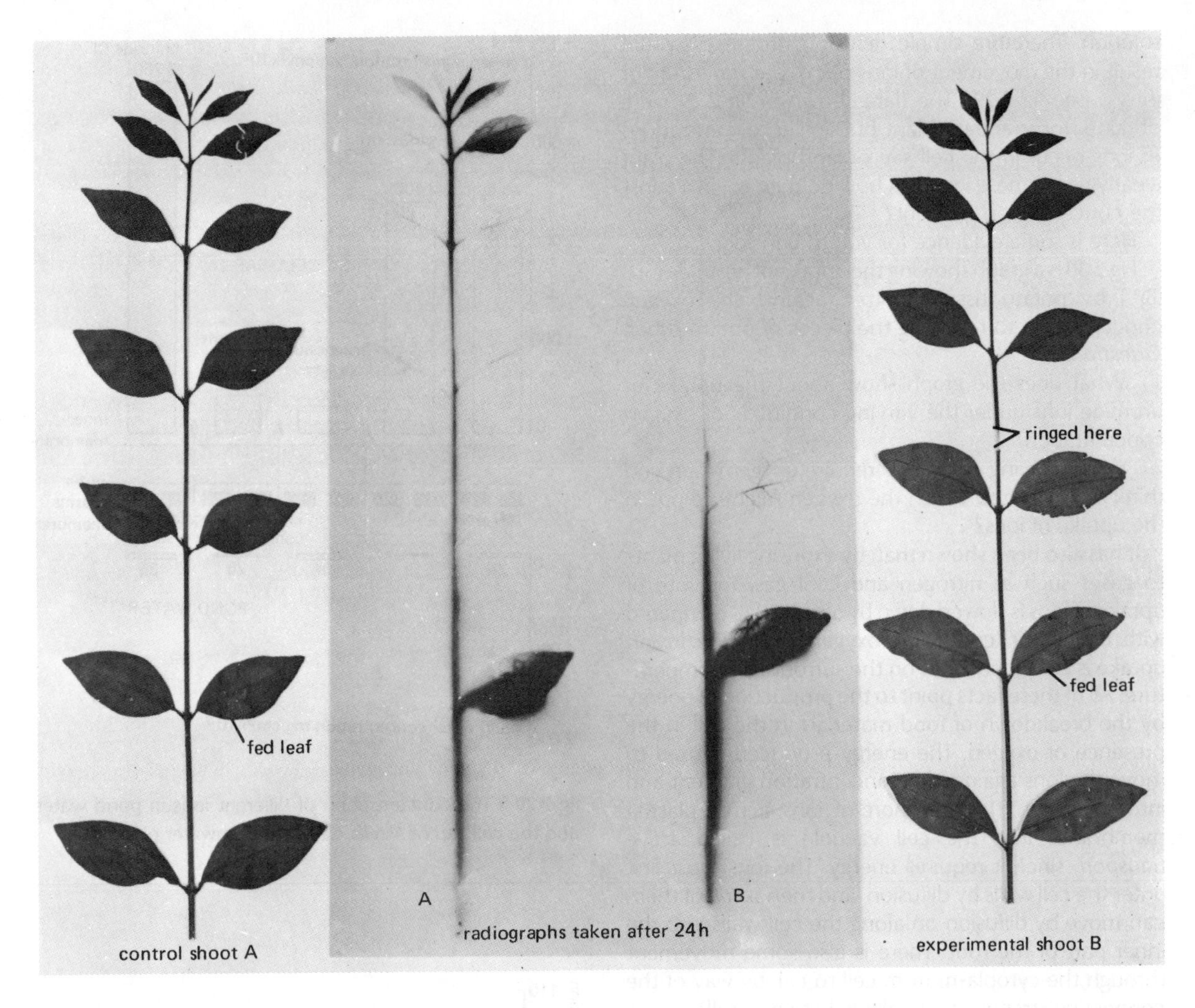

fig 2.28 Experimental and control shoots treated with radioactive sucrose

and dried. Each shoot was laid upon X-ray photographic film in a light proof paper for several days, with the result that the film was affected by the radioactive particles. This produced an image on the film when developed. (See fig 2.28).

Questions

4 What evidence is there that the sucrose has been transported away from its original position?

5 Is there any evidence of a two-way flow, both up and down the stem?

6 Is there any evidence that the sucrose has been transported by the phloem and not by the xylem?

7 Can you suggest why the sucrose should be transported to the young leaves at the top of the control shoot?

We can now summarise the function of the phloem as follows:

i) The phloem transports dissolved synthesised foods such as sugars and amino-acids from the leaves, where photosynthesis occurs.

ii) The dissolved material passes up and down the stem to the region where it is to be used or stored e.g. leaves not actively photosynthesising, growing points of stems and roots, storage organs below and above ground, and fruits and seeds. (Ringing the shoot can stop this flow of food and could eventually result in the death of a plant. Trees can be killed in this way by 'ring-barking.')

2.64 Uptake of mineral salts

The plant requires certain elements in the form of mineral salts which must be absorbed by the root system. The salts are carried up in solution through the xylem in the stream of water flowing throughout the plant. The experiment using radioactive phosphorus in section 2.60 showed how rapidly these salts can be moved up to the topmost portion of the shoot. How do the salts enter the plant?

The concentrations of ions in the cell sap of the cells of the root is considerably higher than that of the soil solution. Therefore simple diffusion processes would result in the movement of ions out of the root and not into it (see fig 2.29). The data shown in fig 2.29 were obtained for an aquatic plant, but similar concentrations of ions occur in the cell sap of land plants. The plant clearly needs these ions, but how do they get in against the concentration gradient?

Here is some evidence for you to consider.

Fig 2.30 is a graph showing the uptake of bromide ions (Br^-) by potato tissue, plotted against the oxygen concentration surrounding the pieces of potato tuber.

Questions

1 What does the graph show about the uptake of bromide ions under the varying conditions of oxygen concentration?

2 Why do living organisms require oxygen? Can you think of any use to which the oxygen might be put in the uptake of ions?

It has also been shown that, by exposing living plants to gases such as nitrogen and coal gas, the rate of uptake of ions is slowed down by about 90% compared with that in air (containing oxygen). Variation in ion uptake is also dependent on the surrounding temperature. All of these facts point to the production of energy by the breakdown of food materials in the cell in the presence of oxygen. The energy produced is used to move the ions against the concentration gradient and into the plant. This transport of ions across plasma membranes into the cell vacuole is called *active transport*, since it requires energy. The ions must first enter the cell walls by diffusion, and then some of them can move by diffusion on along the cell walls into the inner part of the root. There is also some movement through the cytoplasm from cell to cell by way of the connecting strands of cytoplasm between cells.

Once the ions are inside the root they move across to the xylem and then up through the plant in the moving water. When they reach the leaves and growing points they are used, together with the products of photosynthesis, for the manufacture of proteins and other substances.

2.65 Loss of water by the plant

We have seen that if a plant is not watered regularly, it wilts. This is caused by the cells becoming plasmolysed through water loss. A considerable amount of water is taken up by the plant and it is probable that not all of it is required for photosynthesis and other metabolic activities. Where does it go?

Experiment

Can the amounts of uptake, retention and loss of water by a plant be measured?

Procedure

1 Select a young herbaceous plant (e.g. willowherb or groundsel) with carefully washed roots, and place it

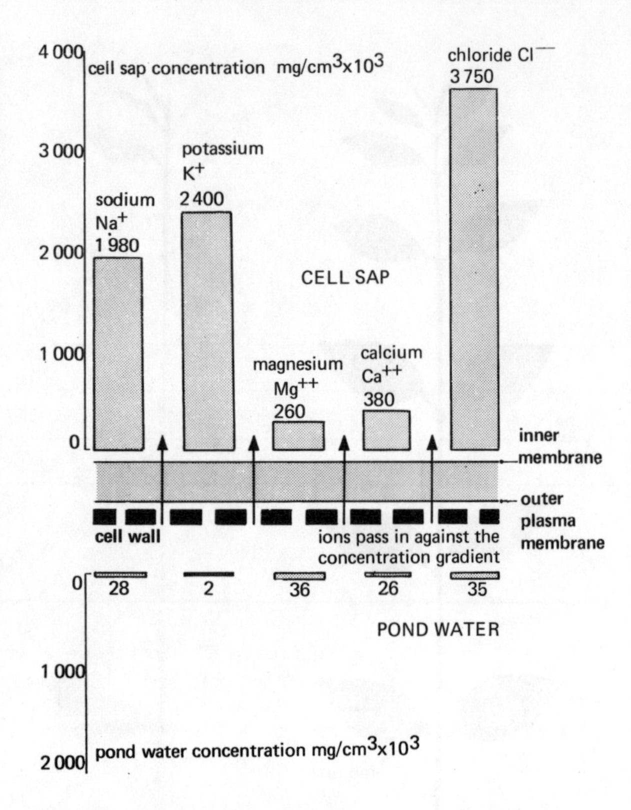

fig 2.29 The concentration of different ions in pond water and the cell sap of *Nitella clavata* (a freshwater plant)

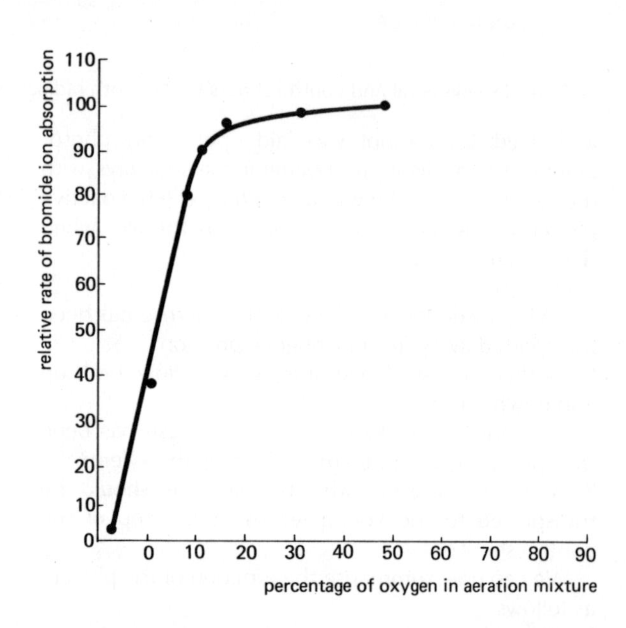

fig 2.30 A graph showing the uptake of bromide ions by pieces of living potato tuber

inside a conical flask (see fig 2.31).

2 Fill the flask with water to about 1 cm from the top and stick a label on the side of the neck to indicate clearly the water level.

3 Add a few drops of thin oil so that the whole water surface is covered (the stem should be free of leaves at this level).

4 To prevent the plant slipping down into the flask, the stem could be supported across the neck by passing it through a piece of cardboard.

5 Weigh the flask and its contents to the nearest gramme on a balance. Record date, time and weight.

6 Set up a control flask in a similar fashion but without a plant. Weigh the flask and its contents. Record the date, time and weight.

7 After 2–3 days, examine the flask with the plant and note that the water level has dropped. Using a burette, syringe or measuring cylinder, add water until the water level is back to its original mark. Record the volume of water poured into the flask. Weigh the flask and plant, recording date, time and weight.

8 Check the control flask and note the water level. Weigh the flask.

9 Record your results in the following table.

Date	Time	Weight of flask with plant (g)	Weight of control flask (g)	Volume of water required to restore level (cm^3)

Questions

1 What purpose was served by the thin layer of oil?

2 Did the level of the water fall in the control flask? Give an explanation of your observations.

3 Give an explanation of why the water level fell in the flask with the plant.

4 What was the difference between the weight of the flask and plant at the beginning and at the end of the experiment?

5 What total volume of water was added to the flask to restore the level?

6 Work out the mass of water absorbed by the plant, the mass of water lost and finally the mass of water retained by the plant.

7 What use is made of the water retained by the plant?

8 Work out the percentage of water lost by the plant compared with its uptake. Compare this percentage with those obtained by other members of your class working on the same experiment. Are your results similar? What variables are operating between the different experiments within the class groups?

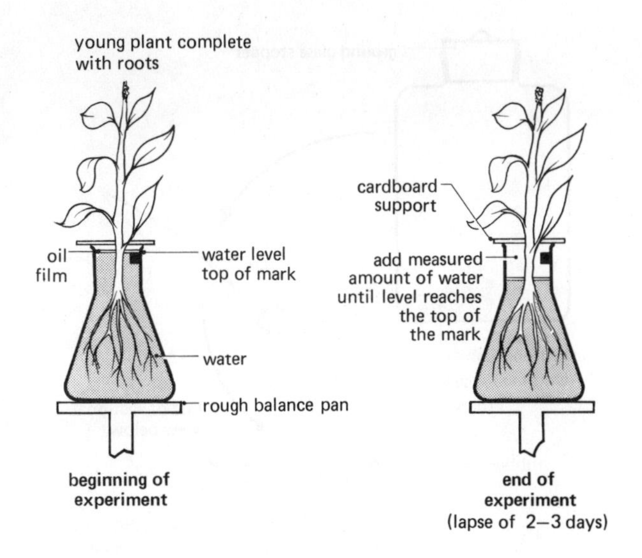

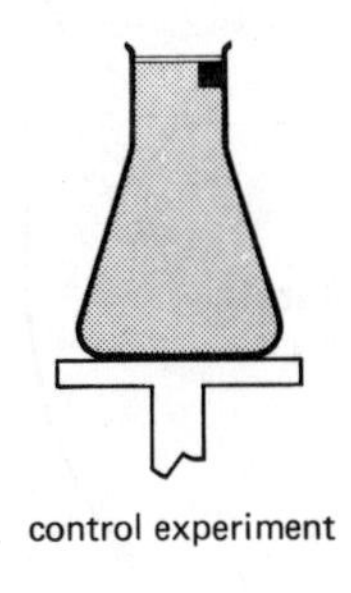

fig 2.31 A weight potometer

The last experiment has shown that plants lose much of the water initially absorbed by the roots. By fairly simple measurements we have calculated the amount absorbed, the amount retained and the amount lost. Where does this loss occur? Again we are led onto another experiment by posing a question on the previous one.

Experiment

From what part of its structure does a plant lose much of the water absorbed from the soil?

Procedure

1 Anhydrous cobalt chloride paper (blue in colour) kept in a well-stoppered jar, away from atmospheric moisture, is very sensitive to water and water vapour (see Introductory Chapter section 5.22).

2 Place one small square (1 cm^3) of this paper (handle with forceps only) on each side of a number of leaves of different plants. The leaves must remain attached to the main shoot.

3 Place glass slides on either side of each leaf, enclosing the square of paper, and hold them with rubber bands (see fig 2.32).

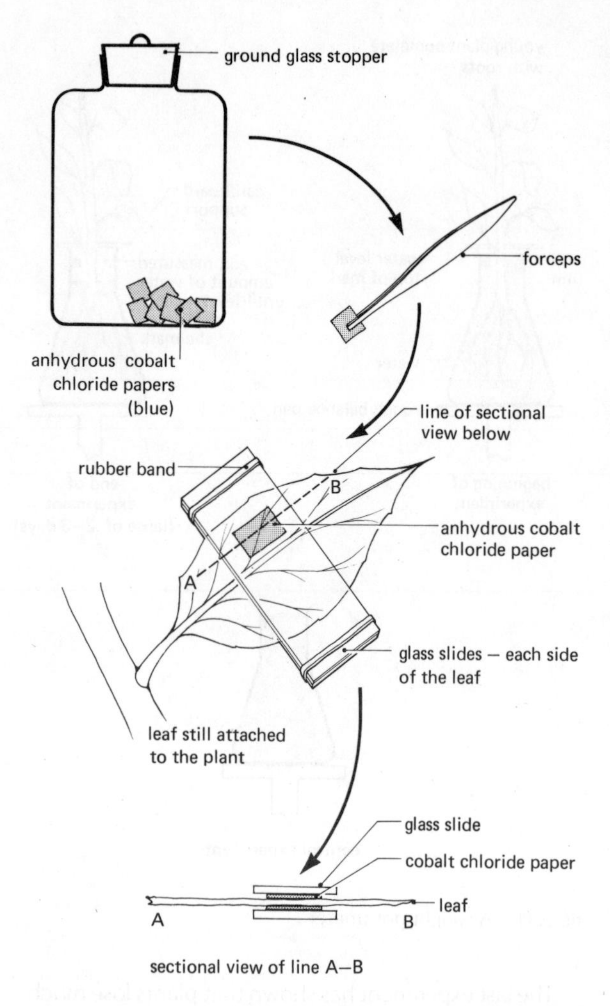

fig 2.32 The experimental apparatus required to demonstrate water loss from a leaf, using cobalt chloride paper

4 For each leaf record the time at which the paper square was put in place, and observe them carefully, noting any colour changes.

5 The paper will turn pink in the presence of water as the anhydrous cobalt chloride becomes hydrated.

6 Record the time when each square of paper becomes entirely pink.

Note: It may be easier to cover each piece of cobalt chloride paper with a large piece of sellotape in place of glass slides and rubber bands.

Questions

9 Why is the cobalt chloride paper handled only with forceps?

10 Why is the paper covered with a glass slide? What control should be set up for this experiment?

11 Is there any difference between the colour changes on the upper and lower surfaces of the leaves? Is there any difference in the timing of colour changes for different types of plant?

12 In what form is water lost from the plant?

13 It would appear that the leaves are important structures with regard to water loss. What other parts of the plant may be involved? How could you investigate their importance?

14 How could you measure the rate of water loss from a potted plant?

The loss of water by the plant through the leaves, and to a certain extent through the stem and flowers, is called *transpiration*. The water passes out in the *form of water vapour*.

The green leaves present an enormous surface area to the sunlight for photosynthesis. Let us now consider the structure of the green leaf and understand how it is enabled to carry out the two processes of photosynthesis and transpiration. What requirements must be fulfilled by the leaf?

The plant can be compared to a wick soaked in water. As water passes out of the leaves by transpiration, and as photosynthesis produces osmotically active sugar molecules, the osmotic pressure of the leaf cells becomes greater. This brings more water to the cells by drawing it out of the xylem. There is a 'pull' on the tiny thread of water in each xylem vessel, and the water moves upwards. These threads of water do not break because of the *cohesion* between the water molecules. The tiny columns in the stems have great tensile strength, and so the water moves in an unbroken stream from root to leaf. This phenomenon is known as the *transpiration stream*.

It has been demonstrated that the water moves across the root cells from the root hair to the xylem along an *osmotic gradient*, and that it does the same across the cells of the leaf from the *xylem to the mesophyll cells*. These events occur in the following sequence:

i) Water molecules diffuse through the stomata from the saturated intercellular space to unsaturated atmosphere.

ii) Water molecules then evaporate from the cell wall of the mesophyll cells into the intercellular space.

iii) Water molecules move from the vacuole of the mesophyll cells through the cytoplasm to the cell wall.

iv) The cell sap in the vacuoles becomes more concentrated and so draws water from an inner cell. This process is repeated across the mesophyll cells.

v) Water is pulled up the xylem vessels in the leaf and stem veins.

Transpiration, although an apparently unavoidable accompaniment to photosynthesis, nevertheless plays a very important part in the functioning of the plant.

Useful functions

1 *Water transport*

The transpiration stream moves water up through the plant to fulfil the six functions outlined in section 2.62.

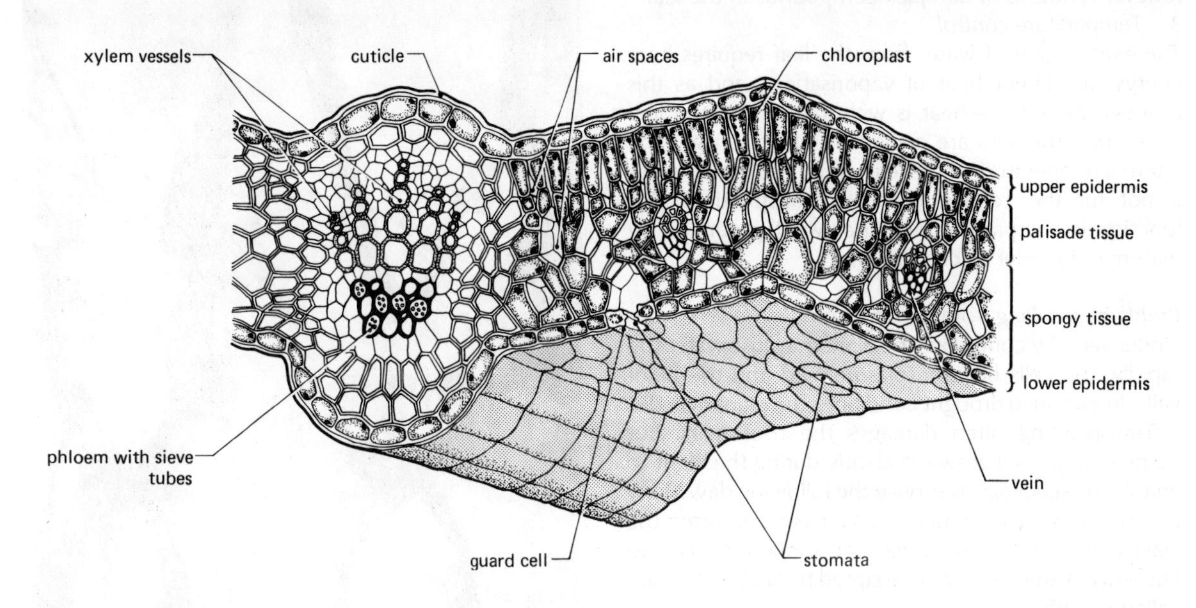

fig 2.33 A stereogram of a leaf *After D. G. Mackean*

Photosynthesis	Transpiration	Leaf structure
1 A large surface area must be presented at right angles to the light.		The leaf is a flat structure, so that a tree with thousands of leaves has an enormous surface area.
2 There must be free movement of gases through its surface.	At the same time there is a free movement of water molecules.	The under surface and the upper surface have pores through the epidermis called stomata – gases diffuse through these holes – CO_2 in and O_2 out in photosynthesis; H_2O molecules out in transpiration – a small amount of gas can diffuse out through the epidermal cells.
3 There must be intercellular spaces through which CO_2 can diffuse towards the cells containing chlorophyll.	The presence of these spaces also permits the evaporation of water from the cell surfaces.	The spongy mesophyll in the underside of the leaf has large intercellular spaces through which the gas molecules can diffuse.
4 Chlorophyll concentrated in chloroplasts must be near the top surface of the leaf so that they can absorb light.		The chloroplasts, containing chlorophyll, are most numerous in the upper palisade cells of the leaf.
5 There must be an ample supply of water by the conducting tissue of the leaf for photosynthesis.	An ample supply of water is needed to replace that lost by transpiration.	Water is supplied through the vascular bundles present in the leaf as veins. (See the leaf structure in fig 2.33.)

Table 2.5 Leaf structure in relation to function

2 *Salt transport*
The transpiration stream carries dissolved salts from the root for synthesis of complex compounds in the leaf.
3 *Temperature control*
The evaporation of water from the leaf requires heat energy (the latent heat of vaporisation), and as this process takes place heat is withdrawn from the leaf cells. Thus, the cells are cooled. In direct sunlight the internal temperature of the leaf would rise rapidly were it not for the cooling effect of transpiration. This function is comparable with the sweat produced by mammals in order to cool the body.

Problems resulting from transpiration
Under very dry conditions the plant may lose water too rapidly, the cells become plasmolysed and the plant wilts. In extended drought conditions the plant may die.

Transplanting often damages the root hairs, but many of them will grow sufficiently during the night to enable the seedlings to survive the following day. Many desert plants have a much *reduced leaf area* (often the leaves are replaced by spines) and *sunken stomata*, to cut down water loss. Plants adapted to dry habitats are called *xerophytes*.

Experiment

Do stomata open and close?

Procedure

1 By means of forceps, strip off a small piece of epidermis from a leaf. Mount this on a slide in water and cover with a cover slip. Examine under a microscope. Make drawings of the stomata and guard cells together with the surrounding epidermal cells.
2 Place a few drops of 50% sucrose solution on the slide next to the cover slip. Draw the sucrose solution under the slip by using a piece of torn blotting paper (or filter paper) to absorb the water from the opposite side of the cover slip. Observe the stomata and guard cells carefully. Make drawings of the guard cells and stomata to the same scale as the first drawing.
3 Compare your drawings. See fig 2.34.

Questions

15 Which liquid results in the stomatal opening being largest?
16 Give an explanation of how these two liquids alter the size of the guard cells and hence the pores.
17 Can you suggest a suitable model for the behaviour of the guard cells?

The stomata tend to *open by day* and *close by night.* This is brought about by the change in turgidity of the guard cells. During the day, the starch is changed into sugar, which increases the osmotic pressure of the cell sap: thus water is drawn into the cell. The guard cells become *more turgid*, with the result that the opening is increased in size. At night the sugar changes to starch, which is not osmotically active. Thus the cell sap is less concentrated, water is lost, and the guard cells become *less turgid* and *collapse inwards, closing the pore* (see fig 2.34).

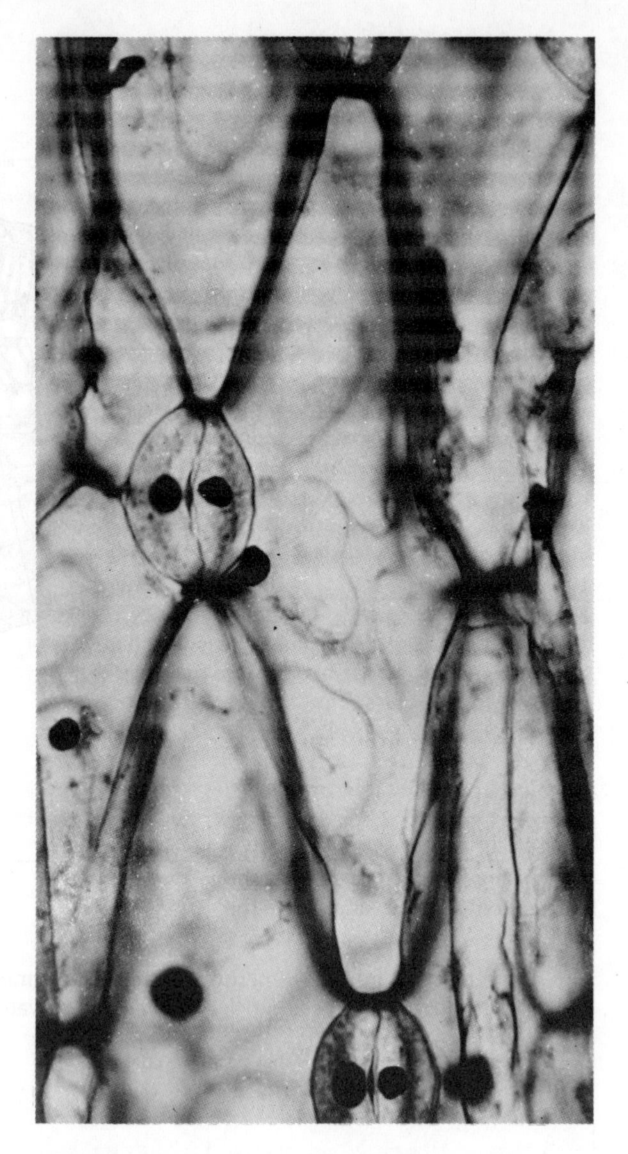

fig 2.34 A photograph of the epidermis in surface view, showing guard cells

Experiment

To compare rates of transpiration under different experimental conditions

Procedure

This experiment involves the use of a *potometer*, an instrument which measures the uptake of water by a leafy shoot. This uptake is indicated by the movement of water along a capillary tube. The capillary tube can be recharged for successive readings from a reservoir of water.

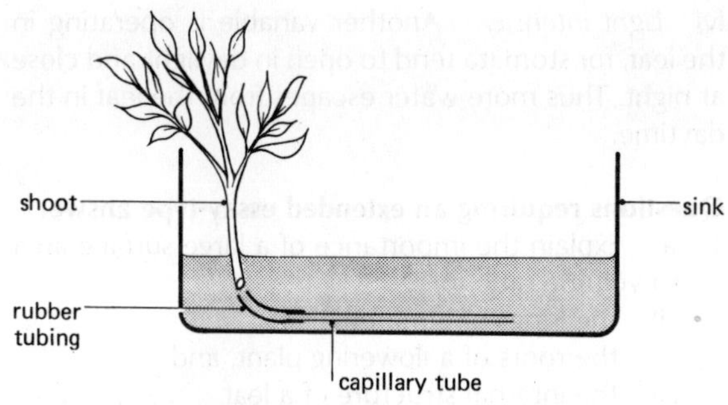

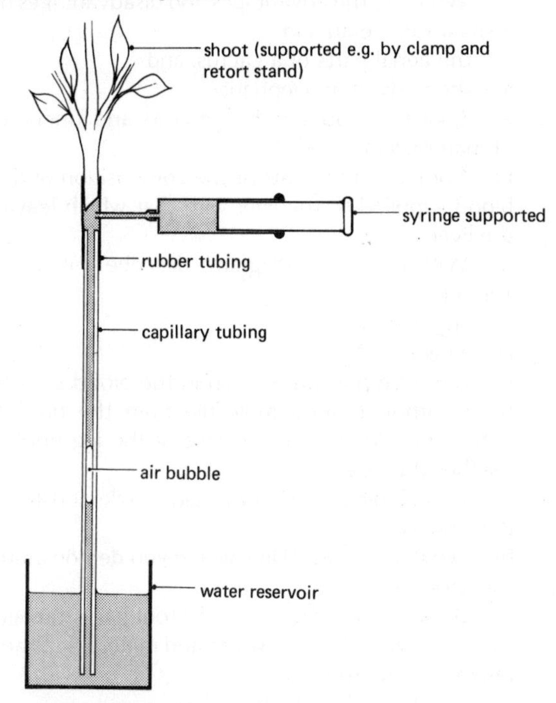

fig 2.35 A bubble potometer, and the method used to insert the shoot into the apparatus

1 Cut a leafy shoot from a tree or shrub as follows. Bend down a branch and place part of the stem under water in a bowl. Cut the shoot under water with a sharp knife. Leave the base of the shoot in water. This procedure prevents air entering the xylem vessels. Ensure that the stem is circular and about the right diameter to fit into the rubber tubing of the apparatus (see fig 2.35).

2 Fill up the capillary tube and rubber tubing with water and insert the shoot into the apparatus. This should be done under water in a large sink if possible, to ensure again that no air enters the xylem. Keep the leaves of the shoot out of water. It is essential that no air pockets are trapped in the rubber tubing. Place a finger over the open end of the capillary tube, take the apparatus out of the sink and place the open end of the tube under water in the reservoir before removing the finger.

3 The water in the capillary tube should now be moving up the tube and so if the apparatus is lifted just clear of the water in the reservoir an air bubble should appear in the capillary tube. When the air bubble has increased in length to about 2 cm replace the capillary tube in the water reservoir.

4 Fill a small plastic syringe with water and insert the needle of the syringe into one side of the rubber tubing; support the syringe in order to keep it horizontal. By means of this instrument water can be released into the tubing and thus send the bubble back down the capillary tube.

5 Time the rise of the bubble along the capillary tube over a fixed distance of 5 cm. Reset the bubble by means of the syringe and time again. Repeat this measurement several times.

Note that the rate of movement of the water in the capillary tube is the *rate of absorption* of the cut shoot, *not the rate of transpiration*. We can, however, *compare* the rates of transpiration if we consider that the water retained under any one set of conditions is constant. Thus we can take the rate of movement of the water along the capillary tube as reflecting the rate of transpiration, assuming that the rate of water loss is directly proportional to the rate of uptake.

6 Now subject the shoot and the potometer to various different environmental conditions and measure the rate of water movement in the capillary tube.

e.g. i) Put the shoot in the dark by covering it with a large black plastic bag.

ii) Place a fan in front of the shoot and move air across the surfaces of the leaves.

iii) Expose the shoot to a brighter light by putting two bench lamps to shine on the leaves.

iv) Take the potometer out of the laboratory into bright sunlight and breeze (the apparatus should be carried carefully to avoid disturbance of the shoot).

v) Cover the shoot with a large transparent plastic bag into which water has been sprinkled. Close the open end by tying it around the base of the shoot.

7 Take several readings under each condition to allow the shoot to adjust to its new environment. When the readings are constant, take the mean of a number of readings. It is not essential to investigate all of these variables, but different groups within a class could each attempt to find out how one of these conditions affects the plant, and results could be compared.

Questions

18 Why will air get into the xylem vessels if the cut end of the shoot is not kept under water?

19 As the shoot absorbs water, why does the bubble move up the capillary tube?

20 As the shoot loses water more quickly or more slowly, what will happen to the bubble in the capillary tube?

21 Under which set of conditions is the rate of transpiration faster than under ordinary laboratory conditions?

22 Under which set of conditions is the rate of transpiration slower than under ordinary laboratory conditions?

Transpiration is due to the evaporation of water in the leaf from the cell walls of the mesophyll cells into the intercellular spaces. Most of this water passes out through the stomata of the leaf. The evaporation of the water is always subject to external conditions. Consider a pool of water left after a rain storm. When the sun reappears the pool of water will gradually disappear as the water evaporates. The conditions that bring about this loss of water must be similar to those in the leaf. These are:

i) *Temperature*. Heat supplied to water (e.g. heat from the sun) causes it to evaporate faster. At the same time the air warms up and is able to absorb more moisture. The warmth of the sunlight absorbed by the leaf, and heat drawn from the cells, will cause water on the surface of the mesophyll cells to evaporate more quickly. It diffuses through the leaf and out of the stomata.

ii) *Wind*. The movement of air over a pool of water will remove the molecules of water vapour more quickly, so that they can be replaced by further molecules. In still air the surroundings of a pool of water or a leaf become saturated with water vapour and thus evaporation is much reduced.

iii) *Humidity*. The air can hold only a certain amount of water at each temperature level. The higher the temperature the more water vapour can be present in the air. When the weather is very humid the air is saturated with moisture and no more can evaporate from a free water surface such as a pool. Dry air can therefore result in a more rapid evaporation of water, from a free water surface or from the cells of a leaf.

iv) *Light intensity*. Another variable is operating in the leaf, for stomata tend to open in daylight and close at night. Thus more water escapes from the leaf in the daytime.

Questions requiring an extended essay-type answer

1 a) Explain the importance of a large surface area to volume ratio in
i) the lungs of a mammal,
ii) the roots of a flowering plant, and
iii) the internal structure of a leaf.
b) What are the advantages and disadvantages of a small surface area in
i) the aerial parts of a cactus, and
ii) the body of an elephant?

2 a) Give an account of the structure and functions of mammalian blood.
b) Compare and contrast the composition of the blood supplied to the liver with that which leaves the liver.

3 a) With the aid of diagrams describe the structure of
i) an artery, and
ii) a vein.
b) Describe the route taken in the blood system by a carbon dioxide molecule from the time it enters the blood in the muscle of the leg until it reaches the lungs.

4 a) What is diffusion? How would you demonstrate this process?
b) What is osmosis? How would you demonstrate this process?
c) Draw a labelled diagram of a root hair attached to a root. Describe how water and mineral salts are taken into the root.

5 a) What is meant by osmosis?
b) Briefly describe and explain
i) one way in which osmosis is important in the life of a freshwater organism, and
ii) one way in which the same process is important in a flowering plant.

3 Other types of nutrition

3.00 Introduction

Green plants are able to manufacture their own food by photosynthesis. This type of nutrition is therefore said to be *holophytic*. Many animals, however, feed by ingesting solid particles of food which are then digested and absorbed; a process referred to as *holozoic* nutrition.

3.01 There are two other types of nutrition:
i) *saprophytic* nutrition – saprophytism
ii) *parasitic* nutrition – parasitism
Symbiotic nutrition is a term applied to certain associations between organisms. The relationship is not always of mutual benefit for nutrition. Indeed, some symbionts are holophytic, some are holozoic and some are saprophytic.

3.10 Saprophytic nutrition

Saprophytes are organisms that obtain their nutrients from *dead organic material*. They are plants (bacteria or fungi) but they lack the pigment chlorophyll. They are unable to manufacture their own food and therefore obtain their nutrients by the process of *extracellular digestion*.

3.11 Saprophytic bacteria
Many bacteria are saprophytes. You have already studied the nitrogen cycle, and in this study you will have met a number of different saprophytic bacteria which bring about the decay of dead plants and animals.
Question
What benefit do bacteria obtain from the process of decay during the nitrogen cycle?

3.12 Saprophytic fungi

Investigation
Where are saprophytic fungi found?
Procedure
1 Put some moist stale bread or pieces of moist, succulent fruit e.g. banana in a petri dish or similar container and place the dish under a bell jar. If a bell jar is not available cover the bread or fruit with anything that will serve to enclose the air space above the material.
2 Leave for a few days, and then remove the cover and examine the surface of the bread or fruit first with the naked eye, and then with the aid of a hand lens (see fig 3.1).

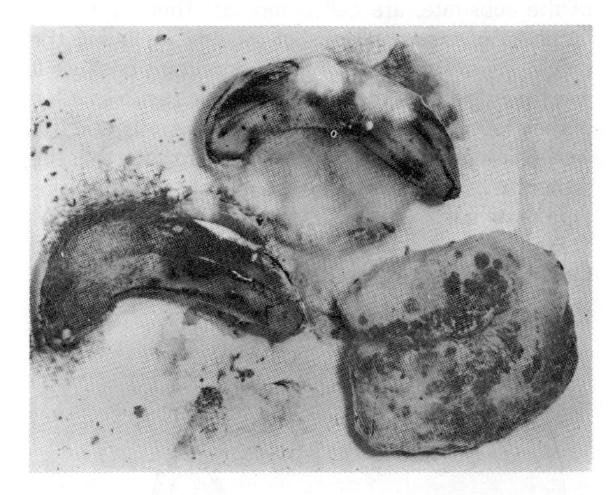

fig 3.1 A fungal growth on pieces of bread and banana

Questions
1 What can you observe on the surface of the bread or fruit?
2 What type of organism is visible on the surface of the bread or fruit?
3 How would you describe the structure of this organism when you use a hand lens?
4 How is it that the organism is able to grow on the bread?
5 How do you think the organism arrived on the bread?

Investigation
What does a saprophytic fungus look like under the microscope?
Procedure
1 Scrape a little of the fungus from the bread and place it on a microscope slide.
2 Add a drop of water and cover with a coverslip. (Methylene blue stain can be used to colour living cells.)
3 Observe the material under the low power of the microscope.

Questions
6 Describe the appearance and shape of the fungus.
7 What evidence do you see to indicate that the fungus you are looking at is a plant and not an animal?

3.13 Moulds

Many fungi have a similar form to the one you have just studied. Saprophytic fungi of this type, where the plant consists of minute threads that spread over the surface of the substrate, are called *moulds*. There are large numbers of different kinds of moulds, including the fungus *Penicillium*. This fungus is important because it was the original source of the antibiotic *penicillin*. One of the most common mould fungi is *Rhizopus*, and this will be described here in more detail. It may be that the saprophyte you have been examining is a *Rhizopus* type of fungus.

Examine fig 3.2. It shows part of a *Rhizopus* mould fungus. The structure consists of a number of filaments called *hyphae*, and these hyphae are known collectively as the *mycelium*. Note that some of the hyphae are specially modified for penetrating the food substratum in order to digest and absorb nutrients. Fungi of this type are able to obtain food by secreting enzymes from the tips of the hyphae. These enzymes digest the food substratum, and the products of digestion are then absorbed by the hyphae, and pass into the main part of the mycelium. This process is called *extracellular digestion*.

Investigation

Fig 3.2 shows an enlarged portion of a hypha. Study it carefully.

Question
1 Suggest ways in which the cellular structure of such a hypha differs from that of a normal cell of a green plant.

Some fungal hyphae have walls containing cellulose, but the majority have walls made up of an organic

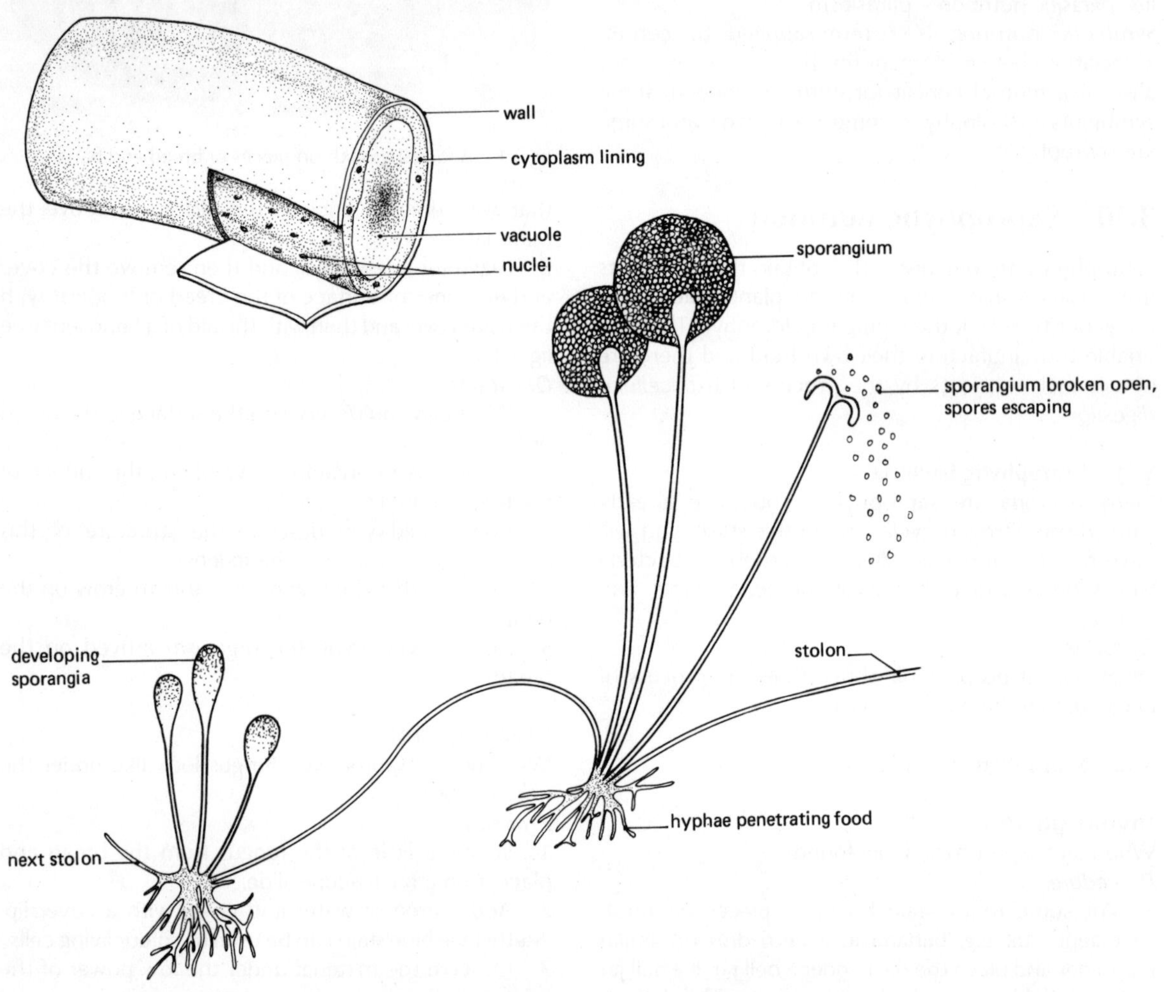

fig 3.2 The structure of the mycelium of *Rhizopus*, and part of a hypha in detail

nitrogenous substance called *chitin*. Like all fungi, *Rhizopus* lacks chlorophyll. Food is not stored in the form of starch, and usually storage material takes the form of oil droplets or glycogen.

3.20 Parasitic nutrition

Parasites are organisms that live on or in other organisms. The organism in close association with the parasite is called the *host*, and the parasite obtains shelter and nutrient from the host. In such an association, only the parasite benefits, and this is always at the expense of the host.

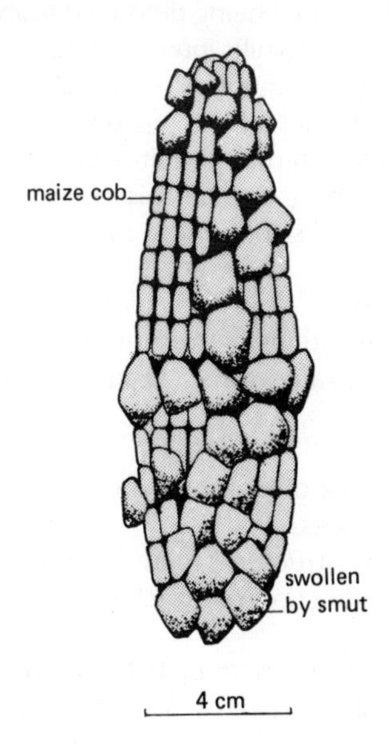

fig 3.3 Maize fruits suffering from an attack of maize smut fungus

Parasitic forms are found in both the animal and plant kingdoms. Some parasites live on the outside of the host, and are called external parasites, or ecto-parasites. Other forms live within the body of the host and are termed internal parasites or endoparasites. Parasitism is in fact a common mode of nutrition, and most free-living organisms harbour many parasites.

3.21 Fungal parasites

Many fungi live parasitically on other plants, and some of these are important to agriculture since they cause serious diseases of specific crops, such as cereals and potatoes. *Phytophthora* is a widespread genus of fungi which causes rotting in fruits and other storage organs in a variety of plants. One species, *P. infestans*, parasitises the potato plant and in wet weather can spread very rapidly. It caused the potato blight responsible for the Irish famine that resulted in many

fig 3.4 A potato plant infected with potato blight (*Phyto-phthora*)

deaths from starvation in the last century, and a great emigration from Ireland to the New World. The fungus is found particularly in the stems and leaves of infected plants and, if the infestation is heavy, in the tubers as well. The hyphae grow within the aerial parts, spreading between the cells. Each hypha is a *coenocyte* (multinucleate), i.e. it has no cross walls, and as it grows the tip produces enzymes which separate and destroy the cells of the host. Branches of the mycelium emerge through stomata and produce asexual reproductive bodies called sporangia. Very rainy conditions are required for the development of sporangia and when released each sporangium produces a number of swimming zoospores. These can swim through water films and attack new plant hosts. Under drier conditions the sporangium develops into a single spore termed a conidium. Both types of asexual reproductive cell can germinate and grow quickly to produce a new mycelium.

In sexual reproduction, fertilisation results in the production of a thick-walled oospore. These spores can overwinter in the ground, causing new outbreaks of potato blight in the following year.

The first signs of disease in a potato crop are brown patches on the leaves. Under moist conditions, the brown coloured patches spread rapidly into the stems and even down to the tubers. Within the tissues, digestion of the cells by fungal enzymes and secondary infection by bacteria causes a general softening and the whole plant may be reduced to a sodden mass. In order to prevent the spread of potato blight it is essential to plant phytophthora-free tubers obtained from reliable sources. In an infected field, all plant material should be burned and no infected tubers left in the ground. Young plants of the potato crop can be sprayed with fungicides such as Bordeaux mixture (containing copper sulphate) and this can often prevent the disease spreading.

3.22 Higher plant parasites

In addition to the fungi, some of the higher plants also live parasitically. Some of the most common higher plant parasites belong to the group of plants known generally as the *dodders*. There are several types of dodder, and one in particular is commonly found as an external parasite on cashew trees. Another series of important parasites belongs to the family *Loranthaceae*. These plants are commonly referred to as mistletoes, and they grow as semi-parasites on a variety of trees. Apple is commonly the host but elm and poplar are also parasitised. The leaves are green and thus the mistletoe can photosynthesise as well as obtain some organic substances from the tree (see fig 3.5).

fig 3.5 Mistletoe (*Loranthaceae*) growing on an apple tree

3.30 Animal parasites

Many animals live parasitically on or in other animals, and, as in the case of plant parasites, they are classified as either exo- or endo-parasites. Some of these parasites are important because Man himself functions as a host during their life cycles. Animal parasites belonging to this group often cause or carry disease, and the more important of them, together with their respective life cycles, will be dealt with in Chapter 12 on disease. One of the most important problems faced by an animal parasite is that of maintaining its position, either on the outside or inside of the host. External parasites need to avoid being dislodged from the body surface of the host, while internal parasites, such as those living in the alimentary canal, are in constant danger of being detached and passed to the outside in the faeces. Some internal parasites live in the bloodstream of their hosts, and, in this environment, they may possess special structures to help in locomotion.

Investigation

Examine the illustrations of the four different parasites; a trypanosome, a tapeworm, a flea and a louse (see fig 3.6).

Questions

1 Which of these organisms are external and which are internal parasites?

2 What specific structures found on the flea and the louse have the function of attaching the parasites to the host?

3 Can you see any part of the trypanosome that might function as an organ of locomotion?

3.31 Method of obtaining nutrition

Most external animal parasites *feed on the blood* of the host. Fleas and lice have biting mouth parts, and they are used to penetrate the skin of the host and to make contact with the blood capillaries just below the surface. Ticks, commonly found in birds and mammals, obtain their nutrients in the same way.

Internal parasites are usually found in the lymph, blood or alimentary canal. They live in a fluid medium and are, therefore, surrounded by nutrient material in a dissolved form. Many internal blood parasites are either microscopic, unicellular forms or small, worm-like structures which, because they are small, have a relatively large surface area to volume ratio. This means that the molecules of dissolved food can be absorbed directly across the body surface and thus most parasites of this type do not need a specialised digestive system. Examples of this mode of life are the protozoans *Plasmodium* and *Trypanosoma*, which cause malaria and sleeping sickness respectively. Parasites such as tapeworms and roundworms are often quite large in size. However, even these forms have reduced

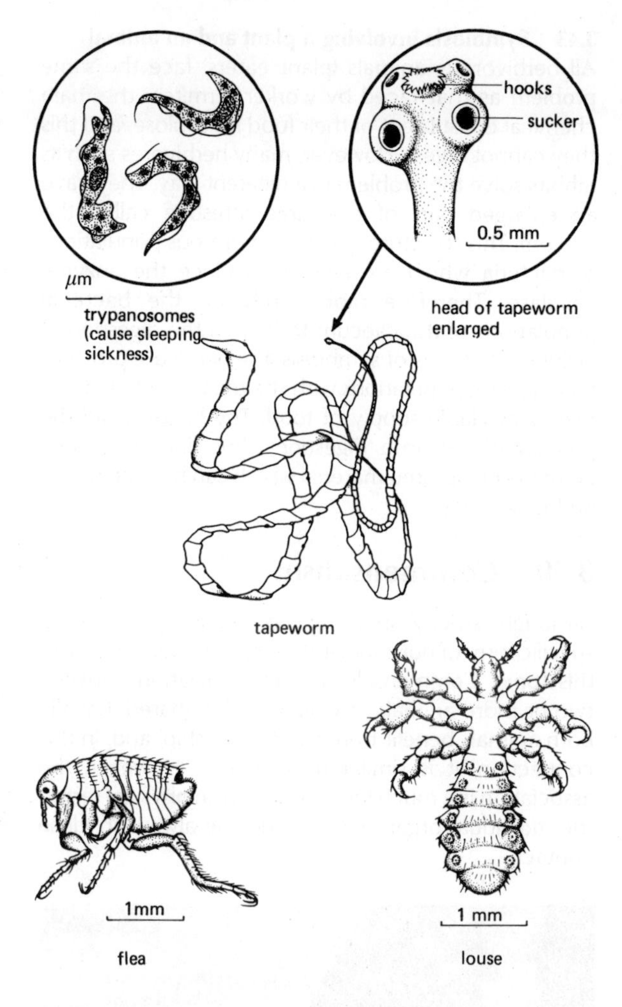

fig 3.6 Four parasitic animals

digestive systems because they live surrounded by a medium of wholly or partly digested food material. This material is absorbed across the body surface.

One of the main problems faced by parasites such as tapeworms is that of living in a medium which is frequently rich in digestive enzymes. All parasites living in the alimentary canal are in constant danger of being digested by the host's enzymes. This danger is overcome, to some extent, by the parasitic organisms secreting large quantities of protective mucus. This material lies on the surface of the parasite's body, forming a protective covering and helping to protect it from the host's digestive enzymes.

In addition, many parasites living in the alimentary canal secrete chemical substances called *anti-enzymes*. These specific materials help to neutralise partially the action of the host's enzymes, thus helping the individual parasites to overcome the danger of digestion by the host.

3.32 Parasites and disease

Many parasites live in close association with their hosts, and little or no harm is done to the host. However, some parasites are responsible for a number of serious diseases of Man, and these will be dealt with in more detail in Chapter 12. Examples were mentioned in 3.31. In addition to these types, some external parasites carry viruses that themselves cause disease. An example of this is a specific form of tick that carries a virus responsible for tick fever in Man.

3.40 Symbiotic nutrition

Symbionts are organisms that live in close association with each other and the relationship is such that *each organism gains some benefit*. The association may be between two plants, between two animals, or between a plant and an animal.

3.41 Symbiosis involving two plants

Investigation

Study carefully the photograph (fig 3.7) of the root system of a bean plant.

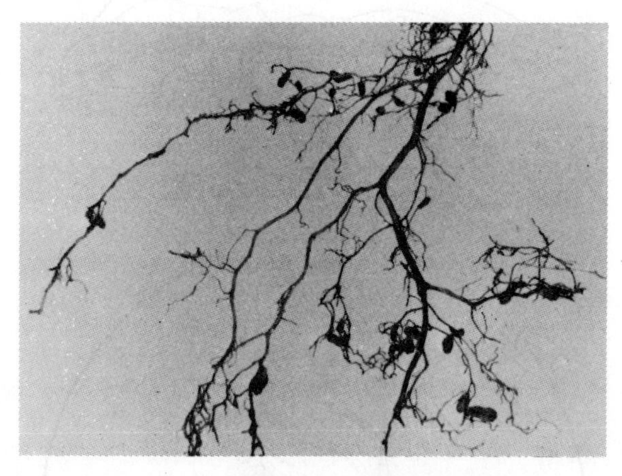

fig 3.7 Part of the root system of a bean plant showing nodules

Questions

1 What type of structures can you see attached to the roots?

2 What are these structures called? You may need to refer back to the nitrogen cycle in Chapter 1 (Nutrition).

3 What do these structures contain?

4 Explain the symbiotic relationship between the roots of a bean plant and the swellings on the roots.

Another example of a symbiotic association between two plants is found in a group of plants called *lichens*. A lichen is an association between a fungus and a green alga. The fungus provides shelter and protection for the delicate algal threads and, at the same time, the respiratory processes of the fungal mycelium produce carbon dioxide which can be used by the alga during photosynthesis. The fungal mycelium is able to absorb nutrient, in a saprophytic way, from the general surroundings upon which the association lives, (trees, stones and buildings).

3.42 Symbiosis involving two animals

Examine fig 3.8. It shows a worker termite and an enlargement of a microscopic organism called *Trichonympha*. *Trichonympha* is a single-celled animal found in large numbers in the alimentary canal of the termite. *Trichonympha* is one of the few animals that is able to produce the enzyme cellulase.

Questions

1 What is the main food material utilised by termites?
2 What is the most important chemical structure that makes up the food material of the termite?
3 What is the function of cellulase?
4 Explain the mutual advantage of the symbiotic relationship between the termite and the large population of *Trichonympha*.

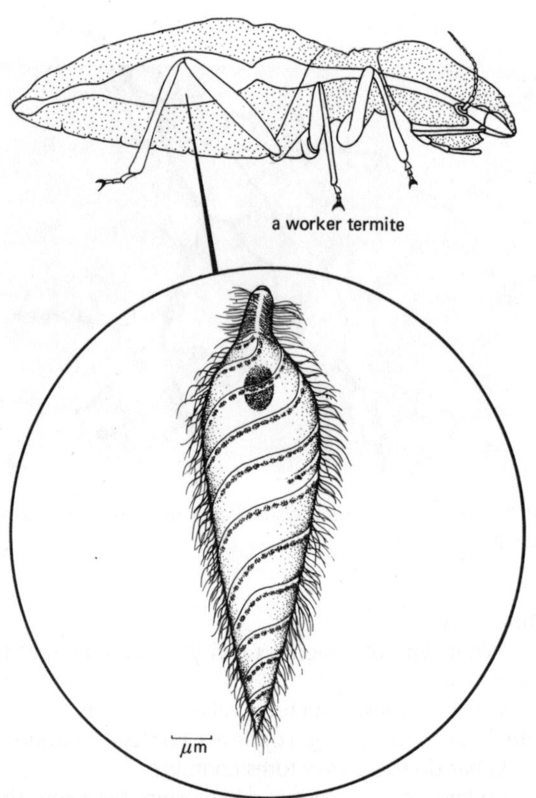

fig 3.8 A worker termite and an enlargement of one of the protozoa (*Trichonympha*) responsible for aiding the digestion of cellulose

3.43 Symbiosis involving a plant and an animal

All herbivorous animals (plant eaters) face the same problem as that faced by worker termites: the main chemical constituent of their food is cellulose, and this they cannot digest. However, many herbivores such as rabbits solve the problem in a different way. They have an enlarged part of the large intestine called the *caecum*. This structure contains enormous populations of bacteria which are able to produce the enzyme cellulase. Therefore rabbits rely on the bacterial populations of the caecum to help in the digestion of cellulose. In terms of symbiosis, the herbivore provides the bacterial populations with shelter, protection and a readily available supply of food. The bacteria aid the process of cellulose digestion, thus liberating large amounts of digested material which can be used by the herbivore body.

3.50 Commensalism

Although, strictly speaking, commensalism is not a specific form of nutrition, it does have certain links with this topic. Commensalism is an association between two or more animals in which food is shared. Usually both animals benefit from the relationship, and, in this connection, it is similar to symbiosis. However, the association is a much looser one and, in physical terms, the individual organisms may not be always in close contact.

fig 3.9 Commensalism between buffalo and egret

Many birds have extremely well-developed senses of smell, hearing and vision. Many of the large ungulates of the African savanna, such as buffaloes have poorly developed senses, and although the elephant has excellent hearing, its sense of vision is poor.

As the large mammals move through the savanna grass, their movements disturb large numbers of insects. In addition, the skins of most large herbivores harbour enormous numbers of external parasites, particularly ticks.

The mutual advantage of this type of association is, therefore, obvious. The large herbivores provide a readily available source of food, either in the form of disturbed insects, or in the form of external parasites. The birds, in turn, provide the herbivore population with greater sensitivity to the surroundings. As the birds themselves are disturbed by changes in the surroundings, they respond by flying away. This causes the larger animals to react, long before their own senses have made them aware of approaching danger.

As stated earlier, most larger mammals are infested with external parasites attached to the skin. The rhinoceros is an example of this. This particular species of mammal forms a commensal association with a bird called an ox-pecker. Look at fig 3.10, and then answer the following questions:

Questions

1 What benefit is gained by the bird from this association?

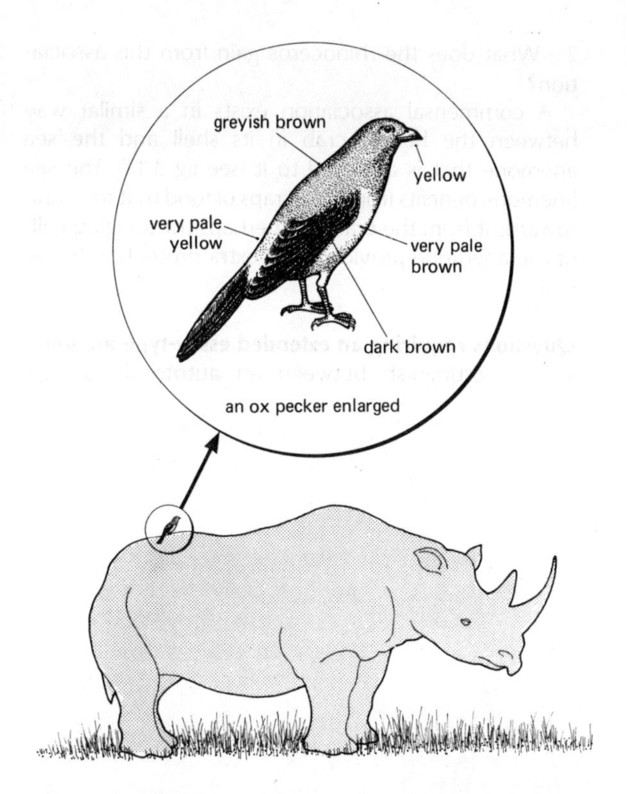

fig 3.10 A black rhinoceros and an associated ox-pecker bird. The ox-pecker is shown in more detail in the circle

fig 3.11 A hermit crab and a sea anemone

2 What does the rhinoceros gain from this association?

A commensal association exists in a similar way between the hermit crab in its shell and the sea anemone that is attached to it (see fig 3.11). The sea anemone benefits from the scraps of food that may drift towards it from the crab and perhaps the stinging cells of the anemone provide some extra protection to the crab.

Questions requiring an extended essay-type answer

1 a) Distinguish between an autotroph, a saprophyte and a parasite.

b) Describe the nutrition of a *named* mould fungus.

c) Describe the nutrition of a filamentous green alga (e.g. *Spirogyra*)

2 a) What is a symbiont and how does it differ from a saprophyte?

b) Describe the structure of a *named* saprophyte.

c) Describe i) an example of symbiosis between a plant and a bacterium and ii) an example of commensalism.

4 Storage

4.00 Storage of food reserves in the plant

In Chapters 1 and 2 we have considered the manufacture of food and its uses in organisms. We have also examined the movement of food around plant and animal bodies. Plants make their own food from raw materials. This food is not all used immediately after formation. In most plants it is transported to some organ in which the organic compounds can be stored. The stored material may remain for a short or long period of time, after which it is used for a variety of purposes. Very commonly it is used during a period of growth when food manufacture is not always possible. In many areas of the world there is a season of the year when plants find environmental conditions so difficult that they cannot go on living actively. In temperate climates this season is the winter, when low temperatures are the limiting factor, but in the tropics it is the dry season when lack of rain is the limiting factor. Plants store up food material before this cold or dry season, and then use it for rapid growth at the end, when conditions are becoming more favourable.

Storage of food material is most important in perennial plants which live on year after year.

Perennials
These are plants which do not die after producing flowers, fruits and seeds, but persist from year to year. The growth is reduced during the dry or cold season, and in some cases, the parts above ground die off and disappear. These are *herbaceous perennials* which are able to survive throughout the difficult season by underground storage organs. The organs are modifications of stem, root or leaf which because of their function are called *perennating organs*.

Woody plants (shrubs and trees) do not die off during the cold or dry season. The trunk and branches persist above ground and continue to grow year after year. During the difficult season, however, the woody plant often loses its leaves to prevent excessive water loss by transpiration; for in dry seasons water would not be available to the root system. Evergreen trees retain their leaves, but are usually specially adapted to restrict water loss. These plants are known as *woody perennials*.

Herbaceous perennials: daisy, dandelion, crocus, mint, *Gladiolus*, iris, daffodil.
Woody perennials: hazel, ash, oak, holly, heather.

A second group of plants which make use of storage of food over the dry or cold season are the biennials.

Biennials
These are plants which grow during their first year, producing root and shoot, and at the end of their growth period store food material. The food is used during the rapid growth at the beginning of their second year and to produce new shoots, flowers, fruits and seeds. The life cycle is completed in two years, after which the plant dies. Man often grows biennials as a food crop and harvests them at the end of the first year of growth, when the plant is swollen with food reserves. He must allow some to grow through into the second year in order to obtain seeds for planting.
Biennials: carrot, radish.

Thirdly, there are the *annuals* a group of plants which do not store food for long-term use, but complete their life cycle within a short space of time.

Annuals
These are plants which grow to maturity, produce flowers, fruits and seeds within the space of one year, and then die. In many annuals, the whole life cycle from germination to seeding lasts only a few weeks. The seeds of these plants germinate and the cycle is repeated again and again within a single growing season. Annuals can only survive the cold or dry season as seeds. Instead of food materials stored in the vegetative structures of the parent plant, the food is present in the seed structure.
Annuals: willowherb, groundsel, poppy, chickweed, shepherd's purse.

4.10 Storage in vegetative organs

In the plant kingdom we can see varying degrees of storage of manufactured food.
i) *Initial storage* of soluble sugars and amino-acids occurs in the cell vacuole during photosynthesis. Temporary storage of starch in the cells of the leaf occurs during the day-time as the products of photosynthesis build up. (*Translocation* of soluble foods

takes place particularly at night, to areas where the food material is required).

ii) *Long term storage* occurs in perennating organs or fruits and seeds, to be used at a much later date for rapid growth. This food is manufactured in excess of normal daily requirements for growth and other living processes.

4.11 Storage in modified stems

The long term storage of food may result in the modification of a plant structure. The stem is often used in perennials, and since the storage structure must be below ground in order to escape the extremes of the cold or dry season, this results in the unusual phenomenon of an *underground stem*. We must look at these storage organs in detail in order to understand why they are stem rather than root structures.

Rhizome

The rhizome is a *horizontal* underground stem. The rhizome of iris, for example, at first sight appears to be a root. Close examination reveals the following typical stem characteristics.

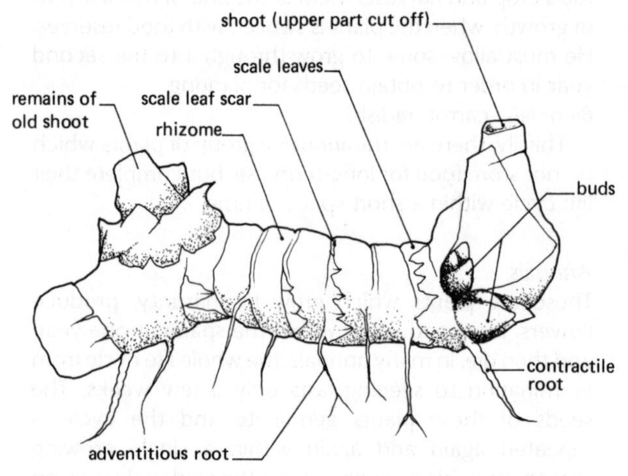

fig 4.1 A rhizome of iris

i) *Scale leaves*, or scars where leaves were attached.
ii) *Buds* growing in the axils of these leaves. The buds develop into green shoots which will grow above the ground.
iii) *Adventitious* roots grow from the rhizome.
iv) *Contractile* roots grow from the rhizome; these by their contractions pull the rhizome down into the soil.
v) A cross-section of the rhizome reveals a distribution of tissues which is *stem-like* rather than root-like.

The rhizome of iris is thick because it is swollen with food reserves. During the summer these are passed down from the leaves which are photosynthesising. Some of the translocated food is used for the growth of lateral buds into new branches of the rhizome. At the end of each branch, the terminal bud grows upwards, and produces the aerial shoot with its leaves and flowers. During the winter the aerial parts die down and the plant exists only as the underground rhizome.

Rhizoms are a common method of perennation, but they are not always swollen to the size of that of the iris. The perennial grasses, e.g. couch grass, have long slender rhizomes, but during the winter they still contain food reserves which enable new green shoots to appear early in spring.

Stem tuber

Stem tubers are the *swollen ends* of underground stems. The plant produces a number of slender rhizomes, and it is the terminal part of these structures which becomes swollen with food reserves. The 'European' or 'Irish' potato (*Solanum*) came originally to Europe from South America. The plant is noted for the large, edible stem tubers that it produces. Examine fig 4.2 and note the stem features shown by each tuber.

i) The presence of *scale leaves* or leaf scars with a bud in the axil of the scale leaf. These are called the 'eyes' of the tuber, because each has a curved mark resembling an eyebrow, and below this a tiny bud in the position of the eye.
ii) The *buds* are arranged spirally around the tuber facing towards one end where a terminal bud is present.
iii) If the tuber remains in the ground the buds will develop *green shoots* which will grow above ground.

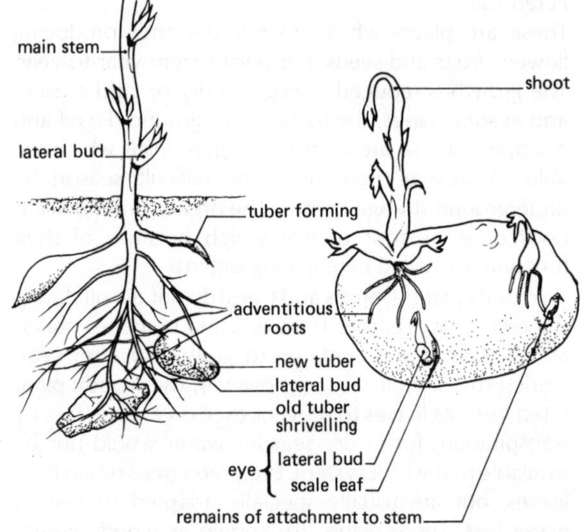

fig 4.2 A stem tuber of potato (*Solanum*)

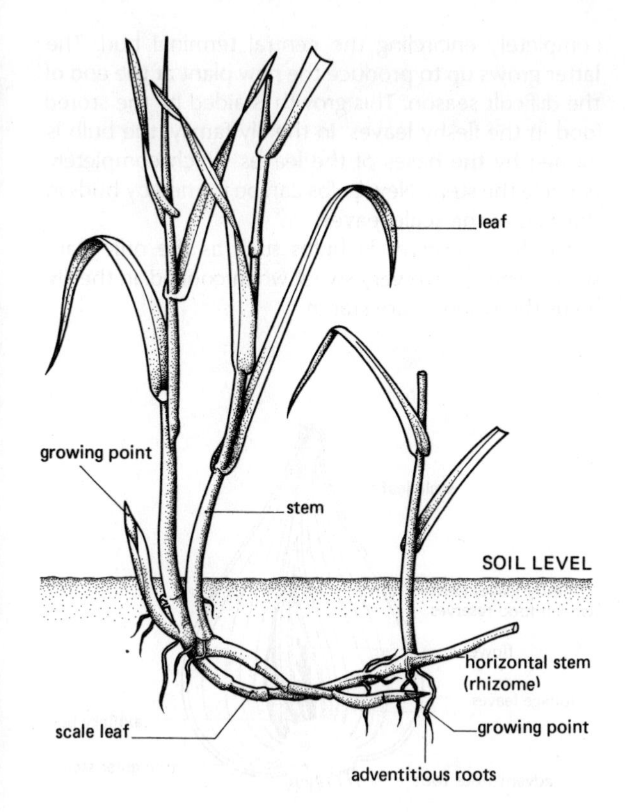

fig 4.3 A rhizome of couch grass (*Agropyron*)

iv) *Adventitious* roots will grow from the base of the shoot.

v) A cross-section of the tuber reveals a *stem-like* distribution of tissues, while on the outside are *lenticels* which allow gaseous exchange. These are normally found on the stems of woody plants. If the tubers are placed in the ground the 'eyes' begin to develop new shoots, using up the starch reserves in the tuber. A new plant is formed and eventually the original tuber withers and decays. New stems from the axils of lower leaves grow down into the soil and begin to form new tubers. In the dry season, the above ground portion dies and the tubers remain in the soil ready to grow when more favourable conditions return.

Corms

The corm is a *short swollen stem base* positioned *vertically* in the soil. It is encircled by leaf bases which form a protective scaly covering, and looks like a condensed vertical rhizome. It can be identified as a stem because it shows the following characteristics.

i) A series of *leaf bases* arises from it.

ii) *Buds* are present in the axils of the leaf bases, and the buds can develop green shoots which give rise to new corms at their base at the end of the summer.

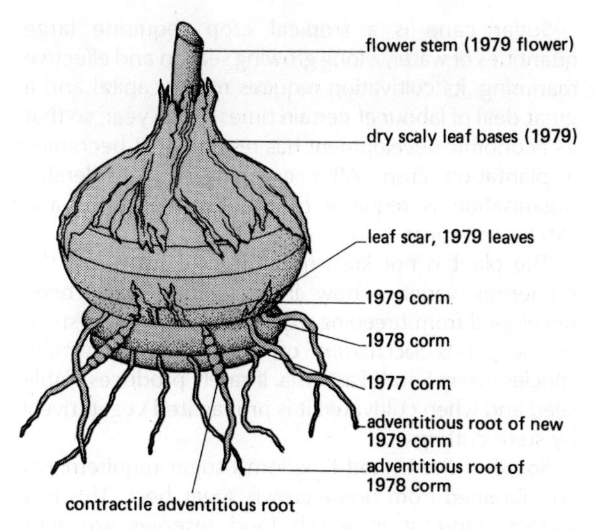

fig 4.4 A series of corms of *Gladiolus*

iii) *Adventitious* roots grow from its base.

iv) *Contractile* roots grow from its base.

v) A cross-section of the corm reveals a stem structure.

There is a terminal bud at the top of the stem which grows upwards in the spring and produces leaves and flowers. Food material is passed down and the new corm forms on top of the old. New lateral corms may also arise from lateral buds. The process is repeated year after year, and the new corm for any one year may have below it old, shrivelled corms dating back several years. As the newer corms reach the surface of the soil the contractile roots can pull them down into the soil.

Well known examples of corms are *Gladiolus*, and *Crocus* which represent horticultural species.

Sugar cane

The storage of sugar in the stems of sugar cane is an example which must be considered, since it is of great commercial importance, being one of the principal export crops of the tropics.

It is obtained from the stem of a member of the grass family, *Saccharum officinarum*. It does not fall into the same class as the other examples in this section of the book, since the stem is not a perennating organ, nor is it underground. Nevertheless, the stem has become enlarged, and it does have much greater reserves of sugar than are normally found in plants.

There are very few plants that store their carbohydrates as sugar, and therefore it was not until the discovery of sugar cane (and sugar beet) that sugar came to be the important commodity that it is today. Until then, sweetening materials had been obtained from honey and sweet secretions from trees.

Sugar cane is a tropical crop requiring large quantities of water, a long growing season and effective manuring. Its cultivation requires much capital and a great deal of labour at certain times of the year, so that its economic development has resulted in it becoming a plantation crop. After cropping, a considerable organisation is required to process the stems and extract the sugar.

The plant is not known in the wild state, but the numerous varieties now in cultivation have been developed from breeding experiments with wild stock of the genus *Saccharum*, of which twelve or more species live in South East Asia. It rarely produces viable seed and when cultivated it is propagated vegetatively by stem cuttings.

Some of the United Kingdom's sugar requirements are obtained from home-grown sugar beet. This is a swollen tap-root in which food reserves are substantially in the form of disaccharide sugar. Modern varieties of sugar beet produce about 16 tonnes per acre, with a yield of 17% to 18% sugar.

Experiment

What foodstuffs are present in perennating organs and how are they distributed?

Procedure

1 Cut thin (2 mm) slices across a rhizome (e.g. iris), a corm (e.g. *Gladiolus*) and a stem tuber (e.g. potato *Solanum*).

2 Flood the surface of the slice with aniline hydrochloride. Leave for five minutes and then wash off with water. Examine the colour distribution in the section. Use a hand lens.

3 Flood the surface of the same slice of each plant with iodine solution. Leave for five minutes and then wash off with water. Examine the colour distribution in the section. Use a hand lens.

4 Make a drawing of the distribution of tissues shown up by the reagents.

Questions

1 What colour is produced in each slide by the aniline hydrochloride?

2 What parts of the section were stained by the iodine solution? What food reserves are present?

4.12 Storage in modified leaves

The leaves of all plants store food temporarily, but in some plants a perennating organ called a *bulb* has developed. This is an underground structure resembling a *large bud* or a *condensed shoot*. The stem is short and triangular in cross-section, while the leaves are of two types: (i) outer leaves which are scaly and dry and protect (ii) the inner leaves which are thick and fleshy containing stored food material.

In the onion bulb the storage leaves are cylindrical, completely encircling the central terminal bud. The latter grows up to produce the new plant at the end of the difficult season. This growth is aided by the stored food in the fleshy leaves. In the lily family, the bulb is formed by the bases of the leaves which completely encircle the stem. New bulbs can be formed by buds in the axils of the scale leaves.

The food reserves in bulbs such as the onion are sugars. These taste very sweet when cooked. In the lily bulbs the reserves are starch.

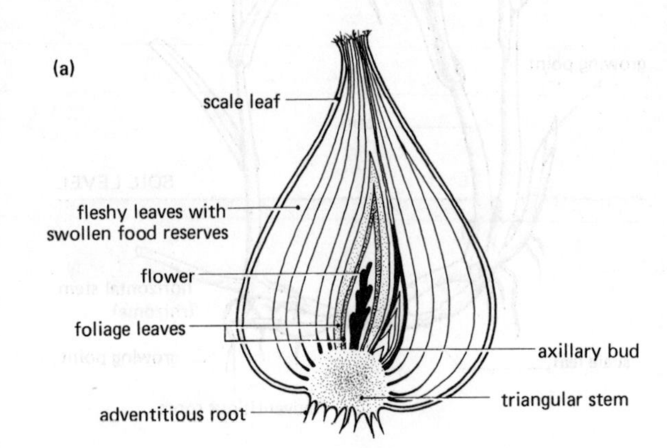

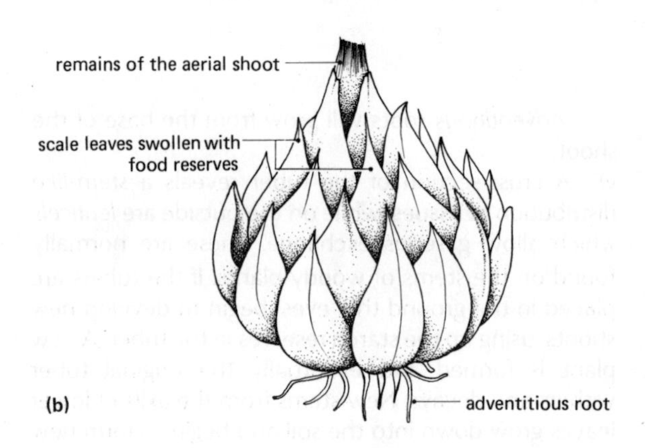

fig 4.5 (a) A vertical section of an onion bulb (*Allium*) (b) A scaly bulb of a lily

4.13 Storage in modified roots

Tap root

The tap root is a swollen main root with a very short stem at the top. A vertical section through a tap root such as a carrot shows the conducting strands, the outer cortex and the central pith areas. The food reserves are sugars which are stored in the cortex of the carrot and in the phloem of the beetroot.

The tap root is a common means of storage for biennials. The main root swells at the end of the growing season of the first year, and these stores are utilised to complete the life cycle in the second year. Taproots can also act as perennating organs in herbaceous perennials.

The radish (*Raphanus*) is grown as an annual plant. It produces a short stem which elongates rapidly and flowers develop amongst the leaves. The tap root grows to some depth, but only the upper portion swells, to form the edible portion. Storage is mainly in the form of starch.

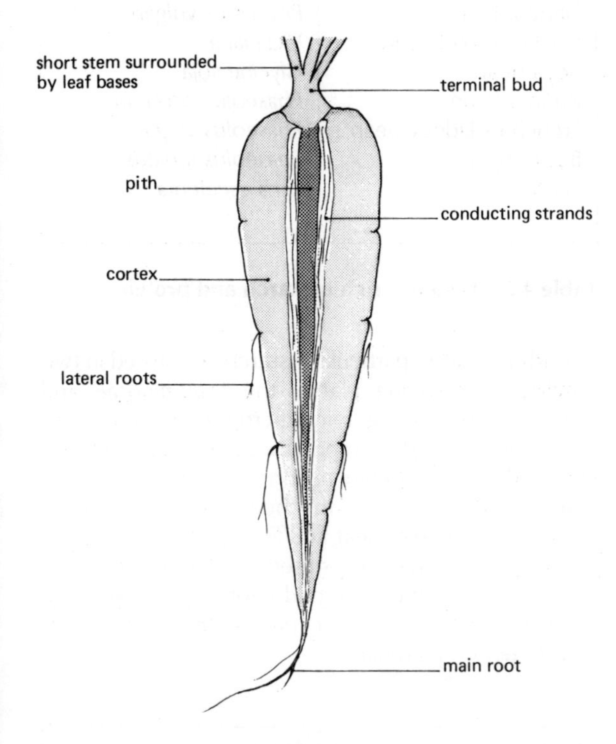

fig 4.6 A vertical section through the taproot of a carrot

Root tuber

Plants can also store food in roots other than tap roots, particularly where they have a *fibrous root system*. The tubers develop in the swollen ends of the roots, and can be distinguished from stem tubers by their lack of buds, scale leaves and scale leaf scars. Their internal anatomy is also different. Food produced by the leaves in the growing season is passed downwards to be stored in the tubers. Since roots do not produce buds they do not usually act as reproductive structures. They are simply food storage organs, e.g. Dahlia.

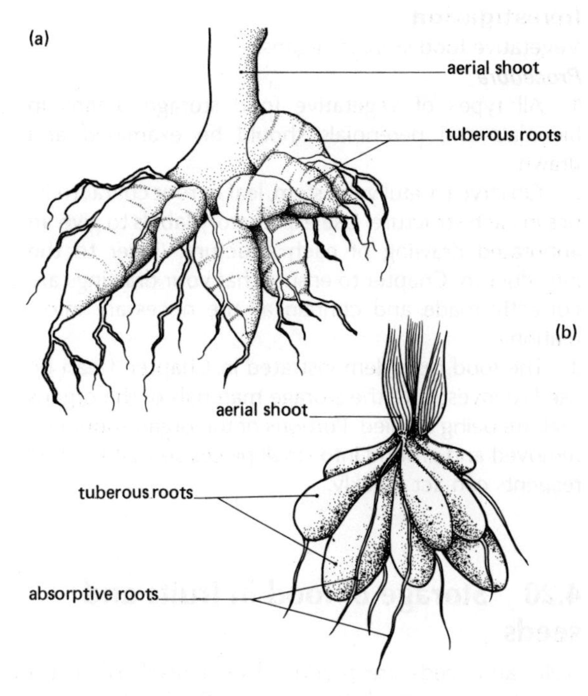

fig 4.7 Root tubers of (a) Dahlia (b) Lesser celandine (*Ranunculus ficaria*)

4.14 Advantages of food storage in perennials and biennials

The *advantages* of food storage to the plants are as follows:

i) It permits the *survival* of the plant over the *unfavourable* season when small quantities of the stores are used for life processes and limited growth.

ii) It permits *rapid growth* at the beginning of favourable conditions, e.g. the spring, when the bulk of the food reserves are used up.

iii) This rapid growth enables these plants to *outgrow their competitors* in the intense struggle for light, air, water and mineral salts, so that flowers, fruits and seeds may be produced.

iv) It permits this rapid growth to occur in the established habitat where the plant is already part of the ecosystem.

As far as Man is concerned, the ability of plants to store food is tremendously important, for by selective breeding and careful cultivation, he has developed strains with relatively huge storage organs which produce correspondingly bigger yields of food for the human population. Examples of this exploitation of plant food storage for human use include potato, carrot, onion, swede, turnip, parsnip and radish.

Investigation

Vegetative food storage organs.

Procedure

1 All types of vegetative food storage organs in biennials and perennials should be examined and drawn.

2 Observe carefully the stem, leaf or root characteristics in each structure. Use explanatory labels to give an annotated drawing of each structure. (Refer to the Introductory Chapter to ensure that your drawings are correctly made and contain all the necessary information.)

3 The food tests demonstrated in Chapter 1 can be used to investigate the storage materials of the organs that are being studied. Portions of the organ should be removed and ground into small pieces so that the test reagents can act quickly.

4.20 Storage of food in fruits and seeds

Fruits and seeds are produced as a result of *sexual reproduction* in the flowering plant. The seeds contain the young plant embryos. In order for these to grow they must be provided with food reserves for their initial development. They will be unable to manufacture their own food until the green leaves are formed. All seeds contain greater or lesser amounts of stored carbohydrates, fat and protein, and indeed many fruits carry large quantities of food stored in the fruit walls. In the case of fruits, the reserves are an attraction to animals, and are thus an aid to dispersal rather than growth.

From the earliest times, Man has used fruits and seeds in his diet, and over the last few centuries he has produced varieties of plants with ever-increasing food stores to satisfy his needs. Most plants store food in the seed in the form of oil, since it uses space more economically by providing more energy per unit weight than carbohydrate. Plants bearing oil seeds are grown and harvested in vast quantities in tropical countries to provide food for local use and for export to many other parts of the world.

A few plant species are grown in the United Kingdom for the commercial production of oil seed. These include oil rape (*Brassica oleracea*), linseed (*Linum visitatissimum*) and black mustard (*Brassica nigra*). (See table 4.1.)

Oil from seeds	Plant
Soya bean oil	*Glycine soja*
Sesame oil	*Sesamum orientale*
Sunflower seed oil	*Helianthus annua*
Groundnut oil	*Arachis hypogea*
Castor oil	*Ricinus communis*
Linseed	*Linum visitatissimum*

Table 4.1 Oils from seeds

Seed	Plant
Haricot bean	*Phaseolus vulgaris*
Field or broad bean	*Vicia faba*
Soya bean	*Glycine soja*
Runner bean	*Phaseolus coccineus*
French or kidney bean	*Phaseolus vulgaris*
Butter bean	*Phaseolus lunatus*
Lentil	*Lens esculenta*

Table 4.2 Legumes rich in starch and protein

Carbohydrates, particularly starch, are stored in two families of flowering plants, the *Leguminosae* and the *Gramineae*. The leguminous crops of the tropics are many and varied. Their importance as human food is far greater than in temperate regions. They provide a great deal of the protein for the human diet in tropical communities where meat and fish are in short supply.

Not only are these crops used by Man but they also supply the proteinaceous fodder for tropical livestock. In addition to the seeds other parts of the plant are rich in nitrogenous material.

Fruit	Plant
Maize	*Zea mays*
Rye	*Secale cereale*
Barley	*Hordeum sativum*
Oat	*Avena sativa*
Wheat	*Triticum* ssp.

Table 4.3 Cereals rich in starch and protein

The term cereal is given to members of the *Gramineae* which are cultivated for their fruits. The cereal crops of the world form the most important source of the world's food supply and are extensively cultivated. It is surprising therefore that the number of species of cereal crops is small. *Oats* and *rye* are the dominant cereals of the colder parts of the world. *Wheat* and *barley* are most important in warm temperate regions, while in the tropics rice, *maize* and *millet* form the bulk of the diet. In temperate climates, such as that of the United Kingdom and of Europe, they are important in the diet as energy-rich foods. (See table 4.2.)

Many non-cereal plants in the tropics produce edible fruits containing food reserves (mostly sugars), but relatively few are cultivated. Only five or six of those cultivated have been commercially exploited or have received attention from the plant breeder. Attention has been focused mainly on bananas, citrus fruits, dates, mangoes, pineapples and avocadoes to the exclusion of most others. In addition to these, guava, durian, pawpaw and pomegranate are common in the diet of tropical countries, but there is considerable local variation in quality and quantity. Many of these fruits are exported from tropical areas into European markets. Cold storage and gas storage can ensure that they are available throughout the year. Areas with temperate climates also produce their own varieties of fruit during the summer season and modern storage methods ensure that they are readily available in greengrocer's shops throughout the winter. (See table 4.4.)

Seeds are classified broadly into two groups, according to the structure storing the food reserves.

i) Endospermic seeds

These develop an additional tissue in the seed apart from the embryonic tissue. This is the *endosperm*. Its food reserve is used up on germination of the seed. e.g. wheat, oats and maize fruits and coconut.

ii) Non-endospermic seeds

No endosperm develops, and the food is stored in the

Fruit	Plant
Banana	*Musa* spp.
Citrus fruits	*Citrus* spp. (includes oranges, lemons, grapefruits)
Pineapple	*Ananas comosa*
Date	*Phoenix dactylifera*
Tomato	*Solanum melongera*
Red or green peppers	*Capsicum annuum*
Plum	*Prunus domestica*
Cherry	*P. avium*
Peach	*P. persica*
Almond	*P. amygdalus*
Pear	*Pyrus communis*
Apple	*Malus sylvestris*

Table 4.4. Tropical and temperate fruits

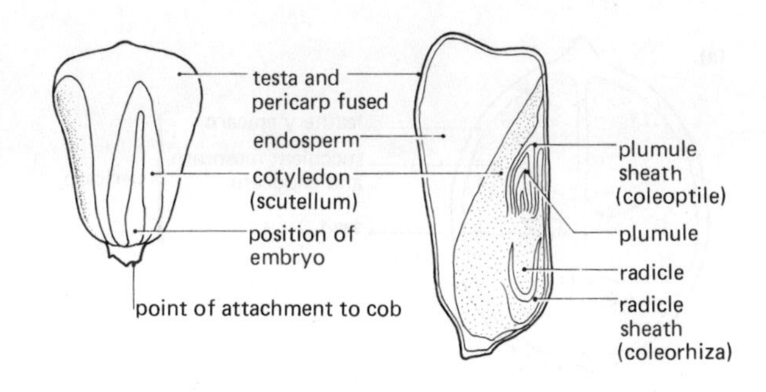

fig 4.8 A fruit of maize (*Zea*) – starch and protein food reserves mainly external to the cotyledon in the endosperm

cotyledons (seed leaves) which are part of the plant embryo. (see fig 4.9), e.g. peas and beans.

In fruits the food reserves can also be found in two different types of tissue.

i) The *fruit wall* or *pericarp* becomes swollen. Different parts of this wall can be utilised e.g. the mesocarp in fruits of the drupe type, or the mesocarp and endocarp in fruits of the berry type. (See fig 4.10).
e.g. lemon, orange, tomato, cucumber, marrow.

ii) The *receptacle*, on which the flower develops, becomes swollen and surrounds the fruit proper. The fruit is then known as a 'false fruit' since its food store is not part of the fruit wall. (See fig 4.10).
e.g. apple.

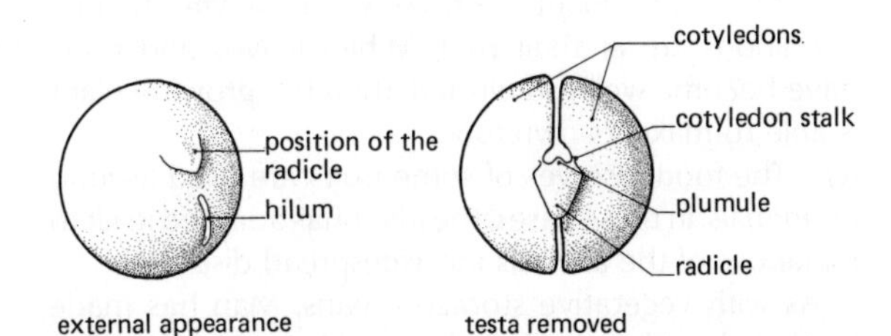

fig 4.9 A seed of pea (*Pisum*) – food mainly in the cotyledons as no endosperm is present

4.21 Advantages of food storage in fruits and seeds

As with food storage in biennials and perennials we must consider the advantage to the plant of this method of storing up reserves.

i) It permits the *survival* of the plant, in that seeds remain dormant over the dry season or even for several years if they are in a dry state. Wheat can survive for as long as fifteen years and then germinate given satisfactory conditions, whereas rubber seeds must germinate a few days after being shed.

ii) After dispersal, the seed may reach an unfavourable habitat, but even so growth can proceed for a while until all the reserves have been used, or until favourable conditions return.

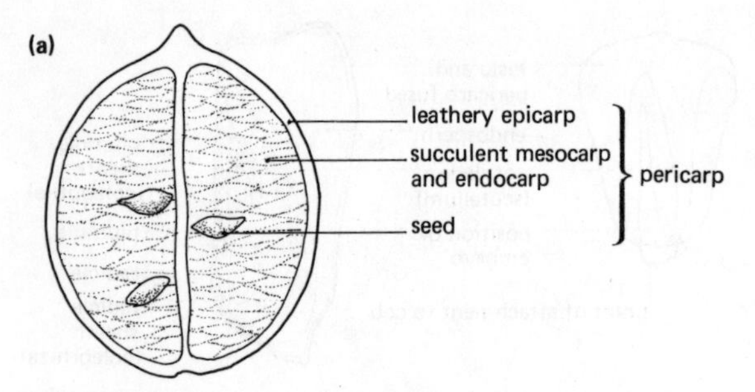

Radial longitudinal section through a lemon (fruit – a berry)
Food reserves in the mesocarp and endocarp of the fruit wall

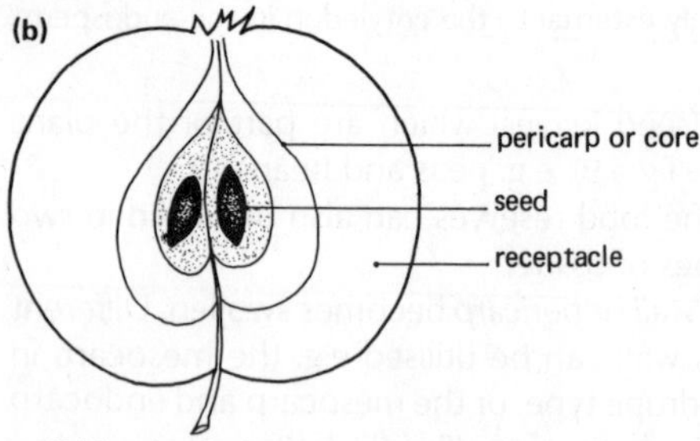

Radial longitudinal section through an apple (fruit – a pome)
Food reserves in the receptacles, not the pericarp, hence a 'false-fruit'

fig 4.10 (a) A radial L.S. through a lemon (fruit a berry) (b) A radial L.S. through an apple (fruit a pome)

iii) The *reserves* of the seed are used in the *rapid growth* of the plant embryo when environmental conditions are at their best. When leaves and roots have become well-established, then the growing plant is able to make its own food.
iv) The food reserves of some fruits are used as *food by animals*. In these cases, the plant has adapted itself to make use of the animals for widespread dispersal.

As with vegetative storage organs, Man has made fruits and seeds part of his diet, and in recent years has greatly improved the productivity of many plants that he now cultivates.

4.30 Storage of food reserves in animals

We have seen in Chapter 1 that the *lipids* (fats) are important to animals as a source of energy, and furthermore that lipids can be stored as a food reserve in many groups of animals. After digestion of fat, the fatty acids and glycerol formed are carried around the body. These building blocks may be used to form new fats which can be stored under the skin and around the internal organs. Each animal stores a form of fat characteristic of its species, so that fat from sheep for example is different from beef fat. Not all of the fat in mammals and other vertebrates comes from dietary lipids, for fat can be synthesised in the body from carbohydrates. Excess carbohydrate, not used immediately, is converted to fats for long term storage. In Man, the fattening properties of a diet over-rich in sugars and starch, are well known. Herbivorous animals synthesise their fat from cellulose which is broken down into its constituent sugar molecules by bacteria in the gut. These molecules are then either utilised by the body for metabolism, or, if unused, are changed into fat.

Glycogen, a complex carbohydrate, is another important storage substance. It has great significance in the immediate supply of energy when required by an animal. It is stored in the liver and the muscles. Certain vitamins can also be stored in the liver. Thus the liver has a most important role to perform as controller of the level of a number of substances in the body.

Animals that live in cold climates have fat reserves under the skin which provide insulation against heat loss. Birds and mammals in particular with their high body temperatures, lose heat continuously to their surroundings. In Arctic or Antarctic regions, their insulating coverings of hair or feathers are supplemented by the fat layers below the skin. Large mammals, such as seals and whales, which live in icy waters have an extremely thick layer of fat called blubber below their skin. It is this fat which has resulted in their being hunted and consequently the considerable depletion of their numbers.

Tropical animals do not generally store large quantities of fat. There are, however, a few interesting exceptions to this rule. The camel (*Camelus dromedarius*) is well known for its hump, which is not a water store as is often thought, but a fat store. This fat is not distributed widely under the skin, where it would form insulation and cause overheating, but is restricted to one large fat deposit. Fatty humps are also present in some African cattle, whereas in desert sheep, the fat store is in the tail. By this localisation of fat, the rest of the body can radiate large quantities of heat and keep itself relatively cool.

fig 4.11 Zebu cattle showing the fat hump

4.31 The liver

The liver, in spite of its relatively simple structure, has a multiplicity of functions. Above all it is an organ which controls the levels of many of the organic substances circulating in the blood, the tissue fluids and the lymph. In this function it is aided by the kidneys.

It is the largest gland in the body of Man, having an average weight of 1.25 kg, and is situated on the right side of the upper abdomen. The upper surface is convex, fitting closely under the diaphragm, while the lower surface is concave. It has a dark red colour and is divided into lobes. There is a large amount of blood circulating through the organ from the hepatic artery and the hepatic portal vein. The latter collects all the blood from the intestines and passes it through the liver.

We are concerned here with the *storage functions* of the liver. The stored material lies in the liver cells and exchanges take place between these cells and the blood circulating.

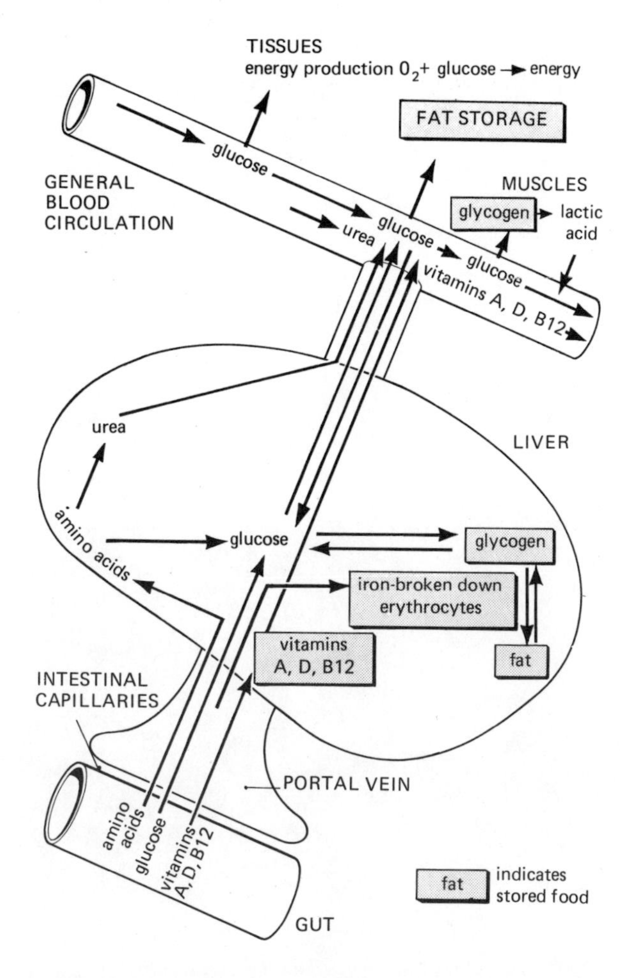

fig 4.12 The relationship between the liver as a storage organ, and the blood circulation

i) The liver has a special role in carbohydrate metabolism. It contains a substantial reserve of glycogen which is converted to free glucose in order to maintain the level of glucose circulating in the blood (known as blood sugar). The glycogen, is, therefore, a carbohydrate reserve. After a meal the digested carbohydrate passes into the blood as glucose, the blood sugar level rises, and the liver takes up glucose to convert it to glycogen. Glucose is continually removed from the blood, during the normal activities of the body and is used by the muscles, the nervous system and other organs. In this way blood sugar level falls, and so liver glycogen is changed into glucose to raise the level back to normal. Thus blood sugar levels are kept within fairly narrow limits.

ii) The liver has a special role to play in fat metabolism. It contains about 4% lipid, and if starvation occurs then the fat content of the liver falls. This will only happen after the exhaustion of all other body fats. Excessive eating of fats will result in the liver considerably increasing its fat content.

iii) *Proteins* are *not* stored in the body and so excess amino-acids must be eliminated. The amino-acids are broken down (*deaminated*) by the liver into *carbohydrate* and *urea*. The urea is excreted through the kidneys, but the carbohydrate can contribute to further formation of glycogen or fat in the body.

iv) *Vitamins A and D* are stored in the liver. They are available to carnivores in the livers of animals on which they feed. The livers of fish are relatively much richer in vitamins (particularly D) than the livers of mammals.

v) *Vitamin* B_{12} is found in liver. This is the anti-anaemic factor which is most important for preventing the symptoms known as *pernicious* anaemia.

vi) The liver also stores *iron* from broken down erythrocytes.

Skeletal muscle, as well as the liver, stores a substantial amount of *glycogen*. The muscles of a well-nourished man can contain up to 250 g of glycogen. During vigorous exercise the glycogen breaks down into lactic acid with the consequent release of energy for the immediate muscular needs. The lactic acid is released into the blood and then can either be built up into glycogen or broken down completely to carbon dioxide and water.

Questions requiring an extended essay-type answer

1 a) *Name* a plant with a vegetative storage organ: state the type of organ and the nature of the storage substance.

b) Make a labelled drawing of a *named* cereal grain in section to show the structure of the fruit. How does the stored food become available to the growing embryo as germination begins?

c) Where, and in what forms, is food stored in the mammal?

2 a) Describe the external appearance of a rhizome of a *named* plant.
b) What evidence have you that this is a modified stem and not a root?
c) What are the advantages of vegetative reproduction by green plants?

3 a) A vegetative storage organ of a plant is thought to contain non-reducing sugar, protein and oil. How would you confirm that these substances are present?
b) Where do these substances originate and how are they transported to the storage structure?
c) Describe briefly their eventual use by the plant.

4 a) By reference to *named* examples, explain the difference between vegetative reproduction and perennation.
b) Show, by means of large labelled diagrams only, how the following structures are used for vegetative reproduction
i) runners (a *named* example), and
ii) stem tubers (a *named* example).
c) What are the advantages and disadvantages, of vegetative reproduction compared with sexual reproduction in the flowering plant?

5 a) Make labelled drawings to show the structure of a stem tuber, a swollen tap root and a bulb.
b) How do these organs illustrate the characteristic external structures of stems, roots and leaves?

5 Respiration

5.00 Breathing

All living things depend upon the presence of air, and we know that any organism will die when deprived of air. What is it about air that is so important? Animals familiar to us, can be seen to move air in and out of their bodies by movements of the chest region. Is the air breathed out (*exhaled air*) different from the air breathed in (*inhaled* air or atmospheric air)? Any difference which can be detected between the two must be due to addition or subtraction by the person or animal breathing the air.

In order to examine the two different types of air we must first obtain samples on which experiments can be performed. Gases are easily collected by the method of displacement. (See fig 5.1).

Experiment

To obtain samples of inhaled and exhaled air.

Procedure

1 Use small gas jars with ground glass edges, together with glass plates to cover the jars. Fill a large trough with water and place in it a bee-hive shelf.

2 Fill several small gas jars with water, invert them and leave them to stand in the trough.

3 Obtain a length of rubber tubing and place one end through the side opening of the bee-hive shelf. Move one of the gas jars on to the top of the bee-hive shelf above the central hole.

4 Exhale most of the air from your lungs and, with the last remaining air, breathe out through the rubber tubing. The exhaled breath should displace all the water from the gas jar.

5 Place a glass plate over the open end of the gas jar, invert the jar, remove it from the trough and stand it on the bench. Repeat this process to fill three gas jars.

6 Atmospheric (inhaled) air can be obtained by simply opening a gas jar on the bench and then covering with a glass plate.

This technique should be used for any experiment in this chapter which requires a sample of exhaled air. It is advisable to use the last air from an exhaled breath since this ensures that it comes from deep inside the lungs.

Now that we have samples of both types of air, we can begin our investigations.

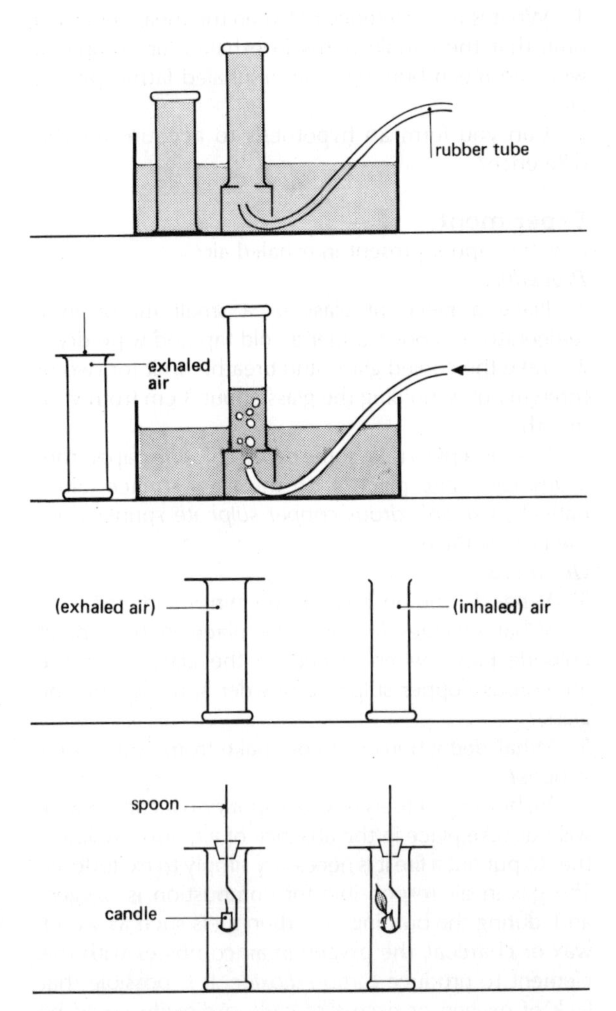

fig 5.1 The method for collecting exhaled gas over water

Experiment

Is the air breathed out different from air breathed in?

Procedure

1 Take two gas jars, one containing a sample of atmospheric air and the other containing exhaled air.

2 Fix a small candle stub on the cup of a *deflagrating spoon*. Light the candle. Obtain a stop watch or a watch with a second hand.

3 Place the burning candle into the jar of exhaled air; time its entry and then observe it carefully. When the

flame goes out record the time again (this is a matter of a few seconds, so very careful observation and timing is important).

4 Repeat this exercise, using the jar of atmospheric air.

5 Carry out both procedures at least three times. Record your results and obtain a mean value for each burning time.

Questions

1 What is the difference between the mean length of time that the candle burns in exhaled air compared with the mean burning time in inhaled (atmospheric) air?

2 Can you form an hypothesis to account for this difference?

Experiment

Is water vapour present in exhaled air?

Procedure

1 Place a piece of glass or a small mirror in a refrigerator or cool it under a cold tap and wipe dry.

2 Take the cooled glass, and breathe on it for two or three minutes, holding the glass about 3 cm from your mouth.

3 Use forceps to take some *cobalt chloride* paper from a *dessicator* and place it on the glass surface (alternatively, use *anhydrous copper sulphate* sprinkled on the glass surface).

Questions

3 What do you observe on the mirror surface?

4 What changes (if any) take place in the cobalt chloride paper when placed on the glass (or in the anhydrous copper sulphate powder sprinkled on the glass)?

5 What deductions can you make from your observations?

The burning of fuels or wax vapour on a candle wick will not take place in the absence of air, also, we know that to put out a fire it is necessary simply to exclude air. The gas in air responsible for combustion is oxygen, and, during the burning of carbon fuels such as wood, wax or charcoal, the *oxygen* in air *combines* with this element to produce *carbon dioxide*. It is possible that lack of oxygen or excess of carbon dioxide could be responsible for the candle flame going out in the air samples. Since we can easily detect the carbon dioxide, let us examine this component of the air samples.

Experiment

Is there any detectable difference between the carbon dioxide content of inhaled and exhaled air?

Procedure

1 Fit up two boiling tubes or specimen tubes with rubber bungs and glass tubing as shown in fig 5.2.

2 Place equal volumes of *bicarbonate indicator* (or lime water) in each tube.

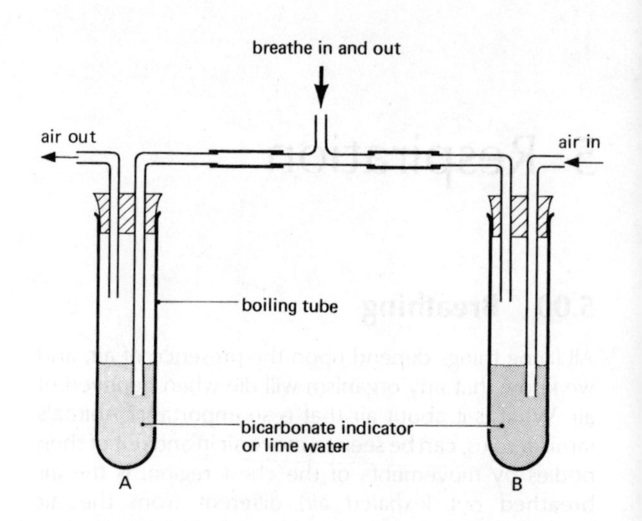

fig 5.2 The experimental apparatus for bubbling inhaled and exhaled air through bicarbonate indicator solution

3 Place the mouthpiece in the mouth and breathe gently in and out through the apparatus.

Questions

6 What happens in the boiling tubes immediately breathing begins, before any colour changes take place?

7 How do you explain these observations?

8 What colour changes take place in the indicator solutions in each tube?

9 What deductions about the inhaled and exhaled air can be made from these colour changes?

10 What function is performed by the boiling tube B?

The experiments have shown that carbon dioxide is present in *greater quantities* in *exhaled* air than in inhaled air. We can now move forward and consider in more detail the gas content of the two types of air. We know that living things need oxygen to survive, and apparently man produces carbon dioxide when breathing out. These two gases are present in atmospheric air, together with the commonest of all gases, *nitrogen*. In order to estimate the quantities of each gas, we can analyse a number of samples of inhaled and exhaled air.

We can use the same apparatus that was used in Chapter 2 in order to analyse the gases given off by photosynthesising plants. The same reagents *potassium hydroxide* and *potassium pyrogallate* solution are used to absorb carbon dioxide and oxygen respectively. The sample of exhaled air should be collected as shown in fig 5.3.

Experiment

How does the gas content of exhaled air differ from that of inhaled air?

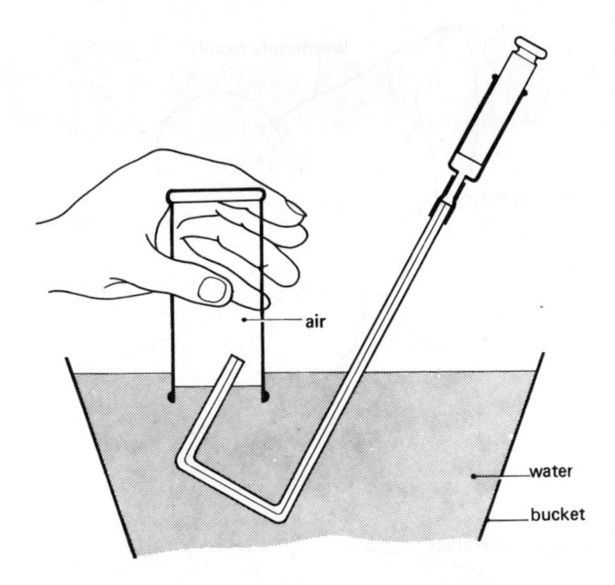

fig 5.3 The method for obtaining a sample of air for analysis

Procedure

Remember that both absorbent reagents are caustic, so take care to avoid drops on the skin, clothing, bench or floor. Be careful when you carry the tube from the reagent to your water trough. Any reagent dropped must be washed away immediately with plenty of water.

2 Collect air samples from the atmosphere and from your exhaled breath as shown in fig 5.1. Smaller samples can be collected in boiling tubes rather than gas jars.

3 Place the container with the gas sample, open and downwards in a bucket of water or a large beaker full of water. The test-tube or gas jars should be held by a clamp and retort stand, but if necessary a person could stand and hold it.

4 Take the J-tube with syringe and place the open end in the water. Draw a thread of water about 5 cm in length into the capillary tube. Then place the open end of the tube up into the gas sample and draw in about 10 cm of gas.

5 Lower the J-tube and draw in another seal of water to enclose the gas bubble. Put the apparatus into a water trough for five minutes. Remove and, without too much handling, measure the length of the bubble (A cm). Record this in your notebook.

6 Push in the plunger of the syringe until the gas bubble is near the open end of the capillary tube. Place the open end of the tube downwards into the potassium hydroxide and draw in about 5 cm of reagent. (See Chapter 1, section 1.23 for further details of precautions to be taken in this experiment.)

7 By pulling out the plunger, move the bubble and reagent round into the long arm of the J-tube, and then move the bubble backwards and forwards in order for the reagent to absorb the carbon dioxide.

8 Place the apparatus into the water trough for 5 minutes. Then remove it and quickly measure the length of the gas bubble (B cm). Record this in your notebook. If the bubble has broken into two or more parts, measure each part and add the lengths together.

9 Push out the potassium hydroxide seal into the sink, so that only a small amount remains next to the bubble. Place the open end of the tube into potassium pyrogallate solution, making sure that it is below the level of the paraffin on the surface.

10 Pull out the syringe plunger to admit about 5 cm of the potassium pyrogallate. The reagent will turn brown as it absorbs oxygen. Pull the bubble into the long arm of the J-tube, and then move it backwards and forwards to absorb the gas.

11 Place the apparatus in the water trough for five minutes. Then remove it and quickly measure the length of the bubble (C cm). Record this in your notebook. Treat a broken bubble as in step 8 above.

12 Work out your results as indicated in Chapter 1 (section 10 of experiment 1.23).

In a class situation, tabulate the class results and obtain the mean values for the percentages of carbon dioxide and oxygen respectively (see table below).

Questions

11 Why was the J-tube placed in the water trough after each part of the experiment?

12 What results were obtained after analysis of (a) inhaled air, and (b) exhaled air? What is their significance?

13 Will exercise have any effect on the results that you have obtained? Suggest how you could investigate this problem.

Name	Inhaled air		Exhaled air	
	% of carbon dioxide	% of oxygen	% of carbon dioxide	% of oxygen
Mean value				

5.01 The content of air

The experimental results in table 5.1, show how the percentage composition of air samples exhaled by humans differs from that of inhaled (atmospheric) air. These figures are constant throughout the world.

Gas	Inhaled air %	Exhaled air %
Carbon dioxide	0.03	3.50
Oxygen	20.93	16.89
Nitrogen	79.04	79.61
	100.00	100.00

Table 5.1 The gaseous content of inhaled and exhaled air

The figures refer to a man in a *resting* condition. After *exercise* the *oxygen consumption rises* slightly and so does the *output* of *carbon dioxide*, but in spite of this it can be seen that the lungs are not very efficient in extracting oxygen from the atmosphere, since exhaled air still has about 17% oxygen.

For this reason, the most effective modern method of artificial respiration is to blow this exhaled air into the lungs of another person. Anyone who has suffered an electric shock, been rescued from gas or smoke, or saved from drowning, may stop breathing. If such a person is to be saved it is most important that the brain is not deprived of oxygen. Therefore, breathing must be restored as quickly as possible, and often the simplest and best method is by mouth to mouth resuscitation. The rescuer blows air from his own lungs into the patient's mouth. In order that this transfer should be effective, the patient's nose must be held closed and the head pulled well back. The rescuer covers the patient's mouth with his own and blows, watching the chest rise as the air enters. He then allows the chest to fall (or presses it down) and then blows in more air. In this way many lives have been saved by the use of the oxygen remaining in exhaled air (see fig 5.4).

The densest part of the atmosphere is nearest to the earth's surface, but as the distance away from the earth increases, the atmosphere becomes thinner and thinner. Although the proportion of oxygen in the atmosphere remains constant at 21%, there are fewer molecules of all gases within a given volume. Thus, a normal inhalation at 6 000 metres will bring only half the amount of oxygen to the lungs brought by a normal inhalation at sea level. Mountain climbing at high altitudes needs a much faster rate of breathing to supply oxygen, and in addition, the breaths become deeper. The conquest of Mount Everest, the highest mountain in the world, has only become possible by the use of a type of oxygen cylinder.

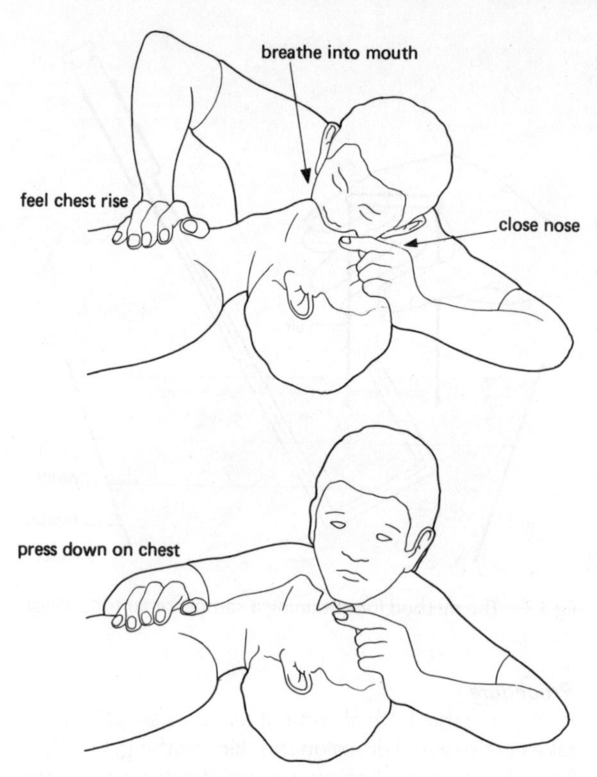

fig 5.4 Mouth to mouth resuscitation

5.02 Gaseous exchange in other living organisms

Experiment

Do all living organisms produce carbon dioxide?

Procedure

1 Take a number of test-tubes and fit them with solid rubber bungs. Each tube should be fitted with a wire or nylon gauze (or cotton wool) platform (see fig 5.5). All pieces of apparatus should be well washed with *distilled water* and then with bicarbonate indicator solution before use.

2 Place about 2 to 3 cm³ of bicarbonate indicator in each tube. Take care not to breathe over the tubes. Put in the rubber bung immediately.

3 Into each tube place some living material so that it rests on the gauze (or cotton wool). Examples of these could be:

i) caterpillars; and	a) piece of potato;
ii) beetles;	b) piece of apple;
iii) millipedes;	c) piece of rhizome;
iv) slugs or snails;	d) piece of corm (crocus).
v) grasshoppers;	(No green plant
vi) cockroaches;	material should
vii) woodlice;	be used).

Replace the rubber bung immediately.

4 Place the tubes in a test-tube rack and allow them to stand. Examine the tubes every thirty minutes. Shake

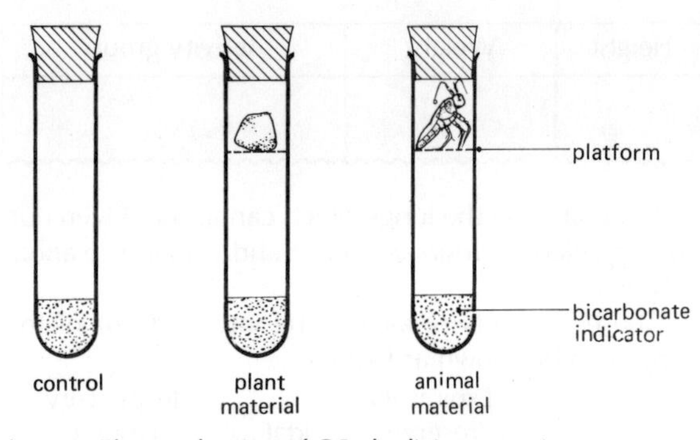

fig 5.5 The production of CO_2 by living organisms

the indicator each time, but do not allow it to splash over the living material.

Questions

1 Why is it essential to have the apparatus clean and washed before commencing the experiment?

2 Why should you not use green plant material?

3 What happens to the bicarbonate indicator in the tubes (a) with animal material, (b) with plant material?

4 What conclusions can you draw from the time taken to change the indicator by the two types of living organism?

Experiment

Do animals remove oxygen from the air?

Procedure

1 Fit up the apparatus as shown in fig 5.6. Insert a glass T-piece. Support the U-tube in the clamp of a retort stand, and use a pipette to put coloured water into it to a depth of about 5 cm.

2 Three-quarters fill a small ignition tube with potassium hydroxide pellets. Attach a thread to the upper end of the tube.

3 Place the living organism (e.g. mouse, insects or snails) inside the flask, and then suspend the tube of pellets in the flask by inserting the rubber bung securely over the thread.

4 Connect the T-piece with the U-tube leaving the screw clip open. Now close the screw clip. Mark the position of the liquid levels in the U-tube.

5 Set up a control experiment in exactly the same way but without the living organism. Place both sets of apparatus in such a position that they are not subject to temperature changes. Observe the liquid levels in each piece of apparatus.

N.B. when using a mouse or other small mammal ensure that the animal does not become distressed as the oxygen is used up.

Questions

5 What is the function of the potassium hydroxide pellets?

6 What do you observe in the U-tube?

7 Is this experiment a direct measurement of oxygen removal? What is the relationship between oxygen use and carbon dioxide emission?

5.03 What volumes of air are moved during breathing?

Experiment

To find the volume of air breathed out during a forced exhalation and to determine if there is any correlation with the level of activity.

Procedure

1 Obtain a large bell jar, turn it upside down and place in it 1 litre of water. Mark the level of the water on the side of the bell jar with a marking pencil. Pour in a second litre of water, and mark the second level. Continue in this way until seven litres have been marked on the jar (or 5 litres if it is a smaller bell jar).

2 Fill a large sink with water, fill the bell jar with water (by lying it on its side) and then stand the bell jar on three supports (metal blocks, evaporating dishes or mortars). Push a rubber tube into the bell jar, between the blocks. Insert a glass mouth piece into the outer end of the rubber tubing (see fig 5.7).

3 Each pupil should take a deep breath and then exhale fully into the bell jar. Measure the volume of air

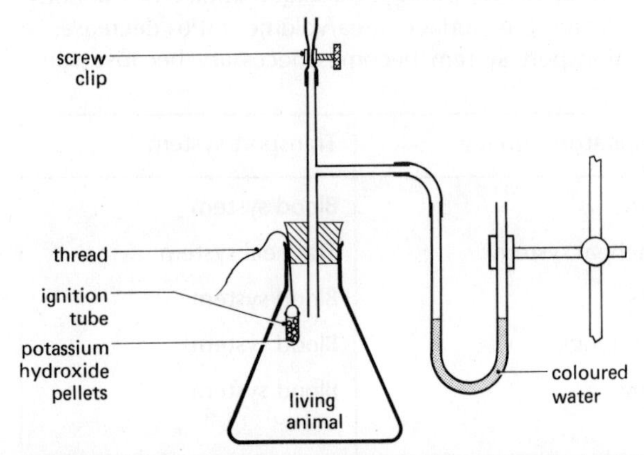

fig 5.6 The experimental apparatus to investigate the removal of oxygen from the air by animals

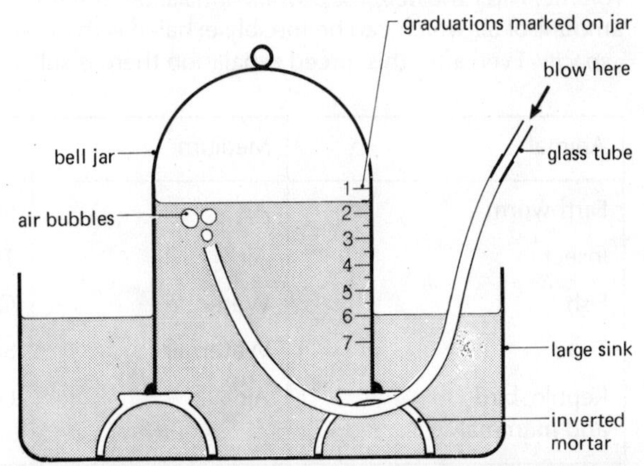

fig 5.7 The experimental apparatus for measuring the vital capacity

Name	Vital capacity	Age	Height	Weight	Activity group

breathed out. An estimate can be made to the nearest 100 cm³, that is a tenth division of the distance between each litre mark.

4 Refill the bell jar, and sterilise the mouth piece with alcohol or a disinfectant.

5 Tabulate the results, indicating volume of air breathed out in a forced exhalation (known as the *vital capacity*) together with other information for correlation purposes.

Classify individuals as follows according to their ability group:

Very active – always in training for athletics (particularly long distance running)

Active – plays games (football, hockey, swimming, netball)

Average – Does some walking and cycling

Inactive – has no interest in games, dislikes walking or cycling

Questions

1 Analyse all class results and any other data that can be obtained. Classify individuals into age, weight, and height groups and work out the average vital capacity for each group. Draw histograms of the results.

2 Do you consider that there is any correlation (relationship) between vital capacity and these factors?

The body under normal quiet conditions moves only a small volume of air into and out of the lungs. This is called the resting *tidal volume* and is about 500 cm³. A *forced inhalation*, to its fullest extent, can take in another 2 000 cm³. This amount is the *inspiratory reserve volume* and together with the tidal volume forms the *inspiratory capacity*. After a normal gentle exhalation it is possible to force out approximately an additional 1 300 cm³: this is the *expiratory reserve volume*. After the *deepest possible inhalation*, the total amount of air which can be forcibly exhaled is the *vital capacity*. Even after this forced exhalation there is still a volume of air in the lungs which cannot be driven out. This is called the *residual volume* and amounts to about 1 500 cm³.

These amounts of air moving into and out of the lungs can be shown as follows:

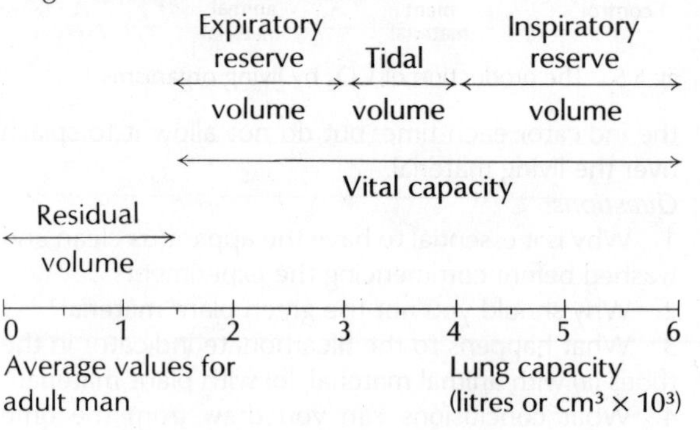

5.10 The structures of the mammal concerned in gaseous exchange

The exchange of carbon dioxide and oxygen in animals demands three conditions.

1 **Medium** – a medium in which the gases can be present e.g. air or water.

2 **Respiratory surface** – a large, moist surface area where the gases are dissolved before diffusing into or diffusing out of the cells.

3 **Transport system** – a system to move the dissolved gases around the body to all the cells, and to collect gases which need to be eliminated.

In simple organisms the exchange of gases takes place through the skin and if the surface area/volume ratio is high, no transport system is required. Gases simply diffuse through the outer surface of the body. When the surface area/volume ratio decreases a transport system becomes necessary because some

Animal	Medium	Respiratory surface	Transport system
Earthworm	Air	Skin	Blood system
Insect	Air	Tracheal system	Tracheal system
Fish	Water	Gills	Blood system
Toad	Water/air	Skin/lungs	Blood system
Reptile, bird and mammal	Air	Lungs	Blood system

Table 5.2 Gaseous exchange mechanisms of different animals

cells will be a relatively long way from the source of oxygen. Also, the outer surface is no longer large enough to absorb sufficient oxygen for the respiratory needs of the proportionately larger bodies. Thus, special structures with a very large area must be developed to absorb oxygen and to eliminate carbon dioxide. These are the gills and lungs of vertebrate animals.

The structures concerned with gaseous exchange in a mammal are two elastic organs called *lungs*, which occupy most of the space inside the *thoracic cavity*. Each lung consists of a series of finely branched tubes (*bronchioles*) terminating in tiny air sacs called *alveoli*. The bronchioles lead into a single wide tube or *bronchus*, and the bronchi from right and left lungs unite with the windpipe or *trachea*. The bronchi and the trachea are strengthened by incomplete hoops of *cartilage* which prevent them from collapsing when bent. (See fig 5.8.)

The trachea opens into the *larynx* (voice box) which contains the *vocal cords*. The vibration of these structures provides sound which is modified by the buccal cavity, lips and tongue. In the evolution of Man, these sounds have become organised into the complex patterns of language. The larynx opens into the *pharynx* by an opening called the *glottis*. This is closed by a flap, the *epiglottis*, as food passes over it (and down the oesophagus) during swallowing. The pharynx is connected to the nasal and buccal cavities, so that air can pass in and out of the lungs during breathing. With the mouth closed, the air passes into the nostrils, through the *nasal cavity*, and into the pharynx. Thus when the buccal cavity is full of food the mammal can still breathe. The bronchi and trachea have a lining of cells covered with minute protoplasmic projections called *cilia*. *Mucus* is also secreted by this lining so that any dust, bacteria and other foreign particles are collected and swept upwards by the cilia into the pharynx, being finally swallowed into the oesophagus.

The thin, elastic walls of the alveoli consist of a single layer of cells interspersed with elastic cells covered by a dense network of capillaries. In one human lung there are approximately 300 million alveoli which provide a *very large surface area*, about the size of a tennis court. (See fig 5.8.)

Obtain the trachea, bronchi and lungs of a mammal either by dissection of a rabbit or rat, or by purchasing them from an abbatoir (where animals are slaughtered).

Experiment

To examine the breathing organs of a mammal

Procedure

1 Put a glass tube into the trachea and blow hard into the tube. what happens when you are blowing, and when you stop blowing? How do you account for this?

2 Observe the size and shape of the larynx and trachea. Note the rings of cartilage.

3 Observe the two bronchi and their points of insertion into the lungs.

4 Observe the thorax and the bones in the thoracic walls. Observe the *diaphragm* which separates the

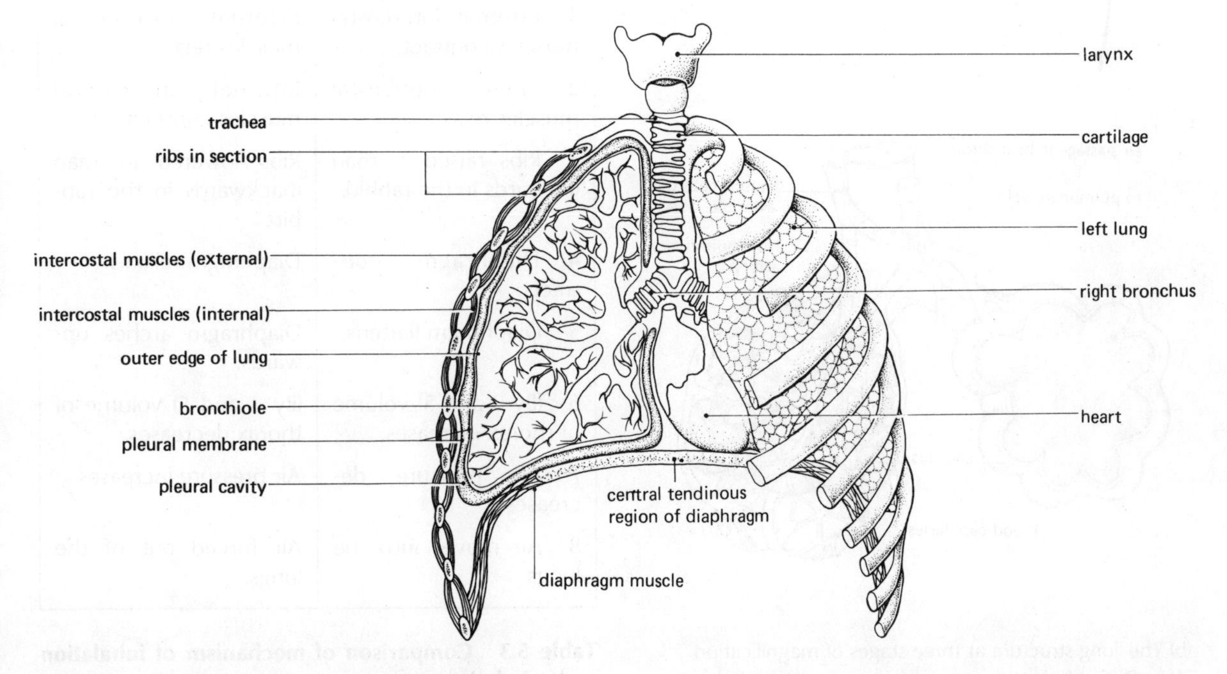

fig 5.8 (a) The lungs, heart and associated organs

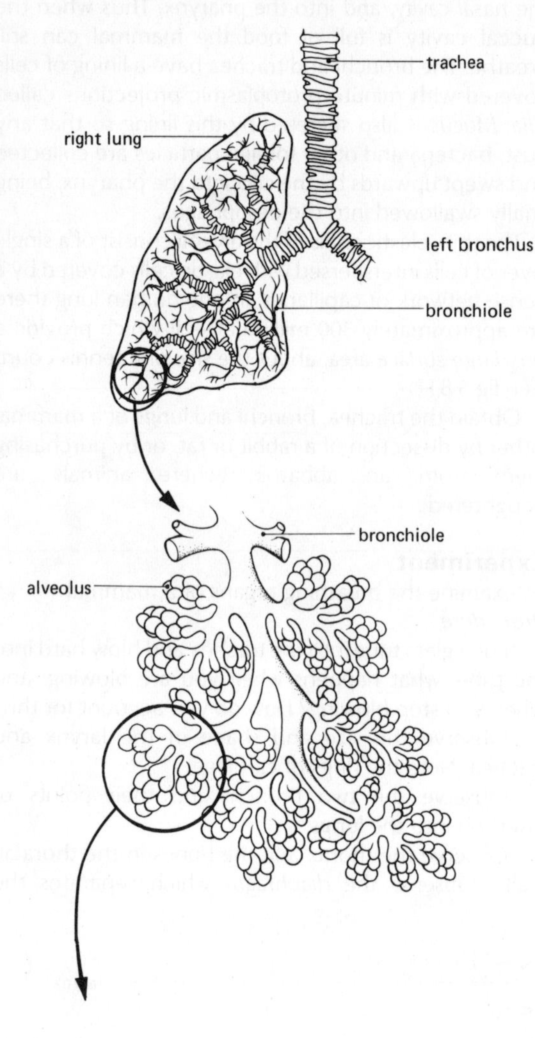

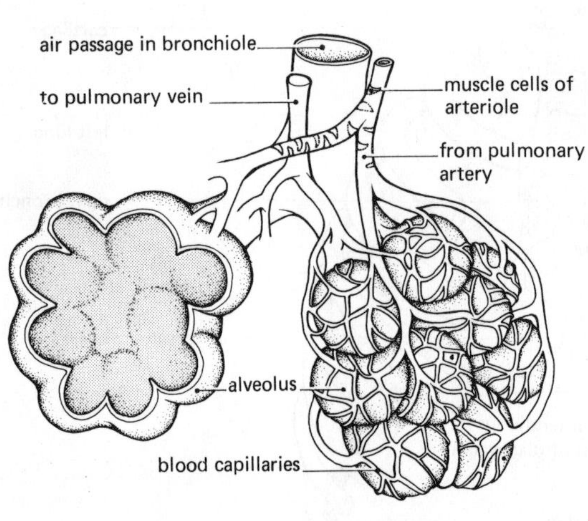

(b) The lung structure at three stages of magnification
After D. G. Mackean

thoracic cavity from the abdominal cavity.

5 Observe the inner walls of the thorax and the surface of the lungs.

6 Cut open the lungs and observe the structure with a hand lens.

Questions

1 Are the surfaces of the lungs and inner walls of the thorax dull or shiny? Are they dry or moist?

2 Notice the large blood vessels. Trace their path. Do they supply the lungs?

3 Can you describe the structure of the diaphragm? Has it a uniform structure throughout?

5.11 Mechanism of breathing

Two groups of muscles are used in breathing, and these are aided by the elasticity of the lungs.

i) The muscles between the ribs are the internal and external *intercostal muscles*. These act antagonistically, for when one set contracts the other relaxes and vice versa. *External* intercostal muscles are arranged diagonally across the ribs, so that by contraction they pivot the ribs on the backbone and thus *lift* the ribcage (in man). *Internal* intercostal muscles are also arranged diagonally, but at right angles to the external muscles, so that their contraction pulls the ribcage *downwards*.

ii) The muscular sheet attaching the fibrous centre portion of the diaphragm to the body wall *contracts* and *flattens* the diaphragm from its relaxed, domed position.

Inhalation	Exhalation
1 External intercostal muscles contract.	External intercostal muscles relax.
2 Internal intercostal muscles relax.	Internal intercostal muscles contract.
3 Ribs raised in man (forwards in the rabbit).	Ribs lowered in man (backwards in the rabbit).
4 Diaphragm contracts.	Diaphragm relaxes.
5 Diaphragm flattens.	Diaphragm arches upwards.
6 (By 3 and 5) volume of thorax increases.	(By 3 and 5) volume of thorax decreases.
7 Air pressure decreases.	Air pressure increases.
8 Air moves into the lungs.	Air forced out of the lungs.

Table 5.3 Comparison of mechanism of inhalation and exhalation

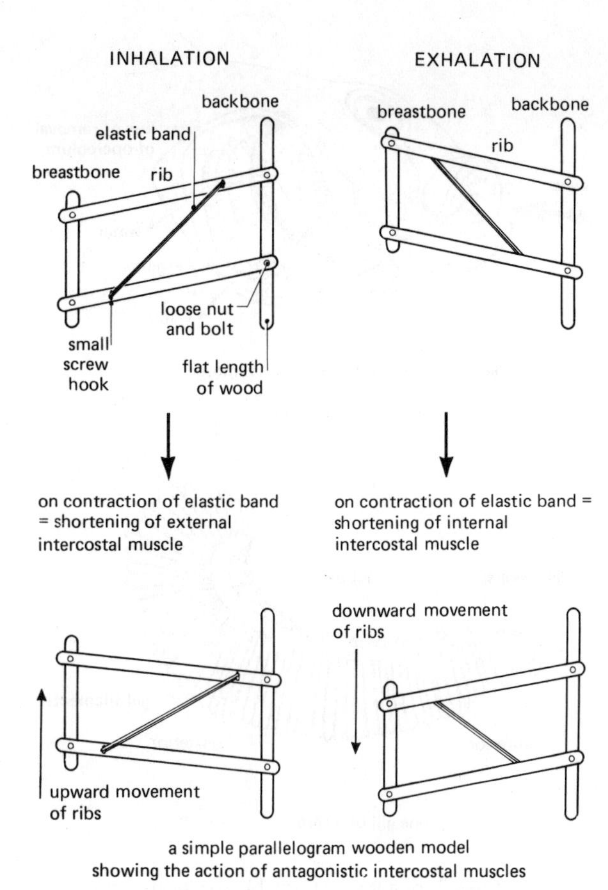

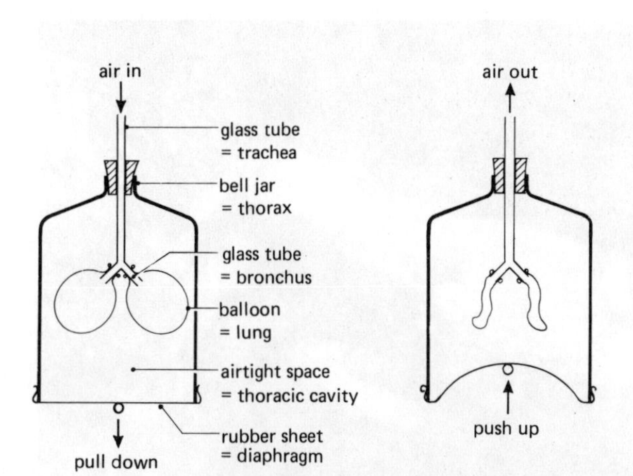

fig 5.9 Models to demonstrate breathing mechanisms

These sets of muscles work to increase or decrease the volume of the thorax, so that pressure is decreased or increased and air moves into or out of the lungs. The action of the two sets of muscles can be demonstrated by the two pieces of apparatus in fig 5.9.

Breathing is normally controlled automatically by the *medulla oblongata* of the brain, but of course we can control our own breathing when singing and talking; and even stop it for a while when we are swimming under water. The medulla is very sensitive to the *concentration of carbon dioxide* in the blood and when the concentration rises (e.g. in a closed room or during exercise), the brain sends messages to the intercostal muscles and the diaphragm. The breathing movements become faster and this causes the carbon dioxide to be removed more quickly. As the carbon dioxide concentration in the blood goes down the breathing rate returns to normal. This automatic control is called a *homeostatic mechanism*.

There are other movements of the thorax and the diaphragm which are variations of the breathing mechanism.

1 *Hiccough* – caused by a spasmodic contraction of the diaphragm forcing air into the lungs.
2 *Cough* – a sudden forced exhalation to clear the trachea and larynx of mucus that contains bacteria and dust.
3 *Yawn* – a long inhalation of air due to a variety of causes which may have increased carbon dioxide concentration in the blood.
4 *Sigh* – a longer inhalation and exhalation than is usual in breathing, involving the thorax and the diaphragm.

fig 5.10 The exchange of gases in the alveoli

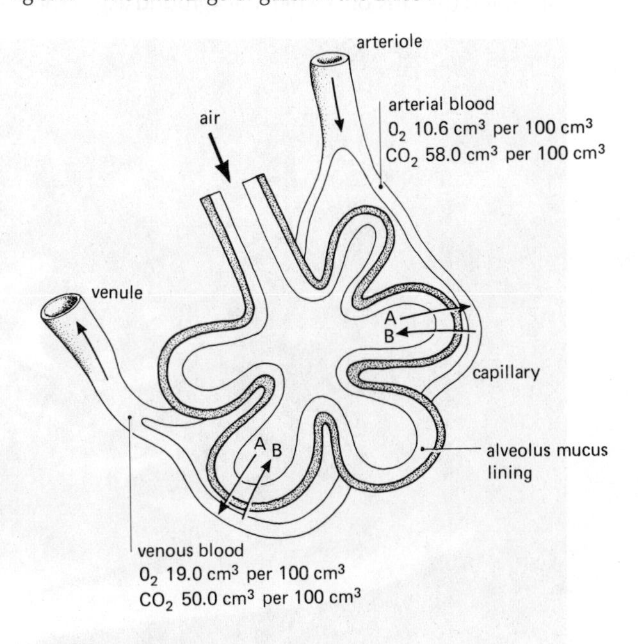

At A, O_2 diffuses into the blood through mucus, alveolar wall and capillary wall
At B, CO_2 diffuses into the alveolus through capillary wall, alveolar wall and mucus

5.12 Gas exchange in the lungs

When the alveoli are filled with air there is an exchange of gases between the air and the capillaries (see fig 5.10). The blood entering the lungs through the pulmonary artery reaches the capillaries. Oxygen and carbon dioxide are present at the levels shown in the diagram. Diffusion of gases takes place between the alveolar cavity and the blood. When the blood leaves the capillary to be collected up by the venules and pulmonary veins the oxygen concentration has increased and carbon dioxide concentration has decreased.

5.13 Gas exchange in a fish

The gills of a fish must extract oxygen from the water. They consist of a series of *gill arches* each with a large number of *filamentous outgrowths*. The outgrowths are well supplied with *blood vessels* and are covered by very thin skin. They provide a large surface area through which oxygen can be absorbed from the water flowing over them. The blood takes up the oxygen and carries it around the body returning with carbon dioxide which is given up to the water.

Investigation

To examine the breathing mechanisms in a fish

Procedure

1 Obtain a large glass beaker or an aquarium tank. Fill it with water. Place in it a small bony fish such as a goldfish. Stand the tank on a white background (e.g. a sheet

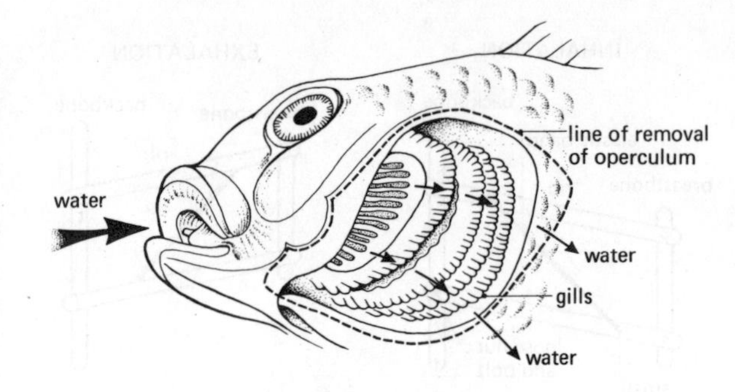

head of a carp with operculum removed

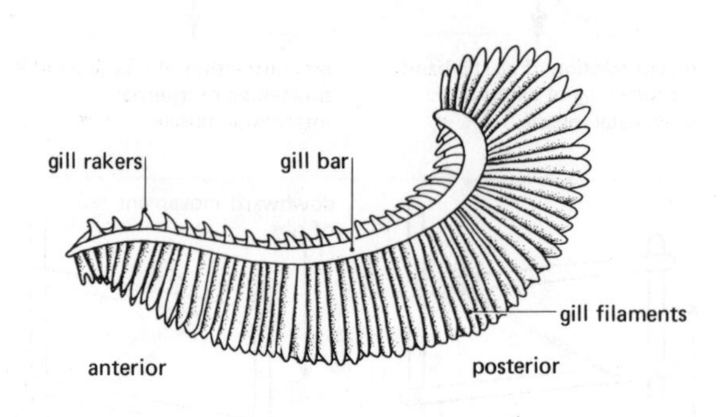

one gill of a carp

fig 5.11 The gills of a fish (a) Carp (b) *Esox lucius* – pike

of drawing paper). Allow the fish to settle in the tank.
2 With a bulb pipette take up a little Indian ink, and expel a drop in front of the mouth of the fish. Complete this slowly and carefully so that the fish is not disturbed.
3 Observe the movement of the Indian ink particles in relation to the breathing movements of the fish. Observe the movements of the mouth and the operculum.

Questions

1 What do you observe about the movement of the Indian ink particles near the mouth of the fish?
2 What relationship exists between the mouth movements and those of the opercular flaps?

Investigation

To examine the structure of the breathing organs of a fish

Procedure

1 Obtain a freshly killed or preserved fish. Lift the operculum and examine the structures underneath.
2 Cut off the operculum to expose the gills. Place the fish under water in a beaker and note how the gill structures float up, but collapse again when the fish is brought out into the air.
3 Push a small glass rod through the mouth and out through the gills to show a continuous channel.
4 Remove one complete gill (see fig 5.11). Put the gill in water. Draw and label it.
5 Examine the structures on the gill bar internal to the gill filaments. These are called *gill rakers*.

Questions

3 How many gill arches are there on each side of the fish?
4 What is the function of the gill rakers?

5.14 Gas exchange in an insect

When examined externally by a hand lens a large insect (e.g. cockroach) can be seen to have an opening on each segment of the abdomen. Each opening is called a *spiracle* and it connects internally with a series of branching tubes called *tracheae*. The branches become smaller and smaller forming minute tubules called *tracheoles,* which reach *every tissue* in the insect. Gases can diffuse along the tracheae and tracheoles eventually reaching the tissues. In this way oxygen is obtained from the atmosphere by all cells, while carbon dioxide moves in the reverse direction. Therefore the blood does *not* contain an *oxygen carrying pigment* comparable with the haemoglobin of the mammal.

The movement of gas along the tubes is aided by the contraction and relaxation of the abdomen. Nevertheless diffusion of gas along minute tubules is only efficient up to a distance of about 1 cm. Insects are *restricted in size* by their breathing systems, and thus insects generally can never grow to the size of birds or mammals. (See fig 5.12.)

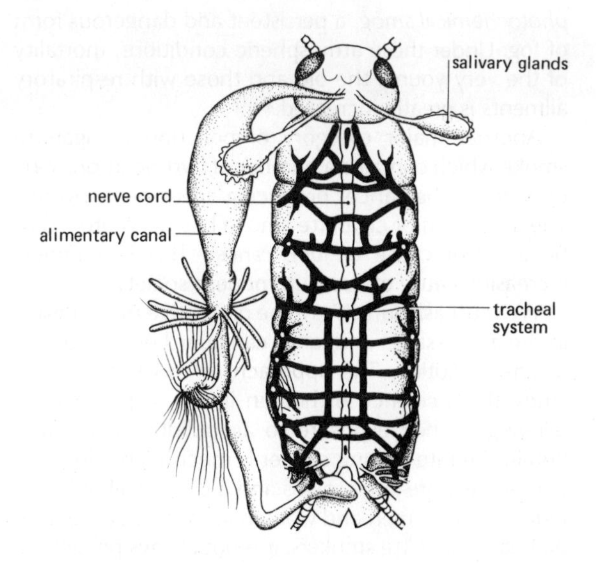

fig 5.12 The tracheal system of an insect

5.20 Air pollution and health

During a period of twenty four hours, each one of us breathes in about 15 000 litres of air. Many people are concerned that in large cities and industrial areas the air is not as clean as it should be in order to maintain the health of the population. Mainly due to man's activities a number of 'foreign' substances find their way into the air, causing pollution of the atmosphere. There are two main sources of atmospheric pollution:
i) *Industrial sources* which include the iron and steel industry, chemical industry, refineries of petroleum spirits, oil and coal-fired generating stations.
ii) The *internal combustion engines* of cars and lorries which emit vast quantities of pollutants from their exhaust systems.
The primary pollutants include:
i) *Soot* from unburned fuel used in domestic fires, industrial furnaces and power stations. The older industrial cities of Europe have their buildings blackened by years of exposure to a soot-laden atmosphere.
ii) *Gases,* such as *sulphur dioxide* from the oxidation of sulphur compounds in fuels such as coal and oil; *carbon monoxide, nitrogen oxides* and other gases are poured into the atmosphere from industrial chimneys and car exhausts.
iii) *Hydrocarbons* produced by the combustion of petrol and engine oil.

In the United States more than 140 million tons of these pollutants are poured into the atmosphere every year. For many human beings air pollution has already proved lethal. Secondary pollutants, produced by the *action of sunlight* on primary pollutants are exceedingly damaging to plant life, and lead to the *formation of photochemical smog*, a persistent and dangerous form of fog. Under these atmospheric conditions, mortality of the very young, the old and those with respiratory ailments is greatly increased.

Another major category of pollution is cigarette smoke which contains a number of hydrocarbons, one of which is the cancer producing agent *benzopyrene*. The inhalation of cigarette smoke has been proved to be a major cause of *lung cancer*. This is a rapidly increasing cause of death in modern society.

We must ask ourselves if the protective mechanisms in the air passages and lungs can guard us adequately against pollution. The approach to this has been to study the incidence of human illness in populations differing in their exposure to air pollutants. In Great Britain the rate of lung cancer is much higher in areas where there are large towns and a great deal of heavy industry. Since men and women in these areas tend to be heavy cigarette smokers, it is not always possible to point to one particular cause of this disease.

As more and more data become available, it is apparent that a wide variety of human diseases are more common amongst urban populations than in rural communities, especially where the town dwellers are exposed to a good deal of air pollution. These diseases include *chronic bronchitis, heart diseases* and other circulatory problems and lung cancer. Also it has been found that death rates are directly proportional to the number of cigarettes smoked each day, extent of inhaling and the duration of the habit. (See fig 5.13).

Hot cigarette smoke destroys the cilia lining the respiratory tract. This interferes with the filter mechanism which protects against microbes. Smokers are therefore prone to respiratory infections.

It is clear that the human respiratory system is not capable of overcoming the extremes of air pollution to which society and the individual subject it.

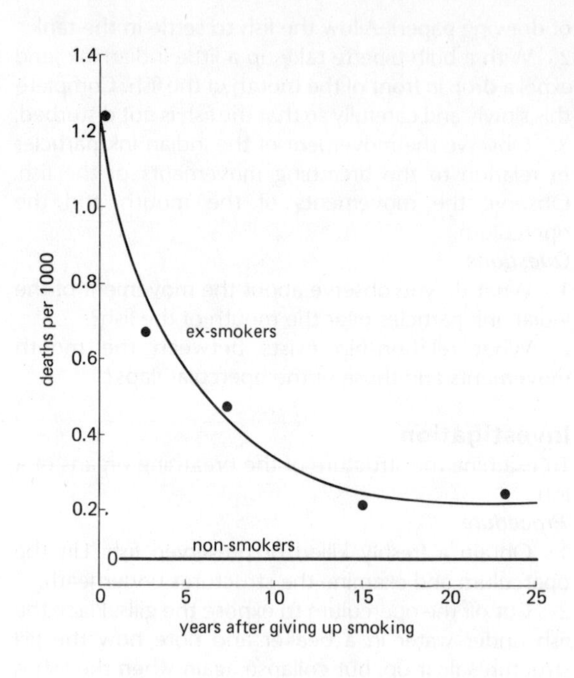

fig 5.13 The death rate from lung cancer, after giving up smoking, compared with the death rate from lung cancer in non-smokers

5.30 Obtaining energy

The work done by a living animal can be compared with that done by an internal combustion engine, such as that of a car. The same is true for plants, but their activities are not always so easily observable as those of an animal. Their energy for living processes is obtained in a similar way, but it is not used for movement and increasing body heat as in animals. Oxygen is not actively breathed in, but diffuses in through pores.

	Engine	Animal
Preparation of fuel	Crude oil purified	Food digested
Fuel utilised	Carbon compounds of C, H and O – petrol	Carbon compounds of C, H and O – carbohydrates and fats
Supply of oxygen	From air through the carburettor	From air through body surface, lungs or gills into blood stream
Combustion	Oxidised fuel in a chamber	Oxidised food in a cell
Waste products	Carbon dioxide, water and other gases	Carbon dioxide and water
Energy production	Heat and mechanical energy	Heat and mechanical energy
Transfer of energy	By a system of cranks and gears	By a system of muscles, bones and joints (in mammals)

Table 5.4 Comparison of an internal combustion engine and an animal

By using *radioactive tracers* in pure foods fed to laboratory animals, it can be shown that the carbon dioxide produced during breathing has come from the foods. On feeding glucose with radioactive carbon ($^{14}C_6H_{12}O_6$) to rats and collecting the carbon dioxide breathed out, it has been discovered that it is the carbon element of this gas that is radioactive ($^{14}CO_2$). The carbon from the sugar has been transferred to the carbon dioxide. Similarly the hydrogen of the sugar can be labelled, and the water present in the urine and the breath shown to contain the radioactive hydrogen.

We may ask, 'but why is the sugar broken down?'. Chemically, it can be shown as follows:

$$C_6H_{12}O_6 + 6O_2 \rightarrow 6CO_2 + 6H_2O$$

Clearly this is not the whole answer, since sugar is not broken down solely to produce waste products of carbon dioxide and water. The water could be very useful to the organism, but most animals obtain the water they require by ingestion. Pursuing our analogy with the internal combustion engine, the most important result of the breakdown of sugar (fuel) is the *release of energy*.

5.31 Types of energy

In the steam engine the *chemical bond energy* of the fuel, which may be coal, oil or wood, is converted to heat energy which can turn a turbine (*mechanical energy*), rotate a dynamo (*electrical energy*), and eventually light a whole town (*light energy*).

The chemical bond energy of the sugar molecules can be converted eventually to other chemical energy or mechanical energy. These different forms of energy are used to work, in the same way that the chemical energy of oil is used to do the work of moving a car up a hill. The chemical bond energy of the sugar and fuel oils has been obtained as a result of the conversion of the *sun's light energy* by plants. In the case of sugars this conversion was carried out recently, but in the case of fuel oils, coal etc. it was carried out millions of years ago.

Experiment

To make a simple steam engine

Procedure

1 Cut a circular piece of tin from the bottom of a tin can. Punch a hole in the centre of the tin. Fix small pieces of material (e.g. cover slips or pieces of tin) on the rim, at angles of 60 degrees to each other.

2 Hold a nail horizontally in the clamp of a retort stand, and put the nail through the hole in the tin wheel. Wrap a rubber band around the end of the nail to prevent the wheel from falling off the nail.

3 Fit a flask with a glass tube and a rubber bung as shown in fig 5.14. The glass tube should be heated and drawn out into a fine point.

4 Put water in the flask and bring it to the boil over a flame. When the water boils, the steam will come out of the glass tube which should be directed onto the vanes of the wheel.

Questions

1 Where does the energy come from to increase the temperature of the water and make it boil?

2 Record the series of changes in energy from the heating of the water to the turning of the wheel.

Living organisms can use and produce all of these types of energy. The following are examples:

Heat energy – used to keep the body temperature constant in birds and mammals.

Mechanical energy – produced during the movement of animals by their limbs.

Light energy – used to produce light in phosphorescent organisms.

Electrical energy – the production of electricity in the electric catfish (*Malapterurus electricus*), and electric eels.

Sound energy – the production of sound in toads and grasshoppers.

Chemical energy – used to bring about muscle contraction.

The heat produced by the bodies of birds and mammals is necessary to maintain the body temperature at a constant level. These animals are *homoiothermic* (warm-blooded). If one places a hand on these creatures one can feel the warmth of their bodies, especially that of a bird which has a body temperature 2 to 3 degrees C above that of Man. It is not so obvious, however, that other animals or plants give off heat. Animals other than birds and mammals are *poikilothermic* (cold-blooded) so that their body temperature *fluctuates* with that of the environment, nevertheless, their bodies do produce heat energy as a result of muscle action and other physiological processes.

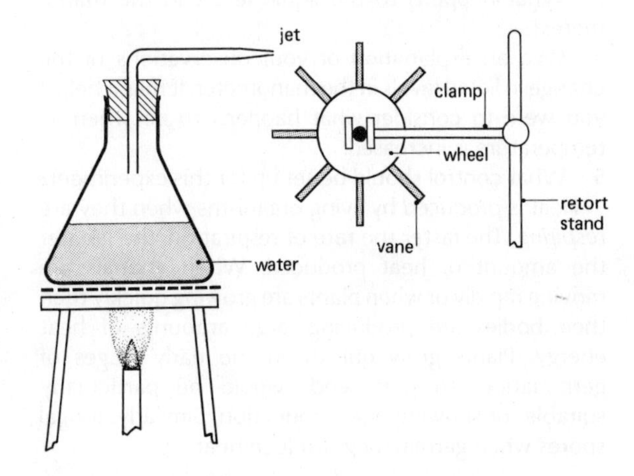

fig 5.14 A simple machine used to demonstrate the transference of energy

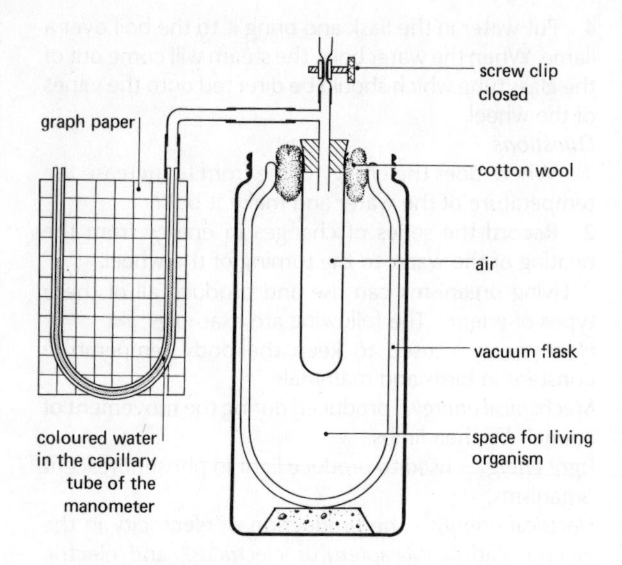

fig 5.15 The production of heat by small animals

Experiment

Do small cold-blooded animals produce heat?

Procedure

1 Set up the apparatus as shown in fig 5.15. (Since the amount of heat produced is very small the vacuum flask is necessary to prevent a large proportion leaking away.)

2 Introduce into the vacuum flask a number of small organisms such as caterpillars or fly larvae.

3 Replace the test-tube and the screw clip; open the clip to equalise the pressure. Close the clip; the liquid levels in the manometer should then be equal. Mark the levels.

4 Leave the apparatus and inspect it at 5 minute intervals.

Questions

3 What happens to the liquid levels in the manometer?

4 Give an explanation of your observations of the change in liquid levels in the manometer. It might help if you were to consider what happens to air when its temperature is increased.

5 What control should be set up for this experiment?

Heat is *produced* by living organisms when they are *respiring*. The faster the rate of respiration, the greater the amount of heat produced. When animals are moving rapidly or when plants are growing quickly then their bodies are producing large amounts of heat energy. Plants grow quickly in the early stages of germination, so that seeds would be particularly suitable for showing heat production. Similarly, fungal spores when germinating produce heat.

Experiment

Do germinating seeds produce heat?

Procedure

1 Soak pea seeds or maize grains for twenty four hours. There should be enough seeds to fill a 50 cm^3 beaker. Divide the seeds into two equal portions.

2 Half the seeds are killed by placing them in boiling water for about five minutes, allowing them to cool and then washing them in 10% formalin.

3 Obtain two vacuum flasks (thermos flasks). Place the living peas in flask A and the dead peas in flask B. Into each flask place a thermometer (0°C to 110°C). Wedge the thermometer with cotton wool, and then carefully *invert* the flasks so that the bulb of each thermometer is surrounded by seeds. (See fig 5.16). This method allows the reading of the thermometers to be taken more easily.

4 Record the temperature of the seeds in each flask morning and evening every day for one week.

Day	Flask temperature in °C	
	Flask A respiring seeds	Flask B dead seeds
1 morning		
evening		

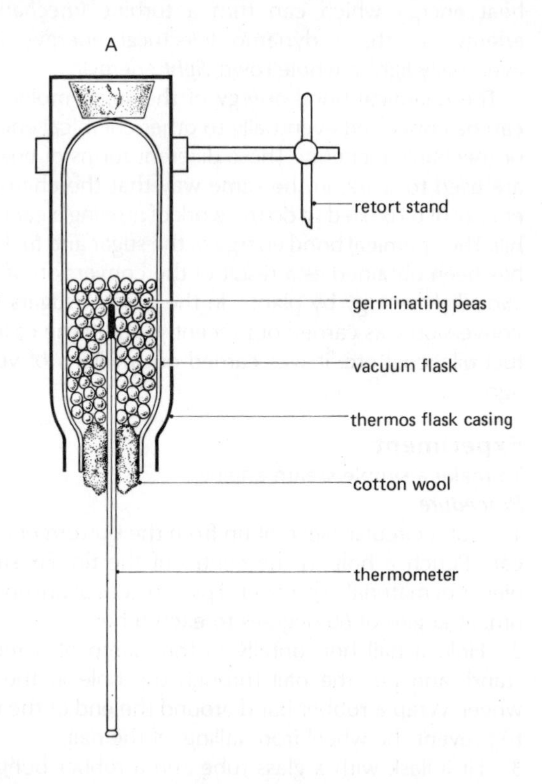

fig 5.16 The production of heat by germinating seeds

5 Draw a graph showing the variations in temperature of the pea seeds in flask A and flask B.

Questions

6 Which flask is the control flask?

7 Why are the boiled seeds washed in 10% formalin?

8 Account for the temperature of the seeds in each flask from the beginning to the end of the experiment.

9 How could you show graphically that living seeds increase in temperature faster than dead seeds?

5.32 Cellular respiration

In section 5.30 it was suggested that the important *breakdown product* of glucose is *energy*. In the presence of oxygen the sugar molecule is broken down completely to the waste products, carbon dioxide and water. This goes on slowly, controlled by many different enzymes, in the cells of living organisms. There is a step by step breakdown of the fuel molecule such that energy is released in small amounts.

Each small quantity of energy is stored in a chemical compound known as ATP (*adenosine triphosphate*). This is formed by the joining of a related molecule ADP (*adenosine diphosphate*) with *phosphate*. The binding of chemical energy into these molecules can be shown as

$$ADP + phosphate + energy \rightarrow ATP$$

Energy in the ATP molecule can be stored in the form of chemical energy and then released restoring the ADP and phosphate.

ATP ⇄ P ⇄ ADP (cycle)

The equation for respiration written in section 5.30 can now be extended as follows:

Aerobic respiration

$$6O_2 + C_6H_{12}O_6 \rightarrow 6\ CO_2 + 6H_2O + Energy$$

$$Energy + ADP + phosphate \rightarrow ATP\ (\times 36)$$

For the oxidation of each glucose molecule *36 ATP molecules* are produced.

In the rapid, uncontrolled oxidation of burning, the sugar molecule releases all of its energy as heat and light. The controlled, step by step, oxidation of glucose in cellular respiration permits most of the energy to be conserved in the form of chemical energy. The high-energy phosphate bonds of ATP can release this energy for the many activities of the cell.

The 36 ATP molecules represent about *39% of the energy available* in the glucose molecule. The conversion figure from the chemical energy of the glucose molecule to the chemical energy of ATP compares very *favourably* with the internal combustion engine. This converts chemical energy in petrol into mechanical energy, and has an efficiency of only 15–30%.

The living cell breaks down complex, energy-rich molecules into simpler energy-poor molecules and this breakdown process is called *catabolism*. In contrast, in Chapter 1, it was seen that small organic molecules enter the cytoplasm and can be used as building blocks to assemble large molecules such as polysaccharides, fats and proteins. This phase of metabolism is called *anabolism*.

The energy released from the ATP can be converted to the different types of energy given in 5.31 (page 123).

Cellular respiration in plants and animals results in the production of carbon dioxide and water. The carbon dioxide liberated by green plants in the presence of light is quickly used up by photosynthesis, and therefore the gaseous exchange of plant respiration can only be investigated in the dark. Photosynthesis is the faster process of the two.

Experiment

Do green plants produce carbon dioxide during respiration?

Procedure

1 Set up the apparatus shown in fig 5.17.

a) Air is drawn through by means of a filter pump (or aspirator).

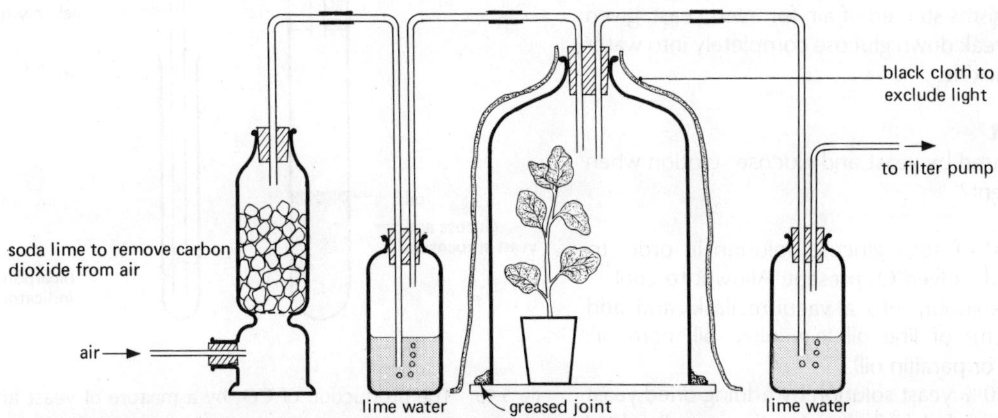

fig 5.17 The experimental apparatus to investigate respiration in green plants

b) The incoming air is passed through soda lime, and then through a flask containing limewater or bicarbonate indicator, before it reaches the plant in the bell jar.
c) After the air has passed over the plant it bubbles through a second flask of limewater (or bicarbonate indicator).
2 Start the air flowing through the apparatus, and observe the indicators.

Questions

1 Why is the bell jar covered with opaque paper or cloth?
2 What is the function of the soda lime?
3 What changes take place in the two flasks of indicator?
4 Do you consider that the plant is respiring?
5 How could you modify the apparatus to show that a small animal such as a mouse is respiring?

5.40 Is energy released without oxygen?

As we have seen in section 5.32, energy can be obtained from complex organic molecules by breaking them into smaller molecules in the presence of oxygen. Where oxygen is not available glucose can be broken down into *ethyl alcohol* (ethanol) and carbon dioxide. This process is called *alcoholic fermentation* and since oxygen is not required, the process is described as *anaerobic*. Fermentation, therefore, is a form of *anaerobic respiration*.

Glucose ⟶ ethyl alcohol + carbon dioxide + energy

Although the process of alcoholic fermentation has been used by man throughout history, it was only about 100 years ago that its true nature was first understood. The French scientist Louis Pasteur found that the process was always associated with the presence of a simple fungus, yeast. He concluded that fermentation was an energy-producing mechanism used by organisms starved of air, for even yeast, given oxygen, will break down glucose completely into water and carbon dioxide.

Experiment

Is energy released by yeast and glucose solution when oxygen is absent?

Procedure

1 Boil 50 cm³ of 10% glucose solution in order to drive out any dissolved O_2 present. Allow it to cool.
2 Pour the solution into a vacuum flask, and add about 5–10 cm³ of fine oil (e.g. olive oil, corn oil, groundnut oil or paraffin oil).
3 Mix up a 10% yeast solution by adding dried yeast to 50 cm³ of cooled boiled water (free of dissolved oxygen). By means of a pipette add the yeast solution, below the oil layer, to the sugar solution.
4 Place a thermometer into the flask and plug the opening with cotton wool. The thermometer should read to one fifth of a degree between 0 and 50°C. (See fig 5.18).
5 Set up a control flask with 100 cm³ of cooled boiled water and an oil layer. Insert a thermometer into the flask, and plug with a cotton wool bung.
6 Record the temperature of each flask at the beginning of the experiment. Record the flask temperatures every hour for one day.

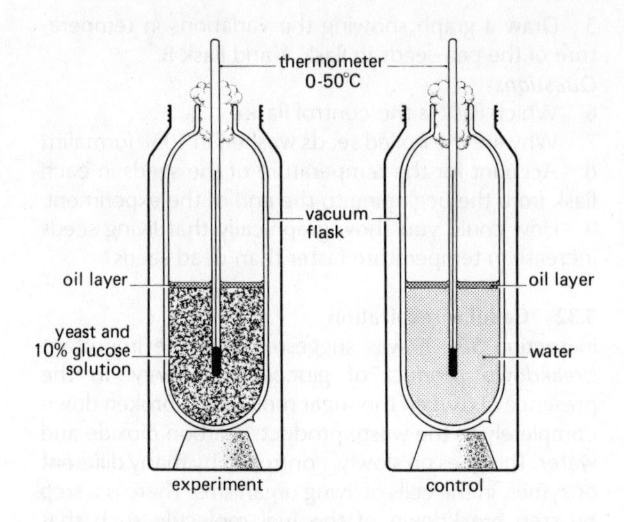

fig 5.18 The production of energy by a mixture of yeast and glucose

Questions

1 Why is it necessary to carry out this experiment in vacuum flasks?

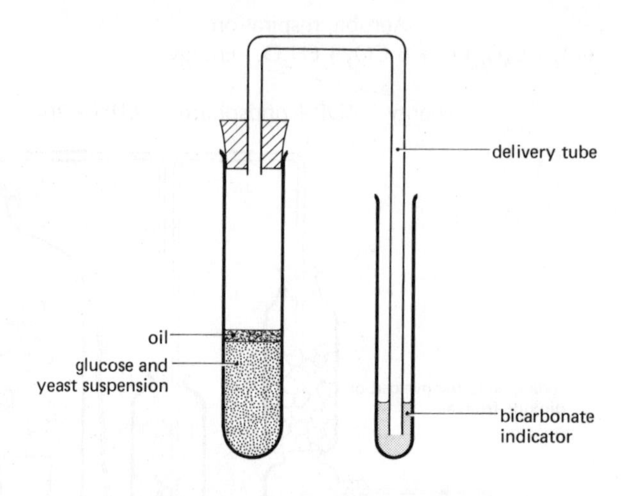

fig 5.19 The production of CO_2 by a mixture of yeast and glucose

2 Is there any evidence that energy is released by activity of the yeast?
3 Why is the surface of the liquid in each flask covered with a light oil?

Experiment

Does the anaerobic respiration of yeast produce carbon dioxide?

Procedure

1 Boil 20 cm³ of glucose solution to drive off any dissolved O_2. Allow it to cool.
2 Pour the solution into a large test-tube and add a small quantity of light oil.
3 By means of a pipette add 10 cm³ of 10% yeast suspension below the oil layer.
4 Connect the test-tube by means of a glass delivery tube to a smaller test-tube containing bicarbonate indicator. The large test-tube must be completely sealed by a rubber bung (see fig 5.19).
5 Leave the apparatus for 1 to 2 hours, and then observe the two test-tubes.

Questions

4 What do you observe in each test-tube?
5 Has any colour change taken place in the bicarbonate indicator? What does this indicate?
6 What other product could be present in the fermenting glucose?

Yeast is used commercially for *both* of the chemical products of anaerobic respiration.

i) **Alcohol production** – The alcohol present in beers and wines is produced with the aid of yeast. This small fungus occurs naturally on leaves and fruits. Any crushed fruits containing sugars or starches, if left to stand in a warm atmosphere, will ferment. In the commercial production of alcoholic drinks, yeast is added to the cereal or fruit, and the fermentation is carefully controlled.

ii) **Bread production** – The carbon dioxide produced by yeast added to a dough mixture, (made from starch-containing flour), causes the mixture to rise. The bubbles of carbon dioxide lighten the heavy starch mixture, and when the dough is baked the yeast dies, but the gas bubbles remain as small holes in the bread produced.

Green plants normally have plenty of oxygen for respiratory processes, but germinating seeds in the soil may be flooded with stagnant water or sealed by clay particles so that oxygen is not readily available, and germination will be delayed or prevented.

Experiment

Do germinating seeds respire anaerobically?

Procedure

1 Soak 20 pea or other seeds in water for twenty-four hours. At the end of this time divide them into two groups. Place one group into boiling water for five minutes and then soak in 10% *formalin*.

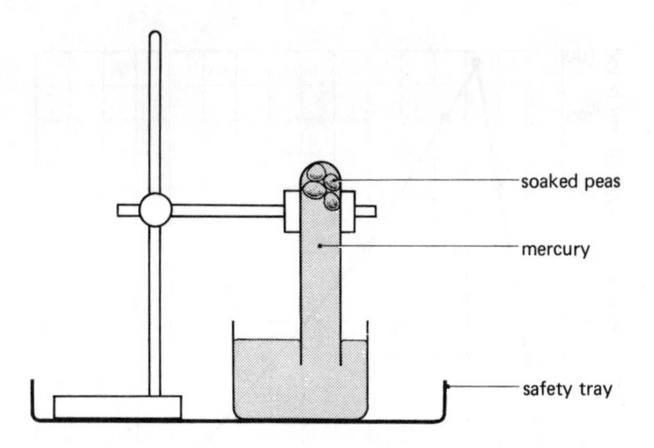

fig 5.20 Anaerobic respiration in germinating seeds

2 Invert a small test-tube of *mercury* into a small dish of mercury placed in a tray (this is essential to avoid loss of spilled mercury). Clamp the tube firmly (see fig 5.20).
3 Introduce 5 or more soaked peas or other seeds into the bottom of the test-tube so that they rise to the top of the mercury in the tube.
4 Set up an identical second apparatus using the boiled (*sterilised*) peas instead of the soaked peas. Leave both sets of apparatus for several days and observe.

Questions

7 Why were the control seeds boiled and sterilised in formalin?
8 What differences between the two tubes were observable after several days?
9 What tests would you perform to attempt identification of the gas produced?
10 What evidence have you that germinating peas respire anaerobically?

5.41 Do animals release energy without using oxygen?

Another type of anaerobic respiration is known in animals. When we take vigorous exercise the activity of the muscles *outstrips* the *supply* of oxygen available for cellular respiration. Under these conditions, glucose is broken down in the muscles into *lactic acid* with the release of a very limited amount of energy. The glucose is derived from glycogen, which is stored in the skeletal muscles.

$$\text{Glucose} \rightarrow \text{lactic acid} + \text{energy}$$

The lactic acid must be removed to avoid harming the body. The removal is slow, but steady, as the lactic acid is rebuilt into glucose and glycogen, or broken down completely to water and carbon dioxide. In the latter case the breakdown uses oxygen which becomes available after the exercise is finished. Examine fig 5.21 which shows the amounts of lactic acid in the blood of

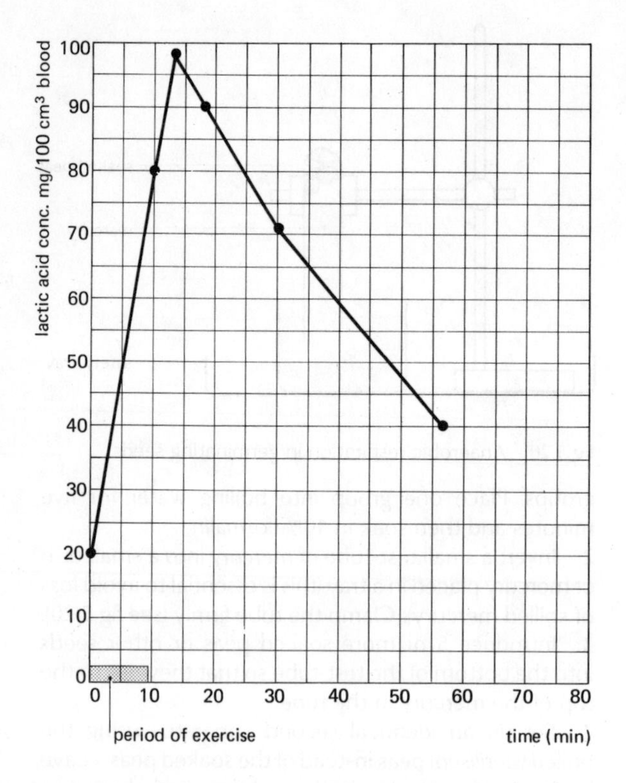

fig 5.21 A graph showing concentration of lactic acid in the blood after exercise *From Nuffield Biology Text III*

a man who has carried on some vigorous exercise for a period of nine minutes.

Questions

1 By how much did the lactic acid increase during the period of exercise?

2 What was the concentration of lactic acid 57 minutes after commencing the exercise?

3 How could you find out the time it takes for the blood concentration of lactic acid to reach its original value of 20 mg/100 cm³ of blood?

During vigorous use of muscles, as in a 100 metres sprint, it is not possible to provide the energy required for contraction by means of aerobic respiration. This is because the muscular work requires more oxygen than the lungs can take in, and more than the blood can carry, during the exercise. The body is then said to be in *oxygen debt*. In this condition we are breathing very deeply and quickly, with the lungs taking in great gulps of air. After the exercise is finished the breathing rate gradually slows, but the extra oxygen needed to oxidise the lactic acid is taken in over a long period (see fig 5.22).

5.50 Amounts of energy released in respiration

Both alcohol and lactic acid production during anaerobic respiration are *inefficient* processes. Only a small proportion of the energy stored in glucose is made available during fermentation. Most of the energy originally in the glucose molecule is still stored in the end products of ethanol or lactic acid. The high energy content of ethanol is demonstrated by the fact that it may be used as a fuel component in racing cars and rocket engines. The amounts of energy released are as follows (see Chapter 1, section 1.40):

Anaerobic respiration

$$C_6H_{12}O_6 \rightarrow \text{ethanol} + CO_2 + 0.22 \times 10^6 \text{J}$$

Aerobic respiration

$$6O_2 + C_6H_{12}O_6 \rightarrow 6CO_2 + 6H_2O + 2.81 \times 10^6 \text{J}$$

$$\text{Thus only } \frac{0.22 \times 10^6 \text{J}}{2.81 \times 10^6 \text{J}} \times 100$$

= 7% approximately of the total energy released by the complete oxidation of glucose is available in anaerobic respiration. Organisms that rely upon fermentation must have large amounts of glucose available to them in order to satisfy their needs. Other carbohydrates could be broken down by enzymes.

Not only is fermentation an inefficient method of obtaining energy but it is potentially a dangerous one. Ethanol is poisonous in moderate quantities. The anaerobic respiration of yeast plants ceases when the concentration of the ethanol reaches about 14%. The yeast can no longer live at this concentration of alcohol, and so wines have a natural limit set on their alcoholic content.

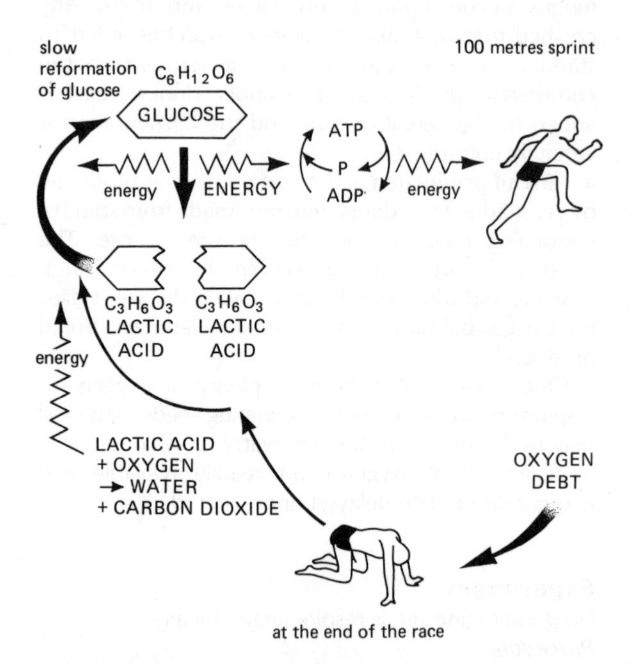

fig 5.22 A diagram of the release of energy in the muscles of a man running a 100 metres sprint

There are very few organisms which rely entirely on anaerobic respiration for their energy requirements. These organisms that can respire in no other way, are very simple in structure and the *tetanus bacterium* which lives in the soil is one such organism.

Summary of energy flow in respiration

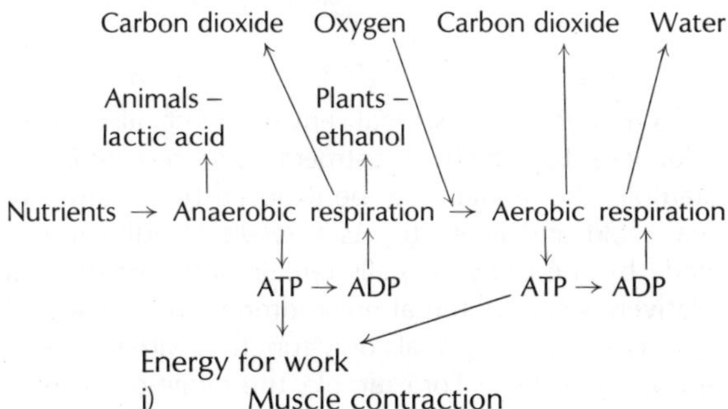

Energy for work
- i) Muscle contraction
- ii) Heat production
- iii) Light production
- iv) Transport
- v) Synthesis of complex compounds – anabolism

Respiration	Photosynthesis
1 Occurs in all living cells of plants and animals.	Occurs only in plants containing the green pigment chlorophyll.
2 Goes on at all times.	Only occurs in the light.
3 Uses oxygen, but the process can occur without this gas.	Carbon dioxide is needed as a raw material.
4 Carbon dioxide is produced.	Oxygen is produced.
5 Water is produced.	More water is used up than is produced. (There is a net gain of water.)
6 Energy is produced.	Energy of sunlight is absorbed by the chlorophyll and stored in complex organic molecules.
7	Proceeds at a much faster rate than respiration in green plants, in terms of gaseous exchange.

Table 5.5 Comparison of Respiration and Photosynthesis

Questions requiring an extended essay-type answer

1 a) Describe an experiment by which you could prove that a green plant or an animal produces carbon dioxide during respiration.
b) Why is there not a continuous increase in carbon dioxide in the atmosphere?
c) What processes in plants and animals balance the quantity of oxygen in the atmosphere?

2 a) Describe, with the aid of diagrams, the structure and position of the trachea, bronchi, lungs and diaphragm of a mammal.
b) How would you confirm experimentally the truth of the following statements
i) exhaled air is warmer than inhaled air, and
ii) exhaled air is moister than inhaled air?

3 a) Explain how the following animals exchange gases with the atmosphere
i) an insect, and
ii) a fish.
b) Compare and contrast the methods by which the following exchange gases with the atmosphere
i) a green plant, and
ii) a woody twig.

4 a) What do you understand by *respiration*?
b) Describe an experiment to show that the process of respiration occurs in micro-organisms.
c) Describe the mechanism by which a fish obtains the oxygen for respiration.

5 a) Give an equation to show that respiration involves the oxidation of a foodstuff.
b) Describe an experiment to show that this process releases energy in the form of heat. Your experiment can use either plant or animal material.

6 a) Define anaerobic respiration.
b) Describe briefly how and where anaerobic respiration occurs in
i) plants, and
ii) animals.
c) Describe briefly an experiment to show that anaerobic respiration in a plant produces carbon dioxide and energy.
d) What is the immediate source of energy in muscles? How are energy-rich molecules rebuilt after muscular contraction?

6 Excretion and regulation

6.00 Regulation of body functions

The ability of complex living organisms to maintain a stable or *constant internal environment* is called *homeostasis*. The concept of the internal environment was first developed in about 1850 by a French physiologist, Claud Bernard. He realised that the ability to maintain a constant internal environment makes animals more independent of the external environment. In order to attain this internal stability, however, the animal must be able to react to any changes in the external environment which could change the internal one. Food, water and oxygen are taken into the body to repair tissues and to provide energy for growth. As a consequence of their intake, regulatory mechanisms come into action which maintain the correct balance of nutrients in the tissues. Waste materials are produced by all metabolic activities and the levels of these substances must also be controlled.

Factors in the external environment also exert influences to which adjustments are required. For example, the animal may be exposed to extremes of heat, cold and light etc. As a result of adjustments made by the body, its cells remain surrounded by a relatively steady internal environment. The bodies of Man and other mammals only function within *narrow physiological limits*. For example, the range of 'normal' body temperature in Man is quite small (see fig 6.2). Small fluctuations outside of this range can result in death.

Let us consider three steady state mechanisms which affect the internal environment of the body.

1 Breathing mechanisms and gaseous exchange (Chapter 5).

2 Temperature control.

3 Water and ionic control – *excretion*.

Each of these body functions are kept at a constant level by a homeostatic system of the kind shown

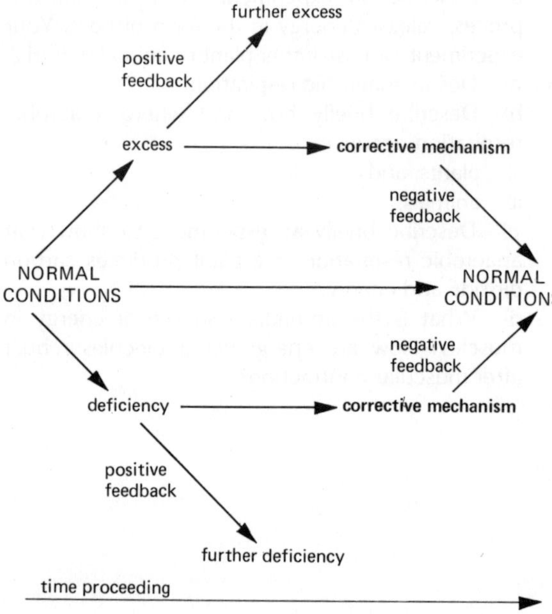

further excess
positive feedback
excess
corrective mechanism
negative feedback
NORMAL CONDITIONS
NORMAL CONDITIONS
negative feedback
deficiency
corrective mechanism
positive feedback
further deficiency
time proceeding

fig 6.1 Homeostasis in its simplest form

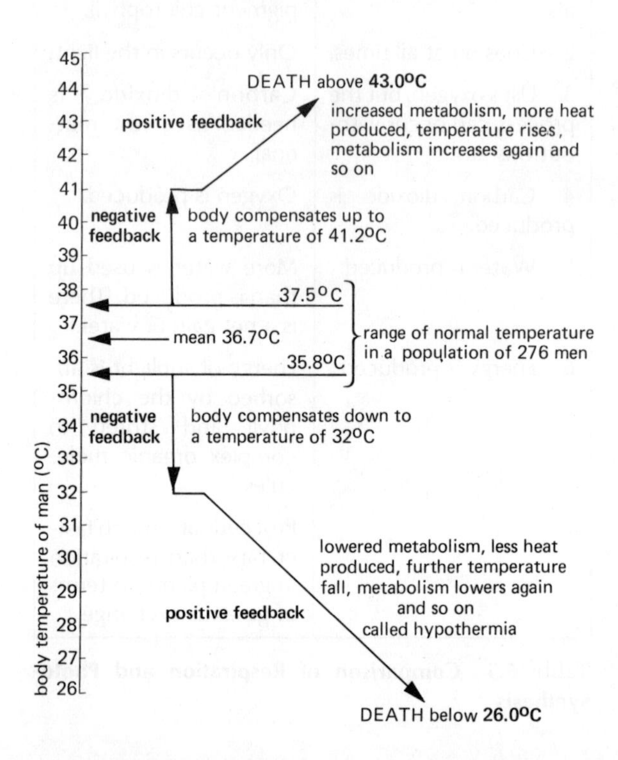

fig 6.2 Normal and abnormal ranges of temperature in Man

diagrammatically in fig 6.1. Any change from the normal operating levels by increase or decrease brings into operation a compensatory change to return the levels to normal. This change is called *negative feed-back*, since it must lessen the effect of the original change. Positive feed-back is rare in living organisms under normal conditions, but when it does occur it accentuates the change and is very difficult to reverse. In fig 6.2 the body of Man, at its danger limits of temperature, is unable to compensate for higher or lower temperatures, and death results.

6.10 Temperature control

A constant warm internal temperature is an advantage to an animal since it permits efficient functioning of the body. In warm-blooded animals with a constant body temperature a heat balance is maintained. Cold-blooded animals have their body temperature determined by the environment. The terms 'cold-blooded' and 'warm-blooded' are not really appropriate because a cold-blooded lizard sitting on a rock in the sun may have a higher body temperature than a warm-blooded man. When night comes, however, the lizard's temperature will fall, but the man's will remain much the same.

The best means of distinguishing between the two types of animal is to use the terms:

homoiothermic – indicating a constant body temperature as in mammals and birds, and
poikilothermic – indicating a variable body temperature as in reptiles, amphibia, fish and invertebrates.

Fig 6.3 indicates the difference between the two states. The graphs show the body temperature of students (homoiotherms) and a frog (poikilotherm) plotted against environmental temperature. Examine the graphs.

Questions

1 What happens to the body temperature of the frog as the external temperature increases?

2 What happens to the body temperature of the students as the external temperature increases?

3 From the graph, what can you deduce about the activity of the students and the frog at night time? What advantage does this give one animal over the other in respect of night activities?

Examine fig 6.3 and its graph.

4 Man is a homoiotherm, and all men belong to a single species. What relevance have these facts with regard to the distribution of Man on the earth's surface?

Examine the graph in fig 6.4.

5 What happens to the body temperature of the insect larvae as the external temperature increases?

6.11 Poikilotherms

Fig 6.5 shows that the sand lizard takes its heat from the environment. Homoiothermic animals work most

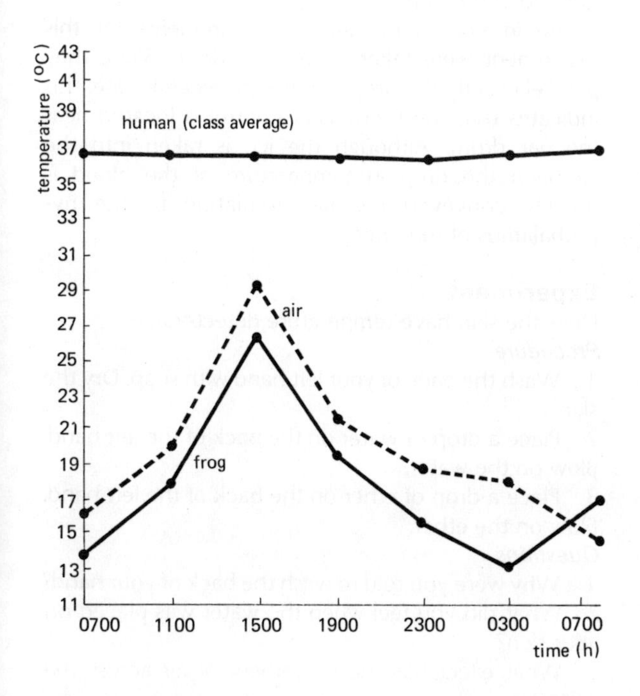

fig 6.3 A graph of the body temperature of a frog and a group of students, measured over a 24-hour period

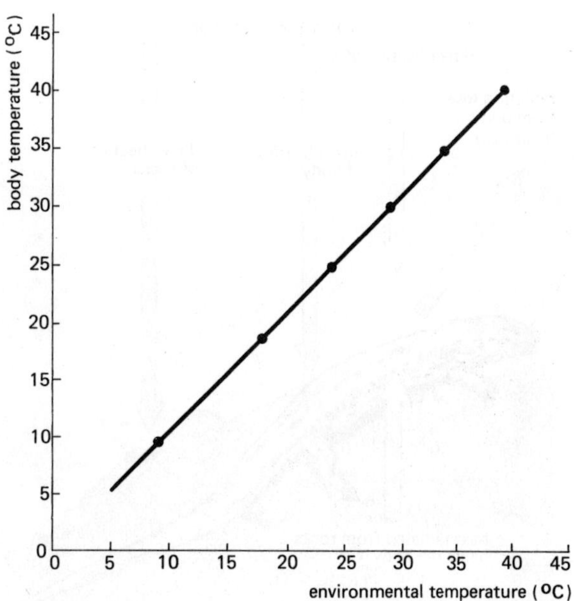

fig 6.4 The temperature of insect larvae, under different temperature conditions

efficiently at a body temperature of 37°C. This is also true of poikilotherms. In cold conditions, where the external temperature is below 20°C, poikilotherms are very sluggish and inactive. At night and in the early morning, the temperature may be too low for the lizard to be active. As soon as the sun appears the animal moves into the sunlight, so that heat flows into the body directly from the sun, and also indirectly from the heated rocks or walls surrounding the animal. The body warms up and the lizard becomes very active.

It could not remain for long exposed to the sun because its temperature would rise, there would be positive feed-back (see fig 6.1) and it would eventually reach a lethal temperature. In the hottest part of the day therefore, the lizard hides in the shade, and heat flows from its body to the cooler rocks.

Fig 6.4 shows that insects also adopt the temperature of the environment. Thus in temperate climates such as Europe and North America with a well-defined winter, few insects are seen at this time. Poikilotherms under such conditions must *hibernate or die*; insects often survive in a developmental stage, the *pupa*. The adult insect hatches from this stage on the return of warmer conditions. In the tropics, butterflies flutter their wings in the morning in order to increase their body temperature sufficiently to fly at an earlier time in the day than would be otherwise possible at that environmental temperature. This could be described as a 'warming-up' exercise.

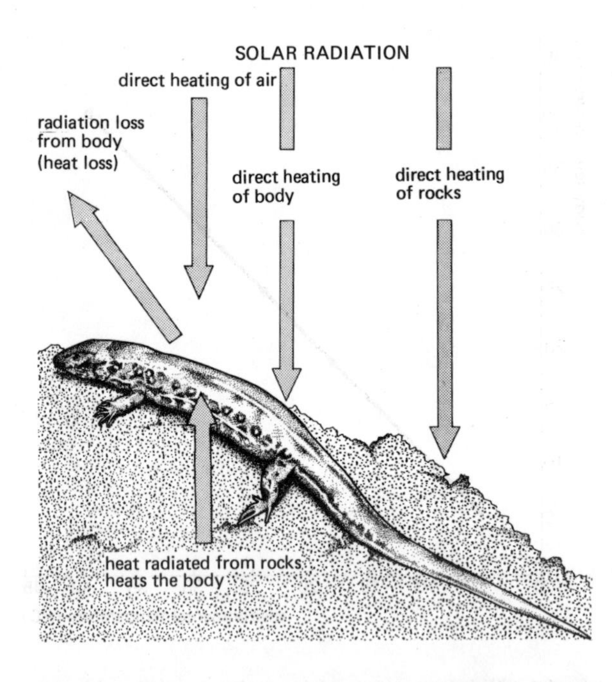

fig 6.5 The sand lizard in the sun increasing its body temperature

Crocodiles are often seen to lie on sand banks in the hot sun with their mouths open. This is a method of cooling the body which is also shown by some lizards. The metabolism of the butterfly and the crocodile, although poikilothermic, is not entirely conditioned to environmental temperatures. Both animals are able to maintain some control over their internal environments and thus show a link between poikilothermy and homoiothermy.

6.12 Homoiotherms

Body temperature in a mammal is controlled by certain cells in the *hypothalamus* of the brain. Very slight rises in temperature result in this part of the brain discharging nerve impulses that set in action mechanisms to *cool the body*. The result is that the blood is cooled and the feed-back 'switches' off the hypothalamus. Examine fig 6.6 and you will see that as the ice is swallowed the internal body temperature falls from 37.6°C to 37°C but regains its value after 15 minutes. At the same time the skin temperature rises steadily to 37.6°C. Notice, however, that the rate of sweating follows the internal temperature, first dropping and then rising. The explanation of these events must be that the rate of sweating falls to compensate for the falling internal temperature, but the skin temperature rises at the same time because of heat absorption from the hot atmosphere (45°C). These facts indicate that the thermostat controlling sweating is not situated in the skin.

The internal temperature measurements for this experiment were taken at the ear drum. Since they parallel exactly the drop and rise in sweating rate, this indicates that the thermostat control is located near the ear drum. Although the ice is taken into the stomach the drop in temperature of the blood is quickly conveyed by the circulation to the hypothalamus of the brain.

Experiment

Does the skin have temperature detectors?

Procedure

1 Wash the back of your left hand with soap. Dry the skin.

2 Place a drop of water on the back of the left hand. Blow on the water.

3 Place a drop of ether on the back of the left hand. Blow on the ether.

Questions

1 Why were you told to wash the back of your hand?

2 What did you feel when the water was placed on your skin?

3 What effect has the movement of air across the water?

4 How do the ether and water differ in their effect on the skin?

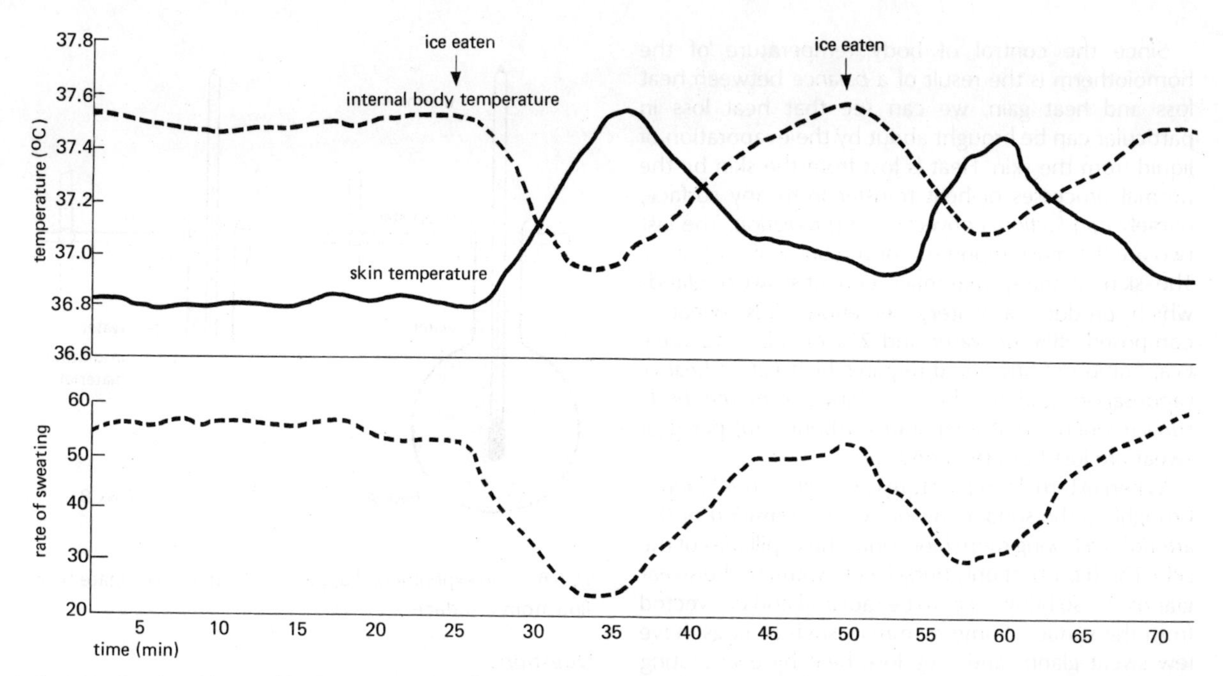

fig 6.6 Graphs of internal body temperature, skin temperature and rate of sweating of a man, in a temperature controlled at 45°C

5 How do you account for this difference?
6 Why is a fan useful in a hot climate?

We can examine this phenomenon more quantitatively by the following experiment.

Experiment

Does the evaporation of a liquid use up heat energy?

Procedure

1 Boil 500 cm³ of water and place it in a round bottom flask clamped onto a retort stand. The flask should be fitted with a rubber bung through which passes a thermometer (0°C to 110°C). The bulb of the thermometer should be in the centre of the flask. See fig 6.7.

2 Record the temperature of the water at the time the thermometer is fixed in the flask.

3 Repeat the temperature readings every minute for 10 minutes, recording the time and temperature on each occasion.

4 Now wipe the surface of the flask with ethanol continually for 5 minutes. Record the time and temperature at one minute intervals during this operation.

5 When the wiping has stopped, continue recording the temperature every minute for another 10 minutes.

6 Using your results, plot a graph of temperature (vertical axis) against time (horizontal axis).

7 Calculate the rate of fall of temperature as follows:

Fall of temperature over 10 minutes $= A°C$

Therefore rate of fall of temperature $= \frac{A°C}{10}$ per minute

Calculate the three rates of temperature fall for (a) the first 10 minutes, (b) the next five minutes when alcohol was on the flask, and (c) the final 10 minutes.

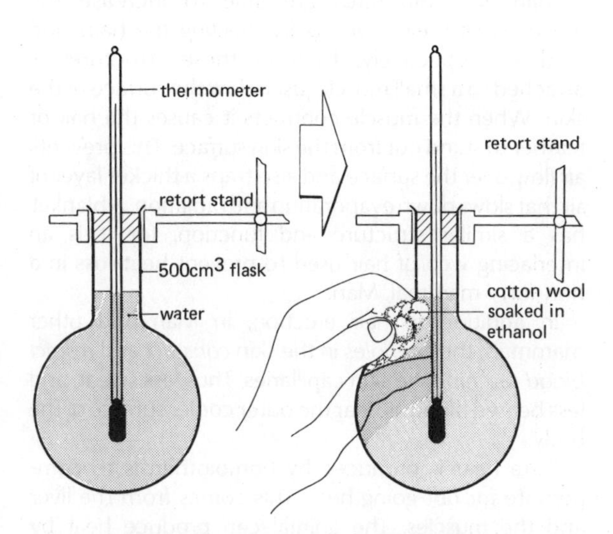

fig 6.7 The experimental apparatus used to investigate the effect of evaporating liquids on cooling water

Questions

7 What was the rate of temperature fall for (a), (b) and (c)?

8 What causes any differences between the rates for (a), (b) and (c)?

9 How could you slow down the rate of fall of temperature in the flask?

Since the control of body temperature of the homoiotherm is the result of a *balance* between heat loss and heat gain, we can see that heat loss in particular can be brought about by the evaporation of liquid from the skin. Heat is lost from the skin by the normal processes of heat transfer from any surface, namely *conduction, convection* and *radiation*. The last two are the most important for the body of an animal. The skin of many mammals contains sweat glands which produce a watery secretion. This sweat is composed 98% of water and 2% of salts and urea. Evaporation of any liquid requires heat (*latent heat of vaporisation*) and the heat is drawn *from the body surface* when sweat evaporates. About 140J per g of sweat are lost from the body.

When external temperatures are high, *more blood* is brought to the surface; owing to the *relaxation* of the *arterioles* allowing more blood into the capillaries of the skin. This blood not only brings more water to the sweat glands, it also brings heat to be radiated and convected from the surface. Some mammals such as dogs, have few sweat glands and they lose heat by evaporating saliva from their tongues during the activity called *panting*. Man can lose heat by:

1 **Behavioural methods** – wearing fewer clothes, taking cold baths or swims, drinking cold drinks, using a fan or air conditioning.

2 **Physiological methods** – producing more sweat which evaporates freely (latent heat loss), and the body can bring more blood to the skin (heat loss by radiation and convection).

6.13 Retaining heat in homoiotherms

We have considered the loss of heat in homoiotherms, but their major problem in conserving a heat balance is to prevent heat passing to the environment. Their surroundings are seldom hotter than body temperature, so that there will be a tendency for heat to flow from the external surfaces of the body.

Experiment

How can heat flow from a surface be slowed down?

Procedure

1 Take two round-bottom 500 cm^3 flasks and cover one (flask B) with an insulating material (e.g. cotton wool, a piece of old blanket or fur).

2 Into each flask, fit a rubber bung and a thermometer as in the experiment in section 6.12 (see fig 6.8).

3 Pour 500 cm^3 of boiling water into each flask. Replace the bung and thermometers, and record the initial temperature of the water.

4 Record the temperature of each flask in the form of a table every minute for twenty minutes. Plot a graph of temperature against time for each flask.

5 Calculate the rate of fall of temperature for the liquid in each flask.

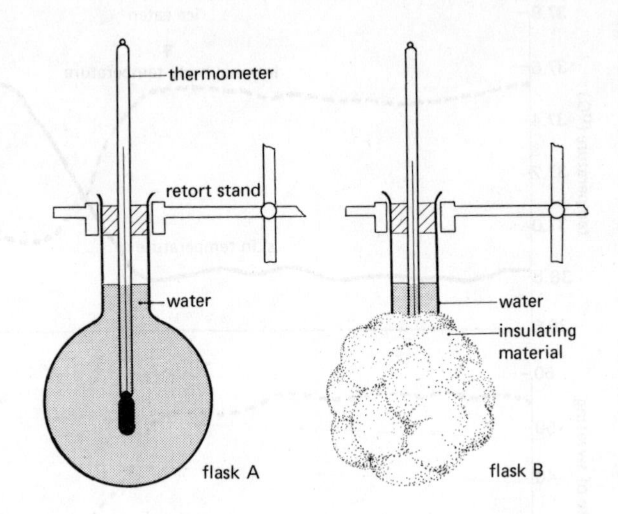

fig 6.8 The experimental apparatus used to investigate heat flow from a surface

Questions

1 What was the rate of fall of temperature for flask A and flask B in fig 6.8?

2 What deductions can you draw from the graph and your calculations?

Mammals and birds, which have constant body temperatures, are insulated against heat loss by the special properties of their skin.

i) Mammals are covered with *hair*.

ii) Birds are covered with *feathers*.

iii) Mammals and birds have an *insulating layer of fat* beneath the skin. This layer becomes extremely thick in homoiotherms living in cold climates e.g. whales, seals and penguins.

Mammals and birds are able to increase the thickness of their covering by *erecting* the hairs and feathers respectively. Each of these structures is attached to a small muscle just below the surface of the skin. When the muscle contracts it causes the hair or feather to stand out from the skin surface. This prevents air flow over the surface and also traps a thicker layer of air that slows down evaporation and radiation. A blanket has a similar structure and function, for it is an interlacing web of hair used to prevent heat loss in a non-furry mammal, Man.

In addition to hair erection, in Man and other mammals, the *arterioles* in the skin *constrict* and *restrict blood reaching* the skin capillaries. Thus less sweat, and less heated blood, is near the outer cooler surface of the body.

Extra heat is produced by homoiotherms to compensate for out-going heat. This comes from the liver and the muscles. The animal can produce heat by *involuntary contraction* of the *skeletal* muscles (*shivering*). Man can take vigorous exercise to gain extra heat.

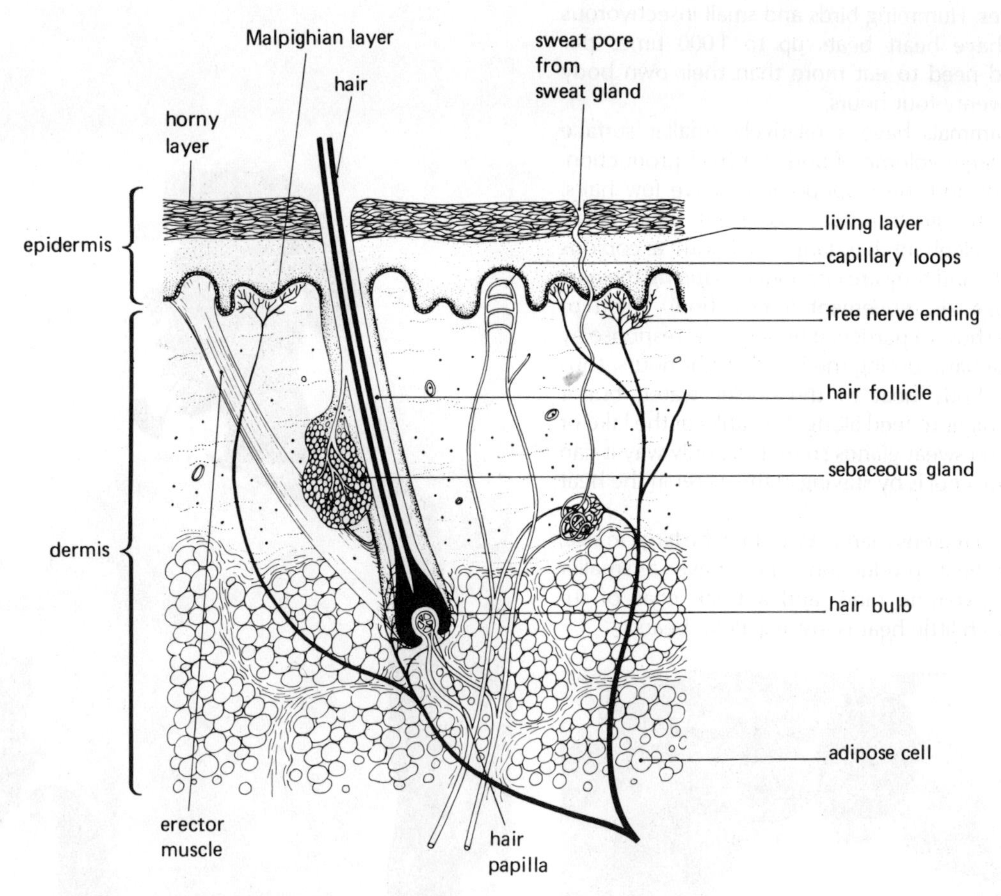

fig 6.9 A vertical section through the skin of a mammal

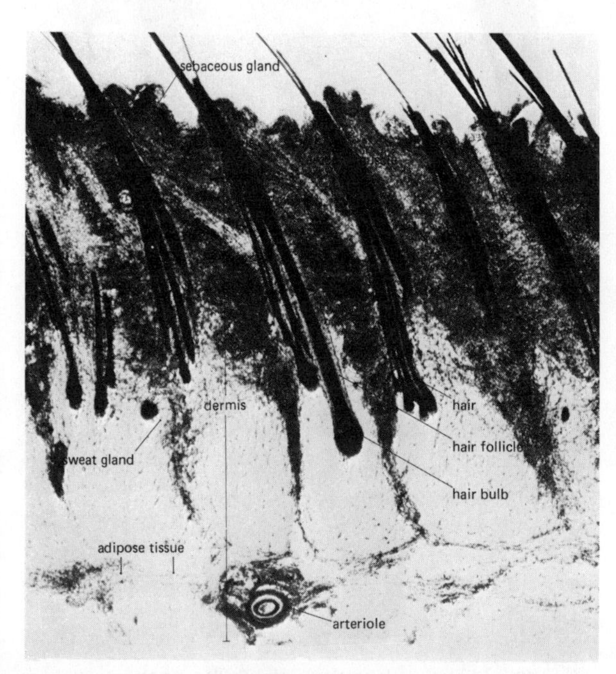

fig 6.10 A vertical section through the skin of a dog

Methods available to Man for slowing down heat loss:

1 **Behavioural methods** – wearing more clothes, drinking hot drinks or eating hot food, heating his house, taking more exercise to produce heat.

2 **Physiological methods** – involuntary body reactions such as shivering, producing less sweat, and bringing less blood to the surface, thus conserving the heat in the inner regions of the organism.

6.14 Heat loss and body size in homoiotherms

In Chapter 2 you have discovered that the larger the organism the smaller the surface area/volume ratio. Since heat loss is through the body surface, and heat production is the result of respiration and energy production from the cells in the total volume of the body, this ratio is of vital importance to the animal. Very small animals have a relatively large surface area (high S.A./V. ratio) and find heat conservation a major problem. *Very large animals* in hot climates have the opposite difficulty. They need to *encourage heat loss* in order to keep cool.

Small mammals and birds have a *high metabolic rate* in order to *produce heat* to maintain their body temperatures. Humming birds and small insectivorous mammals have heart beats up to 1000 times per minute, and need to eat more than their own body weight in twenty-four hours.

Large mammals have a relatively smaller surface area, but a large volume of body for heat production. The elephant and the hippopotamus have few hairs. This allows for greater heat loss from their naked bodies. The elephant has large ears with a *copious blood supply*, and they are continually *flapping* the ears to encourage air movement (convection). The hippopotamus shows a particular behavioural response by remaining aquatic during the hot daylight hours, thus conducting body heat to the cooler water, but it emerges at night to feed along the banks of the lake or river. It has no sweat glands so that the only way it can keep its body cool is by staying in the water in the heat of the day.

In *very cold* regions mammals tend to be large in size, for greater heat production. They have a smaller surface area/volume ratio, and a thick coat of fur through which little heat is lost e.g. Polar bears.

fig 6.11 Climatic adaptation – the Eskimo

fig 6.12 Climatic adaptation – the Sudan negro

In Man the relatively hairless skin enables heat loss to occur by radiation, conduction, and convection under conditions of cold external temperature. We have detailed in section 6.13 how this loss can be decreased, but in spite of these methods, the toes, feet, fingers, hands, ears and nose lose heat very rapidly and thus feel cold. It is these parts of the body where 'frostbite' occurs if they lose heat too quickly over long periods. Blood flow is cut off and the tissues die through lack of oxygen, food and heat. If suitable conditions are not present to revive the cells, then the parts of the body so damaged must be amputated.

The adaptation of animals to differing external temperatures can be seen in many species. Examine photographs fig 6.11 and fig 6.12 showing two races of Man adapted to different environments. The Eskimo weighs about 77 kg and his height is 1.64 m. He lives in a very cold climate and his body shows the typical thickset figure, flat features and short limbs of men *adapted to these conditions*. The Sudan negro, however, weighs 58.8 kg and his height is 1.885 m. He is much taller but weighs less than the Eskimo. His thin body and relatively long limbs are an advantageous adaptation to *tropical climates* permitting rapid heat loss from the body.

Examine fig 6.13 which shows three species of fox, one each from a cold, a temperate and a hot climate.

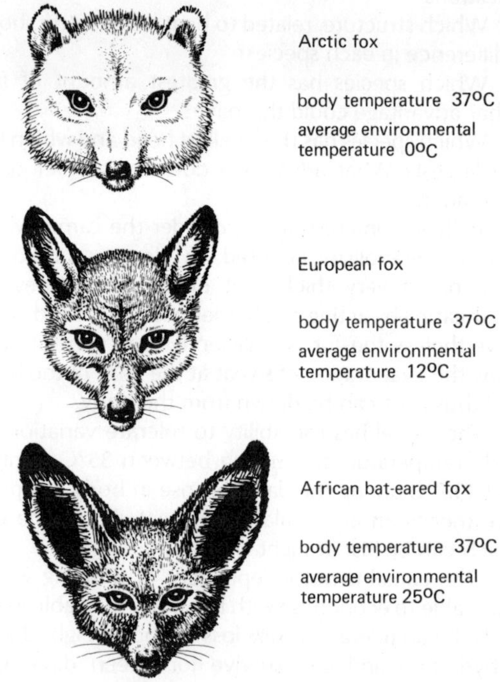

Arctic fox
body temperature 37°C
average environmental temperature 0°C

European fox
body temperature 37°C
average environmental temperature 12°C

African bat-eared fox
body temperature 37°C
average environmental temperature 25°C

fig 6.13 The differences in appearance of three foxes: from the Arctic, Europe and Africa

Methods by which mammals lose heat	Methods by which mammals gain heat
1 Production of sweat increases. The water in the sweat evaporates drawing latent heat from the body.	Production of sweat decreases so that heat lost by evaporation is much less.
2 The water in the saliva evaporates as the animal exhales rapidly – panting.	The jaws are kept closed, breathing is slow and through the nostrils.
3 The whole body is immersed to conduct heat away to the water during the day.	The animal emerges from the water at night when the air is cooler.
4 The hair is lowered making a thinner coat so that heat can escape more easily.	The hair is raised making a thicker coat, thus trapping more air as an insulating layer. In Man the naked skin with its few hairs shows traces of this function. The contracted hair muscles appear as 'goose pimples'.
5 Arterioles relax and more blood enters the capillary network. Extra heat is radiated and convected from the skin – vasodilation.	Arterioles constrict and less blood enters the capillary network. Less heat is lost by radiation and convection – vasoconstriction.
6 A moult occurs making the coat thinner and heat escapes more easily.	A thicker coat of fur is grown in winter. Man can achieve the same result by putting on more clothing.
7 The rate of metabolism decreases so that less heat is produced, and the animal is less active. Man's activity decreases in hot weather.	The rate of metabolism increases producing more heat. Involuntary muscular action also occurs (shivering). Man can exercise specifically to produce heat.

Table 6.1 Methods by which animals lose and gain heat

Questions

1 Which structure, related to heat loss, clearly shows a difference in each species?

2 Which species has the greatest amount of fur? What advantage could this have?

3 Which species has the smallest head and which has the largest? What advantage does this confer upon each animal?

Finally it is interesting to consider the *camel* which has a metabolism adapted to desert conditions.

i) It has a very thick coat of fur which prevents overheating by acting as a barrier to the sun's radiation. Nevertheless the fur is sufficiently well ventilated to allow the evaporation of sweat at the base of the hairs, and thus heat can be drawn from the body.

ii) The camel has the ability to tolerate variations in body temperature fluctuating between 35°C at night and 40.7°C during the day. This rise in body temperature means an accumulation of heat during the day which can be lost at night.

iii) The camel is not independent of drinking water, but is able to economise with the water available in the body. It can tolerate a 30% loss in body weight due to dehydration and can survive for fifteen days at a temperature of 41°C without drinking. Under these conditions a Man would die in two days, and a sheep in six days. The camel is further adapted by having a *low urine flow, little water loss in faeces* and unusually *large drinking capacity*. On one occasion, a dehydrated camel weighing 325 kg drank 103 litres of water in less than ten minutes.

6.15 Other functions of the mammalian skin

1 *Protection* The cornified layer of the epidermis (see fig 6.9) consists of dead cells which provide an outer protection layer. This can perform the following functions

a) Prevention of water loss.

b) Prevention of frictional damage. This is particularly noticeable on the soles of the feet and palms of the hand in Man, where the cornified layer becomes considerably thicker than in the rest of the body.

c) Prevention of entry of bacteria. Micro-organisms can still enter through the skin through sweat pores, hair follicles or skin wounds.

d) Protection against radiation from the sun by the development of a black or dark-brown pigment, melanin. This is hereditary in dark-skinned races, but acquired as a result of exposure by light-skinned races.

e) Waterproofing oil from the sebaceous glands prevents wetting of the fur in aquatic mammals such as the otter.

f) Insulation. Fat layers under the skin prevent heat loss.

g) Camouflage. The hair pigment and its patterns in the fur can protect mammals in their natural habitat by blending with the surroundings e.g. rabbits in ploughed fields or in hedgerows. The colour pattern can be very distinctive but nevertheless break up the outline of the animal e.g. tigers, leopards and zebras. The pigment can change seasonally in certain mammals, such as stoats, which have brown hair in the summer and a white coat in the winter.

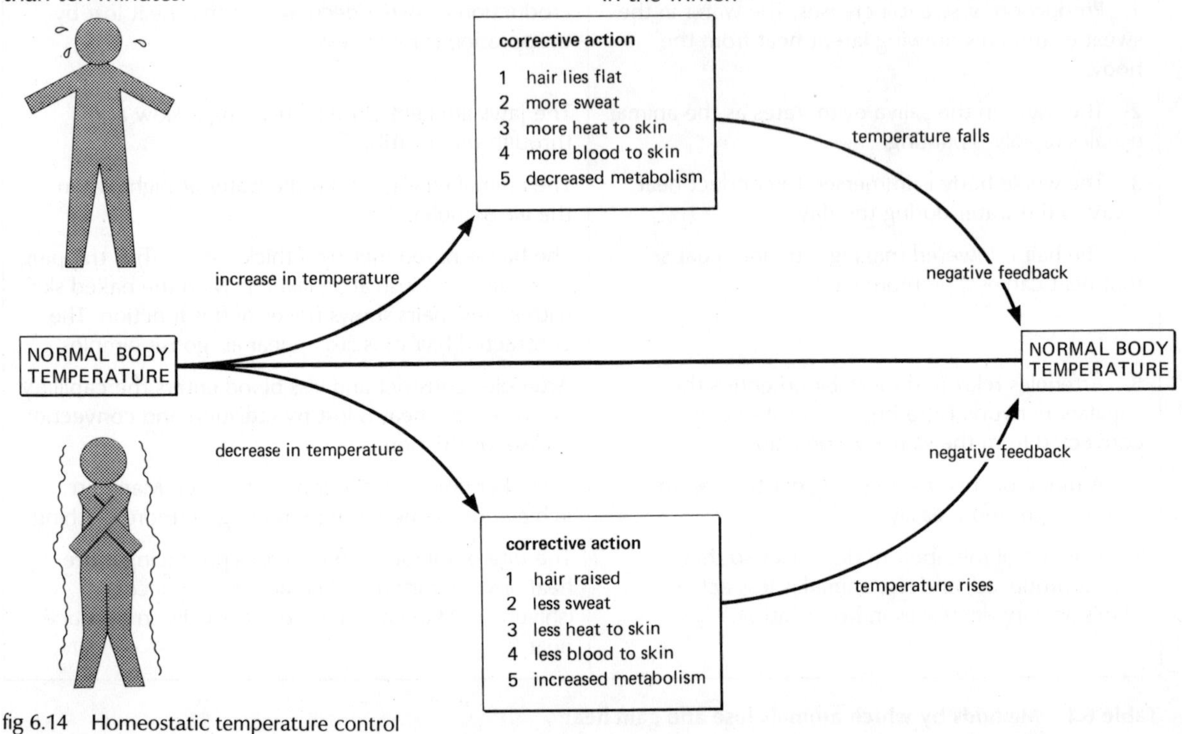

fig 6.14 Homeostatic temperature control

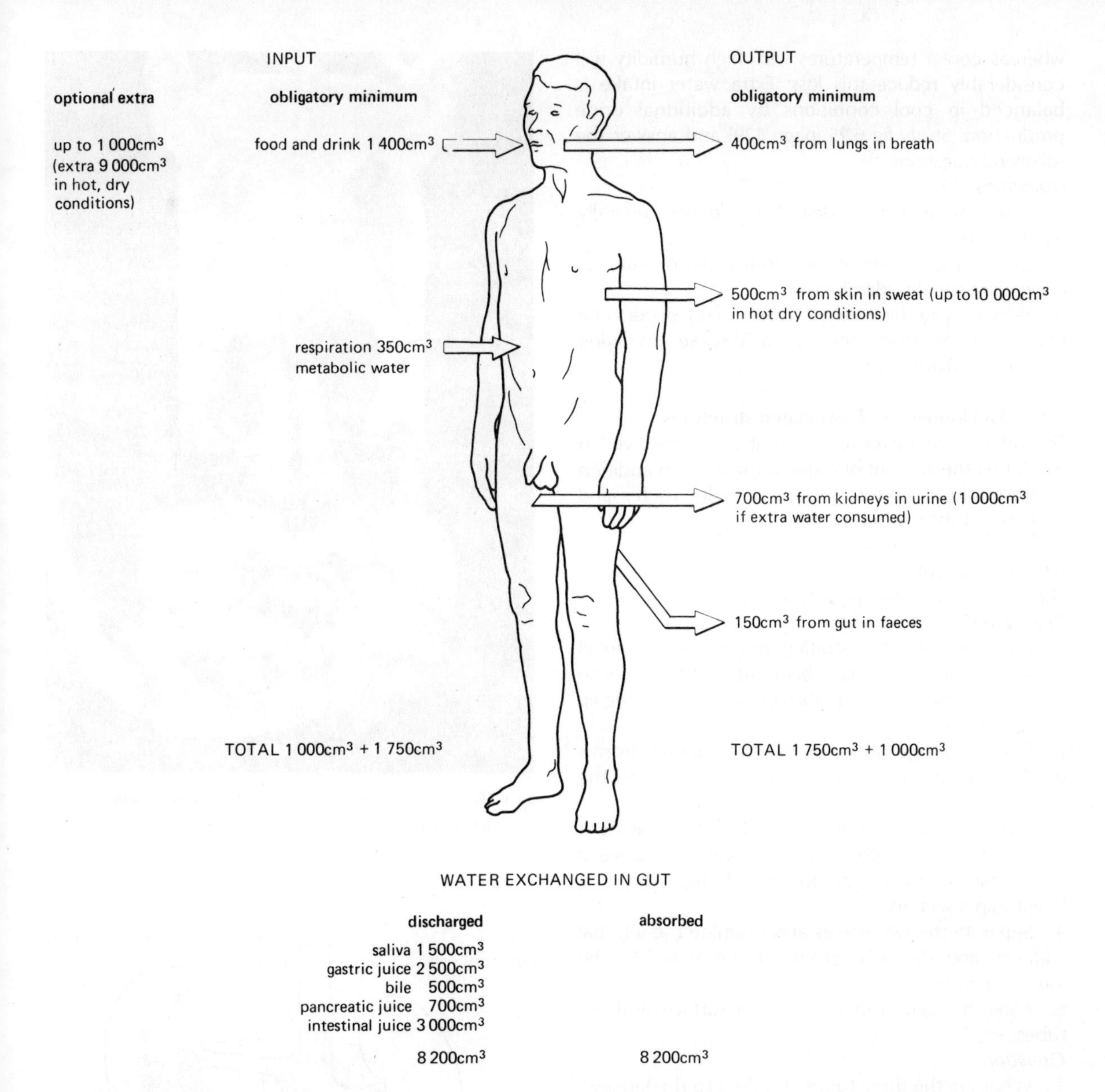

fig 6.15 Water gain and loss in Man

2 *Vitamin D production* Ultra-violet rays of the sun penetrating the skin produce sterols which become converted into vitamin D. Thus in sunny weather some of the requirements of the body can be met in this way, but in dull weather or in industrial regions covered by smoke haze the body has to rely on vitamin D in the diet.

3 *Energy storage* The lower layers of the dermis can store fat. As stated above, this can act as an insulating layer; but in many mammals, especially those that hibernate, the subcutaneous layers of fat are broken down to provide energy.

6.20 Water balance and ionic control

The amount of water in the body of Man and other mammals remains fairly constant due to an approximate balance between water gain and water loss. There is a *minimum* (obligatory) amount which must be *taken in* and a *minimum* (obligatory) amount which must be *eliminated* each day. These obligatory amounts need to be adjusted according to the external temperatures. In hot, dry conditions, the amount of water lost by sweating can amount to 10 litres a day, whereas cooler temperatures with high humidity will considerably reduce this loss. Extra water intake is balanced in cool conditions by additional urine production. Study fig 6.15 (page 139) and answer the following questions.

Questions

1 If we sweat a great deal, how do we generally replace this water?

2 Under what conditions would the loss of water in urine increase considerably?

3 Why do you think dehydration rapidly occurs in a man suffering from cholera, a disease involving continuous diarrhoea?

6.21 The kidneys and associated structures

The kidneys are the most important organs involved in regulating the amount of water in the body in addition to their role as excretory organs, so let us try and understand their structure.

Investigation

The kidney and associated structures

Procedure

1 Examine the demonstration dissection of a small mammal. (The gut will have been removed in order to show the *kidneys, ureters, bladder* and *urethra.* See fig 6.16.)

2 Examine a large kidney removed completely from a sheep, goat or cow. Note any tubes attached to the structure. See fig 6.17.

3 Place the kidney flat on the bench, and by means of a sharp blade, cut it into two equal halves by means of a horizontal cut parallel to the bench top. This is a longitudinal section.

4 Separate the two halves and examine the internal surfaces and the relationship of the tubes to the internal structure.

5 Make a diagram of the internal surface and the tubes.

Questions

1 What are the three tubes attached to the kidney?

2 How can you distinguish between them?

3 How would you describe the structure of the internal face of the kidney?

4 From what part of the kidney does the ureter directly receive its urine?

The *urinary system* can be seen in fig 6.17. It consists of two kidneys lying in the dorsal wall of the abdominal cavity and supplied with blood from *renal arteries.* The blood leaves the kidneys through the *renal veins,* and the urine produced, passes from each kidney into a *ureter.* The left and right ureters join with a muscular sac, the *bladder.* The mouth of this sac connects with a short tube, the *urethra,* which leads to the exterior. There are two small rings of muscle closing the exit of the bladder.

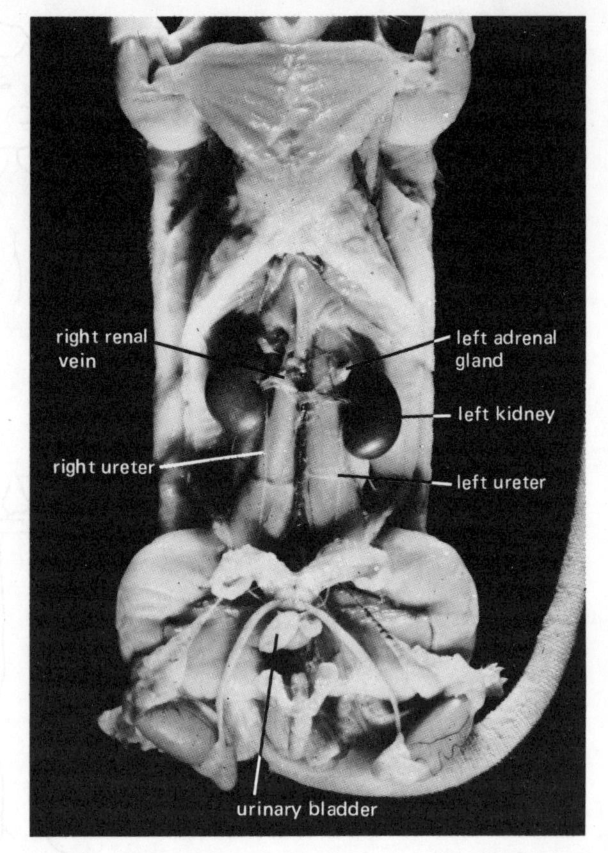

fig 6.16 A dissection of the rabbit to show kidneys and associated structures

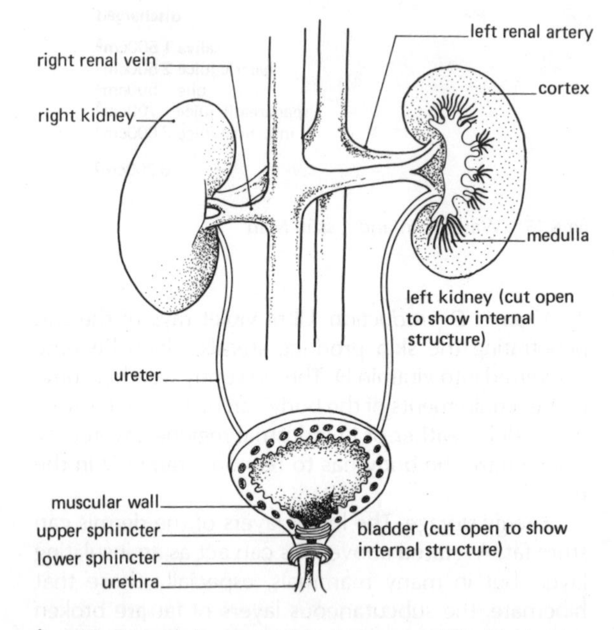

fig 6.17 General structure of the kidneys and bladder in Man

Internally the kidney has an outer portion, the *cortex*, and an inner portion, the *medulla*. The urine leaves the medulla and enters the pelvis, it then enters the ureters.

Investigation

To examine the microscopic structure of the kidney

Procedure

1 Take a prepared slide of a transverse section through a kidney, in which the blood vessels have been injected with a coloured dye and examine the slide under low power of the microscope (see fig 6.18).

2 Examine the slide under high power of the microscope, particularly the cortex (see fig 6.18).

You will note that under the low power the injected dye shows up small knots of blood vessels, but that most of the kidney consists of *tubules*. These are present particularly in the medulla where the tubules radiate from the *pelvis* of the kidney. The interpretation of this complicated structure is difficult from a single section, but by means of serial sections and by microdissection it has been found that each kidney consists of millions of such tubules, called *nephrons*.

The nephron begins in the cortex as a cup-shaped structure, the *Bowman's capsule*, surrounding a small knot of capillaries, the *glomerulus*. These can be seen in the cortical region of the kidney section. The capsule is connected to the tubule of the nephron which is divided into a coiled section, a loop into the medulla, a second coiled section and a duct which finally discharges into the pelvis of the kidney.

fig 6.18 A photomicrograph of kidney tubules

The blood enters the kidney by the renal artery which divides into arterioles supplying the capillaries of the glomeruli. Further capillaries surround each of the coiled sections of the tubule. A large amount of blood reaches each kidney (in Man this is about 500 cm^3 per minute) and slightly less leaves the kidney through the renal vein. The difference between the two amounts is due to the excretory substances and water which is extracted by the kidney tubules.

6.22 The work of the nephron

In the preceding chapters we have seen how living things

i) take up materials and energy from their environment (Chapters 1 and 5),

ii) transport materials through their bodies (Chapter 2), and

iii) transform and store these materials within their cells (Chapter 4).

All of these activities are *anabolic*, involving a building-up process, but to complete our understanding of metabolism we must now examine the ways in which living things return their waste products to the environment. These are produced by the breaking down activities of living cells, activities described as *catabolic*. The most abundant waste products are water, carbon dioxide and ammonia, but they should not be regarded simply as wastes. Remember the importance of carbon dioxide in the regulation of breathing, and even ammonia, a toxic substance, is essential for some amino-acid metabolism.

These substances are dangerous because they can *accumulate* to levels in excess of the organism's needs. The *elimination* of these *products of metabolism* from the body is called *excretion*. The most important organ concerned in this function is the kidney and within this organ the nephron is the working unit. A single human kidney is made up of about one million of these units. How does the nephron function? Many answers to this question were produced by a technical operation which involved sampling, by means of a micropipette, the fluid passing along the tubule. Extremely fine-drawn glass tubes were pushed into the tubule at different points, samples of the fluid withdrawn and then chemically analysed.

The nephron functions as follows. The *exit* capillary of the glomerulus has a *narrower bore* than the capillary which delivers blood (see fig 6.19). The blood is under higher pressure as a consequence and so much of the blood is forced out into the Bowman's capsule. The capsular space and the capillaries are separated only by two thin layers of cells, which act as a filter allowing only the liquid of the plasma and small molecules to pass. Since the filtration is a result of increased blood pressure, it is called pressure filtration or *ultrafiltration*. The fluid which is forced into the capsule is very similar to tissue fluid and lymph (see Chapter 2).

The rate of production of fluid into the tubules is about 130 cm³ per minute (about 100 litres per day). Over a period of twenty-four hours this is equivalent to a volume of fluid more than twice the weight of the body. The final production of urine, however, is only some 1 500 cm³ per day. Thus about 98% of the fluid must be reabsorbed. Examine table 6.2 and answer the following questions.

Questions

1 Why is the protein concentration in urine zero?

2 Glucose is filtered into the tubule. What explanation can you give for its complete absence in urine?

3 How do the concentrations of urea, ammonia and creatinine in the urine compare with those in the plasma?

4 Suggest a reason for any differences in concentration of the three substances.

The values in the table are approximate. There are considerable variations even in healthy people.

6.23 Chemical constancy of the body

The filtrate from the blood collects in the Bowman's capsule and passes down to the first convoluted tubule where considerable *reabsorption* occurs. *Glucose, amino-acids, vitamins, hormones* and a large number of *inorganic ions* (Na^+, K^+, Ca^{2+}, HCO_3^-, Cl^- etc.) are reabsorbed into the blood stream. This reabsorption requires energy for active transport so that the tubule cells have a high rate of metabolism with much ATP being formed. The inner surface of the cells lining the tubule have *microvilli* to increase greatly the area for absorption.

The solutes are returned to the blood in the capillaries surrounding the first convoluted tubule. At

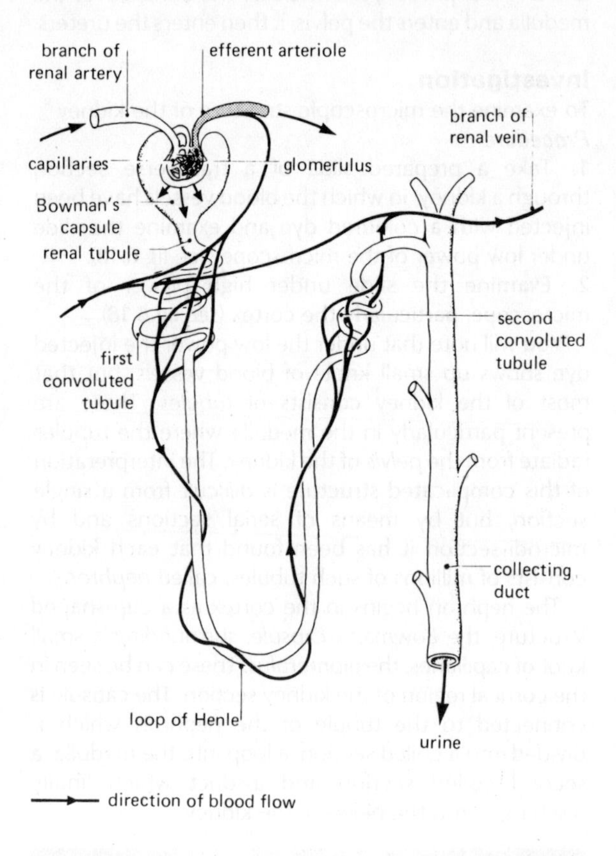

fig 6.19 A diagram of one renal tubule showing the blood supply and the main areas of reabsorption

Substance	% in plasma	% filtrate into nephron	% in urine	Concentration factor
Water	90–93	90–93	95.0	
Protein	7.0	0	0	
Glucose	0.1	0.1	0	
Sodium	0.3	0.3	0.35	× 1.0
Chloride	0.4	0.4	0.6	× 1.5
Urea	0.03	0.03	2.0	× 60.0
Uric acid	0.004	0.004	0.05	× 12.0
Creatinine	0.001	0.001	0.075	× 75.0
Ammonia	0.001	0.001	0.04	× 40.0

Table 6.2 Concentrations of substances in blood plasma and urine

Substance excreted	Excretory process	Animal	Notes
Carbon dioxide	Respiration	All animals	
Water	Respiration and osmo-regulatory processes	All animals	
Mineral salts	Osmo-regulatory processes	All animals	
Nitrogenous compounds			
i) Urea	Deamination of amino-acids to form NH_3 which is converted to urea (non-toxic and soluble in water)	Many aquatic animals; Mammals	Urea removed in solution in cartilaginous fish (sharks and dogfish) bony fish and amphibians
ii) Uric acid	Formed from NH_3 after deamination of amino-acids (non-toxic and insoluble in water)	Terrestrial animals	These animals need to conserve water e.g. insects, reptiles, birds and some mammals. The excretory product is deposited as crystals in the eggs of egg-laying animals, such as reptiles and birds
iii) Ammonia	Deamination of amino-acids and breakdown of urea by the enzyme urease (toxic but very soluble in water)	Many aquatic animals	
iv) Creatinine	Formed from phosphagen in muscles	Some mammals	
v) Bilirubin and biliverdin	Formed from breakdown of haemoglobin excreted in bile	Mammals	

Table 6.3 Substances excreted by animals

the same time a great deal of water follows these solutes back into the blood stream. This reabsorption of *water* is due entirely to *passive osmosis*. The transfer of solutes lowers the osmotic pressure of the filtrate, and so the water passes into the blood to restore the osmotic equilibrium.

In the Loop of Henle and the second convoluted tubule, a more precisely regulated reabsorption of *salts and water* takes place. If the water content of the blood is lower than usual (through heavy sweating) the additional quantities of water needed are taken in through the tubule. The tubule also helps to control the pH of the blood to within the limits of pH 7.3 to 7.4, by the *exchange of ions* when the acidity or alkalinity of the blood tends to rise. The fact that the kidney can produce a urine of pH from 4.5 to 8.5 indicates the flexibility of its adjustment mechanism.

The kidneys are the third example we have chosen of a homeostatic control mechanism. By a delicate adjustment of the urine contents they maintain a *constant level* of water and salts in the *blood* which in turn affects the *tissue fluids* and the *cell contents*. Water uptake in the tubules is controlled by various hormones circulating in the blood stream and these hormones are under the ultimate control of the brain. The *pituitary gland* at the base of the brain secretes the hormone *vasopressin* (ADH). Dehydration of the body is detected as a drop in water content of the circulating blood in the brain. The pituitary gland is stimulated to release vasopressin, which is carried by the blood to the kidneys. Additional water is reabsorbed by the tubules and the water content of the blood is restored to normal levels. Urine production drops and the urine itself becomes more concentrated. If blood becomes too diluted then the secretion of vasopressin is inhibited. The tubules fail to absorb water and large quantities of watery urine are produced (see fig 6.20). Examine fig 6.20.

Questions

1 What period of time passed before the entire litre of water was eliminated from the body?

2 Why do you think that drinking 1 litre of 0.9%

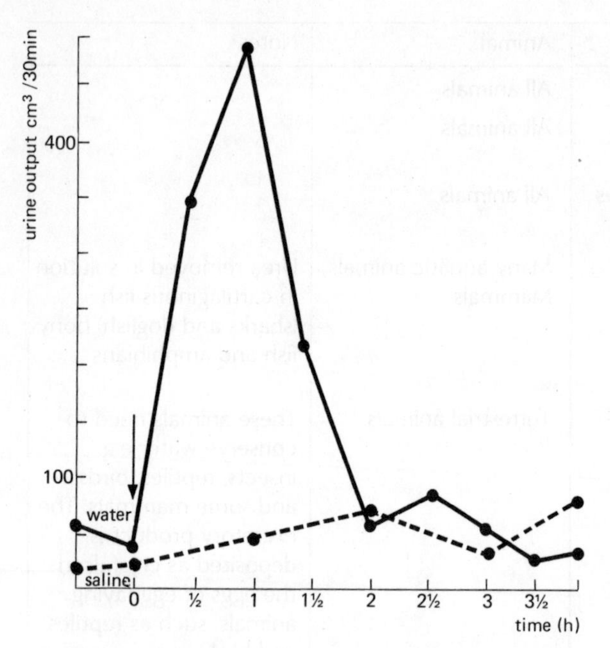

fig 6.20 A graph of response to drinking one litre of water, and on the following day, one litre of 0.96% sodium chloride solution, by a healthy man

sodium chloride solution made little difference to urine output?

If a human is in conditions where no drinking water is available and instead drinks *sea water*, his tissues will dehydrate and he will die. The salt content of the blood is about 1% whereas that of sea water is 3%. Thus when sea water is drunk, the salt content of the blood will rise towards 3%, but the kidney can only eliminate at best about 2% salt in urine. In order to do this, however, the blood drains water even more quickly from the tissues and there is a *net loss of tissue fluid*. As a result, tissue dehydration occurs and this leads quickly to death.

The analysis of urine can give an indication of certain malfunctions of the metabolism of the body. For example, glucose is not normally present in urine, for it is reabsorbed by the tubule after filtration. Evidence of *glucose in the urine* indicates a disease called *diabetes mellitus*, in which the pancreas fails to secrete the hormone *insulin*. Occasionally people lose the ability to secrete vasopressin and they then excrete an enormous volume of urine. This is accompanied by a terrible thirst and the person *drinks large quantities of water* to replace that lost in the urine (up to 20 litres daily). This disease is called *diabetes insipidus*, or *water* diabetes as opposed to *sugar* diabetes.

Kidney failure may occur and, although we can live with only one kidney, the failure of both is fatal. The simple phenomena of diffusion and permeability of membranes are used in *artificial kidneys* in order to help patients over a period of danger. Blood is passed through cellulose tubing (visking tubing) which is immersed in a bath of carefully prepared fluid. The fluid can be made of a precise composition to eliminate any particular ion from the blood. The efficiency is very great, removing urea for example much more quickly than normal kidneys.

Lungs	Skin	Kidneys
Water	Water	Water
Carbon dioxide	Sodium chloride	Sodium chloride
	Urea	Urea
		Uric acid
		Creatinine
		Ammonia

Table 6.4 Summary of substances excreted from body organs

6.30 The liver as a regulator

The body of a mammal is unable to store proteins or amino-acids. *Excess amino-acids* in the blood are taken to the *liver* where they are *deaminated*. The nitrogen-containing amino ($-NH_2$) portion of the molecule is removed and converted to nitrogenous waste, *urea*. The non-nitrogen containing residue from the amino-acid can be converted to glycogen for storage in the liver, or it can enter the blood as glucose and be used in cellular respiration.

In this connection the liver is concerned with the control of *blood sugar levels*. As these levels *rise* in the blood, the pancreas secretes *insulin* which changes the *glucose to glycogen* (an insoluble polysaccharide), for storage in the liver. During heavy exercise, blood sugar levels *fall* and the liver compensates by changing back *glycogen to glucose*.

Fat is also stored in the liver. As fat stores elsewhere in the body decrease during starvation, stored fat is released from the liver. The fat in the liver is very active metabolically for it can be changed to glycerol and fatty acids.

All the blood from the intestine, containing absorbed food materials, reaches the liver via the hepatic portal vein. The liver screens the blood passing through it so that the composition of the blood leaving the liver will be that which is correct for the organism. For example even after a meal rich in carbohydrate, the blood leaves the liver with its blood sugar level at the normal 0.1%. In the reverse process, blood sugar levels below 0.1%

will be restored by the release of glucose obtained from glycogen. Thus we can see that the liver is an organ of prime importance in the maintenance of a constant internal environment.

6.40 Excretion in plants

Excretion in plants does not pose such serious problems as it does in animals, for the following reasons:

1 The *rate* and *amount* of *catabolism* in plants is much *slower* and much less than in animals of similar weights. As a result waste products accumulate more slowly.

2 Green plants can *use* much of their waste products in their anabolic processes. Water and carbon dioxide produced by respiration can be used during photosynthesis. Certain plants have glands that secrete liquid water (*guttation*).

3 *Nitrogenous compounds* produced as waste substances can be used by plants in the *synthesis of new proteins*. It is nitrogenous excretion which makes such demands on the homeostatic mechanisms of animals.

4 The catabolism of plants is based on carbohydrates rather than proteins, and as a result the waste products are *less poisonous* than the nitrogenous wastes.

The metabolic by-products of terrestrial plants are simply stored in the plant. They may be present in the cell vacuoles or form insoluble crystals. These substances are shed in the leaves or the fruits as they drop from trees and shrubs. Commonly occurring crystals in plants are those of *calcium oxalate*. It must be noted, however, that any excretory function is purely *secondary* to the true functions of leaf fall or fruit drop. Some excretory products are stored in the heart wood of trees and shrubs. In herbaceous species the wastes stay in the portions above ground and these die off and fall at the end of the growing season or continuously throughout the year.

In aquatic species it is much easier for the by-products of metabolism to diffuse away into the surrounding water. Water produced as a result of metabolic activity remains in the cells to add to the water which diffuses in and so to maintain the turgidity of the tissues.

Questions requiring an extended essay-type answer

1 a) Define the process of excretion in living organisms.
b) The water content of the body of man remains constant. Show briefly how this balance is achieved.

2 a) What is meant by *homeostasis*? Give three examples in mammals.
b) Discuss the phenomenon of *homiothermy*.

3 a) Compare the methods by which mammals gain heat and lose heat.
b) How are the processes mentioned in part (a) related to water loss and ionic balance in a mammal?

4 a) Describe a *controlled* experiment to investigate heat loss through evaporation of a liquid from a hot surface.
b) How are your findings in part (a) relevant to control of temperature in a mammal?

5 a) Describe the similarities and differences between the processes of sweating in a mammal (e.g. man) and transpiration in a flowering plant.
b) How does the kidney of a mammal function as an osmo-regulatory organ?

6 a) *List* the organs (other than the kidneys) concerned in the process of excretion in a mammal.
b) What substances are excreted by plants? What processes are involved in this excretion?
c) Why is excretion a more important process in animals than in plants?

7 a) What substances are excreted by
i) the lungs,
ii) the skin, and
iii) the kidneys?
b) Describe the part played by the liver and kidneys in the process of excretion in a *named* mammal.

7 Response

7.00 Irritability in organisms

All organisms are in contact with their environment, and in any natural environment, external conditions are always changing. Living things are influenced by these changes; as a result of which their homeostatic mechanisms adjust, to ensure that the organisms are not affected adversely. In addition, they may need to take advantage of the opportunities arising from these changes. The changes in the environment act as *stimuli*, and these are detected by animals through *receptors* or sense organs. The ability to respond to stimuli is known as *irritability*, and is characteristic of all living things.

In plants, responses to changes in the environment are generally slower and less dramatic than those in animals. They may involve changes in the water content of cells (as is the case when daisies 'close up' during the night) or variations in growth rate, as discussed in Chapter 10.

Receptors are specialised sensitive cells which are sometimes combined with other types of cells to form an organ such as the *eye* or the *ear*. The specialised cells detect the environmental changes and so enable the organism to respond in the appropriate manner (see table 7.1). Although sense organs differ throughout the animal kingdom, the stimuli to which they respond are the same. Insects, for example, respond to light, sound, chemicals and touch but their sense organs are different from those of Man and other vertebrates.

Highly developed social insects such as bees and wasps show very complex behaviour, but although much still needs to be learned, enough is known to indicate that this behaviour can be explained as a number of set responses to external and internal stimuli. Such behaviour is termed *instinctive*, and is applied to *inborn behavioural patterns* which may result in apparently intelligent actions, but which are without conscious direction (see section 7.70).

7.10 Taxes

The response of lower organisms such as insects to certain stimuli can often be seen as movement of the whole organism from one place to another. This locomotion may be towards or away from the source of the stimulus i.e. *positive* or *negative*. Such a locomotive reaction of a *complete organism*, which is directly related to the line of action of the stimulus, is called a *tactic response* or *taxis*.

Consider the habitat selection of blowfly larvae. One can see that any dead animal is quickly visited by adult blowflies (*Calliphora* species). These lay eggs which hatch to produce white larvae or maggots. They are not easily visible, but if the carcass is moved, the larvae inside or underneath it are exposed, and quickly disappear from view by burrowing. What stimulus brings about this rapid disappearance of the larvae? We can put forward various hypotheses such as:

a) they move away from the light;
b) they move into the moist parts of the carcass;
c) they move towards their food;
d) they move into small spaces to become enclosed;

or perhaps various combinations of (a) to (d).

We must now select or design apparatus and experiments to test these hypotheses. One such apparatus is called a *choice chamber*, since it provides the test animals with a choice of two different environments, one of which is the control (see fig 7.1).

Experiment

Do blowfly larvae react to light stimuli?

Procedure

1 Set up a choice chamber (see fig 7.1) and leave each of the bottom dishes empty. Cover one chamber with a black cloth or polythene bag.

2 Introduce 10 larvae, placing five in each half of the apparatus. Allow the animals time for dispersal and then count their distribution in each chamber.

3 Repeat the experiment several times and record your results in the form of a table (see below).

Experiment	Distribution of 10 larvae introduced	
	In the dark	In the light
1		
2		

Totals		
Percentages		

4 Calculate the percentage, of the total number of larvae introduced, in each side of the choice chamber.

Experiment

Do blowfly larvae react to a moist or dry environment?

Procedure

1 Place in the bottom dishes of a choice chamber the following substances: in the left side water and in the right side, calcium chloride or silica gel. The substance in the right hand chamber will absorb water and provide a dry environment above it.

2 Introduce 10 larvae, placing five in each side of the choice chamber. Cover the apparatus with a black cloth or polythene so that light is excluded. Allow the animals time for dispersal and then count their distribution in each side.

3 Repeat the experiment several times and record your results in the form of a table.

4 Calculate the percentage, of the total number of larvae introduced, in each side of the choice chamber. Repeat the experiment with moist raw meat on one side and dry meat on the other side of the chamber (dried at 100°C in an oven). The apparatus should be covered to exclude light.

Questions

1 What do your experiments tell you about the tactic responses of the larvae?

welded here or joined by short plastic tube

plastic petri dishes forming upper half of apparatus

tunnel allowing larvae to move from one chamber to another

perforated zinc platform

petri dishes – lower halves (containers for substances controlling humidity)

components of choice chamber in relative positions

fig 7.1 The construction of a choice chamber for blow-fly larvae

2 In the first experiment for the light-dark situation, what other factor could be introduced into the apparatus in order to provide the larvae with a more normal environment?

3 How could you relate your results to the natural conditions in which the larvae are normally found?

4 What other factors could be operating to stimulate the movement of the larvae?

Stimulus	Receptors
Light	Eyes of mammals, compound eyes and simple eyes of insects, eyespot of *Euglena*
Sound	Ears of mammals
Chemical	Nose and tongue of mammals, olfactory organs of fish, skin of toad, tongue of snake
Gravity	Ear of mammal and other vertebrates
Temperature	Skin of mammal and other animals such as snakes
Texture	Skin of Man
Pressure	Skin of Man

Table 7.1 Types of stimulus and their detection by animals

7.20 Light receptor – the mammalian eye

Your experiments have shown that blowfly larvae react to light, but they have no eyes and certainly cannot 'see' objects as a man can. Let us consider what we know about our eyes by answering the following questions.

Question

5a) Are the eyes fixed in their sockets? If not how do they move?

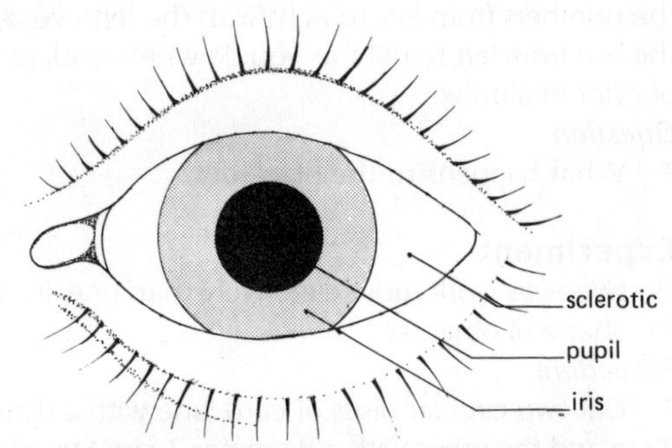

fig 7.2 A diagram of the front view of the eye of Man

• 1 2 3 4 5 6 7 8 9 10 11 12

b) Can we judge how far an object is in front of us?
c) Can we see the shape of objects?
d) Can we see the colour of objects?
e) Can we see in light only, or twilight in addition or even in the dark?

Investigation

The external appearance of the eyes

Procedure

1 Examine your eyes in a mirror. Observe the external appearance of the eye as shown in fig 7.2.

Question

6 How is the eye protected (a) from physical damage, (b) from dust?

Experiment

How do the eyes react to light and dark?

Procedure

1 Look in a mirror. Place the left hand over the left eye and observe any changes that take place in the right eye. Write down any changes that occur.

2 Remove the left hand from the eye and allow both eyes to adjust to light. Continue to look in the mirror, cover the left eye again with the left hand, count slowly up to 10 then uncover the left eye. Observe and record any changes that occur in the left eye immediately after it has been uncovered.

Experiment

Is there any part of the eye which does not react to light?

Procedure

1 Place the page of this book so that the black dot on the left of the line of figures above is opposite the left eye. The page of the book should be at the normal reading distance of about 25 cm. Close the right eye. Without moving the book, or turning the head, read the numbers from left to right with the left eye. Swivel the eye from left to right as you do when reading a line of print in a book.

Question

7 What happens to the black dot?

Experiment

Do two eyes work more effectively than one, in seeing the shape of objects?

Procedure

1 Cut two circular discs of card, one with a diameter 2 cm and the other with a diameter 3 cm. Mount each disc on a long pin or a piece of wire.

2 Hold the 2 cm disc in the left hand in front of the eyes, at a distance of about 45 cm. Hold the 3 cm disc in the right hand, at the same distance.

3 Close the right eye and move the right hand disc further from the eye until the two discs appear exactly the same size.

4 Open the right eye and observe the disc sizes.

Questions

8 When viewing the two discs in stage 3 the discs appear to be the same size. Can you explain this?

9 What seems to happen to the size of the discs when the eyes are open in stage 4 of the experiment?

10 What conclusions can you draw from the above observations about the operation of two eyes as distinct from one?

fig 7.3 The front view of the head of a man

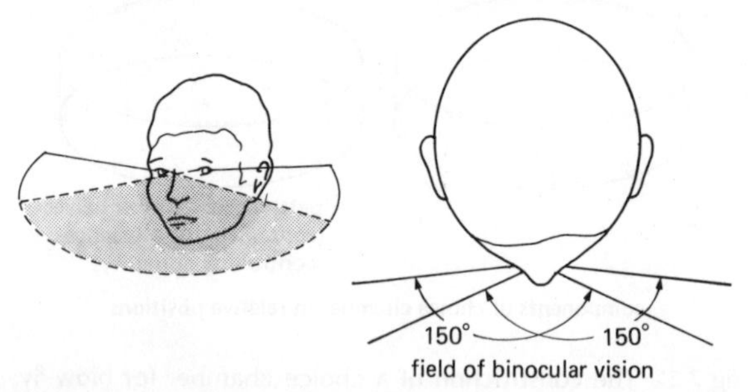

fig 7.4 A view of the head of a man showing his field of vision

In Man, the *visual fields* of the two eyes *overlap* considerably (see fig 7.4). They are in the front of the head and are separated only by the width of the nose in such a way that they receive slightly different images of an object. The brain is able to merge these two pictures into one, and in this way we can appreciate the depth of an object and its distance from us. This is called binocular or *stereoscopic* vision.

Question

11 What types of animal would be aided by this kind of vision?

In herbivorous mammals, such as ungulates (deer, horses and cattle), the eyes are most important for detecting the presence of predators. The eyes are *large* and *set high* on the sides of the head, with the result that an ungulate such as a horse has a very *wide visual field* (see fig 7.6). The visual fields of the eyes overlap to a small extent only, in front of the head: also the pupil has a *rectangular* shape with its long axis horizontal. Both of these factors contribute to increased visual field; and thus by the smallest movement of its head the animal can detect the approach of a predator from almost any direction. The ungulates are, therefore, well-adapted for daylight use of their eyes on open grassland: they can also see better than Man at dawn and dusk.

fig 7.5 A front view of the head of a horse

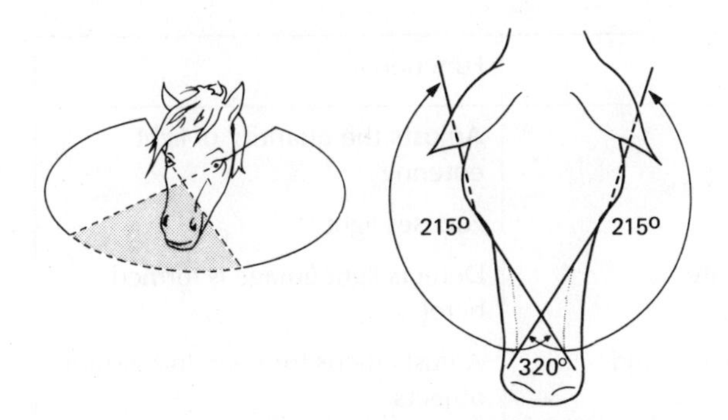

fig 7.6 A view of the head of a horse showing its field of vision

The experiments have shown you some of the functions of the eye, but a knowledge of its internal structure will aid understanding of its working. A sheep, pig, or cow's eye is very similar in structure to that of your own, and therefore can be used for this investigation.

Investigation

To examine the structure of the eye of a mammal

Procedure

1 You are provided with the eye of a sheep (or bullock). Use forceps and scissors to trim away the yellowish-white fat at the back of the eye. Do not cut away any of the structures that are firmly fixed to the eyeball. The fat cushions the eye in its bony socket or orbit.

2 Removal of fat exposes a number of strips of pink tissue attached to the eyeball (examine fig 7.7).

Questions

12 What do you think is the function of this tissue?

13 How many strips are there?

14 Where would you expect the other end of each strip to be attached?

3 Remove the strips of tissue and expose the optic nerve. This is a white cord-like structure attached to the back of the eyeball.

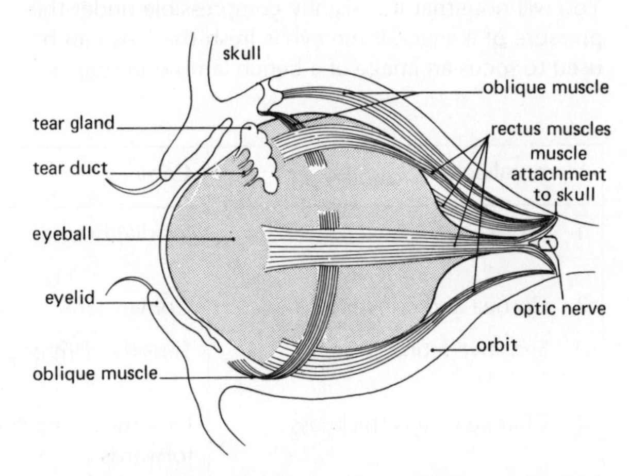

fig 7.7 A diagram of the side view of the eyeball in the socket, showing eye muscles

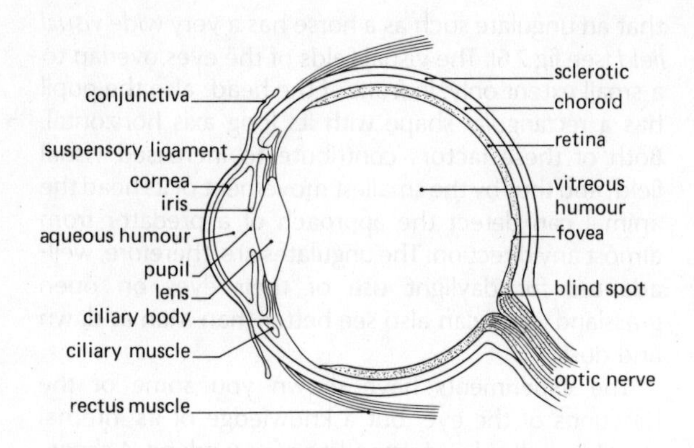

fig 7.8 A diagram of a vertical section through the eyeball

Question

15 What do you think is the function of this structure?

4 Using sharp scissors, make a circular cut around the eyeball with the optic nerve as the circle centre. Remove the disc attached to the optic nerve. The outer white layer (the *sclerotic*) maintains the shape of the eye and it is very tough to cut. Within the sclerotic is the pigmented *choroid,* while inside this is a thin transparent layer, the light-sensitive *retina.* A jelly-like transparent substance, the *vitreous humour,* fills the main cavity of the eye. It probably began to emerge when you made the cut.

Questions

16 What do you think is the function of the choroid?
17 What do you think is the function of the vitreous humour?

5 In the front half of the eyeball can now be seen the *lens.* Remove it from the eye and place it in a petri dish. You will note that it is slightly compressible under the pressure of a finger. If the eye is fresh the lens can be used to focus an image of a bench lamp onto paper.

6 Look again into the front portion of the eye. The black choroid continues forward to form the *iris* surrounding the aperture (the *pupil*), through which light enters. Behind the iris is the ciliary muscle and suspensory ligament, to which the lens was attached.

The sclerotic is continued to the front of the eye as a transparent layer, the *cornea.* The jelly-like substance in front of the iris and lens is the *aqueous humor.*

Question

18 What shape is the pupil? Compare it with your own pupil. Why has the pupil this shape in the eye of this dead mammal?

7.21 Focusing the eye-Accommodation

The eye functions in much the same way as a camera.

Examine table 7.2 and fig 7.9, and you can see how similar they are in structure and in function.

Question

1 State one way in which the eye is fundamentally different in function from the camera.

Experiment

How does the lens focus an image?

Procedure

1 Obtain an empty 500 cm³ flask and fill it with water. Fix a sheet of white paper on a drawing board or stiff cardboard and stand it upright on the bench.

2 Place the filled flask between the paper and the window. Move the flask away from the paper, at the same time observing the paper.

Question

2 What do you observe on the paper?

The flask filled with water behaves as a *convex lens.* The light from the window and the outside scene, passing into a more dense medium such as glass and water, is bent or *refracted* and an image is thus formed (*focused*) on the screen of white paper. The image is upside down or *inverted.* In this case, the light is coming from a distant object, but as objects approach closer to the lens the image moves further back behind the screen.

Mammalian eye	Camera	Function
1 Iris	Iris diaphragm	Adjusts the quantity of light entering
2 Cornea and convex lens	Convex lens	Focuses light
3 Sensitive retina	Sensitive film or plate	Detects light (image is formed here)
4 Change in lens thickness	Lens moves backwards and forwards	Adjusts focus for near and distant objects

Table 7.2 Comparison of the mammalian eye with the camera

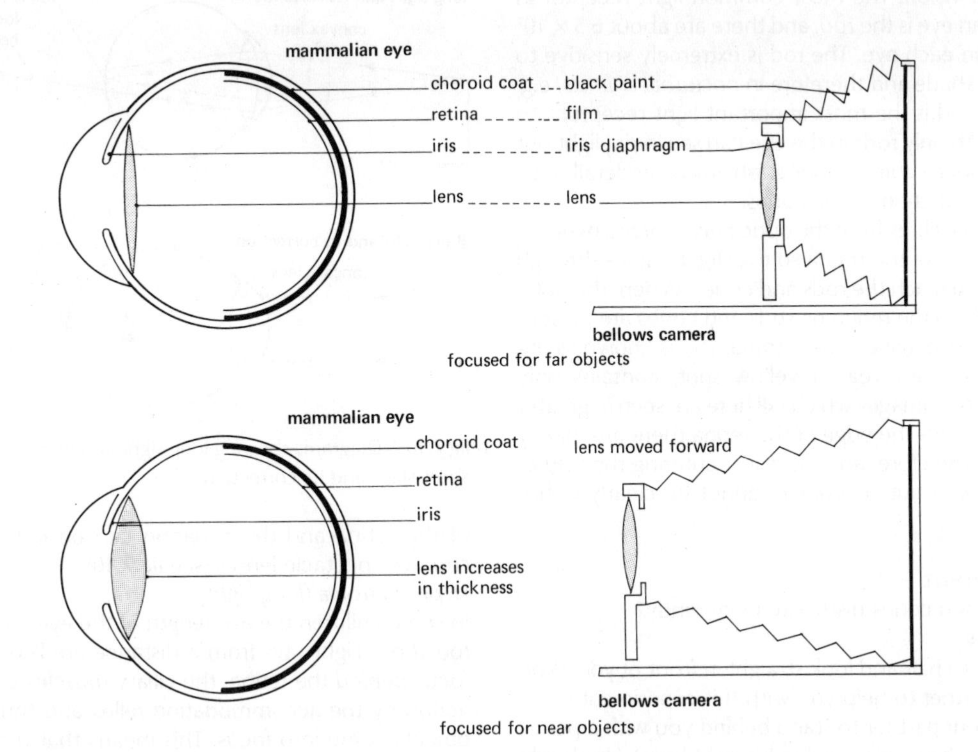

fig 7.9 A comparison of the eye and the camera, focused for near and far objects

Thus in a camera the lens must be pulled or screwed further away from the film in order to keep the image in focus on the film.

The mammalian eye resembles very closely a camera except in one respect; the focusing of a near object onto the retina. Most refraction takes place as the light passes through the transparent cornea in the front of the eye, but the final adjustment is brought about by the lens so that the image is focused clearly on the retina. Any object *more than 10 metres* from the eye is in focus on the retina with the lens at its *thinnest*. Objects closer than this distance are kept in focus by the eye lens bulging and becoming thicker, thus increasing its power to bend the light rays so that they converge on the retina.

The lens is suspended by small ligaments (*suspensory ligaments*) attached to *ciliary muscles*. When the eye lens is thin and focused for distant objects, the ciliary muscle is *relaxed*. In order to see close objects the ciliary muscle, *contracts*, pulling the wall of the eye inwards, *slackening* off the suspensory ligaments and allowing the lens to *bulge*.

Imagine you are sitting outside in the shade of a tree reading a book; you look up from the book to a distant sunlit mountain and then down to the book again. Table 7.3 shows the sequence of events in your eyes as the two actions take place.

Looking up from the book to the distant scene.	Looking back to the book from the distant scene.
1 Circular muscles of the iris contract, radial muscles relax.	Radial muscles of the iris contract, circular muscles relax.
Pupil gets smaller.	Pupil gets larger.
Light intensity adjusted.	Light intensity adjusted.
2 Ciliary muscles relax.	Ciliary muscles contract.
Lens becomes thinner and less refractive.	Lens becomes thicker and more refractive.
Distant scene focused on the retina.	Book print focused on the retina.

Table 7.3 Accommodation of the eye to near and distant objects

The mechanism by which the eye focuses accurately is known as accommodation. The accuracy of the picture seen is due to receptors called *cones* which only function well in high light intensity. There are about 3.5×10^5 cones in each retina. They are also responsible for *colour vision*. The most common light receptor in the human eye is the *rod*, and there are about 6.5×10^7 of these in each eye. The rod is extremely sensitive to light and shade and therefore in *nocturnal animals*, e.g. bats, the rod is the most important light receptor. An animal with only rods in the eye can see in daylight but it would not be able to distinguish any great detail since this is the function of the cones.

The nerve fibres from the optic nerve spread over the inner surface of the retina so that light passes through the fibres to reach the rods and cones. Where the optic nerve pierces the retina no rods and cones are present and thus no image can be formed. This is known as the blind spot. The *fovea*, or yellow spot, contains only cones, so that images which fall here are seen in greater detail. Towards the edge of the retina there are mainly rods and, therefore, an object just entering the field of vision is seen but its colour cannot be clearly distinguished.

Experiment

Are rods and cones distributed differently?

Procedure

1 Sit on a chair and look straight in front of you. (You need a partner to help you with this experiment.)

2 Ask your partner to stand behind you with a pencil (which you have not seen) in his right hand. He should move it forward until it just comes within your field of vision, when you say 'stop'.

3 Without moving the eyes or head, try to determine the colour of the pencil.

Question

3 Can you decide on the colour of the pencil? If not, try to give an explanation of this.

7.22 Defects of the eye

If the eye is to function correctly, there must be a balance between the length of the eyeball and the refractive power of the cornea and the lens. When a child grows, the axis length of the eyeball increases from 16 mm in the newborn to 24 mm in the adult. During this time, refractive power must be adjusted to the increasing axis length.

In extreme cases, the axis length of the eyeball falls outside normal limits, and the eyes are unable to focus by cornea and lens alone. They are aided by *external lenses* (glasses) to focus light rays on to the retina.

Myopia (short sight)

The axis length of the eyeball may be *too long* due to defective nutrition or some genetic reason, and *short sight* or *myopia* results. The term short sight refers to the fact that only objects *near* to the eye can be seen clearly. Children with this defect (detectable by their reading small print close to the eyes) can become progressively worse and the condition must be treated. Light rays from a distance are brought to a focus *in front*

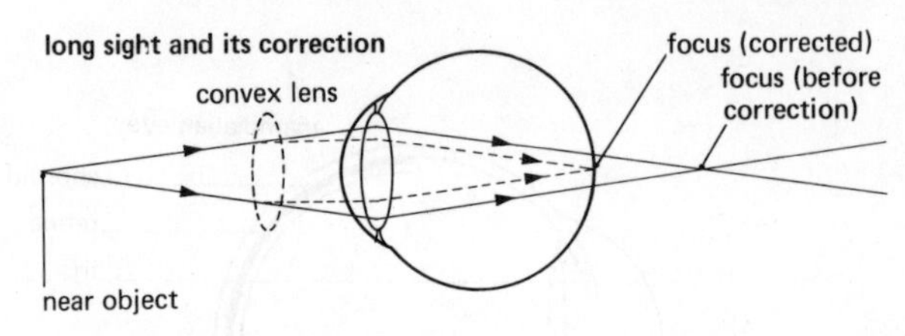

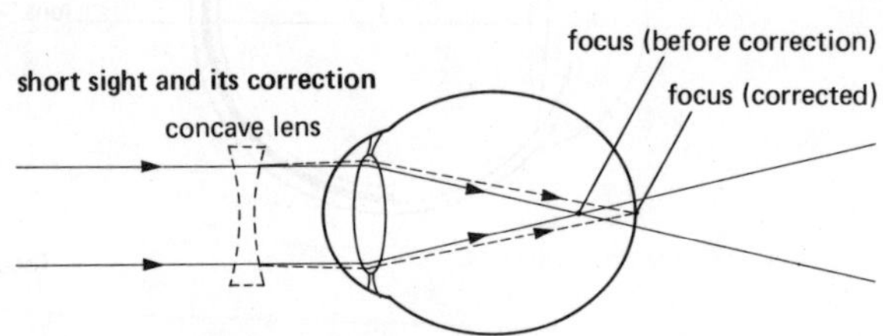

fig 7.10 Diagrams showing long sight and its correction and short sight and its correction

of the retina and this situation can be corrected by *concave* spectacle lenses (see fig 7.10).

Hypermetropia (long sight)

In some children the axis length of the eyeball remains *too short*. Light rays from a distance are brought to a focus *behind* the retina, the ciliary muscles come into action by the accommodation reflex and thus distant objects come into focus. This means that the nearest point of distinct vision may be well away from the eyes. Near objects cannot be seen clearly. This condition is corrected by a *convex* spectacle lens.

Long sight is characteristic of increasing age and in older people this is due to a decrease in the refractive power of the lens of the eye. Again this is corrected by spectacles with convex lenses.

7.30 Sound and gravity receptor – the mammalian ear

The *pinna* or external ear together with the outer ear tube serves to concentrate sound waves. In many animals, but not in Man, the pinna can be turned to the direction from which the sound is coming. The pinna is connected by the *auditory canal* to the *tympanum*.

The middle ear is an air-filled chamber connected by the *eustachian tube* to the back of the throat (the *pharynx*), and thus to the outside air by the *buccal cavity* and mouth.

Question

1 What is the function of the eustachian tube?

The tympanum is connected across the middle ear to the *oval window* by the three tiny, linked bones called ear *ossicles* (hammer, anvil and stirrup). The hammer is attached to the tympanum, and the anvil passes on the movements of this bone to the stirrup which touches a flexible membrane, the oval window. Just below is another membrane, the *round window*.

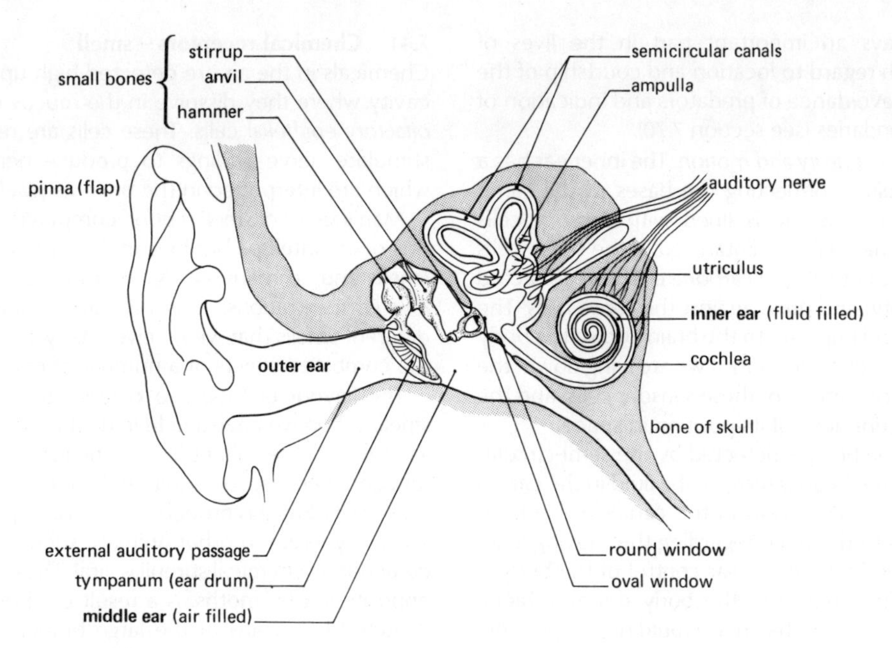

fig 7.11 The structure of the ear

The inner ear is fluid-filled. It contains two structures.

i) The *cochlea*. This is a tube about 3 cm in length, coiled like a snail shell and filled with lymph. Lying within the cochlea is the *organ of Corti* containing thousands of sensitive cells which are the receptors of the vibrations conducted from the tympanum.

ii) Three *semi-circular canals*. These are fluid-filled tubes each one at right angles to the others so that they lie in the three planes of space. At one end of each canal is a small chamber, the *ampulla*, containing sensory hairs. Both structures are connected to the brain by the auditory nerve.

7.31 Functions of the ear

i) Detection of sound (hearing) Sound waves (vibrations of gas molecules in the air), having been concentrated by the pinna, pass down the auditory canal of the outer ear and strike the tympanum causing it to vibrate. To understand this, the following facts about transmission of sound should be known.

a) Sound can be transmitted through a solid, liquid or gas, but not through a vacuum. Light and heat from the sun can travel across outer space but not sound.

b) Transmission of sound in air is due to alternate *compression* and *rarefaction* of the air at the front of the sound wave. These compressions and rarefactions strike the tympanum in succession causing it to move in and out. In a more dense medium, such as wood, the movement of sound is like the rapid transference of energy from one group of molecules to the next.

c) Sound, such as a musical note, has a measurable *frequency*, and this frequency can be transmitted through a medium to cause another object to vibrate at the same frequency. Frequency is expressed in *cycles per sec (c/s)*. The 'pitch' of a sound is its frequency.

d) Sound travels relatively slowly in air at about 330 m/s. Thus the sound of distant thunder arrives long after the lightning flash.

The vibrations of the tympanum are passed across the middle ear by the ossicles and so the oval window vibrates with the same frequency as the tympanum. The vibrations are magnified × 22 when they reach the oval window from the tympanum. This is because the tympanum has over twenty times the diameter of the oval window.

Questions

1 Why should the perforation of the tympanum cause deafness?

2 Remembering that the inner ear is fluid filled and fluids are practically incompressible, can you suggest a function of the round window below the oval window?

The vibrations of the oval window are transmitted through the cochlear fluid, and cause stimulation of the sensitive cells. Impulses are sent from these cells through the auditory nerve to the brain. *Low frequencies* stimulate the organ of Corti near its *tip* and *high frequencies* are detected near its *base*. The ear is a very versatile sound receptor and young people can detect frequencies from 20 to 20 000 c/s. As the body ages the ability to hear higher frequencies is lost and the upper limit may fall to as low as 5 000 c/s. Some mammals, such as dogs, can hear higher frequencies than Man. Bats can detect frequencies as high as 100 000 c/s.

The range of volume of audible sound is also very great since the loudest sound is about one trillion times as loud as the faintest we can detect.

Hearing plays an important part in the lives of mammals with regard to location and courtship of the opposite sex, avoidance of predators and indication of territorial boundaries (see section 7.70).

ii) Detection of gravity and motion. The inner ear has a sac, the *utriculus*, connecting the bases of the semi-circular canals. This sac is lined with sensory cells connected to nerve fibres. Entangled in the sense cells are tiny spheres of *calcium carbonate*. These are acted upon by gravity, and press against the sense cells. The nerve fibres send impulses to the brain and so the body becomes aware of its position. If we are lying down, the spheres press on certain of these sensory cells, and the body is thus conscious of its position in space.

Motion of the body is detected by the semi-circular canals. When the head is moved, the fluid in the canals moves also. The sense cells in the canals are able to send impulses to the brain regarding the moving fluid, and consequently the muscular control of the body is maintained. The control of the body during athletic activity, particularly for balance, would not be possible without this mechanism.

7.40 Chemical receptors – taste

One of the organs of the mammal that detects *chemicals* in the environment is the *tongue*. In order that a chemical may be tasted it must dissolve in the moisture of the buccal cavity, then, when in solution, it can stimulate the taste buds. There are only four primary tastes: sweet, sour, salt and bitter.

Experiment

Are the four primary tastes detected in different regions of the tongue?

Procedure

1 Prepare four solutions giving different tastes, e.g. (a) sugar solution (sweet), (b) very dilute hydrochloric acid, citric acid or vinegar (sour), (c) common salt solution (salt), and (d) quinine sulphate (bitter).

2 Working with the help of a partner, explore the areas of the tongue for sensitivity to the four liquids. Use small glass tubes to place drops of each liquid on different parts of the tongue. The mouth and the tongue should be rinsed well between tests to eliminate the previous tastes.

3 For each liquid, record, on a drawing of the tongue, where it can be tasted (pay particular attention to the tip, sides and back of the tongue). Complete the tests for each liquid before passing on to the next. Leave the bitter substance until last.

Questions

1 Do your tests show any clear pattern of taste areas over the tongue?

2 Are the taste areas separated or do they overlap?

3 Do individuals differ or is there a basic pattern?

7.41 Chemical receptors – smell

Chemicals in the air are detected high up in the nasal cavity where they dissolve in the *mucus* covering the *olfactory epithelial* cells. These cells are receptors and stimulate nerve endings to produce nerve impulses which are interpreted in the brain as smell.

Man's sense of smell is poor, compared with that of a dog or an antelope but he can detect a wide variety of smells and, in many cases, the molecules are at very low concentrations. It is difficult to analyse all the different smells that we receive. Many smells represent the combined effects of a number of chemicals.

The flavour of food is a combination of taste and smell. When we have a cold in the head, food has little flavour, as the olfactory epithelium is unable to function because of additional mucus covering the sense cells. Smell is probably the least important of our senses, whereas in other animals, such as insects, the detection of chemical stimuli is vital. The location of the opposite sex by moths, is a result of chemical stimuli detected by means of the large branched antennae. Many snakes sample the air for molecules of odorous substances by means of their tongues which they continually flick in and out.

Sense cells can become adapted. When we first step into a room where there is a very obvious smell of newly cooked food or new paint we are immediately aware of the smell, but if we stay there long, our perception of the smell diminishes almost to vanishing point. There is a 'conditioning' of the olfactory cells for that particular smell. New smells, however, are detected quickly.

7.50 Receptors in the skin of Man

See Chapter 6 fig 6.9 for the structure of the skin. The skin is an important sensory organ. In the dermis are many nerves which carry information about changes in the external world to the central nervous system. Areas of skin which are hairless, such as the finger tips of Man or the lips of most mammals, tend to be more sensitive than hairy skin. There are receptors at the ends of the nerves in the dermis but it is not always possible to differentiate those responsible for receiving particular stimuli. The following stimuli are detected by skin receptors: (a) heat, (b) cold, (c) touch (roughness, smoothness, etc), and (d) pain.

Experiment

Can the skin distinguish different temperatures?

Procedure

1 Prepare three beakers of water as follows:

i) hot water (about 50°C)

ii) water at room temperature (about 22°C)

iii) ice water (about 5°C).

2 Immerse the index fingers of both hands in beaker (ii) for 10 seconds, then immerse the left index finger in beaker (i) and the right index finger in beaker (iii) for about 30 seconds.
3 Transfer both fingers simultaneously into beaker (ii).

Questions

1 Does the temperature of the water in beaker (ii) feel the same to both fingers at the beginning of the experiment? Why were the fingers both placed in the water of beaker (ii) at the beginning?
2 Do you notice any difference in the response of the two fingers after transfer back to beaker (ii) from the other beakers?
3 What conclusions can you make from your observations regarding the response of the skin to temperature?

Experiment

How do different parts of the skin differ in touch sensitivity?

Procedure

1 Push two pins through a cork so that the points are 2 mm apart.
2 Working in pairs, test different surfaces of the body by pressing gently on the skin with the two pins. First, test the upper and lower surfaces of the hand and then move to fingertips, upper arm and back of the neck.
3 Each time the pins are pressed on to the skin the student should indicate whether he/she can feel two points or one.

Questions

4 Are all areas of the skin equally sensitive to the touch stimuli of the pins?
5 What can you conclude about the distribution of receptors under the skin?

7.60 Co-ordination

The stimuli received by animal sense organs bring about *responses*. These responses are quicker than those of plants which generally take several hours to respond by a growth change (see Chapter 10). The quickness of response is due to the presence of a nervous system which ensures that the effect of a stimulus on sense cells is carried to all other parts of the animal. Information about an animal's environment is collected by sense cells and transmitted to the *central nervous system* by nerve *impulses*. As a result muscles systems work together, the breathing mechanisms exchange gases, the alimentary canals digest food, enemies are avoided and countless other responses enable the body to function as a complete whole. The central nervous system is assisted in co-ordinating the body functions by the *endocrine system*. This consists of a number of glands that discharge chemicals called *hormones* into the blood circulation. These secretions have a particular effect on an organ or on the body as a whole. The nervous system controls these glands by sending appropriate messages to increase or decrease secretion either directly or through the release of intermediate hormones.

The response of the organism to its environment as a result of stimuli received can be called its *behaviour*. The mechanisms underlying behaviour are nervous or glandular in origin and are extremely complex. Nevertheless, we can understand some simple forms of these mechanisms, one of which is shown below.

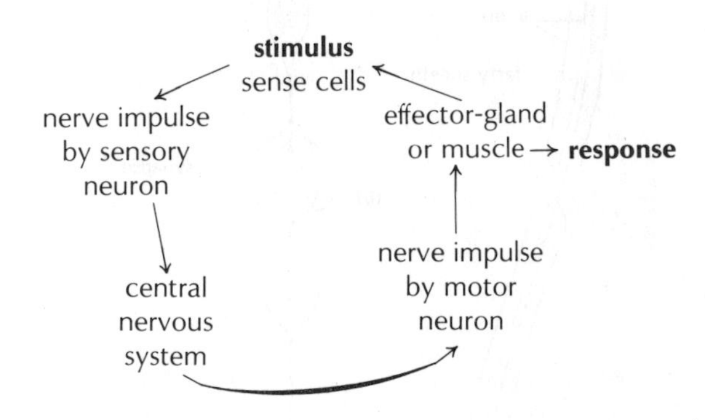

The central nervous system (brain and spinal cord) can be compared to a telephone exchange, sorting incoming messages and transmitting outgoing messages. The nerves represent the telephone wires.

7.61 Neurons

Nerves are made up of units called *neurons*, highly specialised for their function of receiving and conducting impulses. A typical neuron has a very long *axon* or fibre covered and *insulated* by a layer of *fat*. Attached to the axon is the *cell body* containing a nucleus and having many threadlike extensions of the cytoplasm called *dendrites* (see fig 7.12).

The dendrites are in close contact with those from cell bodies of other neurons or with sense cells from which impulses originate. The dendrites do not connect completely with each other but are separated by a minute gap called a *synapse*. The brain and spinal cord have several millions of neurons, each with numerous dendrites: the number of synapses in the nervous system is very great indeed.

Neurons carry nerve impulses which are electrochemical in nature and spread rapidly along the fibres. The impulse is produced by minute electrical changes that normally travel only in one direction, and during this time use up small amounts of energy. The dendrites are not in contact with each other and an impulse reaching the end of a fibre causes the secretion of minute amounts of a chemical substance called *acetylcholine*. This chemical moves across the synapse,

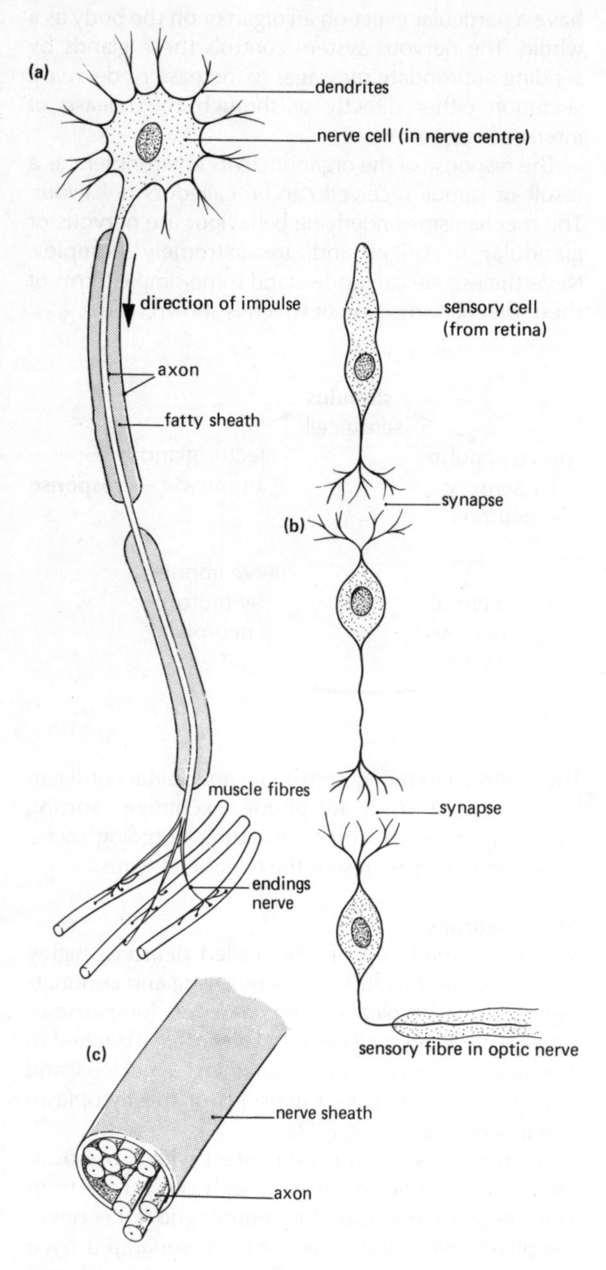

fig 7.12 Diagrams showing (a) a single motor neuron and its connection with muscle fibres (b) synapses between a sensory neuron and a sense cell of the retina of the eye (c) fibres of neurons bound together to form a nerve
After D. G. Mackean

causing a new electrical impulse to be produced and travel along the next neuron. Acetylcholine is unstable and is quickly broken down by an enzyme to prevent the continuous production of new impulses.

7.62 Nervous system

The nervous system is made up of the following parts:

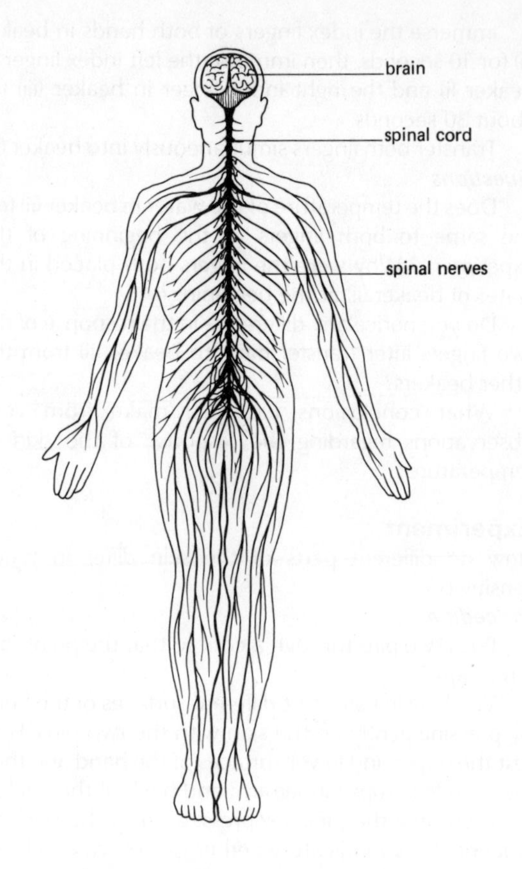

fig 7.13 A diagram of the nervous system of Man

Central Nervous System Peripheral Nervous System

Brain ⟷ Cranial nerves
↓ ↑
Spinal cord ⟷ Spinal nerves
⟷ Autonomic nerves

The mammalian brain is much larger in relation to body size than the brain of other animals. This is due to the proportionately greater size of the cerebrum and the cerebellum associated with the increased variation in behavioural patterns and the greater muscular control that goes with them. The brain of Man is particularly notable for the large *cerebrum*, which extends over all other areas of the brain.
The main regions of the brain are as follows:
Olfactory lobes: Connected by sensory neurons to the organ of smell, which also sends sensory neurons to the cerebrum. These lobes are proportionately larger in animals such as fish which depend heavily on a sense of smell for survival.
Cerebrum (cerebral hemispheres): They have a much folded and wrinkled surface giving the outer cortex a very large surface area. The *cortex* is formed of *grey matter* (cell bodies of neurons) as distinct from the inner

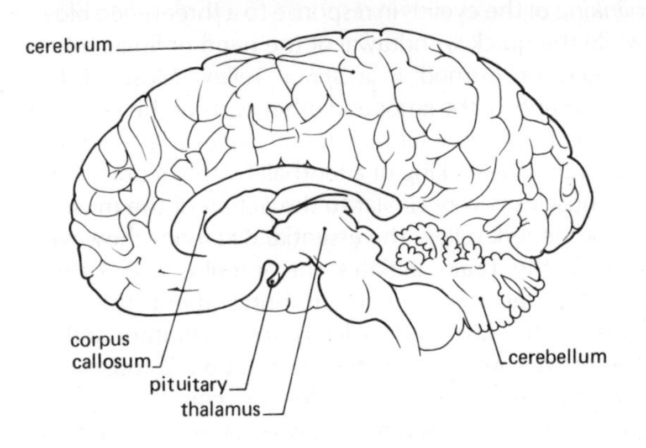

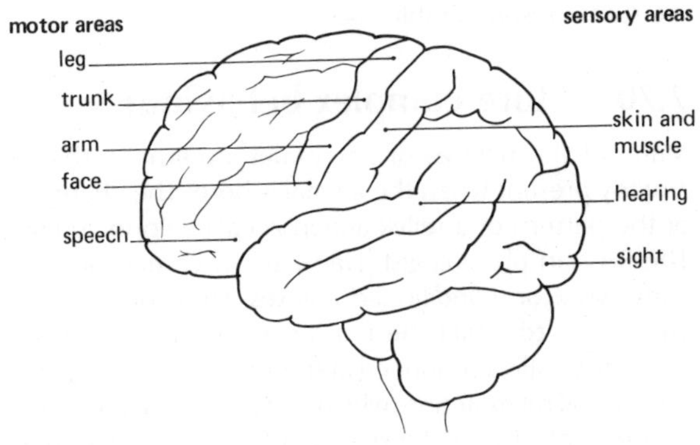

fig 7.14 A longitudinal median section of the brain of Man
Areas of control in the cerebrum of the brain of Man

part which is composed of *white matter* (axons of neurons). Different areas of the cortex control distinct functions. Sensory areas control sight, hearing, smell and skin sensation, while motor areas control muscles of the legs, arms, face, eyes and head. Large areas of the human cerebral cortex are not concerned with sensation or motor control. These areas are missing in all other mammals, and thus it could be that they are concerned with Man's intellectual functions including speech, music, mathematics and other activities requiring the use of symbols and abstract thought.

Hypothalamus: This is the reflex centre concerned with a number of homeostatic mechanisms such as temperature control, water balance and carbon dioxide levels in the blood.

Optic lobes: These receive sensory neurons from the eyes. They are small in Man compared with other animals since the cerebral cortex has taken over much of their role.

Cerebellum: This is a large structure in mammals, as it is concerned with the maintenance of balance, locomotion and positioning of the body, and is thus involved in the co-ordination of muscular activity. It receives sensory impulses from the skeletal muscles and sends motor impulses out to them.

Medulla Oblongata: This is a reflex centre of the brain controlling blood pressure, coughing, swallowing, sneezing, yawning and vomiting (see fig 7.14).

From the brain, twelve pairs of *cranial nerves* leave through small holes in the cranium. They are concerned with sensory impulses (from organs of smell, sight and hearing) and motor impulses (to jaw muscles, eye muscles and tongue muscles). The cranial nerves cross over in the brain so that nerves serving the left side of the head originate in the right side of the brain and vice versa.

Spinal nerves are pairs of *mixed* nerves (motor and sensory) arising at regular intervals along the spinal cord (see fig 7.13). Each nerve splits near the cord into a dorsal sensory branch and a ventral motor branch. The spinal nerves connect with short internuncial neurons, or longer secondary neurons passing up and down the spinal cord. These cross over, mainly in the medulla, so that nerves from the left side of the body reach the right side of the brain and vice versa (see fig 7.13).

The *autonomic* nerves control the *internal* activities of which the individual is not normally aware. Peristalsis of the gut and glandular activity are controlled by the autonomic system.

7.63 Reflex action

Experiment

To demonstrate a spinal reflex

Procedure

1 Sit on a stool with your right leg crossed over the left knee. Tap the patella tendon (just below the knee cap) of your right knee sharply with the edge of a ruler.

2 Now place your left hand on top of your right thigh muscle. Repeat the tapping of the patella tendon.

Questions

1 What happens to the lower part of your right leg and foot when the tendon is tapped?

2 What do you feel with your left hand?

3 What explanation can you give for the response of your lower leg?

The tap on the tendon causes the *stretch receptors* of the thigh muscle to be stimulated and the muscle contracts. This muscle is an *extensor* muscle of the knee joint and the lower part of the limb jerks forward. Thus we have a simple behavioural pattern called a *reflex action*. It results from a sequence of impulses in the nerves and the spinal cord constituting a *reflex arc*. The reflex arc sequence is as follows: the tap on the tendon stimulates the receptors in the muscle; the receptors generate an impulse which is transmitted along the sensory neuron to the spinal cord, where it is passed via an internuncial neuron to an outgoing motor neuron. The impulse then proceeds along the axon of the motor neuron and causes the extensor muscle in the top of the thigh to contract.

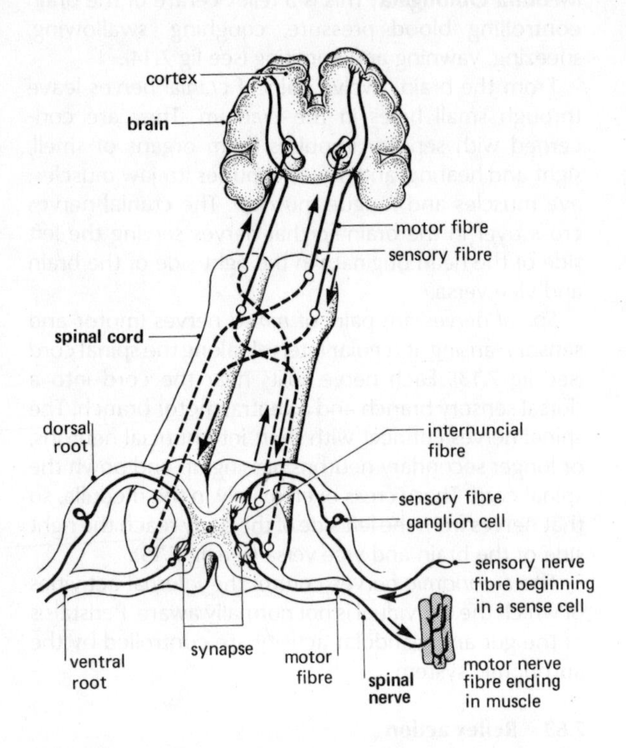

fig 7.15 A diagram of a reflex arc together with a portion of the spinal cord and the nerve connections to the brain

The knee jerk reflex is a spinal reflex action involving only the spinal cord, and like all reflexes, it is not under control of the will. It is not initiated consciously and is difficult to suppress. Nevertheless, when it happens, we are conscious of it because a further impulse is sent along the spinal cord to inform the brain (see fig 7.15). We are then made aware of the response.

Experiment

What happens when we cut up an onion?

Procedure

1 Take an onion and cut it into small pieces.

Question

4 What happens to the nose and eyes?

The vapours arising from the cut onion pass into the nose and eyes causing the olfactory sense cells to react. In addition, the vapour dissolves in the moisture of the eyes. The latter react by discharging fluid from the tear gland to dilute the dissolved vapour. Here is a clear reflex action, *not controllable by the will* and giving a definite response to the chemical stimulus. In this case, however, although quite automatic, the reflex involves the brain rather than the spinal cord, so that it can be called a *cranial reflex action*. The response is a *glandular discharge* as distinct from the muscular action in the case of the knee jerk.

Another example of the cranial reflex action is the *blinking* of the eyelids in response to a threatened blow, while the quick withdrawal of the hand or finger when pricked or burned is a *spinal reflex*. Most of the movements of the limbs and other parts of the skeleton are the result of voluntary action under the control of the brain. The kicking of a football by the action of the leg muscles is very similar to the action of the muscles in a knee jerk reflex. The essential difference, however, is that the brain initiates the impulse and causes muscular action in order to bring about the kick, whereas the knee jerk reflex is an involuntary action. The brain is informed of the contraction of muscles in the knee jerk reflex, but it has no control over the movement. The differences between these two actions are summarised in table 7.4.

7.70 More complex behaviour

Many of the reflexes of an animal are innate, that is already present when the animal is born. The changing of the pattern of a reflex action is called *conditioning*. The Russian physiologist, Ivan Pavlov, carried out the early work on *conditioned reflexes*. He experimented with dogs, to which he fed meat and measured the amount of saliva produced at the sight of the food. This is a normal reflex action, which humans also experience at the sight of food. By ringing a bell every time food was produced, Pavlov conditioned the dogs to expect meat whenever they heard this sound.

Spinal reflex	Voluntary action
Stimulus affects external or internal receptor.	Initiated from the brain at the conscious level.
Spinal cord only involved – not under the control of the will.	Forebrain involved – under the control of the will.
The impulse travels only up or down the spinal cord.	The impulse travels from the brain down the spinal cord.
The path of the nerve impulse is by the shortest route.	The path of the nerve impulse is much longer.
The response is immediate.	The response can be delayed.
The response is in skeletal or internal involuntary muscle.	The response is in skeletal muscle only.

Table 7.4 Comparison of a spinal reflex action and voluntary action.

Eventually after many trials, the dogs salivated in response to the ringing of the *bell only* (i.e. even when no food was produced). Thus the conditioned reflex producing salivation involves nerve pathways *different* from those of the unconditioned reflex.

Experiment

To demonstrate a conditioned reflex in humans

Procedure

1 You will need the cooperation of a group of ten people to act as subjects.
2 Provide each of them with a piece of paper and a pencil.
3 Tell your subjects that you are going to read to them a passage and that they should make a tick on the paper each time that you say the word 'a'.
4 Now read to them the following paragraph at a fairly brisk pace, tapping your pencil on the desk each time that you say 'a'. (You might find it helpful to encircle the letters in pencil, so that they stand out.)

'A farm worker was walking down a country road on a chilly, spring morning, when he saw a white shape immediately ahead of him. A closer look revealed that it was the body of a large sea-bird. It had a pair of webbed feet, a dark band on each wing, a long, orange beak and a crest of black feathers on its head. Death had obviously been caused by a blow on its neck (possibly from a passing car), although there was a smear of oil along its left wing.'

5 Now ask your subjects to draw a line under their ticks before continuing.
6 Read out the second paragraph, this time tapping your pencil on the desk when you say the word 'a' **and** when you say the words printed *in italics*.

'A few days later, he was passing *the* same spot when he saw the body of a second bird. He realised that *it* belonged to a different species (its feet were *not* webbed), but *its* overall size, shape *and* colouring were the same as those of the first bird.'

7 Collect the pieces of paper and count the numbers of ticks above and below the lines. Calculate the mean number of ticks per subject above the line and the mean number of ticks per subject below the line.

Questions

1 What is the ratio of average number of ticks above the line to number of 'a's in the first paragraph (13)?
2 What is the ratio of average number of ticks below the line to number of 'a's in the second paragraph (3)?
3 What feature of the experiment corresponded to the food in Pavlov's experiment?
4 What feature of the experiment corresponded to salivation in Pavlov's experiment?
5 What feature of the experiment corresponded to the bell in Pavlov's experiment?
6 Can you think of a children's game in which the principle of the conditioned reflex is used?

The presentation of the conditional and unconditional stimuli simultaneously or with a short time interval is called *reinforcement*. Many recent investigations into *learning processes* with animals have shown, experimentally, that learning progresses by reinforcement of the response with other stimuli such as food. Skinner's work on rats and pigeons shows how this reinforced stimulus can increase the speed of learning. These studies have been used to help the learning processes in Man especially in the development of teaching machines where each small step in learning is reinforced.

Small children learn many things through the approval or disapproval of their parents. They recognise that good behaviour brings reward in the form of encouraging words. As they grow up, this reward becomes the approval of other children or adults.

Experiment

To demonstrate learning in mice

Procedure

1 Construct a simple maze on a base-board about 0.5 m square. The walls should be at least 5 cm high and the passages at least 7 cm wide. 'Lego' is ideal as construction material, as the shape of the maze can easily be modified to produce patterns of greater or lesser complexity, and the base-board and blocks can be scrubbed and disinfected.
2 Place a sheet of glass or perspex on top of the maze to prevent the mice from jumping over the walls.
3 Place a mouse in the maze and allow it to become familiar with its surroundings for several minutes before removing it.
4 Place a few pellets of dry food at one end of the maze and introduce the mouse at the other end.
5 Record the time taken for the mouse to reach the food.
6 Remove the mouse when it has eaten the food, and carry out the same procedure on the following morning.
7 Repeat the procedure on several successive days and draw a graph of time taken to reach the food plotted against number of repetitions.
8 Repeat the experiment for several mice, washing the maze between runs.

Questions

7 What does your graph show?
8 What is the best time for the experiment to be performed?
9 What modifications to the maze can you suggest in order to prevent the mice from returning to the start?
10 Why should the maze be washed between runs?

11 Why should the food be dry?
12 Why should you leave 24-hour intervals between runs?

Many of the complex activities displayed by animals are not in response to a particular stimulus. They appear to come from within, have a definite aim, and are displayed by young animals where no opportunity for learning has been possible. Such activities are said to be *instinctive* and include the complex behaviour patterns of courtship and nest-building in birds, stalking and hunting in the cat family and territorial behaviour in many types of animal.

Although territorial behaviour is common throughout the Vertebrata, it is particularly well documented in the birds, on which most of the early research was carried out. Usually, this type of behaviour involves the establishment by a male bird of an area of ground on which he will not tolerate the presence of other males of the same species. Such intruders may be chased away by threatening postures or even by fighting: however, it is more usual for the song of the territory 'owner' to prove sufficient discouragement to other males seeking territories of their own. This song may well have a second and complementary function; the attraction of a female mate. Thus territorial behaviour is often most pronounced during the breeding season (see section 14.80).

The size of territory for any given species is closely dependent upon the amount of food that it contains, since it has to support a pair of adults who may not be able to move far from the nest, and a clutch of young

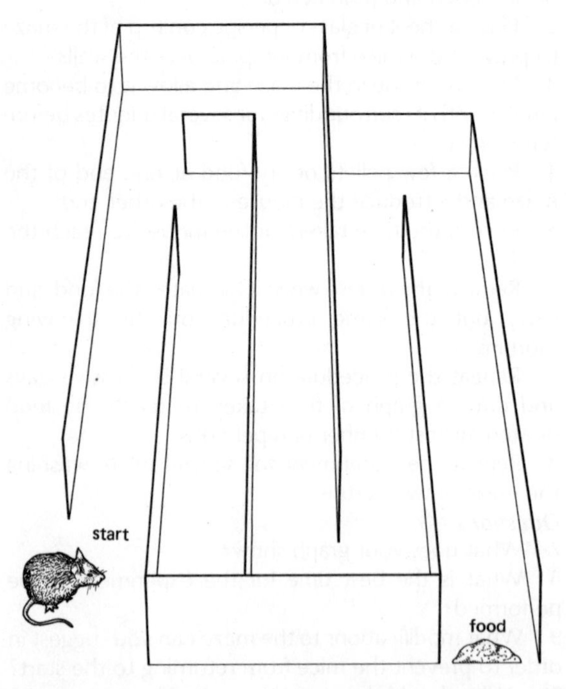

fig 7.16 Maze used to demonstrate learning in mice

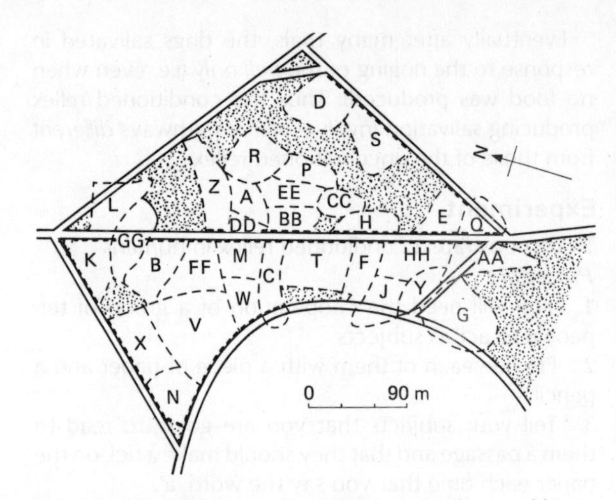

fig 7.17 A territory map of a population of willow-warblers in birch woodland. The size of territories varies from under 850 square metres to more than 4000 square metres. Shaded areas were unoccupied

birds who are completely immobile. The territories of large, carnivorous birds such as hawks may extend over many hectares of moor and mountain, whereas a relatively small garden may support a family of robins. The territories of birds that have virtually no song, such as the guillemots and razorbills, may be only a few square metres in area, since they depend on visual displays to warn off intruders. In these cases a nesting place only is required, since food is obtained by fishing over large areas of sea, and land area on the rocky sea coasts is very limited.

In most animals, contact between adults is restricted to the breeding season. Some species however, notably certain mammals and insects, have evolved highly developed systems of *social organisation*. Perhaps the best known example, apart from that of Man himself, is the honey-bee or hive-bee *Apis mellifera*. Each honey-bee hive contains individuals of three different *castes*, the queen, the workers and the drones. There is a *division of labour* between these three castes, so that each performs its own specific tasks in enabling the community to function efficiently (see Chapter 15). This is termed *social behaviour*.

The workers have a variety of functions. They feed the young larvae, clean the hive, cool it by fanning it with their wings, and collect food in the form of pollen and nectar from flowers. In this last task, they exhibit a very special type of behaviour. On returning to the hive after a successful hunt for food, the worker performs a curious dance which informs her fellow workers of both the direction and the approximate distance at which the food is to be found. Although this behaviour pattern is innate, a definite stimulus (the presence of food) is needed to trigger off the complex chain of events.

The behaviour of higher animals such as mammals is not entirely instinctive however, and they are able to interact with their environment to adjust their behaviour. They profit by their *experience* and so behaviour becomes *modified* as a result of *learning*. It is by virtue of their greatly enlarged cerebrum that such adaptable behaviour becomes possible, unlike birds whose general behaviour is largely instinctive.

7.80 Co-ordination without nerves

Homeostatic mechanisms of the mammalian body (see Chapter 6) are co-ordinated in two ways, by the nervous system and by the *endocrine system* of ductless glands. The nervous system has been likened to a telephone exchange with its telephone wires connected to individual telephones, but the endocrine system is less exact and its action could be compared to a public address system. This means that the products liberated into the *blood stream* by the ductless glands have a much wider effect on the body. The response could be temporary changes produced by a secretion from a ductless gland (hormone) at a particular moment e.g. *adrenalin* (see table 7.5), or they could initiate a complicated cycle of changes e.g. sex hormones preparing the female for ovulation, pregnancy, birth and lactation.

7.81 Evidence of activity of endocrine glands

Man sometimes removes the testes of many of his domestic animals in order to make the males more docile, and also to bring about an increase in weight. This operation, called *castration*, is performed on bulls (bullocks) and stallions (geldings) and male pigs. This operation is relatively simple in mammals, where the testes are external. It is also performed on birds with internal testes. When this is done with young cockerels they grow fat and more suitable for eating. They are known as *capons*.

In 1848, Adolf Berthold performed a series of experiments on the testes of young cockerels (see fig 7.18).

i) He removed the testes, and the bird became a capon.
ii) He disconnected nerve and blood to the testes. A new blood supply developed but not a new nerve supply.
iii) He transferred the testes to another part of the body and the testes developed a new blood supply but not a new nerve supply.

Questions

1 Examine the mature birds in experiments 2 and 3 shown in fig 7.18 on page 163. Do they resemble normal mature cockerels?
2 What controls the development of secondary sexual characters in the cockerel, hormonal secretions or nervous control? Explain the reasons for your decision.

One classical investigation of hormonal activity was made on the pancreas. In 1889, Mering and Minkowski found that on removing the pancreas from dogs, the animals developed a fatal condition called diabetes in which the urine contained sugar. In 1917, Kamimura tied off the pancreatic duct in rabbits, and although the cells secreting pancreatic juice died off, the animals did not discharge sugar in their urine i.e. did not develop the disease. In 1922 Banting and Best injected extracts of pancreas into dogs whose pancreas had been removed. The blood sugar levels of these dogs immediately fell from 0.3% to 0.21%. In their paper, dated 1933, they state:

'. . . in the course of our experiments we have administered over 75 doses of extract from pancreatic tissue to 10 different diabetic animals . . . the extract has always produced a reduction of the percentage of sugar in the blood and of the sugar excreted in the urine.'

Questions

3 Why could the experimenters finally conclude that the chemical secretion of the gland controlling blood sugar levels was secreted into the blood?
4 The pancreas extract does not completely cure diabetes, but has to be given at regular intervals. Why is this?

7.82 Endocrine glands and their functions

Gland	Position in body	Hormone secreted	Response of body to hormone	Abnormal functions
Thyroid	Neck	Thyroxin	Controls basic metabolism and growth rate.	Deficiency causes dwarfism and mental retardation in childhood, myxoedema in adult. Over production causes increased metabolism.
Islets of Langerhans	Pancreas (dual function exocrine and endocrine)	Insulin	Controls balance of sugar in the blood.	Deficiency results in diabetes mellitus.
Adrenal gland	Attached to kidneys	Adrenalin (medulla)	Controls response for 'fight or flight', i.e. increased heart beat, increased blood sugar, dilates coronary artery, dilates pupils etc.	
		Cortisone and other hormones (cortex)	Release glucose from protein in stress. Control salt balance in the body.	
Ovary	Dorsal abdominal wall	Oestrogen	Controls growth of uterus, hip skeleton, underarm hair, pubic hair, breasts, (secondary sexual characteristics).	Deficiency causes delay of appearance of these changes.
(See Chapter 8 section 8.39 – hormones in the sexual cycle)				
Testis	In the scrotum	Testosterone	Controls growth of hair on pubis, under arms and on chest and face, increased muscular development, deepening of voice.	Deficiency causes delay or lack of development of these changes.
Intestinal wall	Duodenum	Secretin	Controls the secretion of digestive juices from the pancreas (exocrine function of this gland). Produced when acid food enters the intestine.	
Pituitary	Beneath the brain	Several hormones	Controls the activity of other ductless glands.	Deficiency causes a type of dwarfism, and inactivity of certain endocrine glands.

Table 7.5 Endocrine glands and their functions

7.83

	Nervous control	Endocrine control
Stimulus	Through receptors, eyes, nose or internal receptors, include light, gravity, sound, temperature, etc.	Through external or internal receptors
Linking mechanism	Central nervous system and nerves	Blood and circulatory system
Effectors	Muscles and glands	Whole body, organs or organ systems
Speed	Rapid reaction – reflex arc or voluntary nerve paths	Slow for some such as growth; rapid for others such as fight or flight hormone, adrenalin.

Table 7.6 Comparison of nervous and endocrine activity

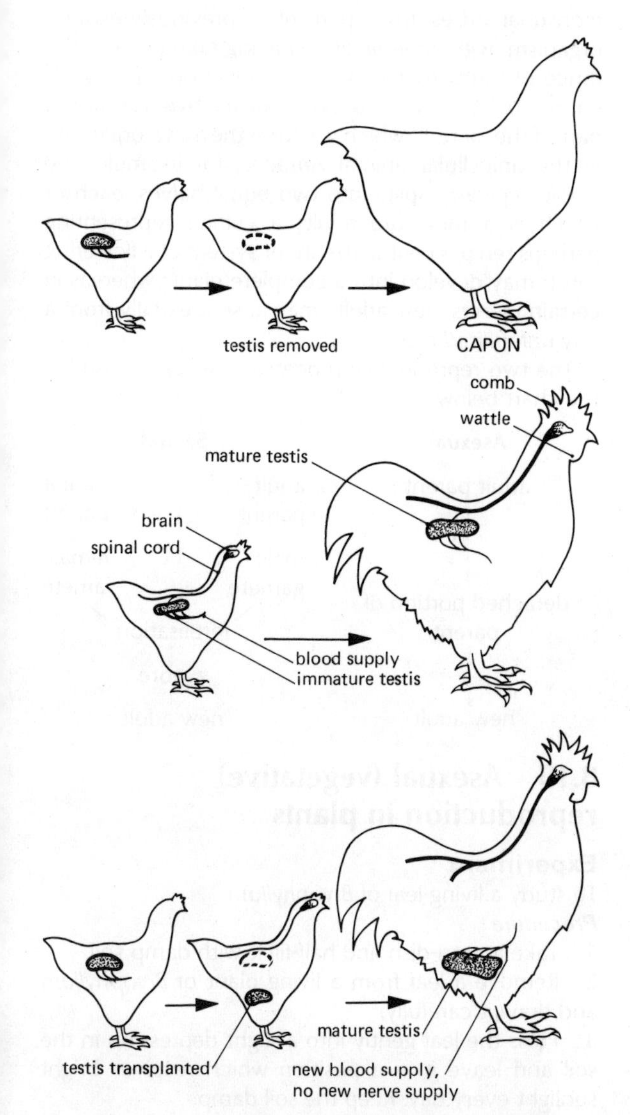

fig 7.18 Three experiments performed by Berthold to investigate the action of the testes as ductless glands

Questions requiring an extended essay-type answer

1 a) What do we mean by the term irritability in living organisms?
b) Describe how each of the following illustrate this phenomenon
i) a deer sniffing the wind with a predator near,
ii) a germinating seed, and
iii) a potted *Geranium* growing near a window.
(See also Chapter 10.)

2 a) Describe how the eyes of a man adjust
i) when he looks down at a newspaper in his hand after watching distant walkers on a hillside, and
ii) when he steps from a well-lit house into the darkness.
b) Describe, briefly, the ways in which a green plant responds to light. (See also Chapter 10.)

3 a) Explain what happens when
i) a bright light from a torch beam is directed into the eye of man, and
ii) light shines on the coleoptile (shoot) of a maize seedling from one side.
b) How do these responses differ from one another? (See also Chapter 10.)

4 a) Compare the structure of the mammalian eye with a camera. In what ways do their workings differ from each other?
b) Describe how the human eye is corrected for long sight and short sight.

5 a) Draw a fully labelled diagram of the mammalian ear.
b) Describe how it is adapted to work efficiently.

6 a) Draw a labelled diagram to show the structure of the inner ear of a mammal.
b) Describe a *controlled* experiment you would carry out to determine the frequency range audible to an old man.
c) How would you expect the auditory range of a boy of fifteen years to differ from that of the old man?

8 Reproduction

8.00 Introduction

Every individual living thing has a limited life span. This may range from a few seconds for certain micro-organisms to thousands of years in the case of large trees. However, all organisms have the ability to pass on life by producing new individuals with the same general characteristics as themselves; this is known as *reproduction*.

Basically, there are two different methods by which reproduction may take place; *sexual* and *asexual*. In sexual reproduction, new individuals are formed by the *fusion* (joining) of small pieces of living material, called *gametes*, from two different organisms. In mammals (including Man) a male gamete or *sperm* fuses with a female gamete or egg. The sperm and the egg (or ovum) are highly specialised cells, completely different from each other in appearance but complementary in function. The details of sexual reproduction are slightly different for plants and for unicellular organisms, but the principle is exactly the same in all cases.

fig 8.1 A leaf of *Bryophyllum*

The process of fusion between male and female gametes is termed *fertilisation*; the male gamete is said to fertilise the female gamete and not vice versa. The fertilised ovum is known as a *zygote*, which will develop to form, eventually, a new adult organism.

Asexual reproduction takes place when a new individual arises from part of a previously-existing organism, *without fertilisation* having taken place. The range of forms that asexual reproduction can take is enormous, as is the proportion or relative size of the part of the 'parent' which will form the 'new' organism. In the unicellular animal *Amoeba*, for example, the *whole organism* splits into two equal halves, each of which is a miniature adult; a *cutting* representing perhaps ten per cent of the shoot system of a flowering shrub may develop into a complete plant; whereas in certain insects, new adults may arise asexually from a tiny *unfertilised* egg.

The two reproductive processes are summarised in the chart below.

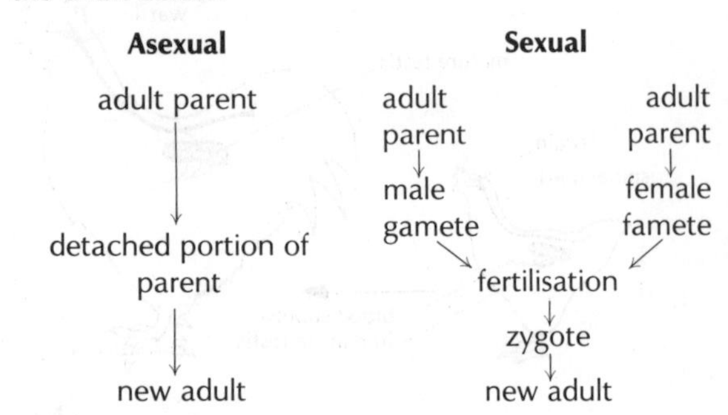

8.10 Asexual (vegetative) reproduction in plants

Experiment

To study a living leaf of *Bryophyllum*

Procedure

1 Take a petri-dish and half-fill it with damp soil.

2 Remove a leaf from a living plant of *Bryophyllum* and draw it carefully.

3 Press the leaf gently into a slight depression in the soil and leave it in a position which receives bright sunlight every day. Keep the soil damp.

4 Examine the leaf carefully every second day, over a period of one month. Record your observations in the

form of a table, and water the leaf after every examination.

5 After a while, small plantlets will become detached from the original leaf. Record their numbers at each examination. Draw one of the plantlets carefully and label your drawing.

6 Remove four or five plantlets and place them in a pot which has been filled to a depth of 5 or 6 cm with soil.

7 Water these plantlets every two or three days, and follow their development over several weeks. Transfer to new pots as necessary. (See fig 8.1.)

Question

1 How can you tell that the structures produced by the leaf are complete young plants?

Other examples of this type of reproduction can be studied in the potato (*Solanum tuberosum*). A potato tuber can be cut into several portions (each with a bud or 'eye' – fig 8.2) which can be planted separately in soil, to produce new potato plants.

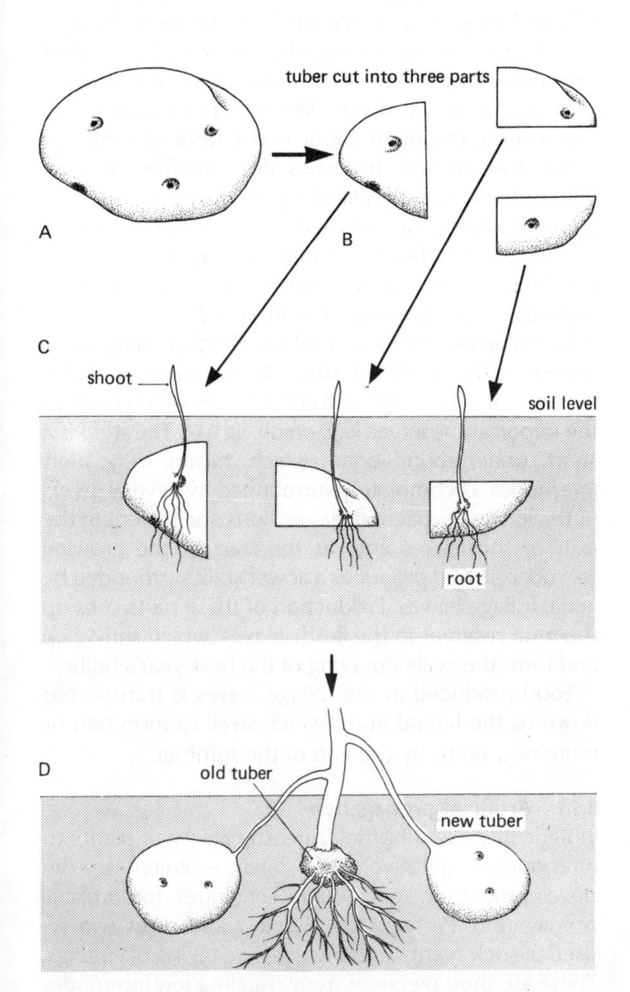

fig 8.2 The artificial propagation of *Solanum*, the potato

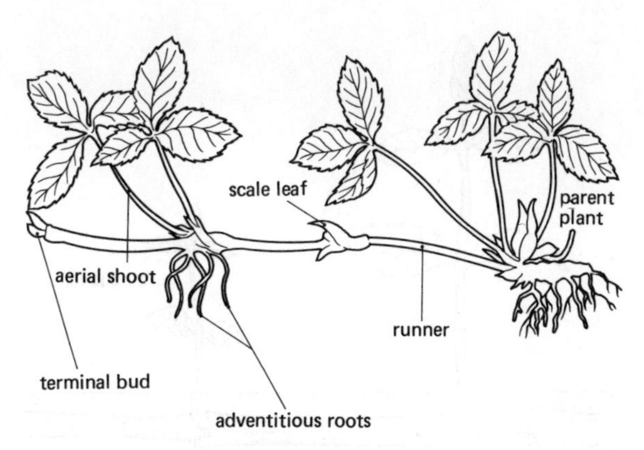

fig 8.3 A runner of strawberry (*Fragaria*)

8.11 Runners and suckers

Many plants have side-branches especially adapted for vegetative reproduction. In the strawberry (fig 8.3), a slender shoot called a *runner* arises from the axil of a leaf at the level of the soil. It grows horizontally along the surface of the soil, and at the nodes it grows *adventitious roots* and *aerial shoots* to form a complete new plant. In this way reproduction is independent of any outside agent, either for fertilisation or for dispersal (although the latter is, of course, restricted by the length of the internode). Also, the young plant receives a supply of nutrients via the runner until it has become established. After establishment of a new plant, it may or may not become separated from the parent plant by the rotting away of the runner.

fig 8.4 Houseleeks (*Sempervivum*). These plants reproduce vegetatively by means of offsets and are planted on cottage roofs to keep slates in position

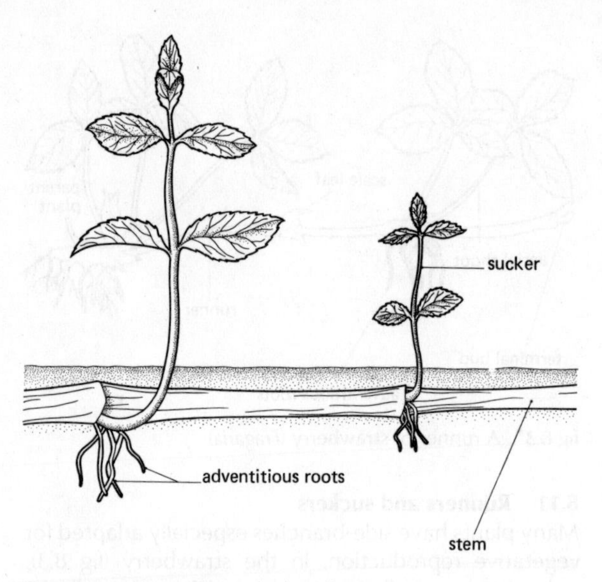

fig 8.5 Suckers of *Mentha* (mint)

The terms *stolon* and *offset* are sometimes applied to short runners of various kinds. A particularly interesting example is found in the houseleek (*Sempervivum*) (fig 8.4).

A *sucker* is similar to a runner except in that it arises *underground* and grows to the surface, where it develops adventitious roots and aerial shoots to produce a new plant. Well-known examples of plants reproducing by means of suckers are the raspberry and mint (fig 8.5).

8.12 Vegetative reproduction and perennation

In a number of well-known cases, the asexual reproductive organs of flowering plants also act as *food stores*. They enable the plant to survive from year to year. The plants may lie dormant over winter and use their stored nutrients in a sudden spurt of rapid growth as soon as favourable conditions return in spring. Survival from year to year in this way is called *perennation* and the organs are known as perennating organs.

Two points are worth emphasising in this context:

1 The food store must remain sufficiently large at the end of the winter to enable the plant to grow rapidly and produce foliage leaves that will take full and immediate advantage of the suitable conditions for photosynthesis.

2 The perennating organ can only be regarded as reproductive if it gives rise to *more than one* 'new' plant.

Important examples of perennating plant organs include rhizomes, corms, tubers and bulbs.

Rhizomes are *underground stems* which grow *horizontally* and then turn upwards to produce leafy shoots forming the terminal bud. The leafy shoots produce food by photosynthesis during the rainy season. This food passes down to the rhizome, which uses it to produce branch rhizomes from lateral buds. By the autumn, when the leafy shoot withers and dies, the rhizomes from the lateral buds have grown for some distance and have stored enough food reserves to produce new leafy shoots at the beginning of the next spring. Well-known examples of rhizomes include those of lily of the valley (*Convallaria*) and couch grass (*Agropyron*).

Another type of modified stem which serves both for food storage and for vegetative reproduction is the *corm*. In this case, the underground stem is *upright* and very *short*. At the beginning of the spring it is swollen with food reserves which enable the terminal bud, containing the foliage leaves and flower stalk, to grow rapidly. The food materials manufactured by the leaves are passed down to the part of the stem immediately above the old, shrinking corm. A new corm forms on top of the old one. In addition to the terminal bud, the corm produces *lateral buds* which develop their own foliage leaves and hence are able to store food in *'daughter' corms* for use over the winter and the start of the next spring. Corms of *Gladiolus* are shown in fig 4.4.

In the case of plants like the potato (*Solanum tuberosum*), the main shoot is not underground but aerial. Certain side branches do, however, develop underground swellings filled with food reserves. These are *stem tubers*, each of which may develop a number of shoots from the lateral buds arising in the axils of small scale leaves and scattered over the surface of the tuber as 'eyes'. (See section 8.10, fig 8.2.)

In rhizomes, corms and tubers, the food material is stored in the modified stem. In the case of *bulbs*, however, it is the large and *fleshy leaves* which contain the important reserves (e.g. onion, fig 4.6). The stem is a short, underground organ which, having very short internodes, is completely surrounded by several layers of thick, closely-packed leaves. Buds are present in the axils of the leaves and, at the start of the growing season, one bud produces a flower stalk surrounded by aerial foliage leaves. Production of these parts uses up the food reserves in the fleshy leaves, which shrivel up and form the scaly covering of the next year's bulbs.

Food produced in the foliage leaves is transported down to the lateral buds, which swell to form two or more new bulbs by the end of the summer.

8.13 Artificial propagation

In agriculture and horticulture, the ability of plants to reproduce vegetatively has been exploited in the development of specialised techniques for artificial propagation. Perhaps the simplest and most widely-used of such techniques involves the taking of *cuttings*. These are short pieces of stem, usually a few internodes long, with two or three leaves at the top end. The

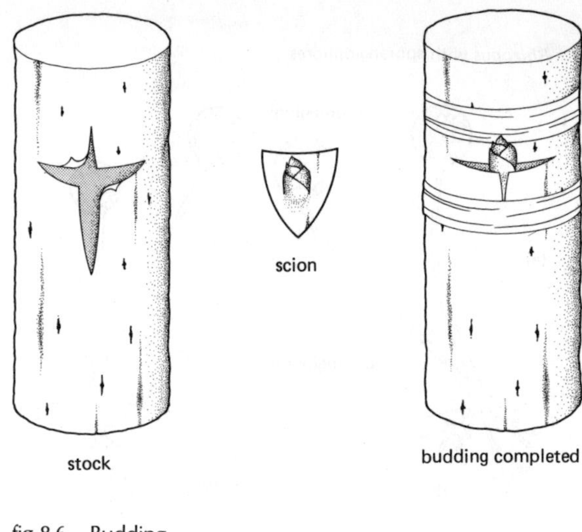

fig 8.6 Budding

bottom end is cut obliquely through or just below a node, from which adventitious roots will arise, and placed immediately in well-watered and aerated soil. Cuttings are used extensively in the propagation of flowering shubs.

Budding is a more highly skilled technique, whereby a dormant bud taken from one woody plant (the *scion*) is made to grow on the well-established root system of another (the *stock*). Usually the two plants involved belong to related species or to different varieties of the same species. (See fig 8.6.)

Grafting is a similar exercise, in which the scion is a twig rather than a bud. In both techniques, the essential point is to bring the exposed *cambia* of the two plants into direct contact. To this end, a cross-shaped cut is made in the bark of the stock, and the flaps are turned back. Into the resulting slot is placed a small piece of bark from the scion, bearing the bud or twig to be grafted, this is then bound in position, using waterproofed raffia, wool or tape.

This type of propagation is used extensively in the growing of roses.

8.14 Use of vegetative reproduction by farmers and gardeners

In addition to artificial vegetative propagation, agriculturalists have more and more come to rely on modifications of natural means of vegetative reproduction for producing the best crops. We have already considered briefly the cutting-up and sowing of pieces of potato tuber (section 8.10). In a similar way, iris rhizomes can be cut into short lengths for planting, and mint suckers can be separated and grown independently. The essential point about these modifications is that they tend to *spread* the daughter plants *more efficiently* than would be likely under natural conditions. Thus they are able to make the best possible use of the land available, while utilising the minimum number of perennating organs. In corms (e.g. *Gladiolus*) and bulbs (e.g. onion), there is very little natural separation, and the young organs must be separated by hand for a worthwhile yield to be obtained.

The main advantages to the farmer of vegetative crop propagation are as follows:

1 *Reliability.* They are less likely to be affected by adverse climatic conditions than are seed-propagated plants, which require a combination of warmth and moisture for germination to occur.

2 *Consistency.* Since it is genetically identical to its parent, a vegetatively propagated plant is likely to produce the same quality and quantity of yield.

3 *Early maturity.* Because of the relatively large food supply present in the perennating organ, shoot systems are able to develop rapidly and take full advantage of the rainy season for photosynthesis. Also, in the case of a budded or grafted scion, the well-established root system of the stock enables a woody shoot system to develop more rapidly, and thus to produce fruit earlier.

4 *Geographical range.* It is often possible to graft scions off foreign plants onto strongly-growing stocks of local related species. This can result in excellent yields of crops which would not grow readily from seed outside their natural environment.

5 Finally, there are certain plants which cannot reproduce sexually, and for them vegetative reproduction is the *only possible means of propagation*: examples include the 'seedless' varieties of orange and other types of citrus fruit.

Of course, along with the advantages of vegetative propagation there is one obvious disadvantage. The consistency referred to in 2 above also implies a lack of variety and flexibility. Therefore, asexually propagated crops are more susceptible than others to changes in climate; and attacks by pests are more likely to spread rapidly, since, if one plant provides a suitable environment for a parasite, the chances are that its neighbours will also. In other words, factors which have an adverse effect on one plant will tend to affect the whole crop in the same way.

8.20 Asexual reproduction in lower organisms

8.21 Binary fission

This is the simplest form of reproduction. It involves the splitting of one parent cell to form two daughter cells and takes place in bacteria and in many unicellular plants and animals. Fig 8.7 shows binary fission in the protozoan *Amoeba*. Division always takes place after

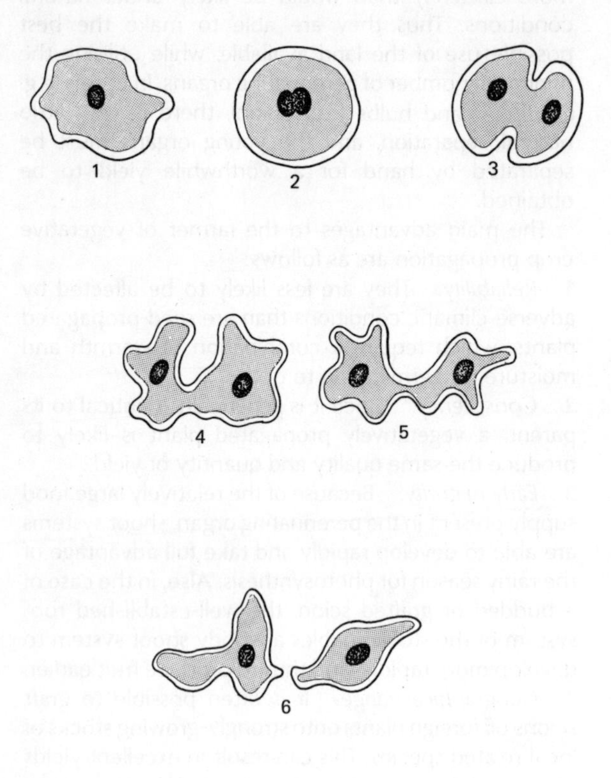

fig 8.7 Binary fission in *Amoeba*

the cell has grown to a certain size. Since all parts of the parent Amoeba are used in the production of the two new cells, the life cycle does not involve the death of the previous generation. Individual amoebae do die, of course, for the usual reasons of starvation and chemical or mechanical damage, but no 'natural' wastage is 'built into' the reproductive process, which may consequently be regarded as efficient in terms of food resources.

8.22 Cell division and fragmentation

The filamentous alga *Spirogyra* represents one step along the evolutionary path that has led from unicellular to multicellular organisms. The plant consists of a series of identical, cylindrically-shaped cells joined end to end in a 'string' or filament. Binary fission of the cells occurs in a transverse plane. However, the daughter cells do not usually separate, but remain joined together to increase the length of the filament. Separation does occur occasionally, possibly as a result of increased mechanical stresses on a greatly elongated filament. This process is called *fragmentation* and results in the formation of two daughter filaments. Thus it is a form of asexual reproduction.

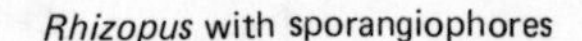

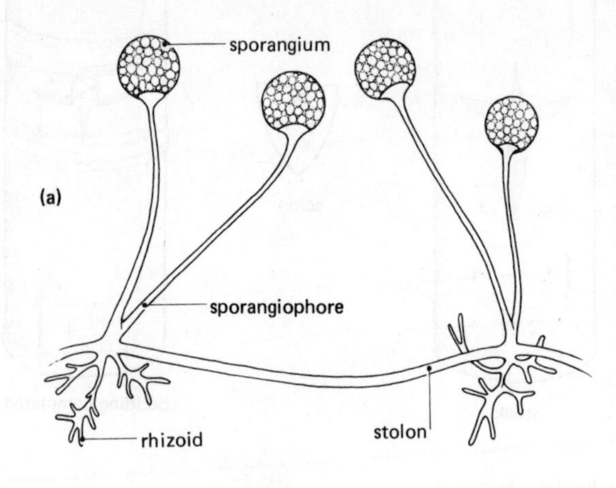

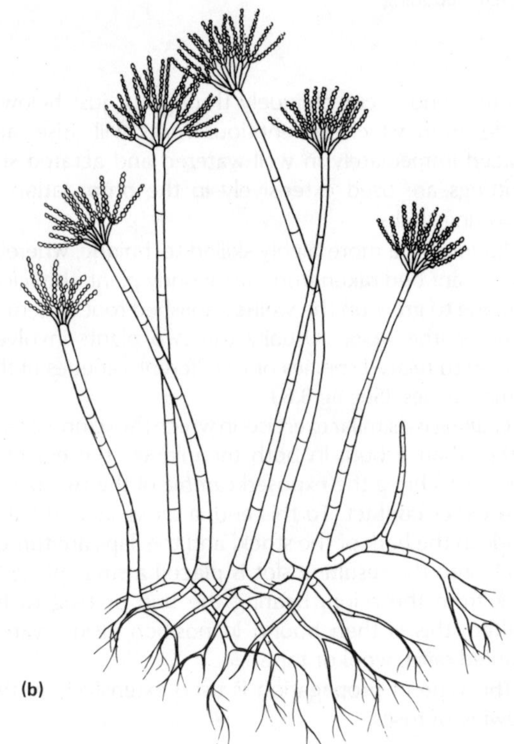

fig 8.8 Spore production in (a) *Rhizopus* (b) *Penicillium*

Spirogyra is also able to reproduce sexually by means of *conjugation*, a process in which the contents of a cell from one filament flow into a cell of another filament.

8.23 Spore production (sporulation)

Many organisms, notably the lower plants, reproduce asexually by means of highly specialised structures called spores. In the mould-fungus *Rhizopus*, found on decaying bread, fruit and other organic material, the 'body' consists of threadlike hyphae of three different kinds: *rhizoids,* which penetrate down into the substratum to provide anchorage and absorb nutrients; *stolons,* which run along the surface of the organic matter; and *sporangiophores,* which grow upwards and bear swollen sporangia at their tips (fig 8.8(a)). The protoplasm of the sporangium is cut off from that of the sporangiophore by a septum and then becomes divided up into thousands of multicellular spores. The spores are released when the sporangium wall ruptures, and are *dispersed* by air currents. Each spore, if it lands in suitable conditions, is capable of producing a new mycelium or network of non-septate hyphae.

Penicillium is a mould which produces spores in a slightly different way. The protoplasm of its hyphae is divided into several *multinucleate* sections by septa. As in the case of *Rhizopus,* the spores (here called *conidia*) are borne on upright hyphae (*conidiophores*). However, they are not enclosed in a capsule or membrane, but arise as constrictions of the branched end of the conidiophore. As they mature, they are dispersed by the wind (fig 8.8(b)).

8.30 Sexual reproduction in mammals

In mammals and other higher animals, *sexual reproduction* is normally the *only* means of producing offspring. Asexual reproduction is virtually unknown under natural conditions, but *parthenogenesis* can be induced artificially in certain amphibians by, for example, pricking an unfertilised egg with a needle. The question of whether such 'virgin birth' is possible in humans has never been satisfactorily resolved.

Two important features of mammalian reproduction stem from the fact that mammals are essentially terrestrial in habit, for even seals and whales arose from land-living forms. These two features are *internal fertilisation* and *viviparity* (the birth of live young).

Most aquatic animals, from sea-snails, crustaceans and marine worms to amphibians and fish, fertilise their eggs *externally.* That is to say, the female deposits the eggs outside the body and the male releases millions of sperms into the water nearby. The sperms swim to the eggs and fuse with them. In terrestrial animals, such a simple system is not possible, since the sperms, in order to swim, must be kept in a liquid environment.

Therefore, most land-based animals have evolved specialised mechanisms for getting the sperms as near as possible to the ova before release. In earthworms, two animals are enclosed together in an elastic sheath called the *clitellum*; in some insects an envelope of sperms, the *spermatophore,* is deposited in the female reproductive tract; and in land snails and mammals the sperms are introduced into the female orifice by a special organ, the *penis.* Once inside the female tract, the sperms can swim along the mucous lining and fertilise the eggs.

Eggs which have been fertilised externally develop, of course, in a watery medium which is relatively free from fluctuations in temperature. If, however, a land-dwelling animal were to release immediately eggs which had been fertilised internally, there would be severe problems of dehydration and of over- and under-heating, since land provides a far less constant environment than water. Thus land animals that are *oviparous* (egg-laying) usually protect their eggs from the external environment by covering them with some form of *waterproof membrane* or *shell* before laying. We can see here how external fertilisation and oviparity would be unlikely to occur together in a terrestrial environment, since a shell sufficiently waterproof to prevent the exit of water molecules would be unlikely to permit the entry of sperms.

The vast majority of mammals do not, however, lay eggs. They carry the idea of protection from the environment one step further and, instead of secreting waterproof membranes around the fertilised eggs, they keep them within the maternal body. In this way they are also able to supply the embryo with all the nutrients required for development. Thus, the eggs of mammals are much smaller than the eggs of birds or even of amphibians, which need to contain very large supplies of food materials, in the form of *yolk,* to last them until hatching.

A comparison of reproduction in the different vertebrate groups is included in table 17.2 (p. 326).

Sexual cycles

In most higher animals, the endocrine system plays a large part in reproduction. The secretion of certain hormones by the pituitary gland determines the time at which eggs can be fertilised successfully. In most mammals, the female will only allow *insemination* (the introduction of sperms into the female reproductive tract) via the penis of the male, at a time when fertilisation is likely to result. At this time, under the control of hormones, the wall of the womb or *uterus* is at its thickest, and eggs are released from the ovaries (a process called *ovulation*). Female mammals in this condition are said to be 'on heat' or 'in season', and they may produce a secretion to arouse and excite the male. If fertilisation does not take place at this time, the wall of the uterus gradually becomes thinner again, the extra material being reabsorbed into the body – usually without bleeding. The process is repeated at regular intervals which vary tremendously, depending upon the species: in mice it is five days; in dogs and cats it is about six months. The period of fertility is known as *oestrus,* and the regular cycle of events is referred to as the *oestrus cycle.*

In humans, the female sexual cycle differs from the above pattern in a number of important respects. The female is receptive to the male *almost equally at all times,* although *ovulation* occurs, from one or other ovary, only once about every *twenty-eight days.* If fertilisation does not occur, the lining of the uterus is not all reabsorbed. Instead, part of it is sloughed off and passes out of the vagina, together with a certain amount of blood and mucus, as the *menstrual flow.* This period of bleeding lasts about five days and is known as *menstruation.* It occurs about midway between one ovulation and the next. These monthly periods take place over the whole of a woman's reproductive life, starting with puberty at the age of about eleven or twelve and ending with the *menopause* or 'change of life', usually during the late forties.

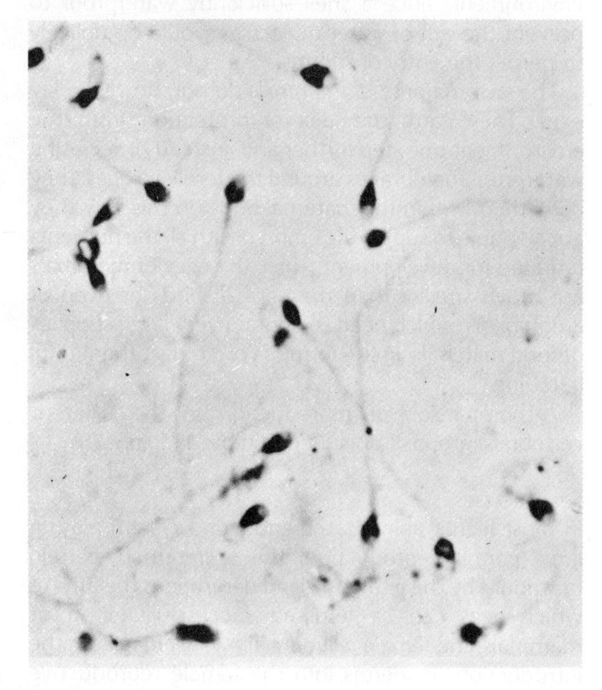

fig 8.9 A photograph of some highly magnified sperm cells (x 1000)

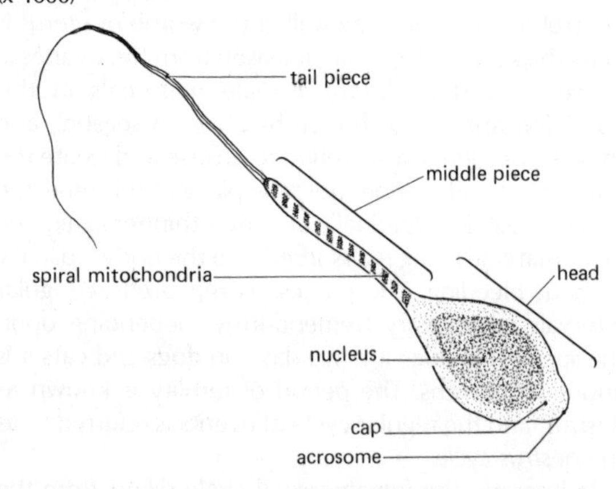

fig 8.10 A drawing of a sperm cell (x 5000)

The gametes

Throughout the whole of the animal kingdom, the structure of spermatozoa and ova is remarkably constant. This is mainly a reflection of the uniformity of their function. The sperm is required:

1 to carry one complete set of genetic information from the male parent;
2 to swim rapidly in an aqueous medium and
3 to fuse with the egg, penetrating any covering membranes that the latter may possess.

To these ends, each sperm (fig 8.10), during the course of its development from an unspecialised body cell:

1 loses one of the two sets of genetic information (see fig 9.6) from its nucleus in the process of *meiosis;*
2 cuts down its weight to an absolute minimum by dispensing with excess water and cytoplasm, and at the same time develops a long tail (flagellum) for swimming ('powered' by organelles called mitochondria which produce the ATP needed for energy conversion);
3 develops at its anterior end an *acrosome,* consisting of a sac of enzymes used for breaking down the egg membranes.

Thus the mature sperm has a head (consisting of a condensed nucleus with an apical acrosome), a middle piece containing mitochondria and a long, slender tail.

The ova, almost invariably spherical or ovoid in shape, are usually the largest cells in an animals's body. This is because they contain reserve foodstuffs in the form of yolk. Even in mammals, where the developing embryo obtains most of its nutritional requirements from the *maternal body,* the egg cell contains some yolk and may be over 100 μm in diameter (fig 8.11).

The ova of most animals are surrounded by one or more egg *membranes.* They usually serve for protection but, in the case of mammals they may also effect the absorption of nutrients from the immediate environment. Mammalian ova are surrounded by a layer of *follicle cells,* arising from the ovary.

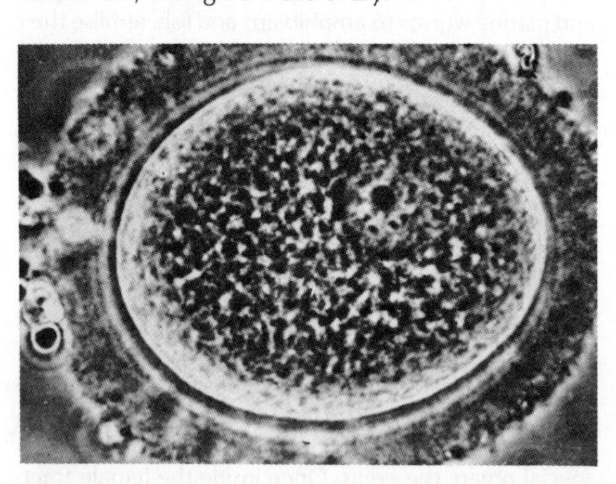

fig 8.11 An ovum of a human (x 500)

Finally, and most important, the egg cell carries one set of *genetic information* (from the female parent) in its nucleus. As in the sperm, one of the two original sets is lost by the process of meiosis.

8.31 The male mammal.

Demonstration

Dissection of the male rat

Procedure

1 A freshly-killed male rat is laid on its back on a dissecting board. Four awls are used to pin its limbs to the board.

2 The loose ventral skin of the abdomen is lifted with forceps away from the body, and a median, ventral incision is made, with scissors or a scalpel, from the rib cage to the pit of the abdomen.

3 · Second and third cuts are made along the legs, and the skin is pinned back.

4 Now the muscular abdominal wall is cut and pinned back in the same way. The coils of the intestine are thus exposed.

5 The intestine is unravelled and the rectum is severed about 6 cm from the anus. The whole of the intestine is then pinned back, well away from the abdominal cavity.

6 The urinogenital system of the male rat is now clearly visible. The skin (*scrotum*) surrounding the testes is carefully removed.

Questions

1 Two tubular structures are visible leaving each testis. What is the function of the more anterior tube (the spermatic cord)?

2 What is the function of the tube leaving the posterior part of the testis?

3 The urethra is the duct inside the penis. How many structures can you see opening into it?

Fig 8.12 above shows a dissection of the reproductive system of a male rat, such as has been demonstrated to you. The *spermatozoa* are produced in the paired ovoid testes, each of which consists of many coiled tubules, with sperms developing in their walls. The sperms pass from the testis into the *epididymis*, which lies against the testis and also consists of a series of coiled and looped tubes. The epididymis may store the sperms temporarily and the walls of its tubules produce a secretion. The *vas deferens* then conducts the sperms to the base of the *penis*, where they receive the secretions of the *prostate* and *Cowper's glands* and the *seminal vesicles*. These secretions, together with those of the epididymis and of the testis, and of course the sperms, make up the *seminal fluid*, or *semen*, which is ejaculated through the *urethra* during copulation. The urethra, which lies inside the penis, also carries the *urine* from the bladder to the exterior and is consequently called a *urino-genital* structure.

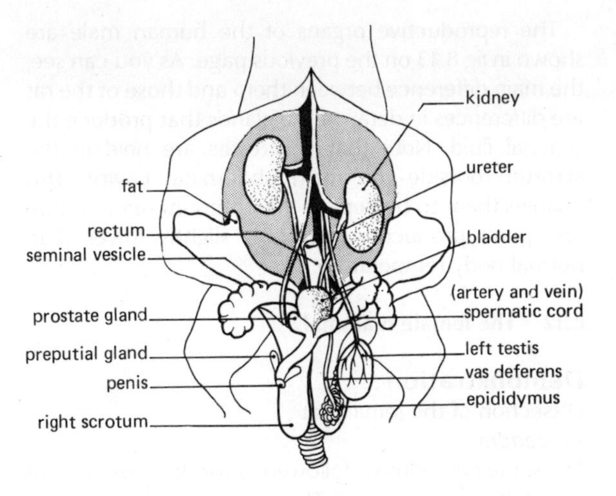

fig 8.12 The reproductive system of a male rat

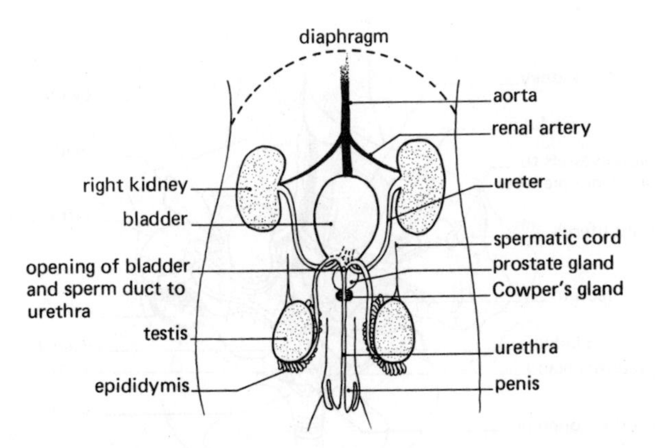

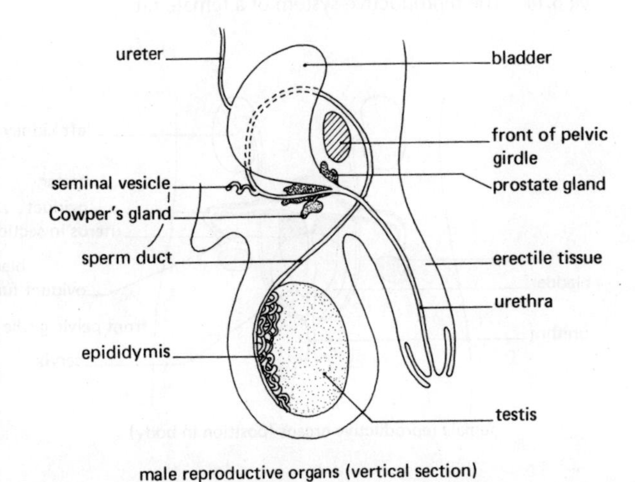

fig 8.13 The reproductive system of a male human

The reproductive organs of the human male are shown in fig 8.13 on the previous page. As you can see, the main difference between them and those of the rat are differences in detail of the glands that produce the seminal fluid. Note that the testes are held in the scrotum outside the main abdominal cavity. This enables them to be kept at the optimum temperature for sperm production, which is slightly lower than normal body temperature.

8.32 The female mammal

Demonstration

Dissection of the female rat

Procedure

The same procedure is followed as for the dissection of the male rat in section 8.31 above. On removal of the alimentary canal, the female urinogenital system is clearly exposed.

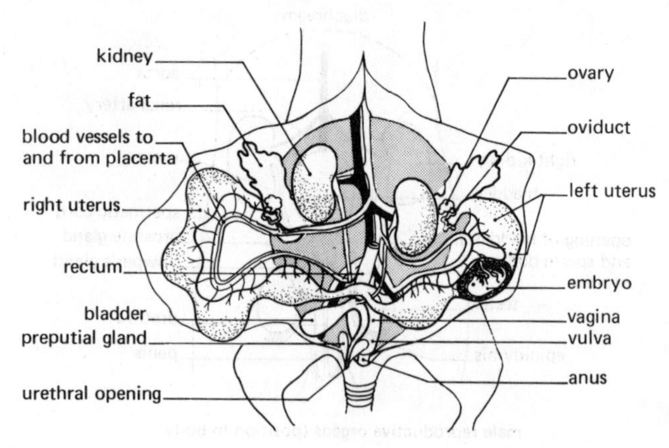

fig 8.14 The reproductive system of a female rat

Questions

1 In what position do the ureters, bearing urine from the kidneys, enter the bladder?
2 Are the two kidneys situated at the same level?
3 What is the function of the urethra?
4 What other differences can you observe between this dissection and the dissection of the male rat?

The stem of a large Y-shaped structure runs below (dorsal to) the urethra, compared with which it is about twice as wide and 50% longer. This stem which opens to the exterior just posterior to the urethral opening, is the rat's *vagina*, a muscular structure which receives the penis during copulation.

The arms of the Y extend to just below the kidney and may be covered with fat, which should be removed. These 'arms' form the twin *uteri* (wombs) of the rat. It is quite possible that the dissected rat may have been pregnant. If this is the case, compare the uteri with those of a non-pregnant rat and note the huge increase in size. Note the positions of the developing embryos and count them.

The 'knobs' seen on the ends of the uteri are coiled *oviducts* or *fallopian tubes* which lie partly around and alongside the small *ovaries*. The fallopian tubes conduct the released eggs from the ovaries to the uteri.

Using a needle and forceps, try to unravel the fallopian tubes as far as possible. It is not an easy task, since the tubes are fairly delicate and are held in position by connective tissue. (See fig 8.14).

The reproductive system of the human female (fig 8.15) differs from that of the rat in that there is only one uterus and the fallopian tubes are not coiled.

8.33 Ovulation in humans

When a baby girl is born, each ovary already contains tens of thousands of potential egg cells (*oogonia*). Of these, only a small fraction (about 450) will be released from her ovaries during her reproductive life. As an

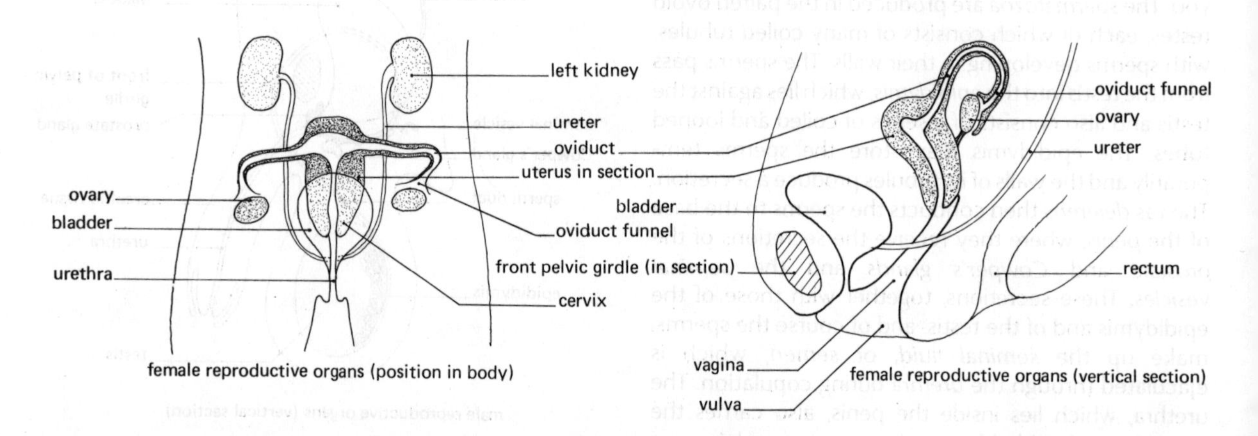

fig 8.15 The reproductive system of a female human

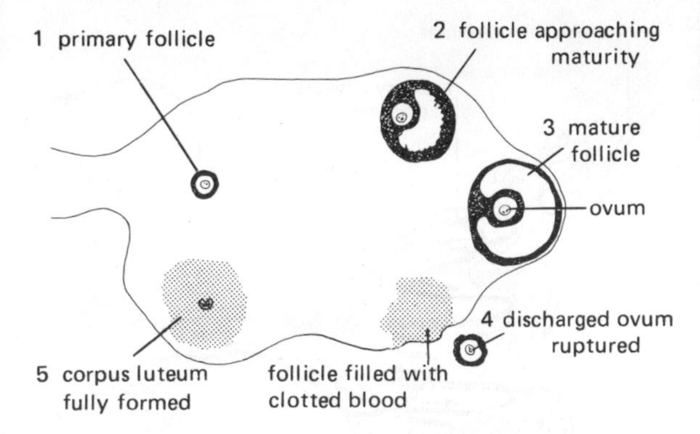

fig 8.16 T.S. of a mammalian ovary

oogonium begins to grow to form a mature egg, it is known as an *oocyte*. The smaller cells around it form a follicle, which eventually develops a large fluid-filled cavity. At this stage, the whole structure is known as a *Graafian follicle*. Its cells help to provide the growing oocyte with nourishment. At the end of the period of growth, the nucleus of the oocyte starts to divide meiotically (see 9.04) in order to reduce its number of chromosomes by half. (See fig 8.16).

The mature Graafian follicle stands out as a bump about 0.5 cm high on the side of the ovary, which is itself about 4 cm in maximum diameter. The follicle bursts and its egg, still surrounded by a layer of inner follicle cells (called the *corona radiata*) is released into the *body cavity*. The cilia on the cells lining the *funnel* of the oviduct help to propel the egg down the fallopian tube on its way to the uterus.

8.34 Sperm production in humans

Sperm production is a continuous process in Man. Spermatozoa are formed from certain cells (*spermatogonia*) which are found in the lining of the seminiferous

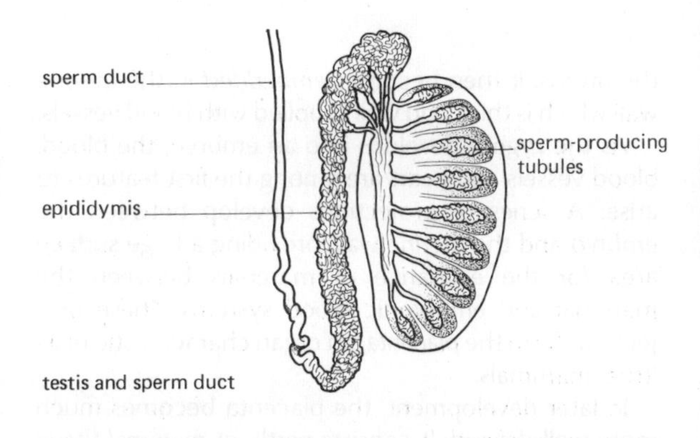

fig 8.17 L.S. of a mammalian testis

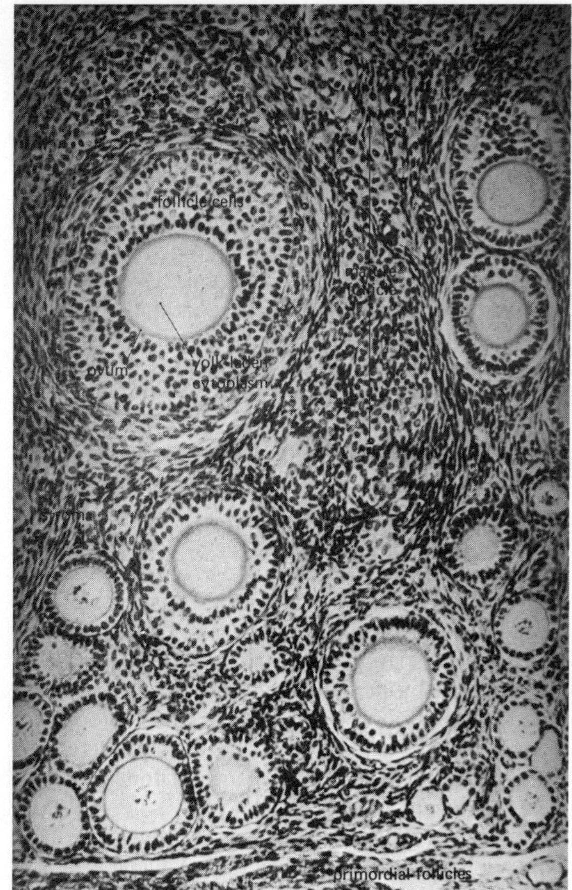

fig 8.18 Photographs of (a) T.S. mammalian ovary (b) T.S. seminiferous tubules

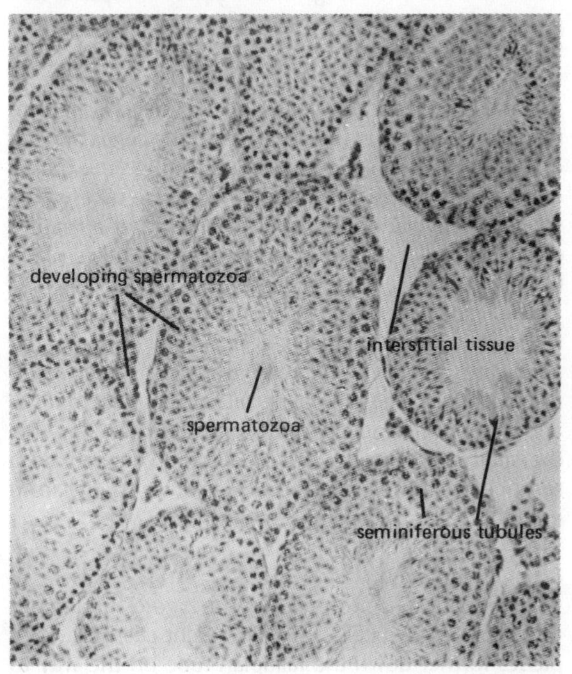

tubules of the testis. (See fig 8.18(b)). Unlike oogonia, spermatogonia are constantly being produced by *mitotic* cell division (see 9.01). The cells thus formed grow to about double their original size (when they are known as *spermatocytes*) and then divide meiotically (see 9.04) to form *spermatids*. The spermatids change shape to develop into the typical spermatozoon form (fig 8.10) and then pass into the epididymis for storage.

During copulation, the sperms, together with the secretions of the prostate and seminal vesicles, are expelled through the urethra by contractions of the duct walls and associated muscles. Sperms which have not been expelled in this way are reabsorbed into the cells of the epididymis wall.

8.35 Copulation and fertilisation

Coition, intercourse, and *mating* are all alternative terms used to describe the act in which the penis of the male is inserted into the vagina of the female and *spermatozoa* are ejaculated into the female reproductive tract. In order for the penis, usually a soft and flaccid organ, to penetrate the vagina, it must become *hard and erect*. In most mammals, including Man, this is accomplished by increasing the amount of blood in the soft tissues around the urethra, thus producing turgidity. The stimulus for erection is largely psychological and depends upon arousal and excitement of the male by the female.

Penetration is also made easier by the secretion of lubricating mucus by cells lining the vaginal wall and the external genital area (the *vulva*). Movement of the penis backwards and forwards inside the vagina stimulates nerve endings in the tip of the penis (*the glans*), which set in motion a reflex action resulting in the ejaculation of semen.

On ejaculation, about 1.5 cm^3 (in humans) of seminal fluid containing about 100 million spermatozoa, is deposited in the vagina, near the *cervix* or neck of the uterus. The sperms *swim* up through the uterus and into the fallopian tubes. If intercourse takes place soon after ovulation, the sperms encounter a mature egg in the fallopian tube and fertilisation takes place there.

The first step in fertilisation is the breaking down of the corona radiata by an enzyme released by the acrosome of the sperm. After these cells have been dispersed, a single sperm penetrates the egg membrane and the sperm head enters the egg cytoplasm, leaving the sperm tail outside. This penetration triggers off a response in the egg, which prevents other sperms from entering. Later the egg and sperm nuclei *fuse*.

8.36 Pregnancy

After fertilisation, the egg, now a *zygote*, undergoes divisions into two cells, four cells, eight cells, etc., as it passes further down the fallopian tube on the way to the uterus. It then becomes *embedded* in the uterine wall which is thick and well supplied with blood vessels.

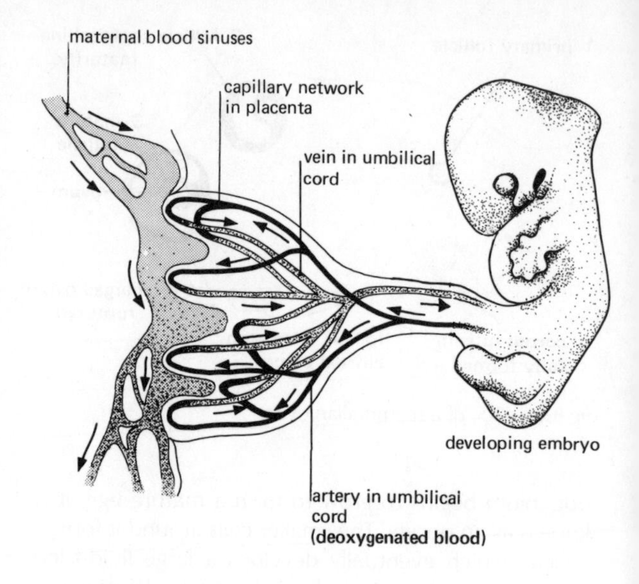

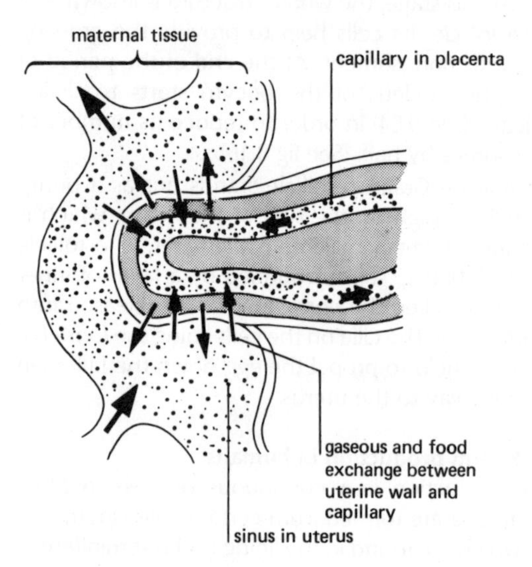

fig 8.19 The structure of the placenta
After D. G. Mackean

As the zygote develops into an embryo, the blood, blood vessels and heart are among the first features to arise. A series of projections develop between the *embryo* and the *uterine wall*, providing a large surface area for the exchange of materials between the maternal and embryonic blood systems. These projections form the *placenta*, an organ characteristic of all 'true' mammals.

In later development, the placenta becomes much more well-defined. It consists partly of *maternal* tissue and partly of *embryonic* tissue and is connected to the body of the embryo by the *umbilical cord*. This

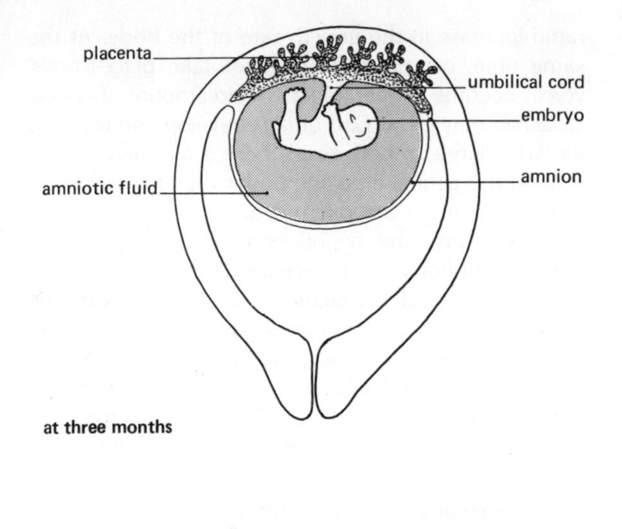

(a)

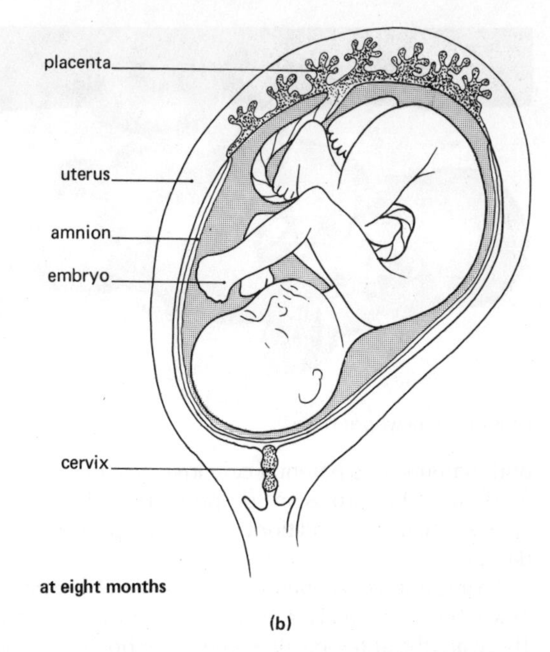

(b)

fig 8.20 A human foetus: (a) at 10 weeks (b) just before birth

structure contains a main vein, which carries oxygenated blood and dissolved nutrients to the embryo, and an artery which takes deoxygenated blood and waste materials to the placenta (fig 8.19).

The embryo is contained in a fluid-filled sac, the *amnion*, which allows it freedom of movement during growth and protects it from mechanical damage.

The period between fertilisation and birth is termed pregnancy or *gestation*, and lasts for about nine months in humans. At the end of that time, the embryo, now called a *foetus*, comes to lie with its head downwards, just above the opening (cervix) of the uterus. (See fig 8.20). With a series of powerful contractions, the muscles of the uterine wall force the foetus, head first, through the dilated cervix and out of the mother's body via the vagina. Shortly after birth, the placenta is ejected in a similar manner. In pigs and certain other mammals, the foetal part of the placenta pulls neatly out of the maternal part, like a hand from a glove. In humans, however, much of the maternal part is also detached, and this results in a certain amount of bleeding.

The umbilical cord is *cut* and *tied* soon after *parturition* (birth) in humans, and the part remaining on the baby withers and drops off within a few days, leaving a small, permanent scar, the *navel*. In many other mammals, the mother bites through the cord, and may eat the placenta.

8.37 Parental care

Immediately after birth, most mammals are almost completely *dependent* for survival on the care of one or

Time after fertilisation	Length	Stage of development
7–10 days	140–160 μm (diameter)	Hollow ball of cells, thickened in one area and implanted in the uterine wall.
3 weeks	1.5 mm	Head region obvious. Spinal cord and heart starting to develop.
6 weeks	10 mm	Brain growing rapidly. Eyes and ears developing. Arm and leg buds forming.
12 weeks	9 cm	The embryo (now called a foetus) has almost the external appearance of a miniature baby.
9 months	50 cm	Birth (parturition) occurs. (See fig 8.21).

Table 8.1 Stages in the development of a human baby

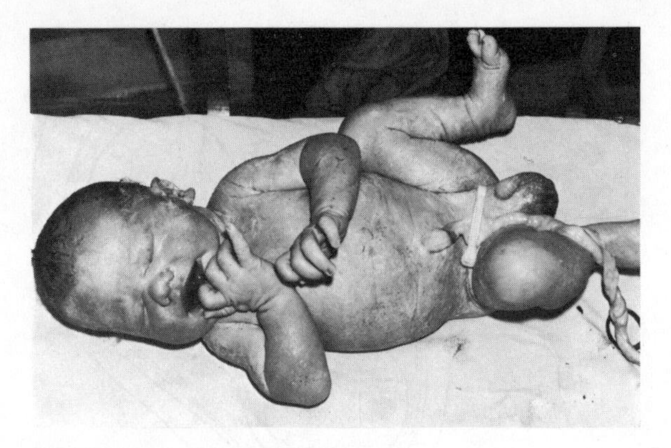

fig 8.21 A newborn baby

both parents. This parental care involves the supply of food and the provision of protection, both from adverse climatic conditions and from potential predators.

An interesting exception to the usual helplessness of newly-born mammals is to be found in the ungulates. These are the antelope, deer and other hoofed animals which form the main diet of large carnivores such as lions. The young of these species are born with disproportionately long legs and with well-developed senses. They are able to run with the herd soon after birth, and thus find safety in numbers.

All mammals supply food to their very young offspring by *suckling* them. The mother's milk contains almost all the constituents of a balanced diet, in approximately the correct proportions. In many cases, the parents may gather or capture food for considerably older offspring. This is particularly true of carnivores and omnivores.

Young mammals, being smaller, have a larger surface area/volume ratio than their parents. Therefore, they will tend to lose and gain heat more readily and are more vulnerable to external changes in temperature. Nest building is a common activity among mammals as well as among birds, and the nests may serve to insulate the offspring and to provide shade. Since young mammals are often born without hair, they may huddle against the mother's body in cool conditions, taking advantage of her body heat.

After female mammals have given birth, they may adopt particularly aggressive behaviour in response to the presence of other animals. This is a behavioural adaptation for the protection of offspring from potential predators.

In humans, the period of parental care is a particularly long one, lasting well over ten years until the onset of puberty. Parental care is also found in other vertebrates (see table 17.2).

8.38 Puberty in humans

At the ages of about 14 in boys and 12 in girls, there is a rapid increase in the growth rate of the body. At the same time, certain other changes take place: boys' *voices* become deeper and a certain amount of coarse *facial hair* may grow. Girls *breasts* develop and their *hip girdles* enlarge; in both sexes there is a growth of hair around the genital area and under the armpits. These physical changes are referred to as *secondary sexual characters*, and the period of their development is known as puberty or *adolescence*. The development of secondary sexual characters is associated with the onset of ovulation and menstruation in girls and of sperm production in boys. At this stage, they are physiologically and anatomically capable of parenthood, although they are usually not ready for it on economic, psychological and emotional grounds.

8.39 Hormones in reproduction

Sperm production and ovulation are triggered by the action on the testes and ovaries respectively of the *gonadotrophic hormones* produced by the *pituitary gland* at the base of the brain. Another effect of these hormones is that they cause the gonads themselves to begin acting as endocrine organs by producing *'sex hormones'*. The testes secrete the male sex hormone (testosterone), which stimulates sperm production and the male secondary sexual characters, while the ovaries secrete the female sex hormones, of which oestrogen produces the female secondary sexual characters and regulates the menstrual cycle and ovulation. Progesterone, the 'pregnancy hormone' is also produced by the ovary.

If pregnancy occurs, the placenta itself becomes an important endocrine organ, interacting with the ovaries and pituitary to bring about the major changes that occur in the mother's body during gestation and birth.

8.40 Sexual reproduction in the flowering plant

Although vegetative reproduction is widespread throughout the plant kingdom, the phenomenon of sexual reproduction is as significant in higher plants as it is in the mammals, where asexual reproduction does not occur.

The sexual reproductive organs of higher plants are known as *flowers*. Since they are in fact specialised shoots, flowers consist of a modified stem and modified leaves. Typically, the stem portion consists of a *pedicel* (the flower stalk) and a *receptacle*, to which are attached the modified leaves.

The modified leaves are arranged in four whorls. The outermost whorl is termed the *calyx* and inside this are, successively, the *corolla*, the *androecium* and the *gynaecium* (or pistil).

Sexual	Asexual
1 Produce variation in offspring.	No variation in offspring.
2 Food reserves restricted to seed	Much larger food reserves available from parent plant of perennating organ.
3 Colonisation of new habitats possible because of dispersal methods.	New plants always arise near parents: results in overcrowding and intra-specific competition.
4 Seeds may land in unfavourable areas.	New plants arise always in areas favourable to the species.

Table 8.2 Comparison of sexual and asexual reproduction in plants

The calyx consists of a number of *sepals*, which are often green in colour and serve to protect the remainder of the whorls during the immature bud stage. The corolla is made up of *petals*, which are particularly large, brightly-coloured and conspicuous in insect-pollinated flowers.

The androecium is the *male* part of the flower. It consists of a number of *stamens*, each of which comprises a *filament* and an *anther*. The filament contains a single vascular bundle and forms a 'stalk' which supports the two pollen-producing sacs of the anther. The *pollen grains* contain the male gametes.

The gynaecium consists of one or more *carpels*. Each carpel has three parts, the *ovary* (which bears one or more ovules), the *stigma* (which receives the pollen grain) and the *style* (which connects the other two). The *ovules* contain the female gametes.

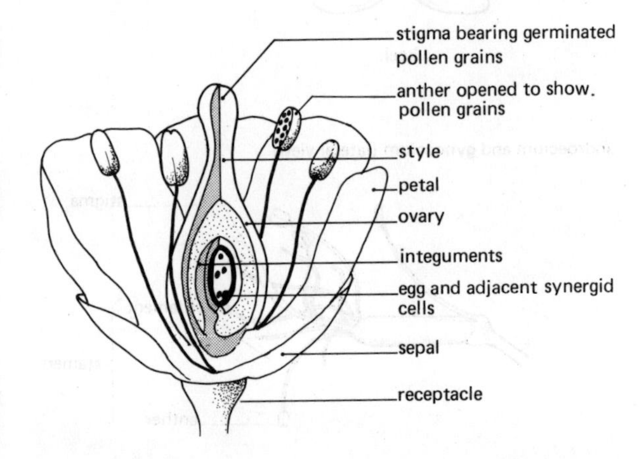

fig 8.22 A longitudinal section of a flower at the time of fertilisation

8.41 Pollination and fertilisation

The function of sexual reproduction in flowering plants is to produce a *fruit*, containing one or more *seeds*, each seed being capable of giving rise to one new plant. Before a seed can be formed, the two processes of pollination and fertilisation must take place.

Pollination involves the placing of *pollen grains* liberated from an anther onto the *stigma* of a carpel. The pollen grains may be from the same plant as the carpel (self-pollination) or from a different plant (cross-pollination): they may be placed on the stigma by air currents (wind pollination) or by bees, butterflies, etc (insect pollination).

After the pollen grain has landed on the stigma, it germinates to produce a narrow tube which grows down the style towards the ovary. (See fig 8.23).

The pollen grain originally contains one haploid nucleus, produced by meiosis (see 9.04). Before the grain leaves the anther, this has divided mitotically to form one *tube nucleus* and one *generative nucleus*. The tube nucleus is concerned with the formation of the *pollen tube* and the generative nucleus divides to form *two male gamete nuclei*. When the pollen tube has grown as far as the ovary, the wall of the tube breaks down and the two gamete nuclei enter the ovule. Inside the ovule, one of the gamete nuclei fertilises the *ovum* or female gamete, thus forming a *zygote*. (See fig 8.23).

The other male gamete nucleus fuses with another (diploid) nucleus in the ovule. The *triploid cell* thus formed gives rise to the *endosperm tissue*, which in some cases acts as a food store for the developing seed.

The seed itself arises from the whole of the ovule, and the ovary which may contain many ovules, develops into the fruit. The *ovary wall*, which may thicken greatly, becomes the *pericarp* of the fruit. In many cases, three distinct layers can be distinguished in the pericarp: an outer skin (the *epicarp*), a middle, fleshy *mesocarp* and an inner *endocarp*.

Investigation

To dissect a flower of sweet-pea (*Lathyrus*)

Procedure

1 Take a sweet-pea flower and examine it carefully.

Question 1 How many sepals can you count?

2 Use forceps to peel back the points of the calyx exposing as much as possible of the corolla.

Question 2 How many petals can you count? Are any of them fused?

3 Using a needle and forceps, remove one of the outer, lateral petals (a wing).

4 Now remove the large, posterior standard petal and the anterior keel (two petals fused). This will expose the

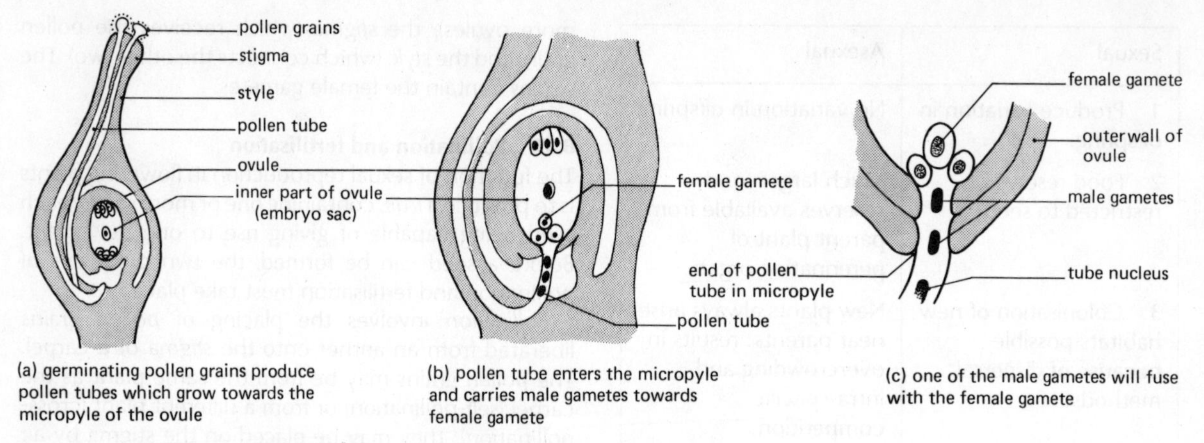

fig 8.23 The process of fertilisation in a flowering plant

androecium and gynaecium. Make a careful drawing of the flower at this stage.

Question 3 How many stamens can you count? Are they uniform in size and shape? Are they separate?

5 Cut the bases of the filaments with the point of a dissecting needle. Remove the single stamen: then slide the remaining nine (fused) stamens over the gynaecium and remove them slowly.

Question 4 How many carpels make up the gynaecium?

Question 5 Can you identify any structure which might serve to trap pollen grains? (See fig 8.24).

Experiment

To investigate the functions of flower parts.

Procedure

1 For this experiment, which is performed out of doors, you will require about thirty-five very young flowers of the same species. Do not pick the flowers, but divide them into seven equal groups by labelling each stem, A, B, C, D, E, F or G with waterproof tags or labels. Choose flowers which are easily accessible.

2 Treat the groups as follows.

A Leave the flowers to develop normally.

B Cover each flower with a plastic bag, tying the neck loosely around the stem.

C Remove the petals from each flower: then cover each flower with a bag as in B.

D Remove the stamens from each flower and cover with a bag.

E Remove the stigma from each flower and cover with a bag.

F Remove the stamens from each flower and place on its stigma pollen obtained from an older flower of the same species.

G Remove the stamens from each flower and place on its stigma pollen obtained from an older flower of different species.

3 Observe the flowers carefully and regularly over a

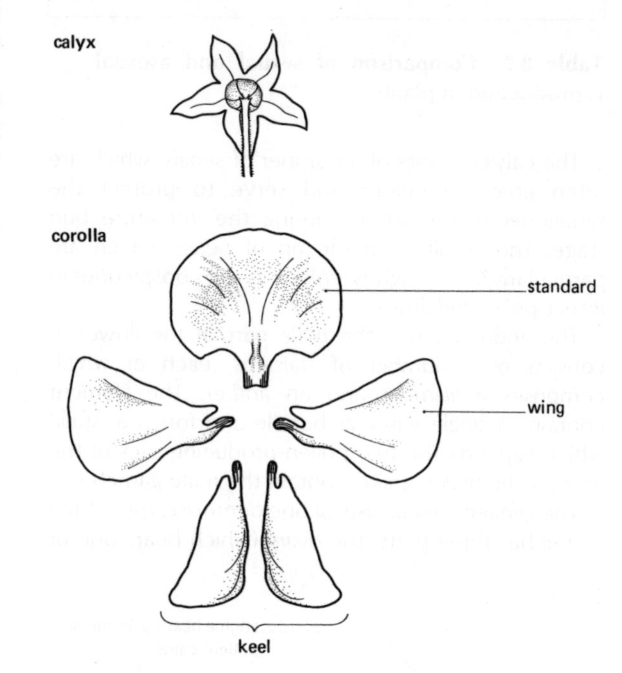

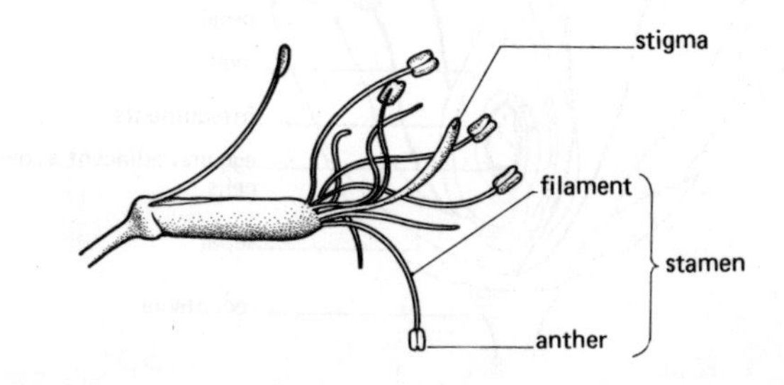

fig 8.24 A dissection of a sweet-pea flower

period of several weeks, paying particular attention to the sizes of the ovaries.

Questions

6 In which of the seven groups do the ovaries swell to form fruits?

7 Which of the groups acts as a control?

8 Why are very young flowers used in this experiment?

9 Which of the floral parts examined are essential in order for fertilisation to take place?

10 What is the function of group B in the experiment?

11 What is the function of group C in the experiment?

Investigation

Variation in flower form

Procedure

1 Collect as many different types of flower as possible. During collection, note the presence of any visiting insects.

2 Study each flower carefully, and record your observations in a table similar to the one shown.

3 From your observations, try to deduce the way in which each flower might be pollinated.

Questions

12 Can you suggest a function for large, brightly-coloured petals and strong scent?

13 If you have discovered species with male and female flowers on separate plants, can you suggest a function for this?

14 Can you suggest, from your observations, another way in which this function might be fulfilled?

15 Referring back to the previous experiment, can you suggest any other ways in which this same function might be fulfilled?

Experiment

To dissect an insect-pollinated flower

Procedure

1 Take a young flower of *Antirrhinum* (snapdragon) and count the sepals, petals, stamens and carpels.

2 Using a razor-blade or scalpel, cut it longitudinally in half. Start your cut in the flower stalk and carry it forward through the middle of the petal corolla tube.

3 If your cut passes slightly to the side of the ovary, make a second cut along the long axis of the ovary to expose the ovules.

4 Make a careful drawing of the floral parts. (See fig 8.25).

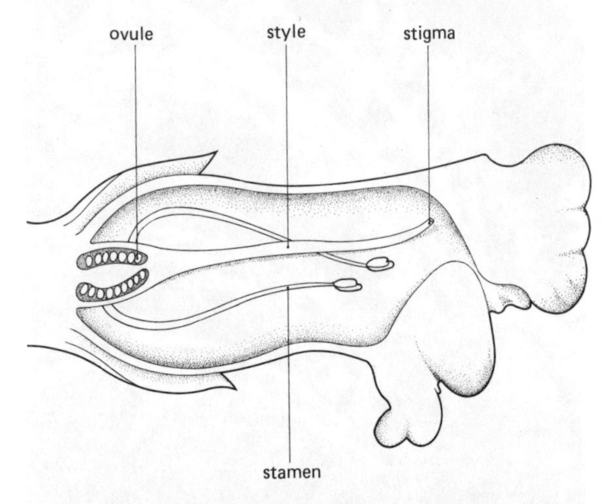

fig 8.25 A half flower of snapdragon (*Antirrhinum*)

Questions

16 How many sepals and petals are there?

17 How many stamens are there?

18 How many ovules are inside the ovary, and how are they attached?

19 What observations can you make to support the view that this plant is insect-pollinated?

Zea mays (maize) a wind-pollinated flower.

Maize plants do not possess sepals and petals like those of *Antirrhinum*: instead they have small, green or yellow, leaf-like structures called bracts. This is characteristic of all grass and cereal plants.

The flowers themselves are grouped in large clusters or *inflorescences*. The sexes of the inflorescences are separate, those at the top of the shoot containing only male flowers (i.e. no gynaecia are present) and those lower down on the sides containing only female flowers.

Name of flower	Number of sepals	Number, colour and size of petals (whether fused or separate)	Number and shape of stamens	Shape of stigma	Position of stigma relative to anthers	Visits by insects?	Presence or absence of scent

fig 8.26 Part of a male inflorescence of *Zea mays*

fig 8.27 Part of a female inflorescence of *Zea mays*

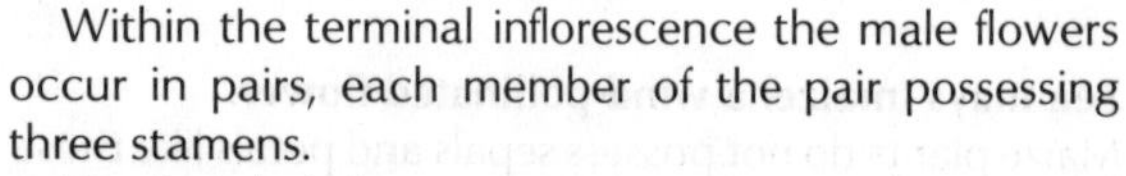

Within the terminal inflorescence the male flowers occur in pairs, each member of the pair possessing three stamens.

The female inflorescences contain carpels arranged spirally on the inflorescence stalk to form the 'cob'. Each cob is completely surrounded by a number of leaves wrapped around each other in layers. The female flowers also occur in pairs, but one of each pair is small and sterile. (See fig 8.28.)

Investigate the structure of both male and female flowers of a maize plant by teasing the flowers apart with a dissecting needle and forceps.

Question

20 What observation can you make to support the view that this plant is wind-pollinated?

8.42 Fruits and seeds

In strict biological terms, a fruit is the *fertilised ovary* of a flower (see section 8.41). In common language, however, the term 'fruit' is often used to describe a variety of structures, from the fused ovaries of several flowers to a swollen receptacle.

The term 'seed' refers to an ovule after fertilisation. It consists of a tough outer coat, the *testa*, surrounding an

Insect pollination	Wind pollination
1 Large, brightly coloured petals.	Small, greenish bracts.
2 Nectar and sweet smell.	No nectar and odourless.
3 Stigma and anthers often at least partly enclosed.	Stigma and anthers exposed to air currents.
4 Filaments fairly rigid and anthers firmly attached.	Filaments flexible and anthers hinged (versatile).
5 Stigma small and rigid.	Stigma flexible and long.
6 Pollen grains relatively large, rough and/or sticky.	Pollen grains small and smooth, may have air bladders.

Table 8.3 Comparison of insect-pollinated and wind-pollinated flowers.

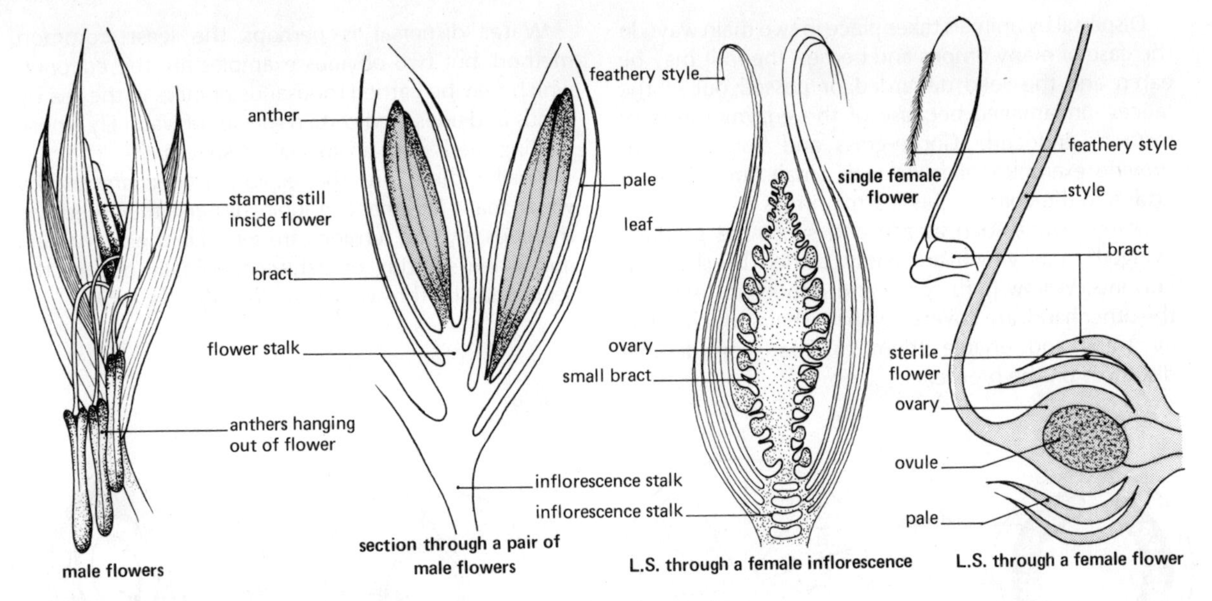

fig 8.28 Details of male and female flowers of *Zea mays* *After D. G. Mackean*

embryo. The embryo comprises a *radicle* (embryonic root), a *plumule* (embryonic shoot) and one or two *cotyledons* (embryonic leaves distinct from the foliage leaves of the plumule). The testa may also enclose food reserves as *endosperm* (see section 8.41). Starch obtained from cereal grains is mainly endosperm.

A true fruit can be recognised as having two scars: the point of attachment of the stigma at one end and the point of attachment of the receptacle at the other. A seed, on the other hand, has a single scar or hilum; the point at which the funicle or ovule stalk was attached.

Classification of fruits

There are a number of ways in which fruits can be classified, all of them more or less artificial. Perhaps the most obvious is to divide fruits first of all into those which are succulent and those which are dry. The dry fruits can be further divided into those which *dehisce* or open to release their seeds and those which do not. A scheme of classification based on these principles is shown here.

- **Fruit**
  - **succulent**
    - pome (e.g. apple)
    - berry (e.g. tomato, orange)
    - drupe (e.g. plum, coconut)
  - **dry**
    - **dehiscent**
      - legume (e.g. pea, bean)
      - capsule (e.g. poppy)
    - **indehiscent**
      - caryopsis (e.g. maize)
      - schizocarp (e.g. garden nasturtium)

A *pome* is a so-called *false fruit*, because the succulent part is a swollen *receptacle*, with the pericarp and seeds (ovary – 'true fruit') forming only the core.

A *berry* has a *succulent pericarp* (divided into epi-, meso- and endocarp). In the orange, the *endocarp* bears hairs which have developed into *succulent* structures.

A *drupe* also has a three-layered pericarp, but only the *outer two layers* are fleshy. The endocarp is hard and forms the 'stone' which encloses the seed.

A *legume* has a relatively thin, *dry* ovary wall, which splits down both sides to release its numerous seeds. It arises from a single carpel.

A *capsule* is formed from *more than one carpel* and may dehisce in one of several ways.

A *caryopsis* consists of a *single seed* surrounded by a dry pericarp to which its testa has become *fused*.

A *schizocarp* originally contains several seeds, but splits into several parts, each containing one seed.

Fruit and seed dispersal

It is undesirable for all the seeds of a plant to fall from it directly onto the ground and germinate there, since (a) it tends to result in competition for light and nutrients between parent and offspring, and (b) it prevents the colonisation of new and favourable habitats. Since plants, unlike most animals, do not have the power of locomotion in the adult state, a number of mechanisms have been evolved for fruit and seed dispersal.

These mechanisms fall into four main categories: *animal* dispersal, *wind* dispersal, *water* dispersal and dispersal by *explosion*.

Dispersal by animals takes place in two main ways. In the case of many drupes and berries, the fruit may be eaten and the seed discarded or passed out in the faeces, undamaged because of the enzyme-resistant testa or endocarp. Goose-grass and burdock both provide examples of fruits with small hooks, which attach to the coats of passing mammals.

Many plants, such as ash and sycamore produce 'winged' fruits which are easily carried by slight air currents. Willow-herb seeds and dandelion fruits, on the other hand, are covered with fine hairs which act as air traps and enable them to float considerable distances on the breeze.

Water dispersal is perhaps the least common method, but two obvious examples are the coconut, which may be carried thousands of miles in the sea by winds and tides, and certain species of water lily whose floating seeds contain special air-spaces.

Finally, there are the explosive mechanisms, by which many legumes (e.g. lupin) disperse their seeds. As the pods dry out, tensions are set up in the ovary wall; and when splitting or dehiscence finally occurs, the seeds may be thrown several feet. (See fig 8.29.)

fig 8.29

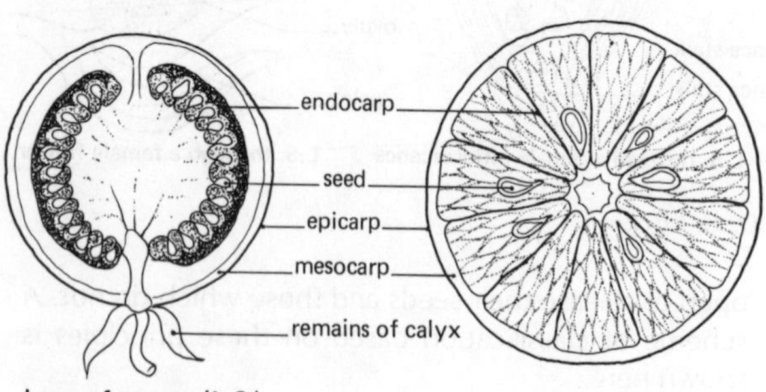
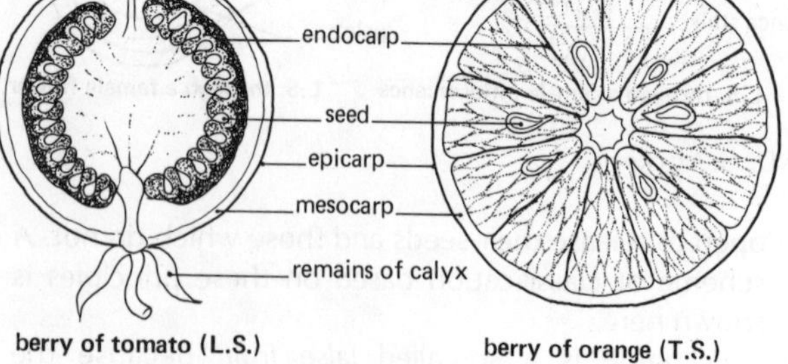

berry of tomato (L.S.) **berry of orange (T.S.)**

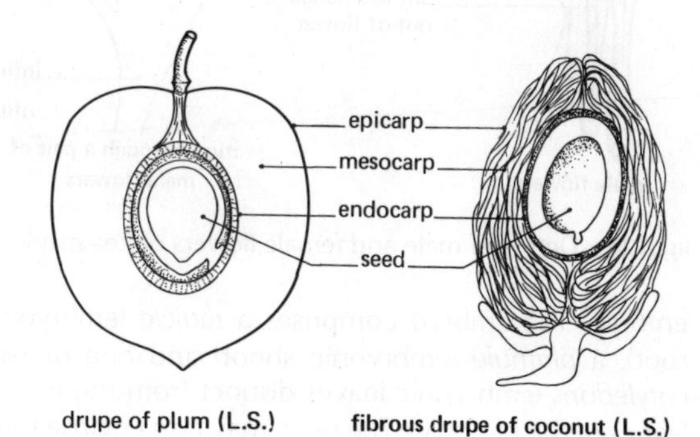

drupe of plum (L.S.) **fibrous drupe of coconut (L.S.)**

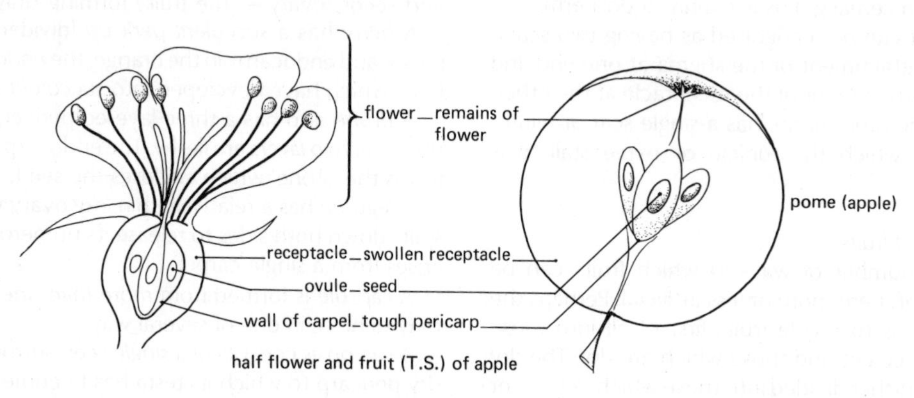

half flower and fruit (T.S.) of apple

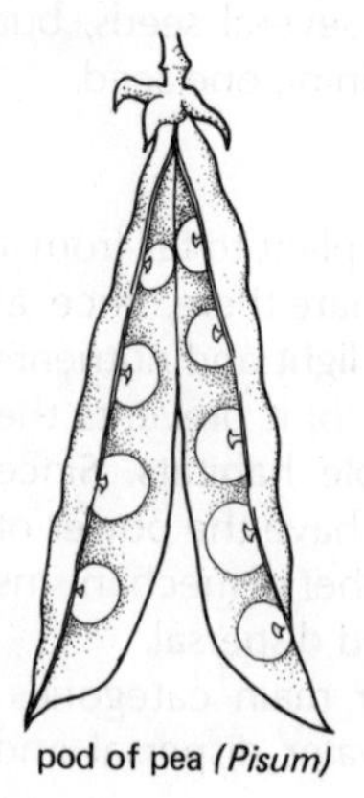

pod of pea *(Pisum)*

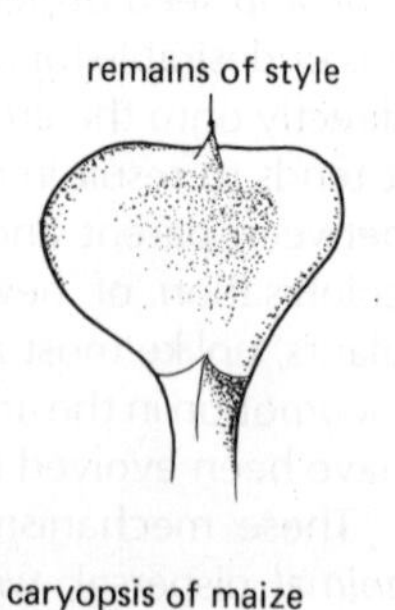

caryopsis of maize

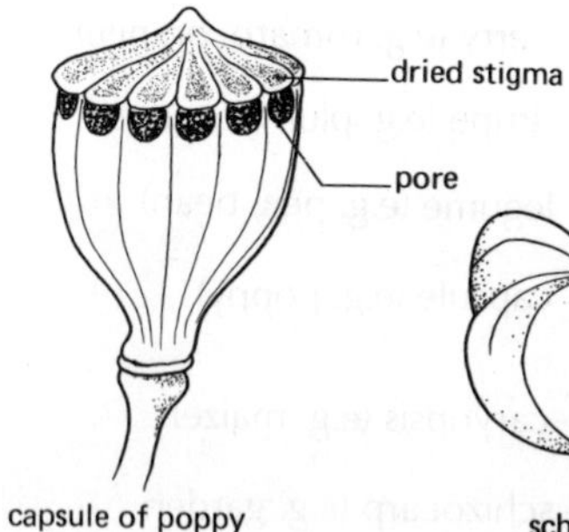

capsule of poppy

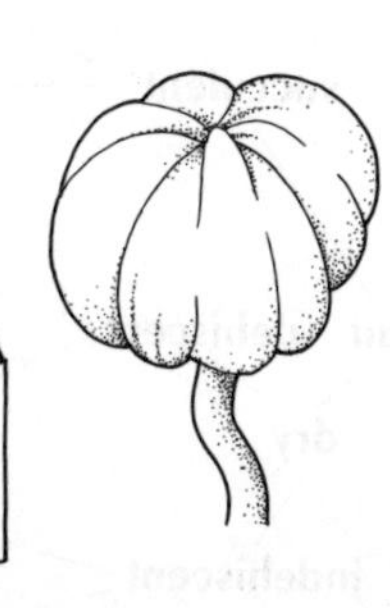

schizocarp of garden nasturtium *(Trapaeolum)*

Investigation

To study a variety of fruits and their dispersal mechanisms.

Procedure

Collect as many different kinds of fruit as possible, and classify them according to their type and their method of dispersal.

Make labelled drawings of selected examples to illustrate the features associated with dispersal.

Questions requiring an extended essay-type answer

1 a) Describe one method of natural vegetative reproduction in a *named* flowering plant.
b) Describe one method of artificial vegetative reproduction of a *named* flowering plant that man uses in order to grow the plant.
(Diagrams should be drawn in part a) and b)).

2 a) Describe the methods of asexual reproduction found in lower organisms.
b) How does asexual reproduction differ from sexual reproduction?

3 a) Draw a diagram of the male urinogenital system of a mammal.
b) Describe briefly the events that occur during the following stages of reproduction in the female mammal
i) the release of eggs and preparation of the uterus, and
ii) fertilisation.

4 a) Draw a diagram of the female urinogenital system of a mammal.
b) Describe, with the aid of diagrams, the main stages in the development of a human baby before birth.

5 a) What are secondary sexual characteristics? Describe their appearance in a human male and female. Give two examples of these in other mammals.
b) Describe the hormonal control of the production of gametes and the development of secondary sexual characteristics.

6 a) Define sexual reproduction.
b) How does gaseous exchange occur in the embryo of a mammal?
c) Describe birth and parental care in a *named* mammal.

7 a) What do you understand by cross-pollination, self-pollination and fertilisation?
b) Describe pollination, fertilisation and fruit formation in a *named* insect-pollinated flower.
c) How is seed dispersal effected in the plant named in part b)?

8 a) Compare the advantages and disadvantages of sexual reproduction and asexual reproduction in flowering plants.
b) Compare the structures and adaptations of flowering plants for insect and wind pollination.

9 a) Define the terms of pollination and fertilisation.
b) Describe how you would investigate the pollination mechanism of an unfamiliar plant. Give details of controls that are set up.

9 Genetics

9.00 Cell division

All living organisms are made up of units called *cells*, and of their products. In spite of the enormous variety in size and structure of living organisms, their cells are relatively uniform. The cells of an elephant do not differ very much from the corresponding cells in a mouse; there are simply more of them.

As an organism grows, its cells increase in number in much the same way as *Amoeba* divides to reproduce (see Chapter 8, section 8.21). The process in which a cell divides to form two identical daughter cells is known as *mitosis*. It differs slightly between animals and plants, mainly because of the structural differences between the cells of the two kingdoms.

Fig 9.1 and fig 9.2 show the appearance of typical animal and plant cells, as seen under a microscope.

Questions

1 What are the main similarities between the two cells?

2 In what respects does the plant cell differ from the animal cell?

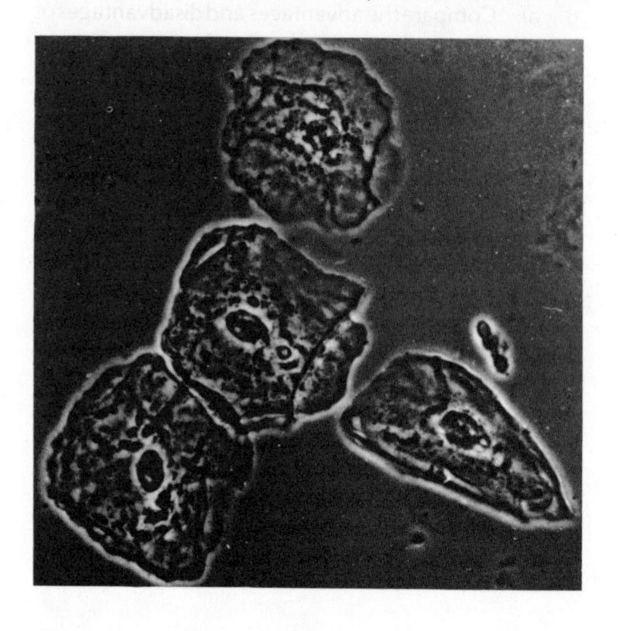

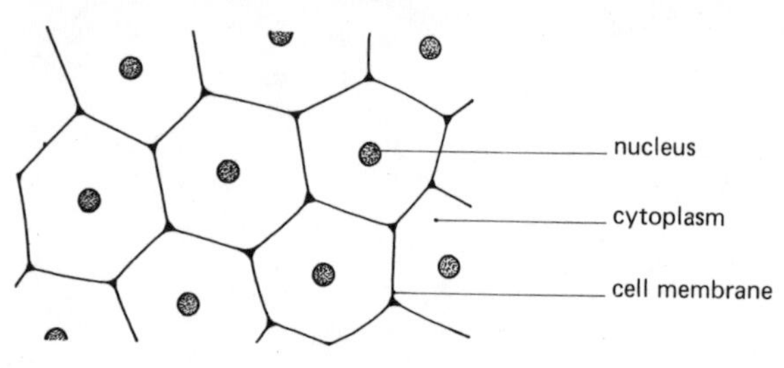

fig 9.1 (a) A photograph of a group of animal cells from the lining of the mouth (b) A diagram of a group of cheek cells

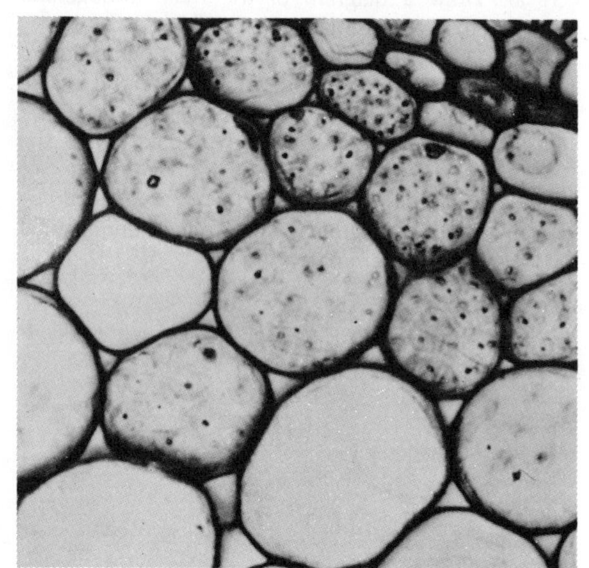

fig 9.2 (a) A photograph of a group of plant cells (b) A diagram of a group of plant cells

9.01 Mitosis in animal cells

In a growing animal, cells are continually dividing: one cell divides to form two, these two divide to form four, then eight, sixteen, thirty-two, and so on in a continuous progression. In between divisions, the daughter cells grow to the size of the parent cells and

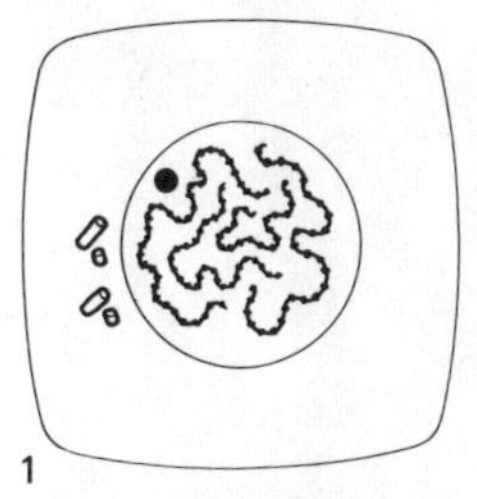

1
EARLY PROPHASE
During the interphase the cell carries out everyday activities. In prophase, the chromosomes begin to shorten and the nucleolus begins to shrink

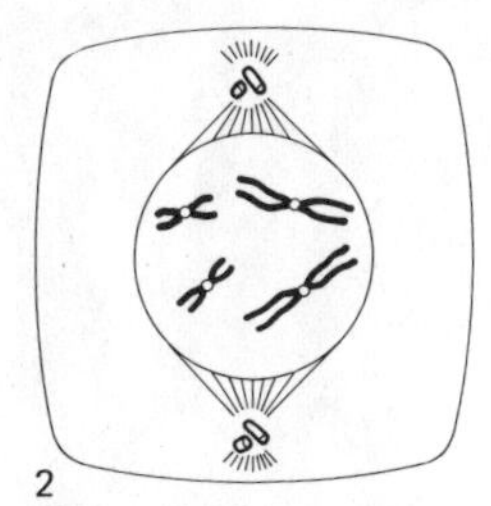

2
LATE PROPHASE
Chromosomes shorten further and duplicate themselves, each one consists of a pair of chromatids joined at the centromere. Centrioles move apart, nucleolus disappears

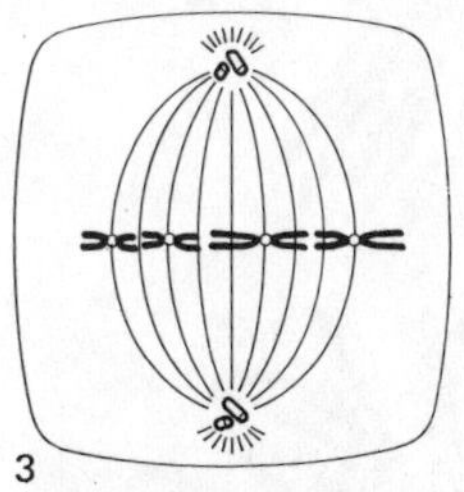

3
METAPHASE
Nuclear membrane breaks down, spindle forms, chromatid pairs line up on equator of cell. (Single chromosome per spindle.

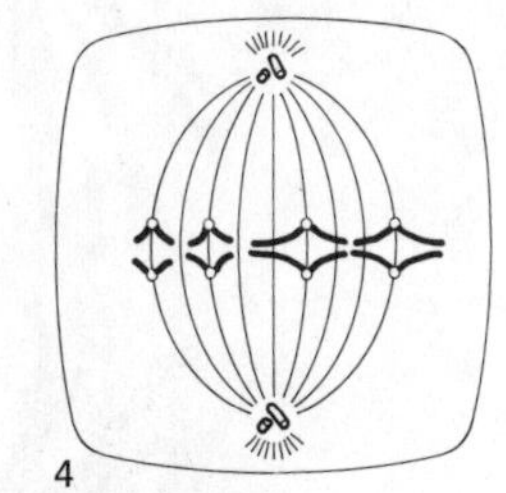

4
ANAPHASE
Chromatid pairs separate and move to opposite ends of cell.

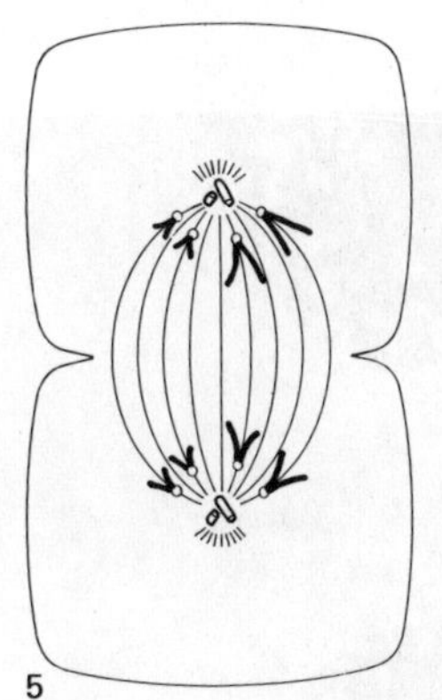

5
EARLY TELOPHASE
Chromatids reach destination and reform into chromosomes. Cell begins to constrict at equator.

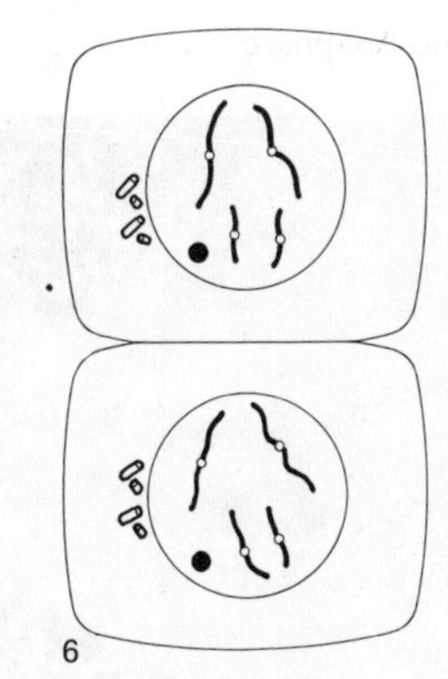

6
LATE TELOPHASE
Constriction completed and nuclear membrane and nucleolus reform in each cell. The chromosomes become invisible threads and each cell enters interphase

fig 9.3 The main stages of mitosis in a generalised animal cell

prepare for the next division in certain other ways. This stage is known as *interphase*. Mitosis consists of four other more or less distinct stages: *prophase, metaphase, anaphase* and *telophase*.

At the beginning of prophase, an examination of the nucleus under a light microscope reveals the presence of a number of fine thread-like structures, the *chromosomes*. Later these are seen to be double-stranded. Each pair of strands (*chromatids*) is connected at one point, the *centromere*. The chromosomes are very important structures, since they contain the coded information that will determine the physical and chemical features of the daughter cells.

Gradually, the chromosomes become shorter, thicker and, as a result, more easily seen. At the same time, two star-shaped *asters* of transparently fine fibres appear on opposite sides of the cell. Towards the end of prophase, a *spindle* of fine fibres appears between the asters, and the membrane surrounding the nucleus breaks up and disappears. The *nucleolus*, seen as a dense spot within the nucleus, also disintegrates.

During metaphase, the chromosomes become arranged on the middle or *'equator'* of the spindle, and each centromere splits into two. The daughter centromeres then begin to move towards opposite ends of the spindle, each one pulling a single chromatid with it.

In anaphase, the chromatids separate completely and travel along the spindle towards the asters.

Finally, during telophase, the cell begins to constrict (narrow) at the equator, and the chromatids become surrounded by two new nuclear membranes.

Two new nucleoli are also formed and when constriction is complete, the two daughter cells are separated completely by a membrane between them. Each new cell passes into interphase again.

Note that *one half* of each chromosome has passed into each daughter cell. By the time these daughter cells are ready for division, the chromosomal material will have doubled and the chromosomes will again be two-stranded.

9.02 Mitosis in plant cells

Experiment

Preparation of an onion root tip squash

Procedure

1 Place two drops of molar hydrochloric acid on a microscope slide and add a drop of *acetic-orcein* stain.
2 Cut about 2 mm from the tip of an onion root and place it in the stain/acid mixture.
3 Warm the slide gently over a Bunsen burner for about 3 minutes using a low flame. Do not boil.
4 Draw off the stain/acid mixture carefully, using a piece of blotting paper, at the edge of the liquid.
5 Add a further drop of stain to the root tip and cover with a coverslip.

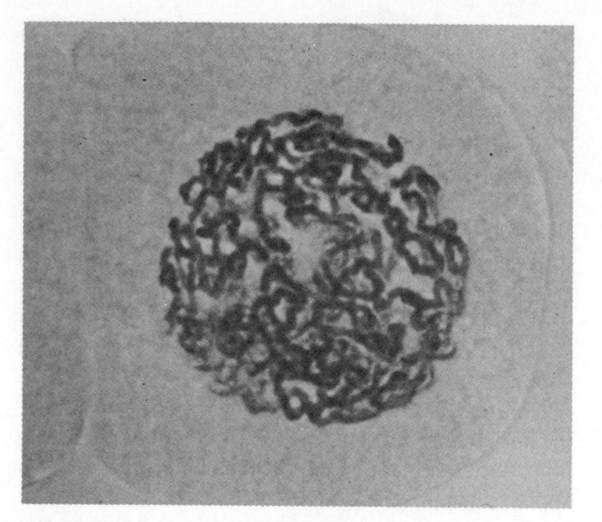
(a) Prophase

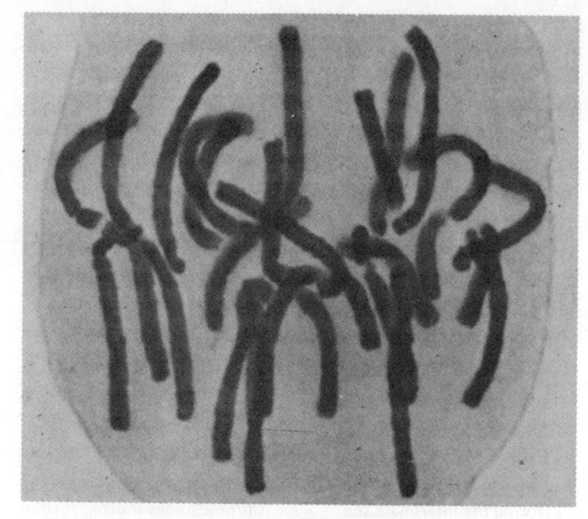
(b) Early metaphase

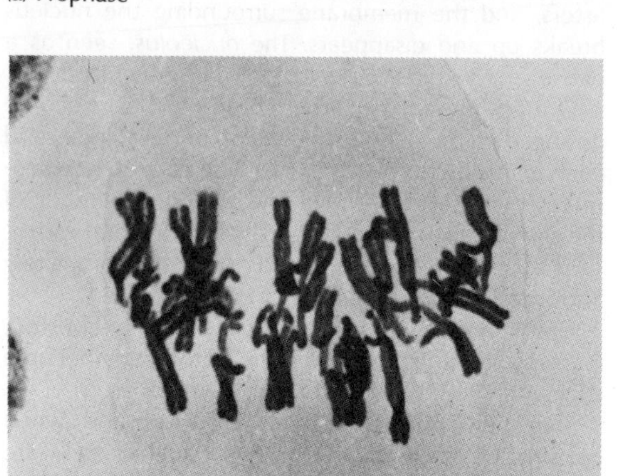
(c) Late metaphase

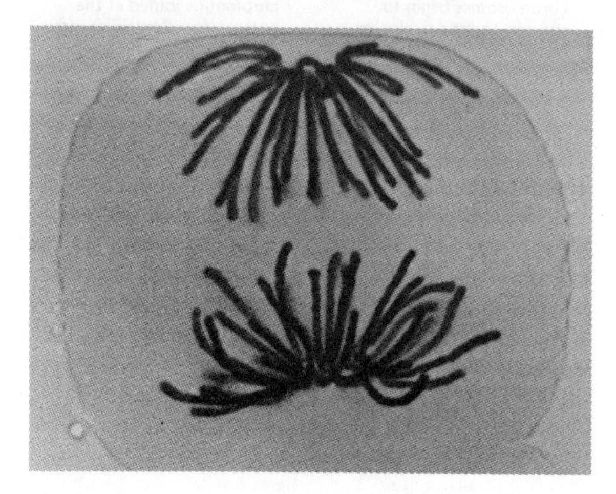
(d) Anaphase

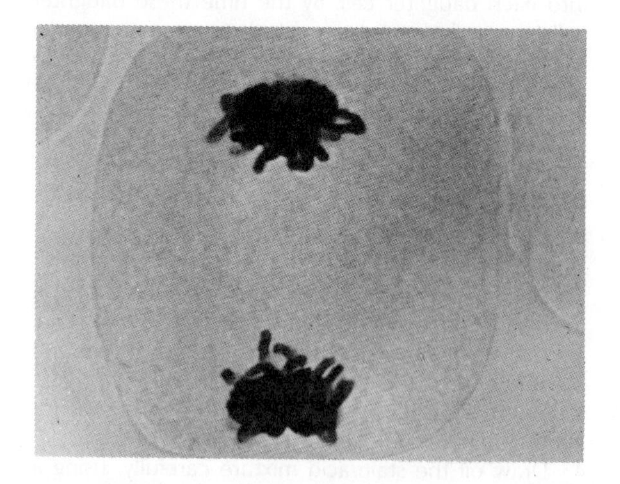
(e) Telophase

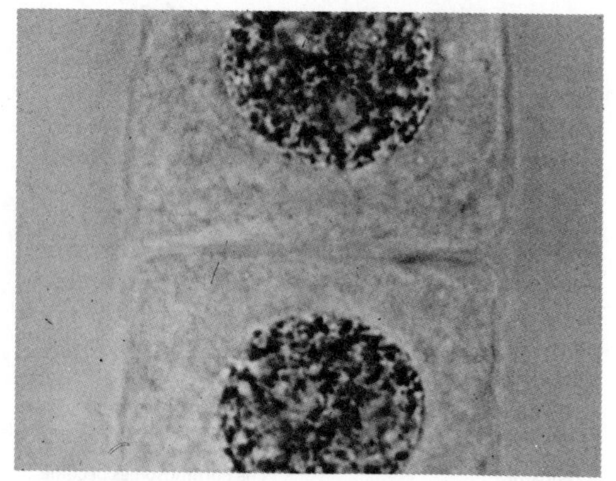
(f) Completed division

fig 9.4 The main stages of mitosis

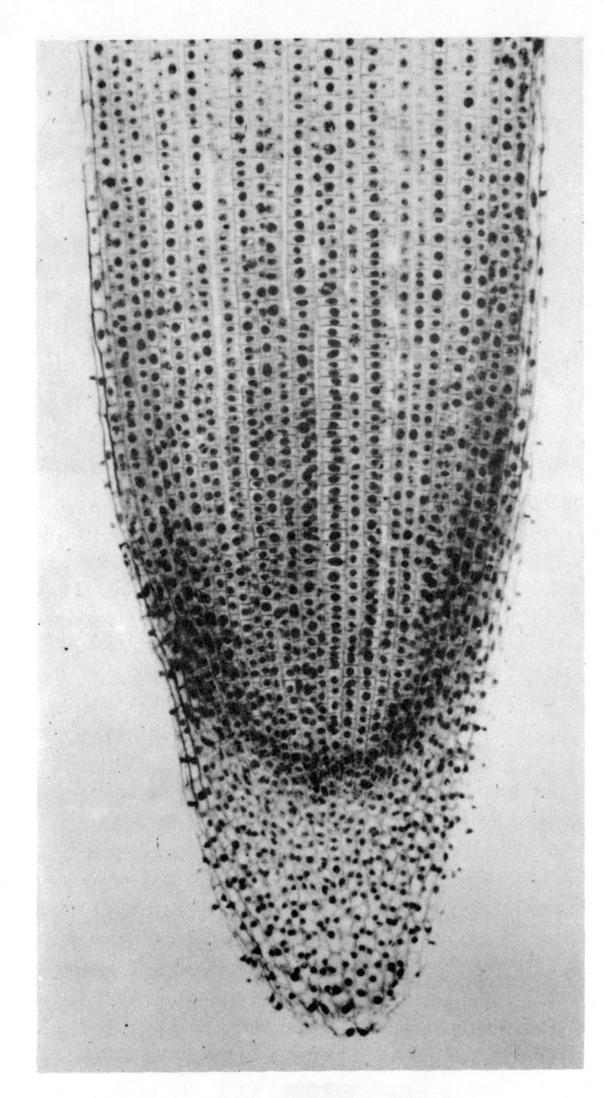

fig 9.5 A group of cells from the tip of an onion root

6 Place a piece of filter paper over the coverslip and apply firm but gentle pressure with your thumb in order to obtain a thin squash.

7 Remove the filter paper and examine the squashed root tip under the microscope.

Questions

1 Study fig 9.5 and your own preparation carefully. In what respects do these two examples of plant cell mitosis differ from the process in animals as described in section 9.01?

2 Why are root tips particularly suitable for the demonstration of mitosis?

9.03 Fertilisation

The number of chromosomes in the nucleus of a *somatic* (body) cell varies considerably throughout the animal and plant kingdoms, but within any one species this number is constant. For example, in humans each somatic cell contains 46 chromosomes.

If we refer back to section 8.35, we can see that sexual reproduction involves the fusion of a male gamete or sex cell with a female gamete in the process of fertilisation. This would lead us to assume that the cells of an organism resulting from this fusion have *double* the number of chromosomes contained by the cells of the parents. We know that this cannot be so, since it has already been pointed out that the number of chromosomes per cell is constant for all individuals in a species.

The apparent contradiction between the above two statements could be explained if the number of chromosomes were to be *halved* during the formation of the gametes. This, in fact, is exactly what does happen in the process known as *meiosis,* which takes place in the reproductive organs of plants and animals during gamete formation.

9.04 Meiosis

When we first considered chromosomes in section 9.01, it was mentioned that they contain the coded information that determines the structure and function of the cells, and of the organisms that they make up. In fact, each body cell possesses two sets of information, one from the male parent and one from the female parent, contained in two similar sets of chromosomes. For every chromosome originating from the mother, there is one of similar length and appearance (containing equivalent information) from the father. These are termed *homologous* chromosomes, and each human body cell has 23 pairs.

The process of meiosis consists of two divisions, during which the pairs of homologous chromosomes are separated. In many ways, these divisions are similar to mitotic divisions, and the same terms are used to describe the stages.

The more important differences between the two processes are to be found in the very first stage: the prophase of the first meiotic division (prophase I). At the beginning of this stage in animals, the chromosomes shorten and thicken, and there is a breakdown of the nucleolus and nuclear membrane, just as in mitosis. Then the pairs of homologous chromosomes come to lie together, each pair forming a *bivalent.* Later, it can be seen that each chromosome consists of two chromatids.

During metaphase I, the bivalents arrange themselves on the equator of the spindle.

At anaphase I, the bivalents split, one homologous chromosome moving to each pole or end of the cell. Note that the chromosomes originating from the male parent do not all move to the same end, so that each daughter cell will probably contain some chromosomes originating from each parent.

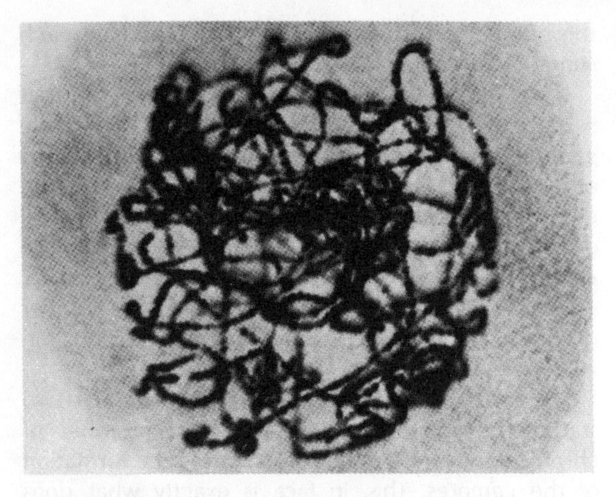
(a) Pachytene

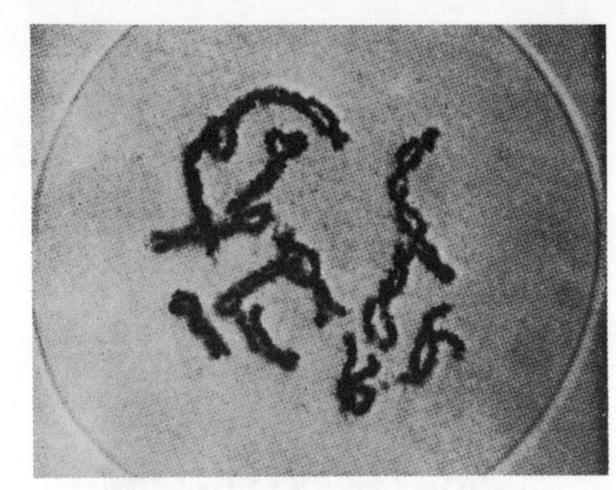
(b) Diplotene

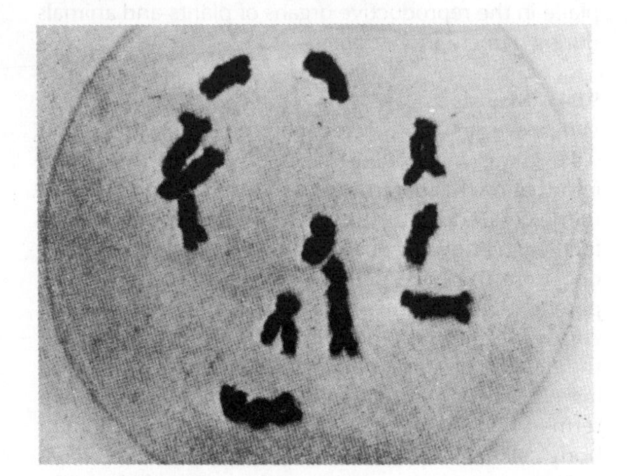
(c) Diakinesis

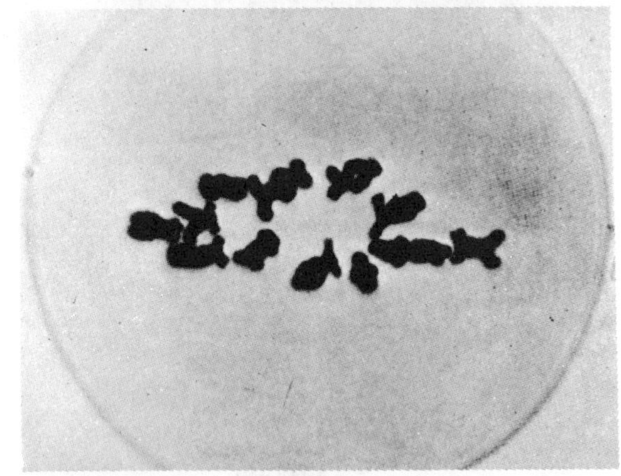
(d) First metaphase

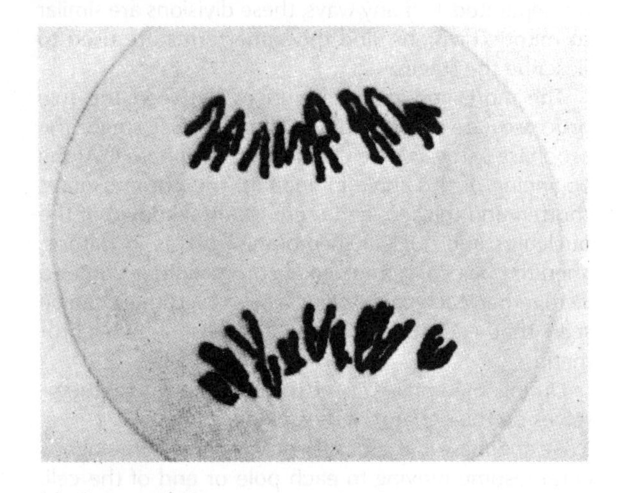
(e) First anaphase

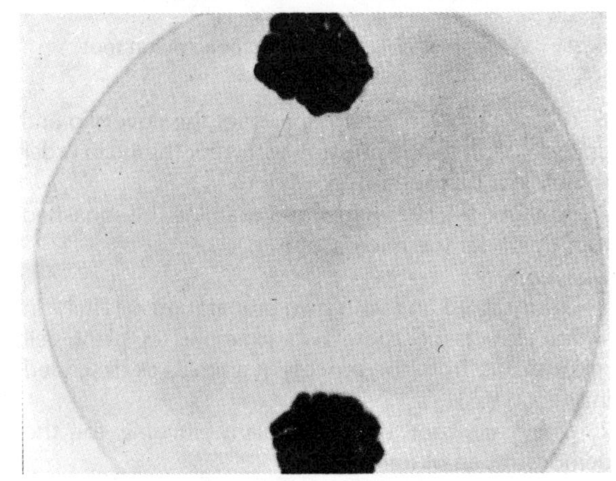
(f) First telophase

fig 9.6 The main stages of meiosis

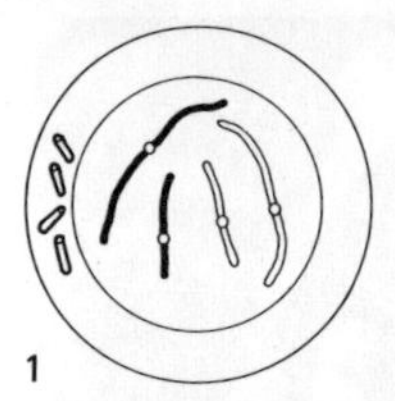

1

EARLY PROPHASE 1
Homologous chromosomes appear in nucleus.

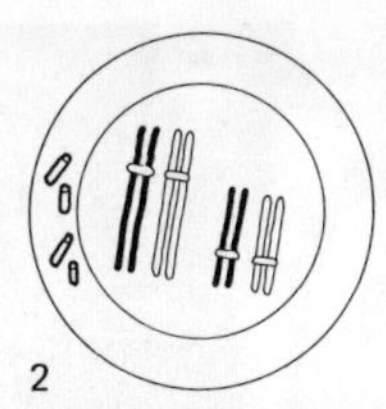

2

MIDDLE PROPHASE 1
Homologous chromosomes pair up, then split into chromatids.

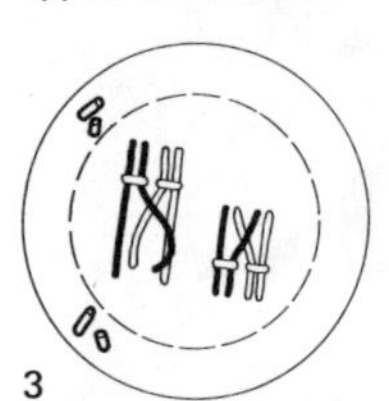

3

LATE PROPHASE 1
Chromatids of homologous chromosomes cross over each other.

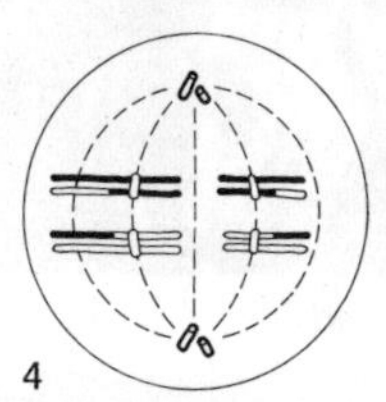

4

METAPHASE 1
Homologous chromosomes arrange themselves on equator of spindle. Segments of crossed chromatids have exchanged by breakage and subsequent rejoining.

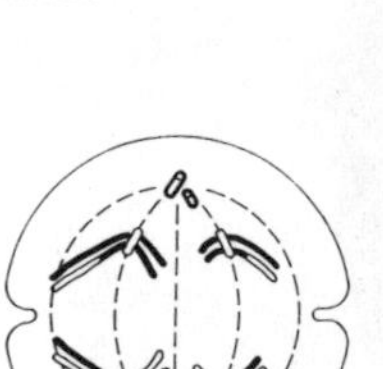

5

ANAPHASE 1
Homologous chromosomes separate.

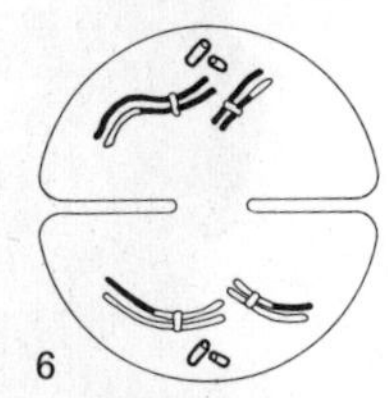

6

TELOPHASE 1
Two new cells form, each with half the number of chromosomes of the original cell. Nuclear membrane may not reform, as metaphase 2 may follow immediately.

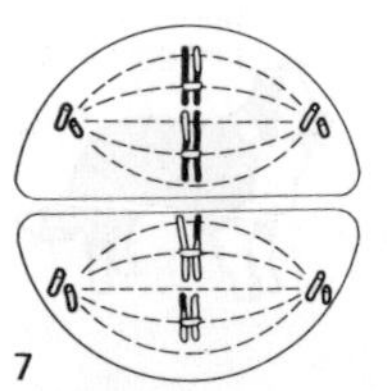

7

METAPHASE 2
A new spindle is formed in each new cell and two chromosomes line up on each equator

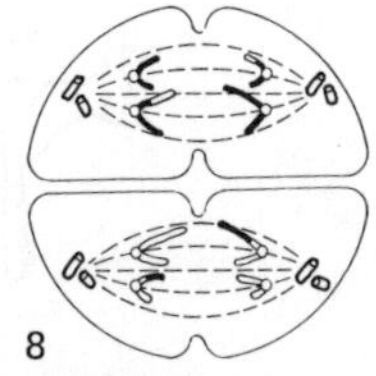

8

ANAPHASE 2
Chromatids of the two chromosomes in each new cell separate. (As in mitosis)

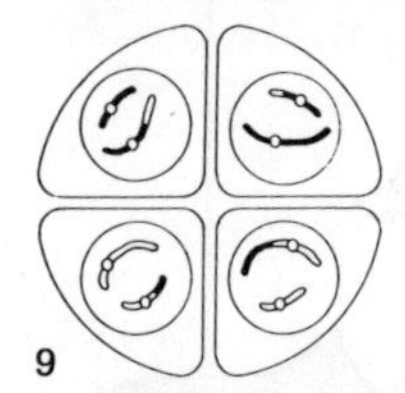

9

TELOPHASE 2
Four new cells form, each with half the number of chromosomes compared to the original cell. The composition of the chromosomes is altered.

fig 9.7 The main stages of meiosis, involving two pairs of homologous chromosomes

Telophase I follows the same pattern as mitotic telophase, and the second meiotic division is very similar to mitosis, where the chromosomes split and the daughter chromatids enter separate cells.

The entire process is summarised in fig 9.7.

Questions

1 How many gametes are produced from one body cell during meiosis?

2 How does the number of chromosomes in a gamete compare with the number in a body cell of the same organism?

3 What would be the effect if, during anaphase I, all the chromosomes originating from the male parent moved to one pole, and all those originating from the female parent moved to the other?

4 What is the purpose of exchange of chromatid segments?

9.10 Variation and heredity

Consider the four people shown in fig 9.8. They come from different parts of the world and represent the four major races of mankind. In spite of the differences in appearance, they all belong to the same species, *Homo sapiens*. This means that it is possible for them to interbreed and to produce children who are themselves fertile. If an African man were to marry a European woman, their children would have some negroid characters, some caucasoid characters and some characters intermediate between those of the two races.

A more striking example of variation can be seen in the case of the dog. For centuries, dogs have been used by Man on every continent for a variety of purposes; for hunting, for guarding property, for herding livestock and even for drawing small carts. As a result, an enormous number of different shapes, sizes and colours of dog have developed throughout the world. Nevertheless, all domestic dogs belong to the same species, *Canis familiaris*, and thus are capable of interbreeding.

Fig 9.9 shows the results of crossing two pedigree animals of different breeds. The offspring, which are shown when fully grown, are all similar, having some characters in common with their dam (mother) and some in common with their sire (father). Crossing of two animals from this generation results in a much wider variety of offspring. You probably have inherited some characteristics from your mother, and some from your father. However, some of your features may appear unlike either parent. These may have been present in your grandparents or in earlier generations.

fig 9.8 The four races of Man

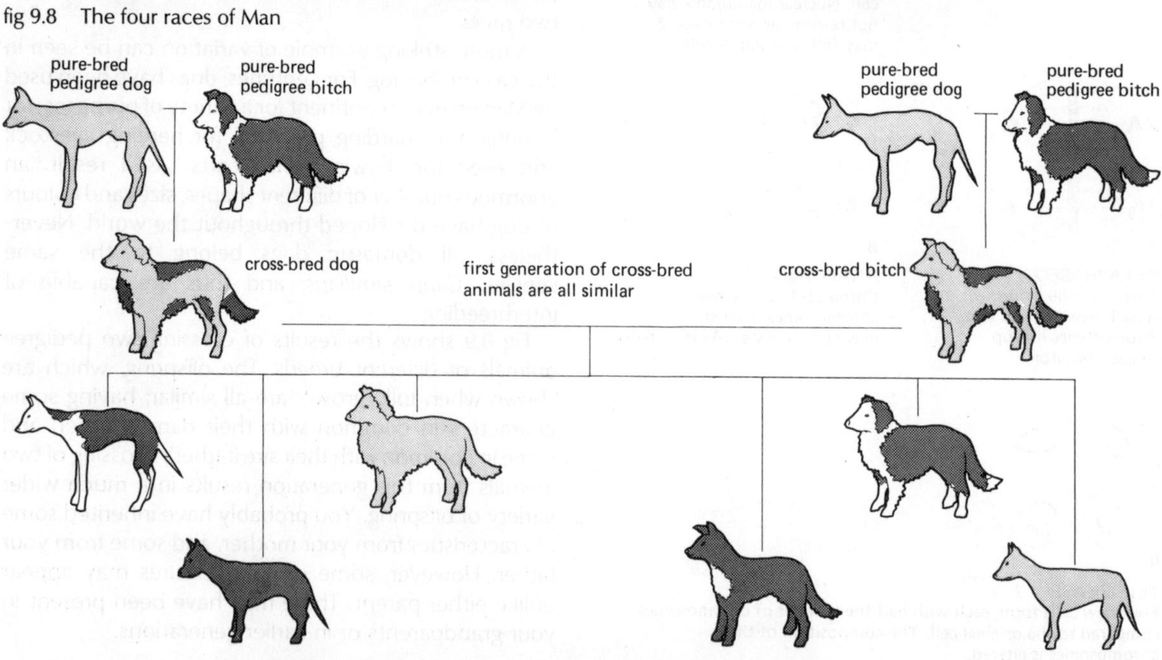

fig 9.9 Variation produced by cross-breeding pedigree dogs over two generations

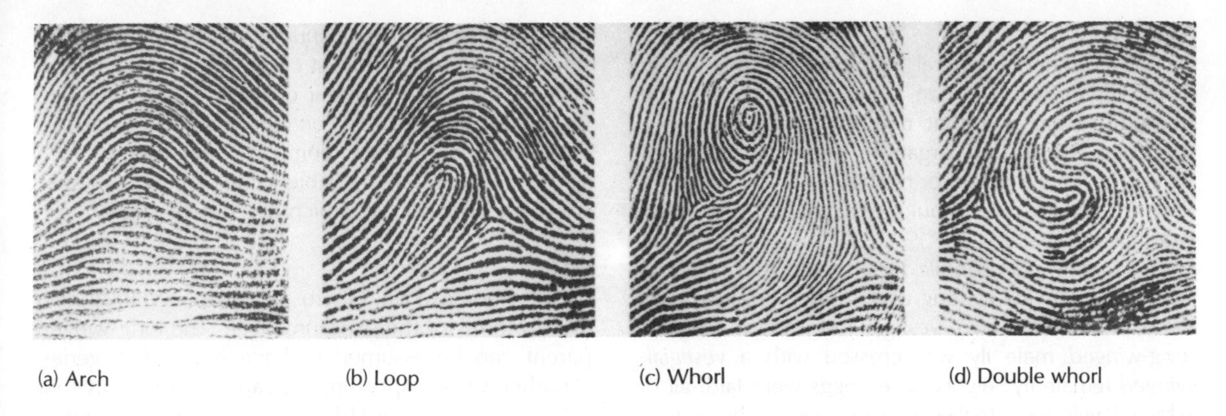

(a) Arch (b) Loop (c) Whorl (d) Double whorl

fig 9.10 The main groups of fingerprints

9.11 Variation in humans

Investigation

Variation in height

Procedure

1 Measure your own height (to the nearest centimetre) against a scale marked on a wall.

2 Measure the heights of all members of your class in the same way. If possible, obtain the heights of as many as a hundred students of about the same age. You will obtain more accurate results with a large sample.

3 Plot your results on graph paper, with height shown on the horizontal axis, and numbers of students in each height group on the vertical axis. You will find the detailed method in the Introductory Chapter (section I 4.10).

Question

1 What is the shape of the curve that you obtain?

Experiment

To investigate variation in fingerprints

Procedure

1 Wash and dry your hands thoroughly.

2 Place the tip of your right index finger lightly but firmly on an office ink pad.

3 Place the inside of your fingertip on a piece of white paper, and roll it carefully to the outside before removing it. Take care not to smudge the print.

4 Note the pattern of loops, arches and whorls (see fig 9.10).

5 Take a large sheet of paper and rule it into 2 cm squares. On each square obtain a fingerprint from the right index finger of a different member of your class.

6 Examine each print with a hand lens and classify it according to the patterns shown in fig 9.10.

Questions

2 Are any of the fingerprints identical?

3 What are the proportions of each type found in your sample?

Investigation

The inheritance of tongue rolling

Procedure

1 Put out your tongue and try to roll it as shown in fig 9.11.

2 Observe other members of your class, noting how many students can roll their tongues and how many cannot.

Questions

4 Do any members of the class show a partial ability to roll their tongues?

5 What explanation can you offer for the results of your investigation?

6 This is one of the few examples of an easily identifiable human characteristic inherited *via* a single gene. (Height and fingerprints are controlled by the action of several genes.) What other factors make humans unsuitable material for studies in genetics? (See section 9.12, question 2.)

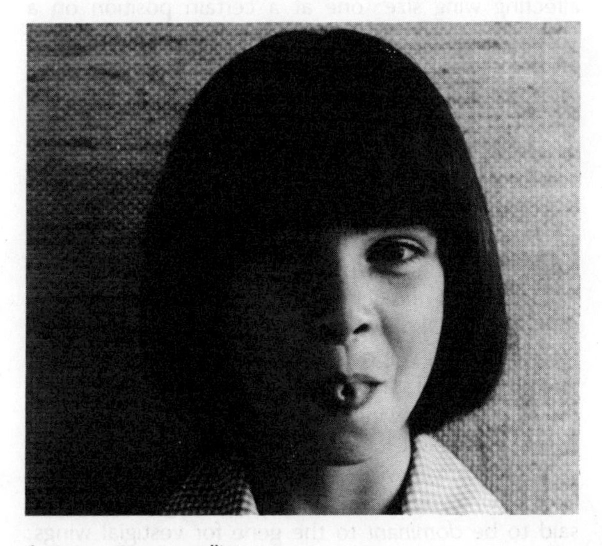

fig 9.11 Tongue rolling

9.12

Having considered several examples of variation in living organisms, we can now ask ourselves the question 'What is it that determines whether, in a particular feature, an organism resembles its male parent, its female parent or neither parent?' In order to find an answer, we should examine as simple an example as possible.

The fruit fly, *Drosophila melanogaster*, usually has wings about twice as long as its abdomen. However, certain individuals have vestigial (very short) wings. A *long-winged* male fly was crossed with a *vestigial-winged* female fly: eighty-seven eggs were laid, all of which developed into flies with long wings. The original pair of flies are referred to as the *parental* (P) generation and the 87 progeny as the *first filial* (F_1) generation. When two members of this long-winged F_1 generation were mated, the resulting *second filial* (F_2) generation consisted of 61 long-winged flies and 20 vestigial-winged flies.

Questions

1 What was the approximate percentage of vestigial-winged flies in the F_2 generation?

2 *Drosophila* is a very suitable animal for use in experiments on heredity. Can you suggest any reasons for this?

9.13

In order to explain the results obtained in 9.12, we must refer back to the chromosomes which we encountered in our study of cell division (see sections 9.01 and 9.04). Every organism arising from sexual reproduction receives from each parent a set of coded 'instructions' for development. These 'instructions' are called *genes* and are found on the chromosomes.

Thus, each fertilised fly egg will have *two* genes affecting wing size: one at a certain position on a chromosome received from its female parent, and one at the corresponding position on the homologous chromosome received, via the sperm, from its male parent. If both genes are 'instructions' for producing long wings, the wings will be long; if both genes are 'instructions' for producing vestigial wings, the wings will be vestigial. In these two cases the flies are said to be *homozygous* for the long-winged and vestigial-winged conditions respectively.

However, if one gene of each type is present, (i.e. if the organism is *heterozygous* for wing length), the resulting wings will not be halfway between the long and the vestigial state: they will be *long*. This is because the gene for long wings (referred to for convenience as L) is able to exert its influence in the presence of the gene for vestigial wings (referred to as l), and prevents l from having any effect. The gene for long wings is thus said to be *dominant* to the gene for vestigial wings: conversely l is *recessive* to L. This is a conveniently simple example of *discontinuous variation*, involving a single pair of genes. In most cases, a given character is affected by a large number of genes acting together, the variation produced is *continuous* and the pattern of inheritance is extremely complex. Nevertheless, such simple examples are valuable in helping us to understand the principles on which heredity operates.

9.14

We can now explain the crosses mentioned in section 9.12 in terms of the genes involved. The long-winged parent can be assumed to have had two L genes. (Another way of expressing the same idea is to say that it had a *genotype* of LL.) The vestigial-winged female parent had a genotype of ll. The F_1 generation all had genotypes of Ll, since they received one gene of each type in the gametes from their parents. However, their *phenotypes* (the physical expression of the genotypes) were long-winged. The conventional way of expressing this cross diagrammatically is in one of the following forms:

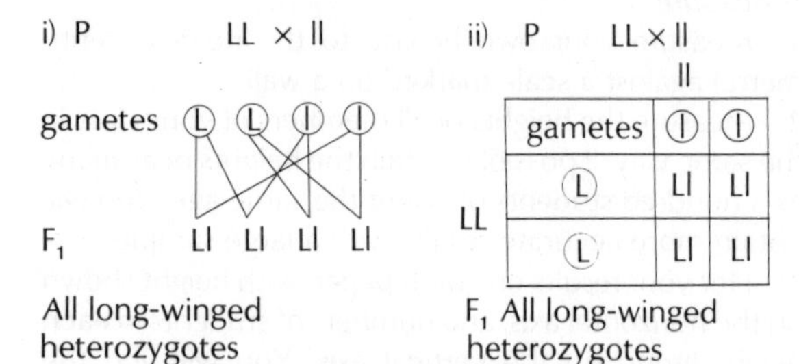

Remember that, during meiosis, the homologous chromosomes separate. Thus the F_1 generation produces gametes of two different types in respect of wing length, those containing the chromosome that bears the L gene and those containing the chromosome that bears the l gene. When two organisms are crossed, any sperm can fertilise any egg: thus any gene can be combined with any other gene. The diagram can now be completed as follows, to show the F_1 cross.

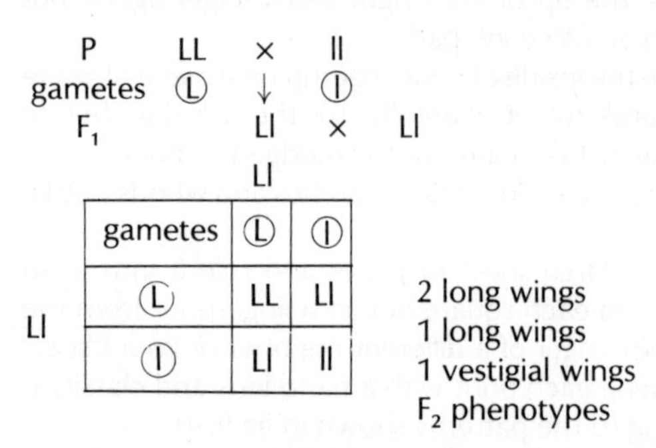

We can see that, for every vestigial-winged fly produced in the F_2 generation, three long-winged flies are produced. This 1:3 ratio corresponds closely with the ratio of 20: 61 obtained in section 9.12.

Experiment	Dominant character	Recessive character	No. of F_2	Ratio of F_2 dominants to F_2 recessives
1	Round seeds	Wrinkled seeds	7324	2.96:1
2	Yellow seeds	Green seeds	8023	3.01:1
3	Red flowers	White flowers	929	3.15:1
4	Inflated pods	Constricted pods	1181	2.95:1
5	Green pods	Yellow pods	580	2.82:1
6	Axial flowers	Terminal flowers	858	3.14:1
7	Long stem	Short stem	1064	2.84:1

Table 9.1 The F_2 results of Mendel's Monohybrid Crosses

9.15

The above experiment helps to emphasise a very important aspect of all statistical work in genetics: it is essential to deal in sufficiently large numbers.

Drosophila is such a useful organism for genetics experiments partly because each female fly lays so many eggs. The fact that the ratio of 61:20 is not exactly the same as 3:1 is not important because we are dealing with such large numbers. However, if the extra one long-winged fly had occurred in a brood of five, the ratio would have changed drastically to 4:1, and the figures would have been impossible to interpret.

Let us consider a more familiar example. The number of boy babies born is statistically equal to the number of girl babies. Thus, in a village where there are 160 children, the expected number of boys is 80 and of girls is also 80. It would be quite unremarkable for a village of this size to have 75 boys and 85 girls, or vice versa, and the ratio would still be almost 1:1. However, if we were to study individual families, it would be quite common to find mothers who had given birth to two, three or four children, all of whom were boys, and others who had only girl babies. If we were to consider only one of these families, then we would obtain a completely misleading picture of the ratio of the sexes.

A further example of the importance of using large numbers in studies on inheritance can be seen in the work of Gregor Mendel. Mendel was an Austrian monk, regarded by many as the father of genetics, who carried out most of his work more than 100 years ago, using pea plants. He investigated the type of cross seen in section 9.12 except that he studied several pairs of contrasting characters. The results of these crosses are shown in the table above. (Since only one pair of characters is being considered, the crosses are described as *monohybrid*.)

Question

1 Draw a graph, showing the difference between the ratio obtained and the ideal ratio (3:1), plotted against the number of plants in the F_2 generation. What deductions can you make from this graph?

Investigation

To study monohybrid crosses, using models

Procedure

1 Into each of two large jars or boxes, place one hundred red beads and one hundred white beads. The jars represent two individuals of the F_1 generation in the experiment, described in section 9.12. The red beads represent the dominant gene for long wings (L) and the white beads represent the recessive gene for vestigial wings (l).

2 Shake each jar thoroughly to mix up the beads.

3 Put one hand into each jar and select a bead at random (do not look at the jars while you are making your selection). Put the two beads together on the bench: they represent a zygote. Note whether the combination is red and red, white and white or red and white.

4 Repeat this process one hundred times.

Questions

2 What wing phenotypes will be produced by the following bead combinations: red and red, red and white, white and white?

3 What is the ratio of long-winged to vestigial-winged flies produced from the hundred zygotes?

9.16

Refer back to the *Drosophila* cross covered in section 9.12. One of the male flies from the F_1 generation was mated with its vestigial-winged parent.

Questions

1 What are the genotypes of the two parents in this cross?

2 What gametes are produced by the male?

3 What gametes are produced by the female?

4 How could the cross be expressed diagrammatically? State the phenotypes of the offspring.

9.17

Let us consider another type of cross, this time involving a species of pea plant which is normally self-pollinating and thus pure-breeding. The anthers were removed from a flower of a tall-stemmed pea plant, and its stigma was dusted with pollen taken from a short-stemmed plant of the same species. The resulting seeds produced F_1 generation plants which were all tall-stemmed.

Several flowers from this F_1 generation had their anthers removed and their stigmas dusted with pollen from the original tall-stemmed parent plant. The resulting seeds produced plants that were all tall-stemmed.

Questions

1 Construct a diagram to illustrate this cross, using suitable symbols for the genes involved.

2 What proportion of the plants produced from this cross are likely to be homozygous (pure-breeding) for the gene for tallness?

3 Refer back to section 9.12. Can you think of any ways in which pea plants are preferable to *Drosophila* for use in genetics experiments?

4 Another organism that is particularly useful in genetics experiments is the maize plant (*Zea mays*). Each ear of maize contains a large number of seeds or kernels, and each seed is the result of a single fertilisation. The seeds possess a number of characters which show discontinuous variation, including colour and shape. For example, the gene for *colour* is dominant to the gene for *lack of colour*, and the gene for *smoothness* is dominant to the gene for *shrunkenness*. Suppose a plant grown from a heterozygous smooth seed were pollinated by a plant grown from a shrunken seed. What proportions of seed phenotypes would you expect to be produced in the resulting F_1 generation?

5 What particular advantage does the use of maize seed characters in genetics experiments have over, for example, the use of stem length and flower colour in peas?

9.20 Incomplete dominance

In a certain plant species, a red-flowered individual when crossed with a white flowered individual, will produce pink-flowered offspring. If self-pollinated, the pink-flowered plants will produce an F_2 generation with flowers in the ratio of 1 red: 2 pink: 1 white. This is an example of incomplete dominance or blending inheritance, where the heterozygote produces a condition intermediate between those produced by the two homozygotes. Neither gene or allele is completely dominant over the other, and so the above crosses can be explained diagrammatically in the following manner.

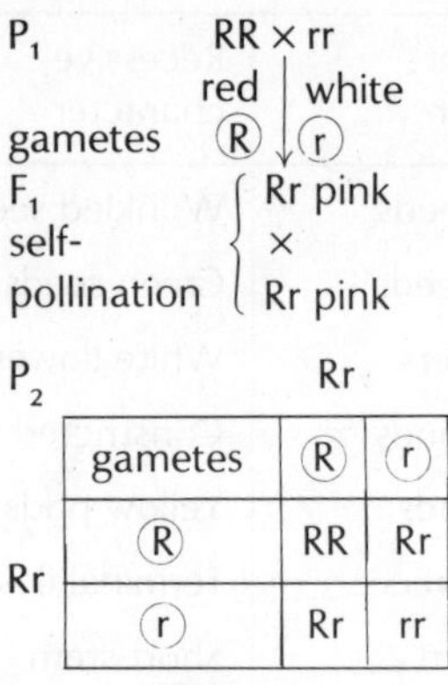

F_2 phenotypes:

1RR red   2Rr pink   1rr white

Although the cells of a single individual cannot possess more than two genes of any one kind, this does not mean that a greater variety of genes is not possible. For example, the inheritance of blood groups in Man is governed by three alleles; G^A, G^B and G^O (see section 2.31). Both G^A and G^B are dominant to G^O, but G^A is not dominant to G^B nor G^B to G^A: in other words, these two genes are *co-dominant*.

Therefore, an individual will have blood of group A if his genotype is G^AG^A or G^AG^O; he will have blood of group B if his genotype is G^BG^B or G^BG^O; but he will have blood of group O only if his genotype is G^OG^O. However, since G^A and G^B are co-dominant, a genotype of G^AG^B will produce a fourth phenotype, blood group AB.

Question

1 Construct a diagram to show the possible phenotypes arising in the children of a marriage between a woman with blood group A and a man with blood group B, where both these parents are heterozygous.

9.30 Sex determination

'Will our new baby be a boy or a girl?' This is the question that all prospective parents ask. We all know that the chances of any one baby belonging to a particular sex are about 1:1 or 50%, but how can this be explained in terms of chromosomes?

To answer this, we must look at the 46 chromosomes that occur in each human cell. In women and girls, they consist of 23 pairs of similar chromosomes: in men and boys, there are 22 pairs of similar chromosomes, but one member of the 23rd pair is much smaller than the other. This smaller one is called the Y chromosome, the larger one is called the X chromosome. Together they form the pair of sex chromosomes. There is no Y chromosome in human females; the pair of sex chromosomes consist of two Xs.

Thus, each one of 50% of the sperms produced by a man (as a result of separation of the homologous chromosomes during meiosis) will contain an X chromosome and each one of the other 50% will contain a Y chromosome. However, every ovum produced by a woman will contain an X chromosome. Therefore, the probability of an offspring having XX sex chromosomes is equal to the probability of its having XY sex chromosomes.

This can be represented diagrammatically as follows:

	Man $\times$ Woman	
Sex chromosomes	XY	XX

Man

	gametes	Ⓧ	Ⓨ
Woman	Ⓧ	XX	XY
	Ⓧ	XX	XY

2 girls 2 boys

9.40 Sex linkage

When the genes controlling a certain character are located on the sex chromosomes, the character is inherited along with the sex. Such characters are described as *sex-linked*. Since the sex chromosomes in the male are the only pair of dissimilar chromosomes, a gene carried on the X chromosome need not have a corresponding gene (or allele) on the Y chromosome (in fact, the human Y chromosome is thought to contain very few genes, if any at all).

Now a boy child receives his X chromosome from his mother and his Y chromosome from his father. Therefore, he cannot inherit his father's sex-linked characters.

However, a man can pass on sex-linked characters, via his daughters, to his grandchildren.

9.41 Haemophilia

One of the best-known examples of sex-linked inheritance is shown by the disease *haemophilia* (bleeder's disease), in which the blood does not have the ability to clot. The disease is caused by a recessive gene carried on the X chromosome.

Female haemophiliacs are rare, since they can only be produced when both parents carry the gene. However, a woman with the recessive gene on one of her X chromosomes will act as a 'carrier' of the disease.

Queen Victoria was such a carrier. Only one of her four sons (Prince Leopold) was a haemophiliac, but she passed on the disease to many of the European Royal Houses through her two carrier daughters, the princesses Alice and Beatrice. Fig 9.12 shows the progress of the disease through four generations of royal families. Note that all affected individuals are males. This indicates that the sex chromosomes are involved. A haemophiliac male of the second generation produces a son who is unaffected by the disease since a boy receives the Y chromosome from his father and the X chromosome from his mother. This confirms that the gene for haemophilia is likely to be carried on the X chromosome.

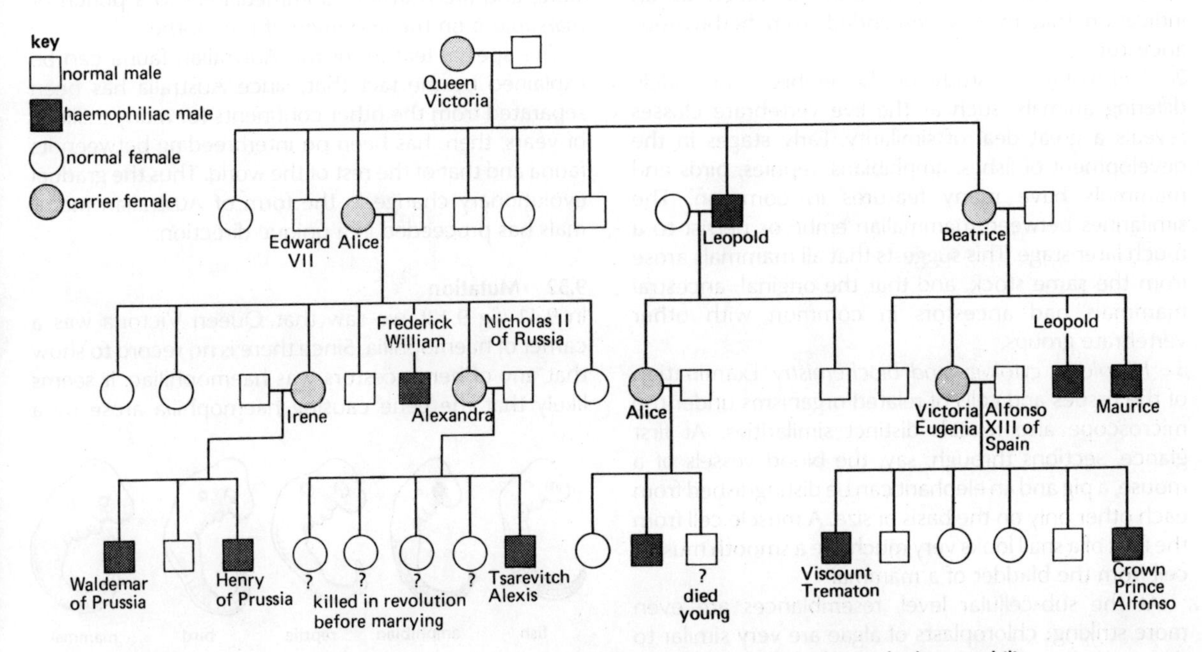

fig 9.12 Queen Victoria's family tree, showing inheritance of the recessive sex-linked gene for haemophilia

9.50 Evolution

The fact that variation can be inherited from one generation to the next helps us to explain the process of evolution – the gradual change in the nature of living organisms over long periods of time. That evolution occurs is almost universally accepted today, but not much more than a century ago Charles Darwin caused great scientific and religious controversy when his book *On the origin of species by means of natural selection* (1859) contradicted the traditional idea of divine creation as described in the *Book of Genesis*.

9.51 The evidence for evolution

Evidence to show that evolution has taken place, and is still taking place, can be obtained from studies of several branches of science.

1 *Anatomy* Certain groups of living animals resemble each other in a way that suggests common ancestry. For example, men, sparrows, lizards and frogs all have feet which bear five toes (or remnants of them). Close examination reveals that birds' wings, sea lions' flippers and horses' hooves are based on the same plan, and that snakes possess remnants of limb girdles. This suggests that all land vertebrates arose from forms that had two pairs of pentadactyl (five-digit) limbs (see Chapter 11).

The limb-girdle remnants of snakes are referred to as *vestigial*. Among our own vestigial structures is the appendix – relatively a very small structure compared with the corresponding structure in a rabbit, which contains symbiotic bacteria for the digestion of cellulose in plant food. This may be taken as an indication that man is descended from herbivorous ancestors.

2 *Embryology* A study of the embryos of widely differing animals, such as the five vertebrate classes reveals a great deal of similarity. Early stages in the development of fishes, amphibians, reptiles, birds and mammals have many features in common. The similarities between mammalian embryos persist to a much later stage. This suggests that all mammals arose from the same stock, and that the original, ancestral mammals had ancestors in common with other vertebrate groups.

3 *Histology, cytology and biochemistry* Examination of the tissues and cells of related organisms under the microscope also reveals distinct similarities. At first glance, sections through, say, the blood vessels of a mouse, a pig and an elephant can be distinguished from each other only on the basis of size. A muscle cell from the foot of a snail looks very much like a smooth muscle cell from the bladder of a mammal.

At the sub-cellular level, resemblances are even more striking: chloroplasts of algae are very similar to those of higher plants, and all mitochondria are built on the same plan.

Most fundamental of all, perhaps, is the point that the molecules upon which the structure and functioning of all living organisms are based show remarkably little variation: the composition of plasma membranes in protozoans differs little from that in mammalian cells, and the biochemical pathways of aerobic respiration are similar in animals and plants.

4 *Palaeontology* Palaeontology is the study of fossils: it is probably the branch of science most closely associated in people's minds with evolution.

The fact that the flora and fauna of the earth many millions of years ago were quite different from those found today can have only two possible explanations. Either the old animals and plants died out completely and were replaced by newly-created forms, or they underwent a gradual change from generation to generation. The first alternative is plainly absurd, whereas the second is supported by the existence of many 'fossil histories', showing progressive changes in form of organisms through succeeding strata of rocks.

In this way, the evolution of man, monkeys and apes has been traced back to a group of small tree-shrews that lived 60 million years ago. Note that monkeys and man arose from a **common ancestor**; it is **not** true to say that man evolved from monkeys.

5 *Geography* Lastly, it is worth noting the effect of *isolation* on flora and fauna. In Australia, the native mammals are quite different from those of other continents. They are marsupials, such as the kangaroo, wallaby, wombat and koala. One of their most striking features is that the young are born in a very immature state, and are transferred immediately to a pouch or *marsupium* on the abdomen of the mother.

This special feature of the Australian fauna can be explained by the fact that, since Australia has been separated from the other continents for many millions of years, there has been no interbreeding between its fauna and that of the rest of the world. Thus the gradual evolutionary change in the form of Australian mammals has proceeded in a unique direction.

9.52 Mutation

In 9.41 (fig 9.12), we saw that Queen Victoria was a carrier of haemophilia. Since there is no record to show that any of her ancestors was haemophiliac, it seems likely that the gene causing haemophilia arose by a

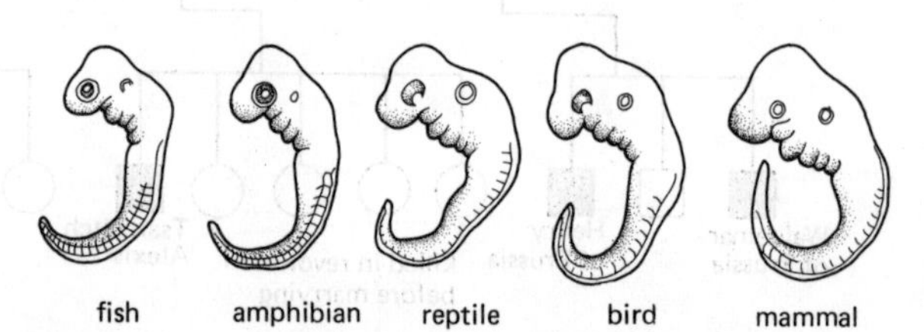

fig 9.13 Early embryos of vertebrates

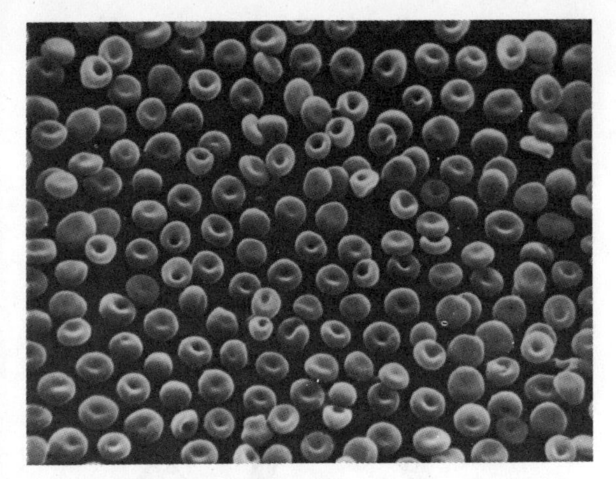
fig 9.14 (a) Normal red blood cells

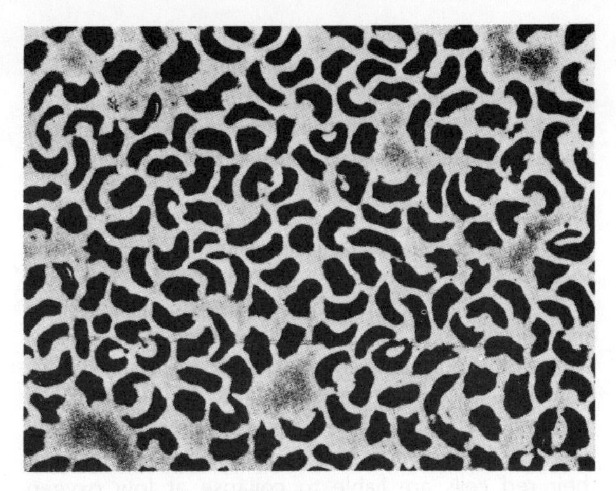
fig 9.14 (b) Sickled red blood cells

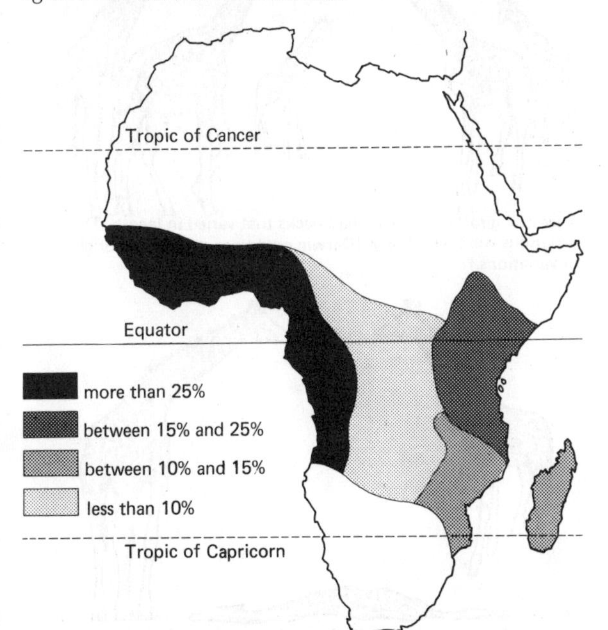

fig 9.15 A map showing the distribution frequency of people carrying the gene for sickle cell haemoglobin

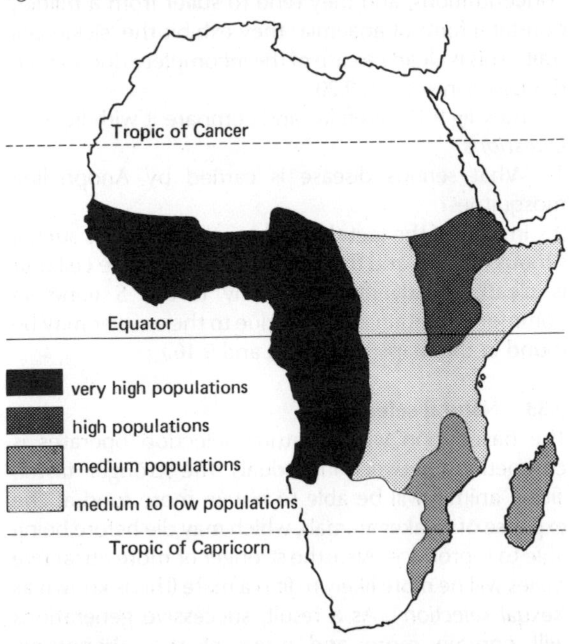

fig 9.16 A map showing the distribution frequency of *Anopheles* mosquitoes which act as vectors for the organisms causing malaria

change in a gene responsible for blood clotting, before she was conceived. This gene change could have occurred in the X chromosome of the egg or of the sperm that fertilised it. Such changes are known as *mutations*: they may affect single genes or relatively large parts of chromosomes.

Sickle cell anaemia

Not all mutations produce disastrous results in the phenotype. In some cases, their effects are beneficial, and they result in an organism being able to compete successfully with other organisms for food and for the production of more offspring. Such a mutation is said to be *selectively advantageous* and it will thus be possessed by more and more individuals in successive generations. In this way a species evolves: mutations are the raw material of evolution.

An interesting and important example of the selective advantage of a mutation can be seen in the condition known as the 'sickle cell trait'. This is a very common blood condition in many parts of the world, and may occur in up to 30% of some populations in West and Central Africa (see fig 9.14).

It arises in the following way. The blood pigment haemoglobin is essentially a protein, containing iron. One particular form of haemoglobin (haemoglobin S) is produced by a mutation of the gene responsible for the production of normal haemoglobin. This mutation causes the substitution of one amino-acid (*valine*) for another (*glutamic acid*) during the formation of the haemoglobin protein, and as a result the red blood cells containing haemoglobin S collapse into the shape of a sickle at low oxygen concentrations.

People who are homozygous for this mutated S gene suffer from an often fatal disease known as *sickle cell anaemia*, where all the haemoglobin is of the S type and all the red blood cells are liable to collapse. People who are homozygous for the non-mutated gene do of course have normal haemoglobin and normal red cells. However, heterozygotes (people who possess both mutated and non-mutated genes) have some normal haemoglobin and some haemoglobin S. Thus some of their red cells are liable to collapse at low oxygen concentrations, and they tend to suffer from a milder, non-fatal form of anaemia: they exhibit the 'sickle cell trait'. This is clearly a case of the incomplete dominance discussed in section 9.20.

Study fig 9.15 carefully, and compare it with fig 9.16.

Questions

1 What serious disease is carried by Anopheline mosquitoes?

2 In view of the fact that sickle cell anaemia is such a serious disease, and that even the milder sickle cell trait is clearly disadvantageous, why is the S gene so common in certain areas? (A clue to the answer may be found in the maps in figs 9.15 and 9.16.)

9.53 Natural selection

The basis upon which natural selection operates is competition between individuals. The stronger, faster, 'fitter' animal will be able to obtain more food at the expense of weaker animals, which may die before being able to reproduce. Also, the stronger or more attractive males will be more likely to find a mate (this is known as *'sexual selection'*). As a result, successive generations will contain more and more of the phenotypic characters of the *'fittest'* animals. Among plants, the 'fittest' are those best able to use the environment in which they live, competing for light, water and nutrients. An example of this is to be found in dense forest, where the fastest growing trees compete more successfully than the slower growing forms. They reach the light more easily and can, therefore, photosynthesise more effectively. A further example can be seen in the evolution of the giraffe from short-necked ancestors (see fig 9.17).

9.54 The effects of a changing environment

During the nineteenth century, butterfly and moth collecting (lepidoptery) was a popular hobby. One particularly common species was *Biston betularia* (the peppered moth), which usually had greyish-white wings with scattered dark markings; from time to time, odd specimens with very dark wings were recorded. Towards the end of the century, lepidopterists noted that more and more dark forms were being collected in the North and Midlands, until the situation shown in fig 9.18 was reached.

Questions

1 Study fig 9.19. What does it tell you about the vulnerability of the moths to predatory birds?

2 How can the results shown on the map in fig 9.18 be related to the changing environment in Britain during the last century?

Ancestral giraffes probably had necks that varied in length. The variations were hereditary. (Darwin could not explain the origin of variations.)

Competition and natural selection led to survival of longer-necked offspring at the expense of shorter-necked ones.

Eventually only long-necked giraffes survived the competition.

fig 9.17 The origin of the giraffe's long neck by natural selection

fig 9.18 The distribution of light and dark forms of the peppered moth in the British Isles. The proportion of the light form in a locality is represented by the white portion of each circle. The dark area of each circle represents the proportion of dark forms present.

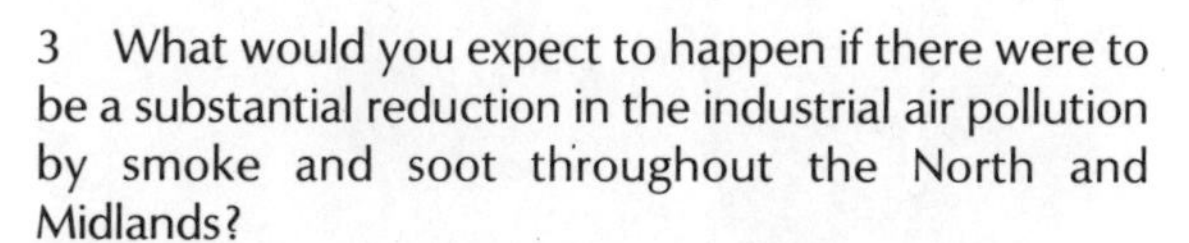

fig 9.19 Light and dark forms of the peppered moth resting on tree trunks in (a) Birmingham and (b) Dorset

3 What would you expect to happen if there were to be a substantial reduction in the industrial air pollution by smoke and soot throughout the North and Midlands?

Over the centuries Man has succeeded in interfering with the process of natural selection in order to produce animals and plants suitable for his own agricultural purposes. He has taken animals which would not necessarily have competed successfully in the wild, but which had phenotypic characters (such as the ability to put on weight quickly or to produce large quantities of milk) which were useful to him, and has bred them over many generations. At each generation he has selected for breeding those animals which possess these qualities to the greatest degree, with the result that the domestic animals have, in many cases, become quite unlike their wild ancestors.

In the case of food plants, *storage organs* are particularly important. The potato, for example, has been produced by breeding from wild South American forms with much smaller stem tubers. Gradually, generation after generation, the plants with the largest tubers have been selected for breeding until the modern potato has been evolved, with tubers which are much larger than would be needed to supply the wild plant with enough food reserves to last over winter.

Other phenotypic characters which have been selected for artificially in the production of strains suitable for modern agricultural purposes are:

i) *early maturity* in both animals and plants (in the case of cereals, for example, this can result in the harvesting of two or more crops per growing season);

ii) *resistance to disease,* as in the case of wheat where strains resistant to the fungi causing rust diseases have been developed;

iii) *increased length* of productive season;

iv) *adaptation* to local or unfavourable conditions;

v) greater efficiency of *conversion* of plant to animal tissues in animals bred for meat;

vi) *higher yields* in terms of the production of milk, eggs, wool, fruits, etc;

vii) *ease of harvesting,* as in the case of thornless blackberries;

viii) improved eating quality, as in the cases of

coreless carrots and seedless oranges;

ix) hardier constitution, such as sweetcorn that can be grown in Britain.

In many cases it has been found profitable to improve the stock by crossing locally bred animals and plants with strains adapted to similar environments in different parts of the world.

fig 9.20 A Hereford bull, bred for beef production

fig 9.21 A Guernsey cow, bred for milk production

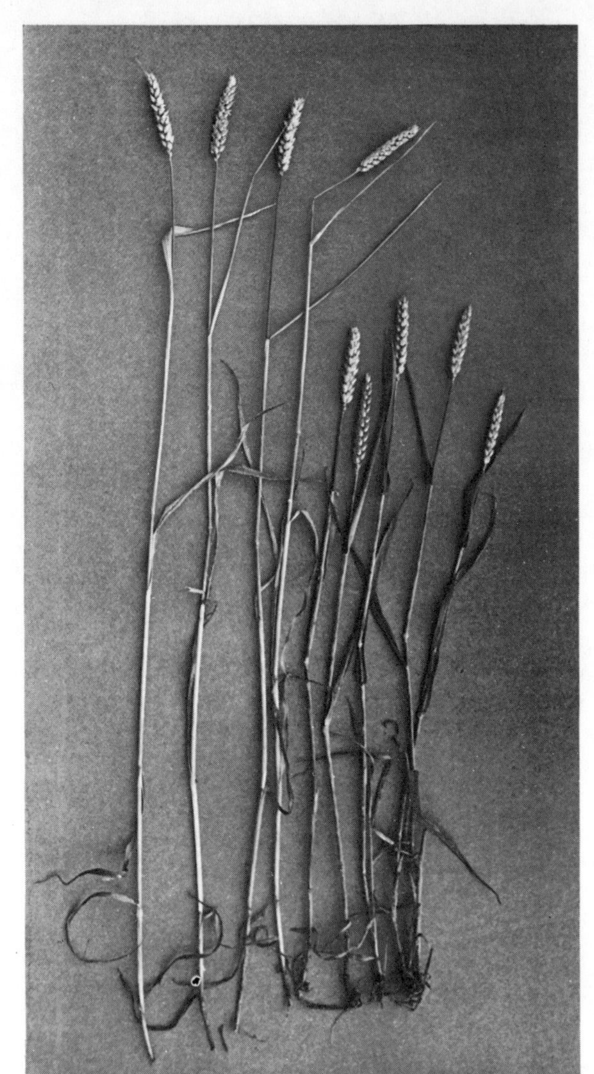

fig 9.23 Winter wheat variety with long stalk together with a short-stalked variety

fig 9.22 A British Fresian bull and cow, an all-round breed with high beef and milk yields

9.70 Glossary of terms used in genetics

Allele (allelomorph): one of a pair (or more) of alternative forms of a gene.

Bivalent: a pair of homologous chromosomes, situated together during meiosis.

fig 9.24 Wild pig

Centromere: a part of the chromosome which has no genes and by which the chromosome becomes attached to the spindle.

Chromatid: one half of the chromosome, when the chromosome splits longitudinally during cell division.

Chromosome: a thread-like structure, bearing genes and located in the nucleus.

Diploid: describes a cell that has two sets of homologous chromosomes.

Dominant: describes an allele whose effect is seen in the phenotype of a heterozygote, in spite of the presence of an alternative allele.

Equator: the plane through the middle of the cell, at right angles to the main axis of the spindle.

Gene: a unit of hereditary material, located on the chromosome.

Genotype: a description of an organism in terms of certain of its genes.

Haploid: describes a cell with a single set of chromosomes.

Heterozygote: an organism which has two different alleles of the same gene.

Homozygote: an organism which has two identical alleles of the same gene.

fig 9.25 English Large White pig, bred for meat production (to same scale as fig 9.24)

Phenotype: a description of an organism in terms of what can be seen or measured.
Recessive: describes an allele whose effect is not seen in the heterozygote, because of the presence of a dominant allele of the same gene.
Spindle: an arrangement of cytoplasmic fibres, between the poles of a cell, along which the chromatids (or chromosomes) move during mitosis (or meiosis).

Questions requiring an extended essay-type answer

1 a) Name an organ in which *meiosis* occurs in
i) a flowering plant, and
ii) a mammal.
b) Name an organ in which *mitosis* occurs in
i) a flowering plant, and
ii) a mammal.
c) How would you prepare plant cells to show mitotic figures? Draw diagrams to show the appearance of the chromosomes seen in your preparation at the four main stages of mitosis (assume that in your plant cells $2n=8$).

2 A breed of dogs has long hair dominant to short hair. A long-haired bitch was first mated with a short-haired dog and produced three long-haired and three short-haired puppies. Her second mating with a long-haired dog produced a litter with all of the puppies long-haired. Use the symbol (L) to represent the allele for long hair and (l) to represent that for short hair.
i) What was the genetical constitution of the long-haired bitch?
ii) How could you determine which of the long-haired puppies of the second mating were homozygous?

3 a) A specimen of *Drosophila* has red eyes and when crossed with a purple-eyed mutant, all of the offsprings were red-eyed. The offspring were mated among themselves and the following proportions of flies were produced: 224 red-eyed and 76 purple-eyed. Define the symbols that you will use to represent the alleles of these flies, and then by means of diagrams explain fully the genetics of these two crosses.
b) Draw diagrams to show the genetic details of a cross between red-eyed males of the parental generation and red-eyed females of the F_2 generation.

10 Growth

10.00 What is growth?

At first glance, the answer to this question may seem obvious. 'Growth is increase in size', would be the first reaction of many people.

Consider the following examples of increase in size, and decide for yourself whether or not they agree with a biologist's idea of growth.

1 A crystal of salt increases in volume and weight by addition to its external surface.

2 A camel increases in volume and weight when it takes a long drink after travelling for several days without access to water (fig 10.1).

3 A wilted plant increases in weight and height immediately after a rainstorm.

4 A butterfly is much longer and wider than the pupa from which it has emerged.

5 A puffer fish increases in volume by blowing up its body with gas when it is disturbed.

6 After weighing yourself, you drink a litre of water: on reweighing yourself, you find that you have gained 1 kg.

7 A 36-year-old man, after following a certain diet for one year, has gained 3 kg in weight.

8 A 12-year-old boy, after following the same diet for one year, has also gained 3 kg in weight.

9 A tree in a well-watered garden increases in height by 10 cm and in trunk circumference by 1 cm in one year.

To help you with this decision, here is a definition of growth given by a leading expert on animal development.

'Growth is the increase in size of an organism or of its parts due to synthesis of protoplasm or of extracellular substances. Protoplasm in this definition includes both the cytoplasm and the nucleus of cells. Extracellular substances are the substances produced by cells and forming a constituent part of the tissues of the organism, such as the fibres of the connective tissue, the matrix of bone and cartilage, and so on, as opposed to substances produced by the cells and subsequently removed from the organisms, such as the secretions of digestive and skin glands, or substances stored as food, such as fat droplets in cells of the adipose tissue. Absorption of water or the taking of food into the alimentary canal before the food is digested and incorporated into the tissues of the animal, although it may increase the weight of the animal, does not constitute growth.'

According to this definition, only examples 8 and 9 qualify as true growth. However, this is only one definition; and it is probable that many people would disagree with it. Some biologists would argue that an increase in complexity alone constitutes growth: others would feel that elongation of plant cells by increasing the size of the vacuole is a form of growth. There is no universal agreement on the precise meaning of this very important term.

fig 10.1 Photographs of a camel (a) dehydrated (b) one hour later after drinking

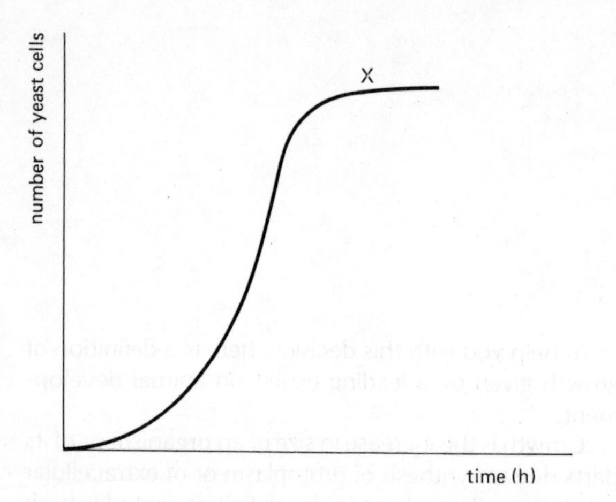

fig 10.2 A sigmoid growth curve of a population of yeast cells

Many small living organisms are studied in terms of the population as a whole, and the changes in number as it increases in size can be thought of as growth. Bacteria, unicellular algae and fungi can all be studied in this way, but a convenient organism is *yeast*. A single yeast cell divides into two cells at regular intervals. The yeast culture will continue to divide and double its numbers as long as none of the cells die. If the growth of this population is analysed it starts slowly, increases rapidly and then slows down again, finally levelling out. These numbers plotted against time give a characteristic S-shaped curve said to have a *sigmoid* form (see fig 10.2).

Hence it can be determined whether numbers are increasing, remaining stationary, or decreasing. The population increases in bulk, and we can apply the term 'growth', and knowing the organism and the environment in which it lives, we may be able to explain the growth changes.

Question

1 With your previous knowledge of yeast cells (see Chapter 5, Section 5.40) give reasons why the sigmoid growth curve (fig 10.2) slows its rate of growth and finally flattens out at point X.

10.10 Germination

This is the process in which a seed starts to develop into a young plant or seedling.

Investigation

The structure of a pea seed

Procedure

1 Soak two or three pea seeds in water for 24 hours, so that their outer coverings (testas) become soft.

2 Examine one seed carefully, locating the scar (hilum) by which the seed was attached to the seed stalk or funicle inside the pod. At one side of the hilum

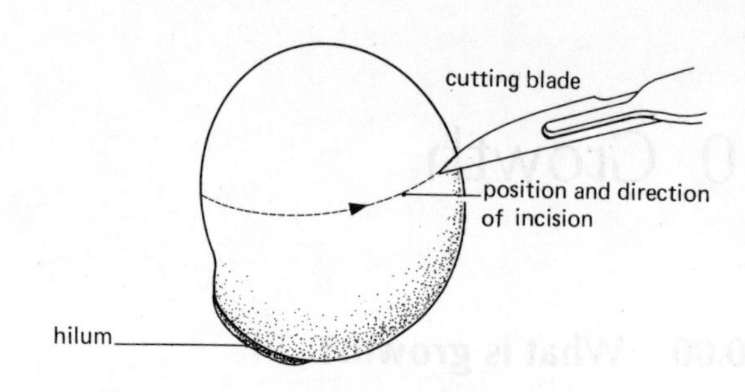

fig 10.3 The removal of the pea testa

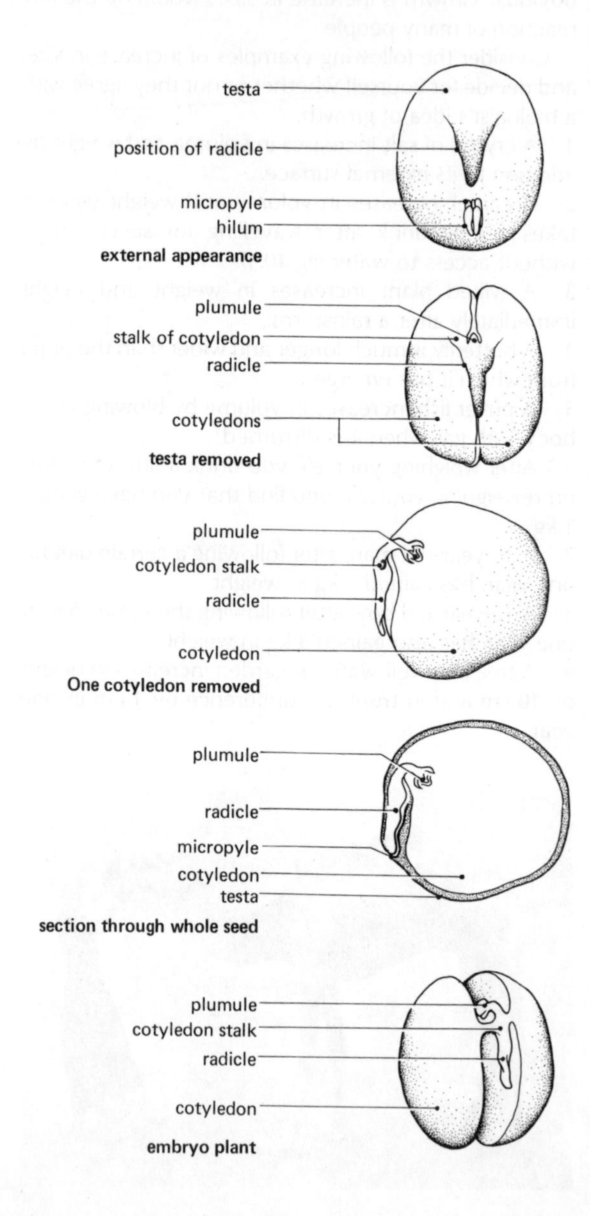

fig 10.4 Structure of a pea seed *After Stone and Cozens*

is a small hole, the micropyle, through which most water is absorbed.
3 Using a scalpel or razor blade, cut carefully around the outside of the testa, so that it can be removed in two halves, without causing damage to the internal structures (fig 10.3).
4 Note that the bulk of the seed is made up of two fleshy leaf-like structures (cotyledons) which contain food reserves. Identify the young shoot (plumule), and the young root (radicle), which is slightly larger.
5 Remove one of the cotyledons carefully, and examine the plumule and radicle in more detail. Can you see the beginnings of leaf structure at the tip of the plumule? (see fig 10.4).

Experiment

To investigate germination in pea seeds

Procedure

1 Place about 1 cm depth of water in a glass jar, and then line the jar with a roll of filter paper or blotting paper (see fig 10.5).

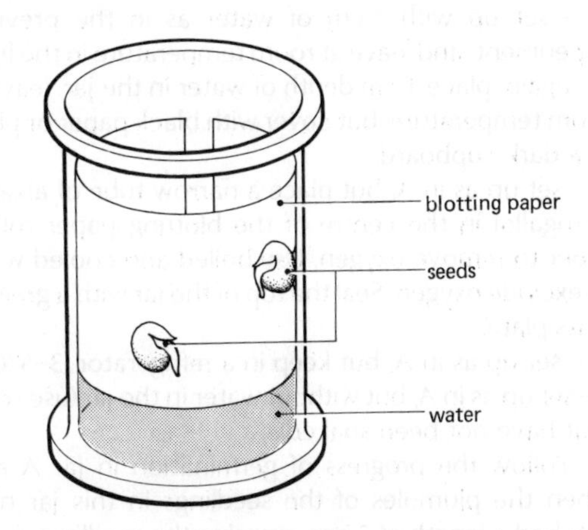

fig 10.5 A method of germinating peas in a gas jar

2 Take two of the pea seeds remaining from the previous investigation and wedge them between the blotting paper and the wall of the jar, about 10 cm from the bottom.
3 Check the appearance of the seeds every day.
4 Keep a record of the progress of germination, noting the lengths of the plumule and radicle at daily intervals, together with any changes in appearance, such as the production of root hairs and leaves. Express your results in the form of a table.

Question

1 From your table, which emerges first, the radicle or the plumule? Give reasons why this structure should always be first to emerge.

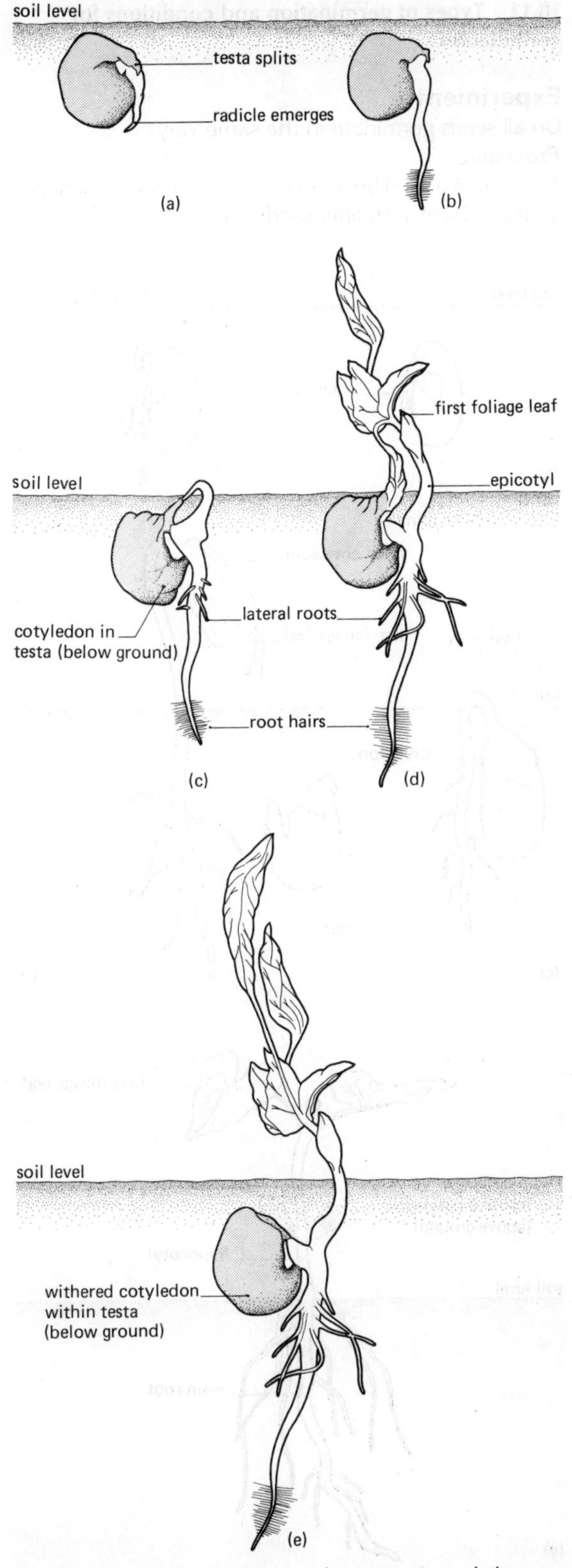

fig 10.6 Stages in the hypogeal germination of the pea, *Pisum spp*

10.11 Types of germination and conditions for germination

Experiment

Do all seeds germinate in the same way?

Procedure

1 Take two seed boxes or trays and fill each to a depth of about 7 cm with fine, sandy soil.

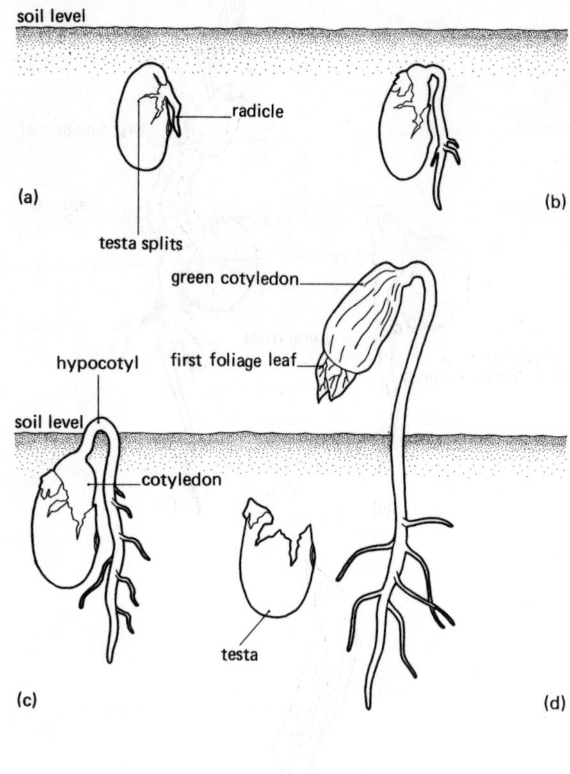

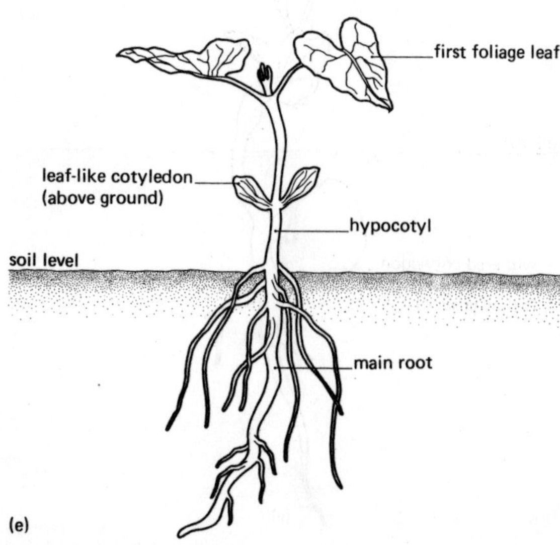

fig 10.7 Stages in the epigeal germination of the kidney bean

2 Plant a row of pea seeds in one box, labelled A, and a row of kidney (French) bean seeds in the other box, labelled B. Planting should be carried out by pressing a hole about 3 cm deep with your finger in the soil and dropping the seeds into it. In the case of the beans, the long axis of the bean must stay vertical.

3 Cover the seeds with soil and leave them at room temperature, watering at regular intervals.

4 Follow the progress of germination in both boxes, keeping a careful record of your observations.

Question

1 What difference do you observe between the mechanisms of germination in the two types of seed with regard to the movement of their main structures?

Experiment

What are the conditions necessary for germination?

Procedure

1 Set up 5 jars, with 3 pea seeds wedged against the wall of each jar, 10 cm from the bottom, with a roll of filter paper. Label the jars A to E and treat as follows.

A – set up with 1 cm of water as in the previous experiment, and leave at room temperature in the light.

B – again, place 1 cm depth of water in the jar, leave at room temperature, but cover with black paper or place in a dark cupboard.

C – set up as in A, but place a narrow tube of alkaline pyrogallol in the centre of the blotting paper roll, in order to remove oxygen. Use boiled and cooled water to exclude oxygen. Seal the top of the jar with a greased glass plate.

D – set up as in A, but keep in a refrigerator (3–5°C).

E – set up as in A, but without water in the jar (use seeds that have not been soaked).

2 Follow the progress of germination in jar A and, when the plumules of the seedlings in this jar have reached a length of 3 cm, examine the seedlings in the other jars.

3 Record your results.

Questions

2 What evidence of advanced germination appears in jars B to E?

3 What conditions do you conclude to be necessary for germination?

The three main conditions required for germination are all essential for a high rate of respiration. The seed becomes a centre of great biochemical activity during germination, with enzymes activated to break down the food reserves. Raw materials are therefore made available for growth.

Apart from observing increase in size in germinating seedlings as an indication of growth, experiments may be carried out to assess the increase in *dry weight* of seedlings. Fig 10.8 is a graph showing the changes in dry weight in pea seedlings from the onset of germination.

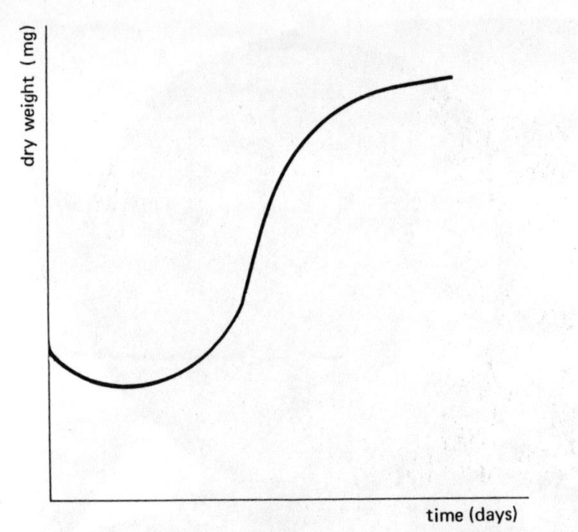

fig 10.8 Dry weight changes in pea seedlings during germination

Questions

4 How does this graph differ from the sigmoid curve obtained from a growing population of yeast cells?

5 How would you account for this difference?

Investigation

Structure of a maize grain

Procedure

1 Soak two or three maize grains in water overnight.

2 The maize grain is in fact a fruit containing a single seed. Its outer covering consists of the pericarp fused to the testa (see section 8.42). On one of these grains, locate the scar which has been left by the style.

3 Using a scalpel or razor blade, slice the grain longitudinally in half, through the style scar.

4 Place the two halves of the grain in a watchglass and cover with dilute iodine solution.

5 After two minutes, wash off the iodine solution in water and examine the half-grains with a hand-lens.

6 Using fig 10.9 as a guide, make a careful drawing of one of the half-grains, shading in the part that has stained blue-black with iodine solution and labelling the following features: style scar, fused pericarp and testa, radicle, plumule, endosperm, cotyledon (scutellum).

7 Take the remaining two grains and germinate them in the same manner as the pea seeds.

Questions

6 Which part of the grain contains starch?

7 What type of germination is shown by this fruit?

10.20 Growth in roots

Experiment

Where does growth take place in the radicle of a bean seedling?

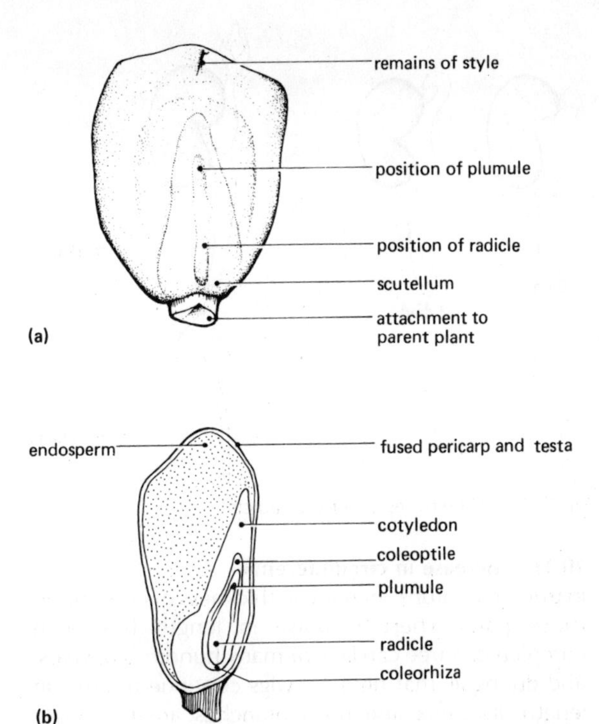

fig 10.9 (a) External view of a maize grain (b) Vertical section through a maize grain

Procedure

1 Soak two or three bean seedlings in water for 24 hours and allow them to germinate in the blotting paper and glass jar apparatus used in section 10.10.

2 When the radicles have reached a length of 1 cm, remove them carefully from the apparatus and make a series of marks along them at 2 mm intervals, using a piece of fine cotton thread soaked in indian ink.

3 Return them to the apparatus, taking care to ensure that the weight of the seedling is supported by the pressure of the blotting paper and jar on the testa and not on the radicle.

4 Measure the intervals between the ink marks every day for six days.

Questions

1 At what part of the radicle have the ink marks moved furthest apart?

2 What explanation can you offer for this, in the light of your observations in Chapter 9, section 9.02 (examination of onion root tip cells)?

10.30 Growth in shoots

The pattern of growth in the plumule is essentially similar to that in the radicle, but is slightly more difficult to follow because of the presence of leaves at the apex.

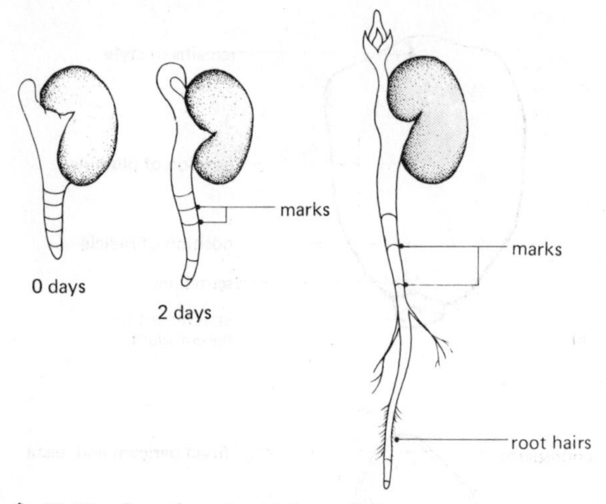

fig 10.10 Growth region of the radicle

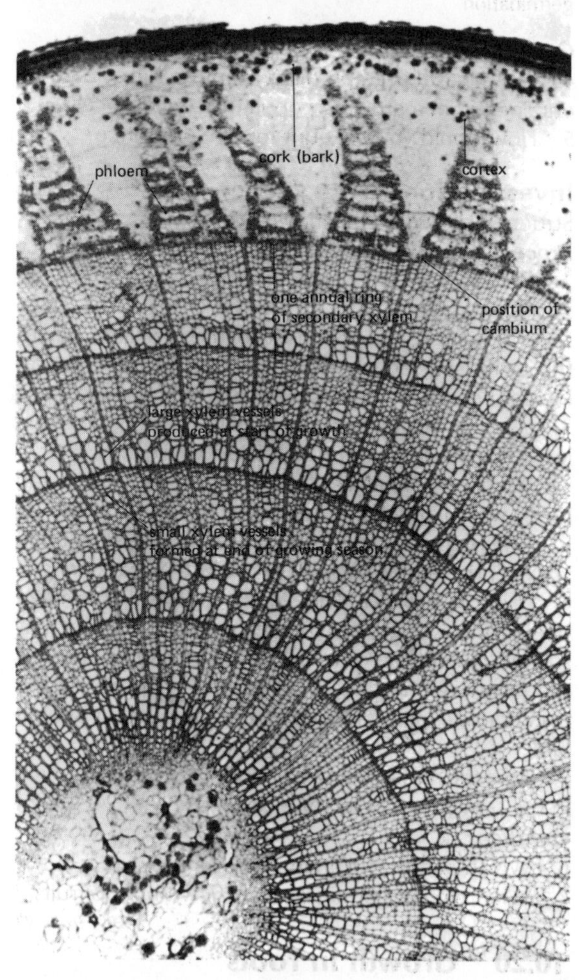

fig 10.11 (a) Photograph of T.S. of a tree trunk showing annual rings (b) Photomicrograph of xylem vessels showing annual rings

10.31 Increase in circumference

In trees and shrubs, increase in thickness occurs only in those parts where increase in length has been completed. A tree can live for many hundreds of years, and during all that time its twigs continue to grow in length. Its trunk and main branches are no longer increasing in length, but they are becoming thicker and thicker by the formation of more and more wood. Wood is the term given to the *xylem tissue* that *conducts water* to the photosynthesising and growing parts of the tree (see section 2.63). As the tree grows, its twigs branch and produce more leaves, which in turn require more water to be conducted up through the xylem vessels. A second function of wood is to provide *support* for the increasing weight of the growing tree. This it is able to do because of the strength of the xylem vessel walls, which are thickened by the deposition of a complex organic compound known as *lignin*.

The additional xylem is formed by the cutting off of new cells on the inside of a ring of meristematic cells, the *cambium*, which also cuts off food-conduction *phloem cells* on the outside. The process is referred to as *secondary thickening*.

It is possible to tell the age of a felled tree by counting the number of wood rings seen in a cross section of the trunk (see fig 10.11 (a)).

This is because the xylem vessels produced in the spring at the beginning of the growing season are always larger than those produced at the end.

Question

1 How old is the tree shown in fig 10.11(b)?

10.40 Plant growth and external stimuli

Experiment

What is the effect of gravity on root and shoot growth?

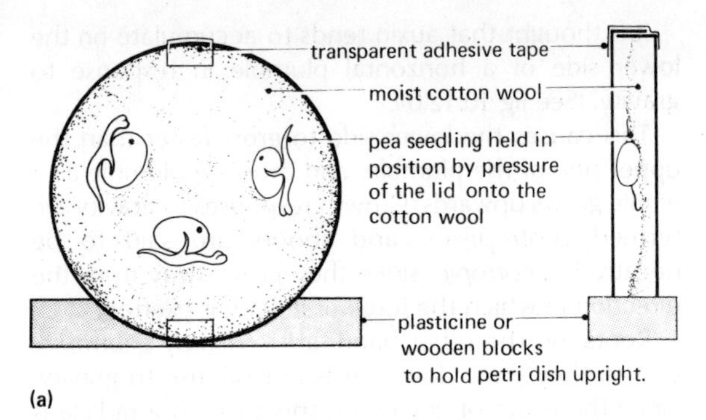

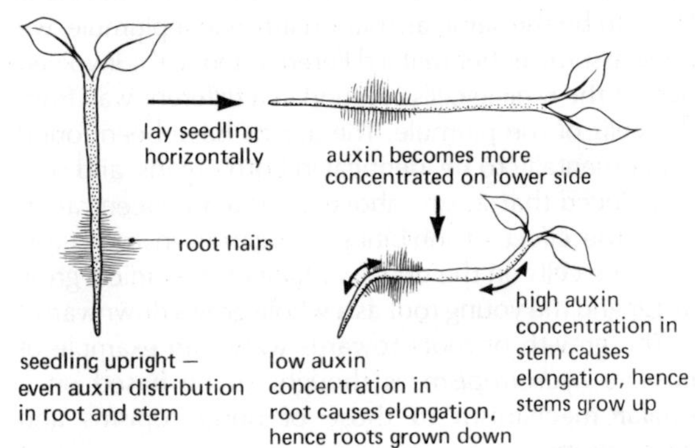

fig 10.12 (a) Apparatus to demonstrate geotropisms (b) The mechanism of geotropic response

Procedure

1 Germinate two or three pea seedlings, using the blotting paper/glass jar method (section 10.10).

2 When the seedlings are 5 or 6 cm long, place them in a petri dish with damp cotton wool or blotting paper as shown in fig 10.12(a). Replace the lid of the dish and secure with transparent tape in such a way that the seeds are firmly pressed against the damp substratum. Fix the dish in an upright position by plasticine or wooden blocks.

3 After three days, re-examine the seedlings and record your observations.

Questions

1 What is the shape of the seedlings after three days?

2 Of what use would these responses be in the life of the plant?

Experiment

What is the effect of light on the growth of shoots?

Procedure

1 Germinate ten or twelve maize grains in a pot filled with damp soil. The grains should be placed just below the soil surface and the pot should be covered by an upturned cardboard box, to exclude light.

2 When the coleoptiles (shoots) have grown to a height of about 2 cm, treat them as follows.

For one third of the seedlings, cut off the top 2 mm carefully, using a razor blade.

For another third of the seedlings, cover the tips with small caps made of aluminium foil.

Leave the remaining third of the seedlings untouched.

3 Cut a hole of about 3 cm diameter in one side of the cardboard box, level with the coleoptiles.

4 Leave for two days at room temperature.

5 After two days, remove the cardboard box and record your observations.

Questions

3 What has happened to the coleoptiles from which the tips had been removed?

4 What has happened to the coleoptiles whose tips had been covered by foil?

5 What has happened to the untreated coleoptiles?

6 Of what advantage would this response be in the life of the plant?

Experiment

What is the effect of water on the growth of roots?

Procedure

1 Set up a floor of wire netting (gauze or fine chicken wire) in a light-proof box, as shown in fig 10.14. Part of the floor should be horizontal and part sloping.

2 On top of the netting place a layer of blotting paper, and cover that with about 2 cm depth of moist soil.

3 Place a number of previously soaked pea seeds just below the surface of the soil, equally spaced about 3 cm from each other.

4 Leave for 3–5 days, keeping the soil moist by frequent watering.

5 When the plumules have grown to a height of about 5 cm, dismantle the box and examine the radicles of the seedlings. Record your results.

Questions

7 What has happened to the radicles of the seedlings grown on the horizontal netting?

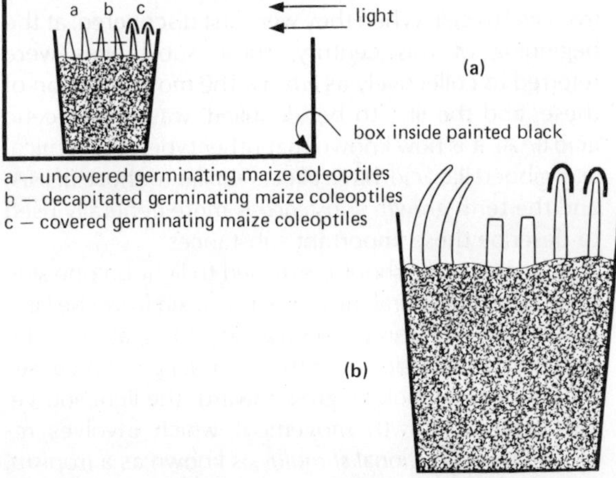

fig 10.13 (a) Coleoptiles receiving unilateral light (b) Coleoptiles after 24 hours of unilateral light

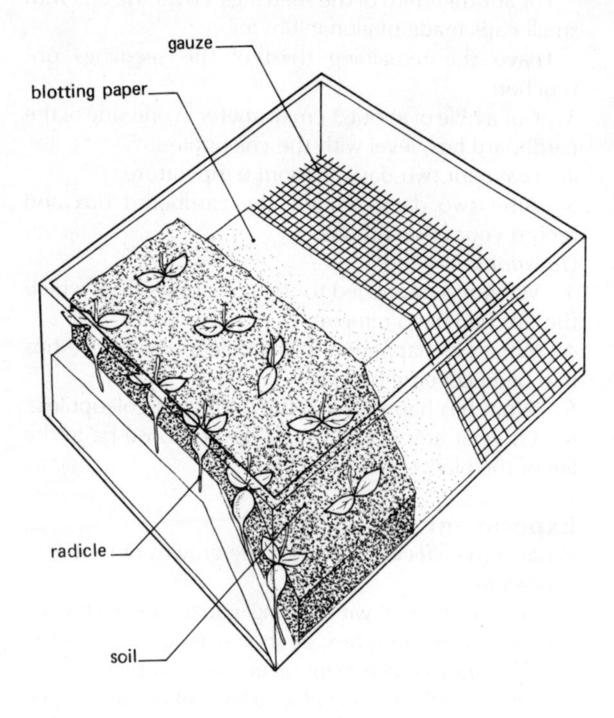

fig 10.14 Apparatus for hydrotropism experiment

8 What has happened to the radicles of the seedlings grown on the sloping netting?
9 Of what use would this response be in the life of the plant?

10.41 The mechanism of growth responses

It has been found that the shoot tips of seedlings produce chemicals which cause elongation of the cells behind the tip, and thus growth of the shoot as a whole. These chemicals are soluble in water and diffuse easily from cell to cell. When they were first discovered, at the beginning of this century, these substances were referred to collectively as *auxins*. The most common of these, and the first to be identified, was indoleacetic acid (IAA). It is now known that other types of chemical (e.g. gibberellic acid) have effects similar to those of IAA, and the term *growth substance* is more generally used to describe these important substances.

When a young shoot is exposed to light on one side only, there is a lateral movement of auxin from the light to the dark side. This causes a greater elongation of the cells on the dark side, and this unequal growth causes the shoot as a whole to grow towards the light source. This type of growth movement which involves *response* to a *directional stimulus*, is known as a *tropism*, and thus shoots are described as being *positively phototropic*.

It is thought that auxin tends to accumulate on the lower side of a horizontal plumule in response to gravity. (See fig 10.12(b)).

This causes the lower side to grow faster than the upper side of the plumule, and thus the plumule as a whole grows upwards. Growth responses to gravity are termed *geotropisms*, and shoots are said to be negatively *geotropic*, since they grow away from the direction in which the force of gravity is exerted.

Roots, on the other hand, are *positively geotropic*: that is, they grow downwards in response to gravity. Since the effect of gravity on the auxin in a radicle is likely to be the same as that on auxin in a plumule, we must assume either that a different chemical is involved or that the radicle cells respond in a different way from the cells of the plumule. The auxins have been found experimentally to be identical in both organs, and so it is deduced that auxins above a certain concentration have the effect of inhibiting elongation in root cells. Thus the cells on the upper surface of the radicle grow faster and the young root as a whole grows downwards.

The growth of roots towards water, an example of *positive hydrotropism*, is thought to be based on a similar mechanism to those of phototropisms and geotropisms.

10.42 Responses of plants to non-directional stimuli

The examples of plant movement discussed so far have involved responses to stimuli of light, gravity and water coming from a particular direction. The response in each case involves growth towards or away from that direction.

Plants also respond to more diffuse *non-directional stimuli*, such as changes in temperature, humidity or light intensity. These are referred to as *nastic* responses.

Examples include the closing up of flower petals at night in response to decreasing temperature and/or light intensity. For this reason, they are sometimes referred to as 'sleep movements'. They are brought about by the growth of the lower surface of the petal at a faster rate than the growth of the upper surface. Not all nastic movements are brought about by growth. Similar movements are brought about in response to humidity fluctuations by changes in turgor pressure of cells found in special structures (pulvini) at the bases of leaves or leaflets of many species, including members of the Leguminosae.

Plants also make movements in response to shock or injury. The best known example is *Mimosa pudica*, the 'sensitive plant'. The folding movements of the leaflets one after the other is also due to changes in turgor pressure of the pulvini cells and not to differential growth.

Movements of insectivorous plants enabling them to trap and digest prey are again due to changes in turgor pressure, not to differential growth.

10.50 Growth in animals

10.51 Growth in Mammals

Experiment

To measure the growth rate of a laboratory mouse.

Procedure

1 Take a litter of new-born mice and weigh each one separately on the day of birth. Weighing should be carried out on a pan balance, with the mouse enclosed in a weighed, covered jar.

2 Repeat the weighings every two days, until there is no further increase in weight. Always carry out the weighings at the same time of the day and before a feed. After the young mice have been weaned, place them at weighing times in a jar with a weighed amount of food. This will discourage them from moving around and causing the balance to waver.

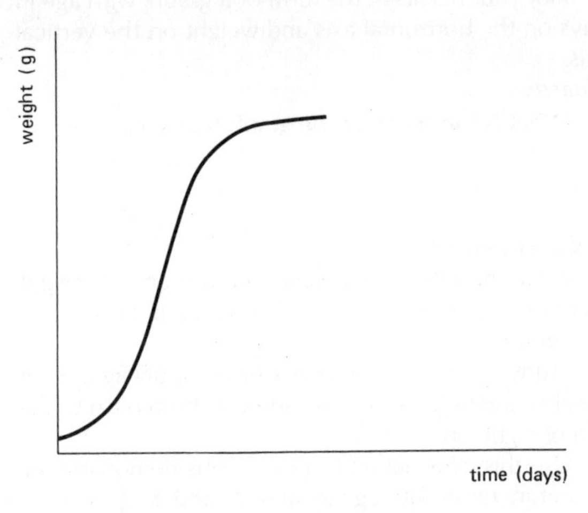

fig 10.15 Idealised sigmoid growth curve of a mouse

Boys			Girls		
Age	Weight (kg)	Total height (cm)	Age	Weight (kg)	Total height (cm)
Birth	3.4	50.6	Birth	3.36	50.2
1 year	10.07	75.2	1 year	9.75	74.2
2 years	12.56	87.5	2 years	12.29	86.6
3 years	14.61	96.2	3 years	14.42	95.7
4 years	16.51	103.4	4 years	16.42	103.2
5 years	18.89	110.0	5 years	18.58	109.4
6 years	21.9	117.5	6 years	21.09	115.9
7 years	24.54	124.1	7 years	23.68	122.3
8 years	27.26	130.0	8 years	26.35	128.0
9 years	29.94	135.5	9 years	28.94	132.9
10 years	32.61	140.3	10 years	31.89	138.6
11 years	35.2	144.2	11 years	35.74	144.7
12 years	38.28	149.6	12 years	39.74	151.9
13 years	42.18	155.0	13 years	44.95	157.1
14 years	48.81	162.7	14 years	49.17	159.6
15 years	54.48	167.8	15 years	51.48	161.1
16 years	58.3	171.6	16 years	53.07	162.2
17 years	61.78	173.7	17 years	54.02	162.5
18 years	63.05	174.5	18 years	54.39	162.5

Table 10.1 Weight and height in boys and girls

3 Plot your results in the form of a graph, with age in days on the horizontal axis and weight on the vertical axis.

Question

1 What is the shape of the graph that you obtain?

Human growth

Investigation

To study the rate of (i) weight increase and (ii) height increase in a population of (a) boys and (b) girls

Procedure

1 Study carefully the charts which supply figures for weights and heights for boys and girls from birth to the age of eighteen.

2 Use this information to plot graphs demonstrating weight increase with age in (a) boys and (b) girls.

3 Similarly plot graphs to show height increase with age in (a) boys and (b) girls.

Questions

2 What are the shapes of the two curves obtained in procedure 2?

3 Which of these curves forms the larger sigma? How do you account for this?

4 What differences are evident in the graphs you have drawn of increasing height in boys and girls?

10.52 Growth in lower vertebrates

Metamorphosis

Amphibians, such as frogs and toads do not show a gradual, uninterrupted increase in size and weight. During the course of their growth there is a marked *change in form*. This is called metamorphosis.

Frogs and toads hatch out of their eggs as *tadpoles*, about two days after fertilisation. These minute forms have external gills for breathing, and long tails for swimming.

Investigation

To observe the changes apparent in different stages of the development of a frog.

Procedure

1 Prepare water to receive frog spawn by allowing it to stand for at least 24 hours and to reach a temperature between 15°C and 18°C.

2 Following introduction of the spawn into the aquarium, grind up pond weeds for the young tadpoles to feed on.

3 Make a drawing of the tadpoles when they first hatch. How do they feed? How do they breathe? How long are they?

4 About 3 days after hatching, examine the tadpoles again, and make another drawing. Note any changes that are apparent, including size.

5 Re-examine the tadpoles about two weeks later. Record their length and make a careful drawing to

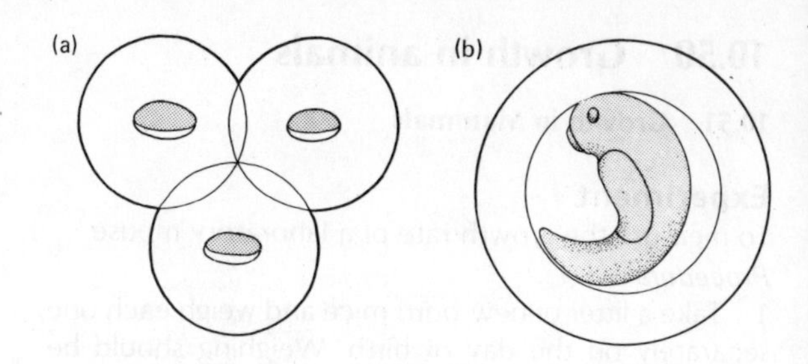

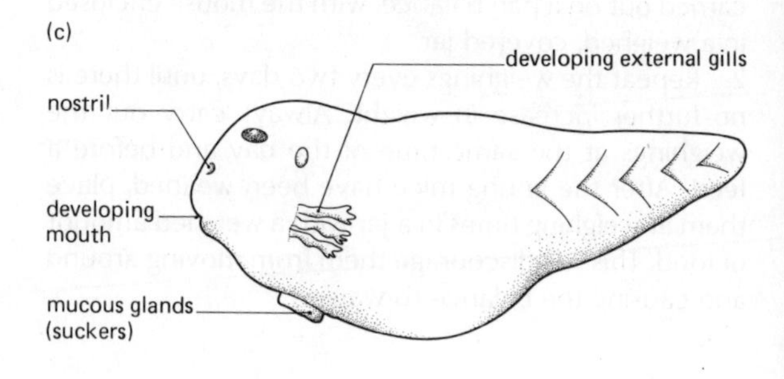

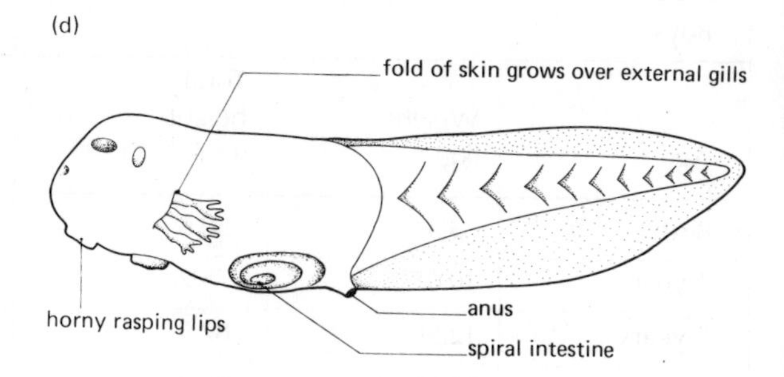

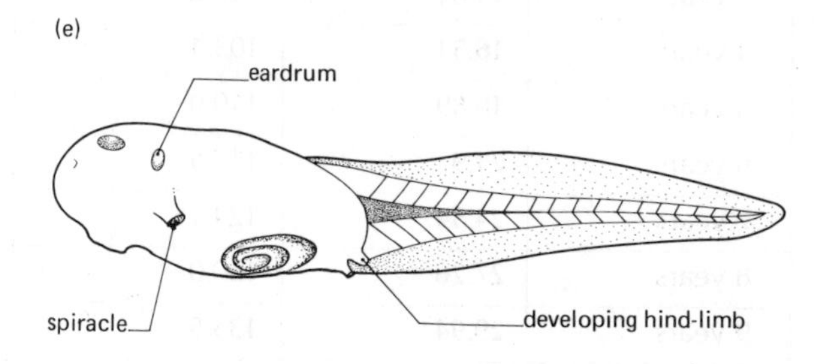

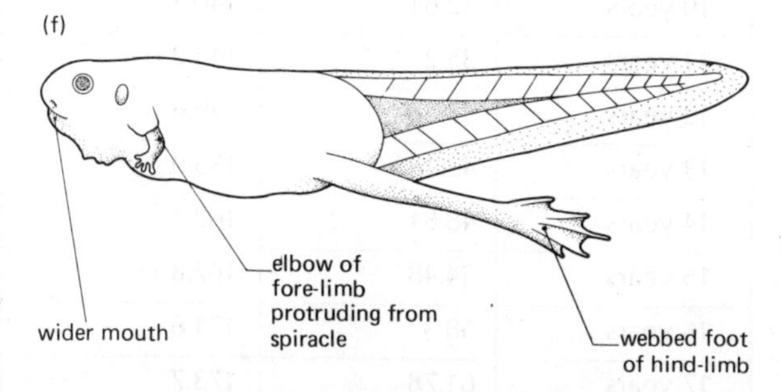

fig 10.16 Stages in the metamorphosis of a frog (a) Frog spawn – jelly mass containing eggs (b) Just before hatching (c) Newly hatched (d) One week after hatching (e) Three weeks after hatching (f) Seven weeks after hatching

show any further changes evident. How do the tadpoles breathe at this stage?
6 In another two weeks, more changes should be seen. Measure and draw a tadpole and list all the changes you can find.
7 About three weeks later, further changes in form should be visible. Metamorphosis is being speeded up by a hormone. What are these changes? Again draw and measure the tadpole.
8 In another month examine the tadpoles again. By this time metamorphosis may be complete. Draw the resulting frog and record its body length. Make a note of all the final external changes.
9 Add chopped worms to the water now: the frog is basically carnivorous.

Following metamorphosis, there is a period of growth: a frog is not fully grown until it is about three years old. If possible find a large frog, and compare its measurements with that of a newly metamorphosed frog.

If it is too difficult to obtain and maintain frog spawn, study fig 10.16 and make a list of all the changes you can see between the stages represented.

10.53 Growth in invertebrates – insects

Growth in insects is limited by their possession of a hard exoskeleton. In some insects, growth from the egg to the adult size or *imago* is accompanied by a series of *gradual* changes. This is known as *incomplete metamorphosis*. From the egg emerges the first *nymph* or *instar* (see fig 10.17(1)). Once the cuticle is hard the instar is unable to grow until it *moults*. Moulting occurs about five or six times before the adult stage is reached. Each moult is followed by a period of rapid growth before the newly-formed cuticle hardens.

Investigation

To analyse the pattern of growth in the locust.

Procedure

1

Stage	Period of no growth – from cuticle hardening to next moult
1st instar	6 days
2nd instar	1 week
3rd instar	1 week
4th instar	8 days
5th instar	8 days

Study the figures in the table above, and the drawings of the instars of the locust.

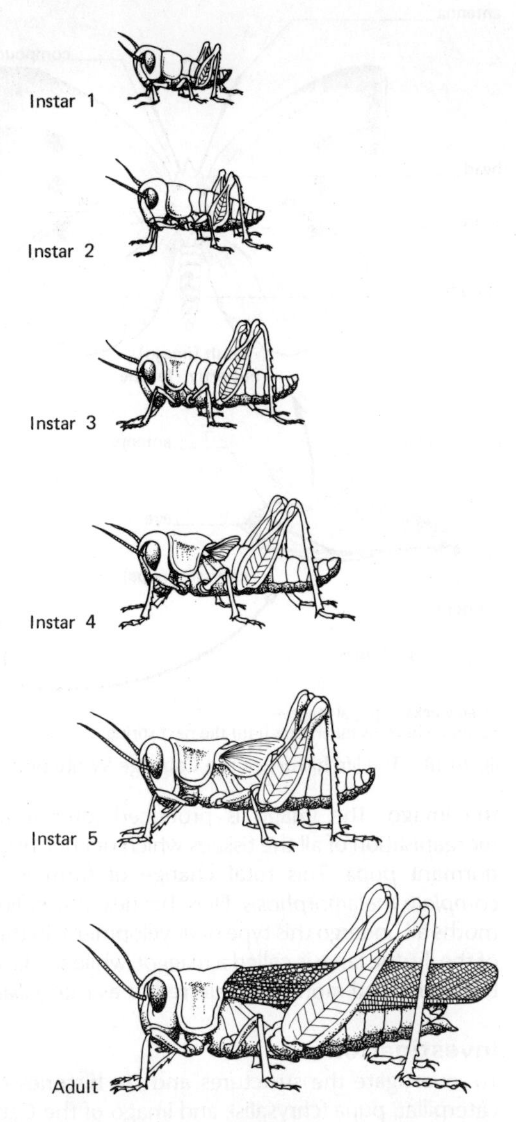

fig 10.17 The development of the locust

2 Measure, on the diagrams, the length of each instar, from the head below the eye to the tip of the abdomen, and record these measurements.
3 Assuming a maximum of one day for each moult and cuticle-hardening, construct a graph to show how growth occurs in the locust. Plot time in 2-day intervals along the horizontal axis, and length in cm along the vertical axis.

Questions

1 How would you describe the pattern of growth in a locust?
2 List the changes you can see, apart from size increase, in each instar.

The majority of insects develop in a radically different way from the locust. The form which hatches from the egg, the *larva*, bears no resemblance at all to

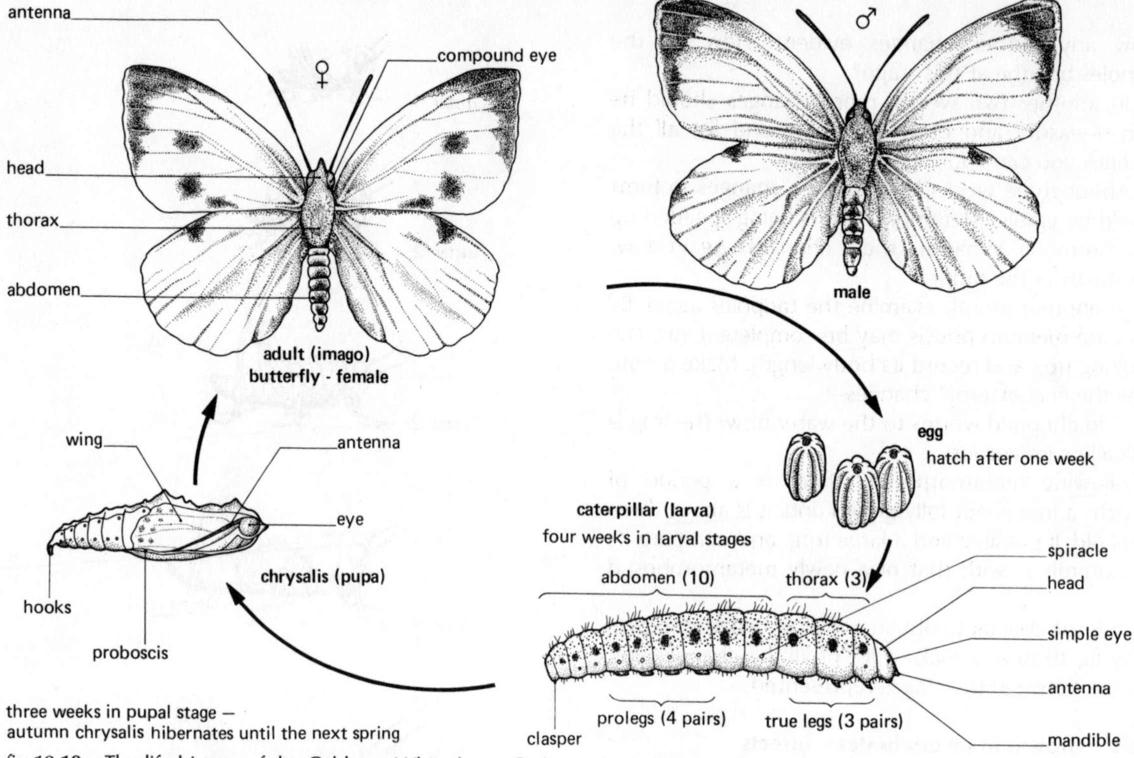

fig 10.18 The life history of the Cabbage White butterfly (*Pieris brassicae*)

the imago. The imago is produced after a drastic reorganisation of all the tissues which occurs inside the dormant *pupa*. This total change of form is called *complete metamorphosis*. Flies, beetles, butterflies and moths all undergo this type of development. In the case of the fly, the larva is called a *maggot*, while the larvae of butterflies and moths are referred to as *caterpillars*.

Investigation

To investigate the structures and life histories of the caterpillar, pupa (chrysalis), and imago of the Cabbage White butterfly.

Procedure

1 Explore the school grounds or your garden at home to find eggs of a butterfly such as the Cabbage White butterfly. These occur commonly under the leaves of cabbage plants. Observe the eggs regularly until the caterpillars emerge.

2 Take a bunch of leaves bearing some caterpillars into the school and make drawings of the larvae. Observe the behaviour of the caterpillars and make notes.

3 Collect some pupae, draw them, and make notes on any observed behaviour.

4 If possible observe the imago feeding, and make relevant drawings. If not, see fig 10.18.

Questions

3 Make a list of all the differences in structure you observed between the caterpillar and the imago.

4 What behaviour did you observe in the pupa?

5 What is the main role of the caterpillar?

6 What is the basic difference between feeding in the caterpillar and feeding in the imago?

Questions requiring an extended essay-type answer

1 Describe the *internal* and *external* changes that occur

a) when a seed grows into a seedling, and

b) when a larva (of a *named* insect) forms a pupa and then finally an imago (adult).

2 a) What is the biological meaning of sensitivity (irritability)?

b) Describe two different examples of sensitivity in

i) the root of a flowering plant, and

ii) the shoot of a flowering plant.

c) How do plants respond to the stimulus of light?

3 a) What *external* conditions are necessary for the germination of a seed?

b) Describe the internal and external changes occurring when a *named* seed is planted and grows above the surface ot the soil. (Do not describe any development *beyond the first pair* of green leaves).

4 a) Describe an experiment to determine the effect of gravity on the growth of the root of a flowering plant.

b) How do your experimental results show that the radicle is adapted for forcing its way through the soil?

11 The skeleton and locomotion

11.00 Introduction

One of the features that enables us to distinguish animals from plants is the ability of animals to move from place to place. This ability – locomotion – is necessary because animals, unlike plants, cannot make their own food but have to move around to obtain it.

Three essential elements occur in almost all animal locomotory systems: a *contractile tissue* usually muscle acting upon a *rigid skeleton* (made of bone in the case of vertebrates) which in turn acts on a supporting and *resisting medium* (earth in the case of most mammals, water in the case of fish and air in the case of birds).

Not all skeletons are internal, and not all are made of bone. Insects, for example, have an *external* skeleton made of a horny material, and annelid worms have a rigid *hydrostatic* skeleton consisting of fluid-filled chambers.

It is noticeable that animals which live completely surrounded by their food (e.g. tapeworms) have very poorly developed skeletal systems.

Of course it is neither possible nor desirable to consider the skeleton in isolation from the other organs and tissues of the body. Most movements occur in response to a particular stimulus, for example the sight of food. In this case, the stimulus is received by the eye and the resulting nerve impulse is transmitted to the brain by the sensory neurons of the optic nerve. Impulses from the brain are conducted via motor neurons to certain muscles, which *contract* to produce movement of parts of the skeleton and hence locomotion of the body towards the food. Thus, the functioning of sense organs, nervous system, muscles and skeleton are closely interlinked. In particular, the skeleton and its associated muscles form a co-ordinated *effector system*.

Investigation

To examine the arm

Procedure

1 Stretch out your right arm horizontally in front of you, palm upwards.
2 Observe the muscles of the upper part of the arm, and feel them carefully with your left hand.
3 Bend your right arm at the elbow, so that the fingers of your right hand are touching your right shoulder.
4 Again, observe the muscles of the upper arm and feel them carefully with your left hand.

Questions

1 What changes have taken place in the muscle on top of the upper arm (the *biceps*)?
2 What changes have taken place in the muscle underneath the upper arm (the *triceps*)?

11.10 The mammalian skeleton

Throughout the class *Mammalia*, in spite of great variation in form and size, the pattern of the skeleton is remarkably constant. It may be divided into two main parts: the *axial* skeleton and the *appendicular* skeleton. The axial skeleton consists of the skull and the vertebral column, together with the ribs and sternum; the appendicular skeleton consists of the limbs and limb girdles (see figs 11.1, 11.2 and 11.3).

11.11 Functions of the skeleton

In addition to locomotion, skeletons have two other main structural functions: support and protection (bones also produce blood cells in the marrow and provide reserves of calcium and phosphorus, but these non-structural functions are dealt with elsewhere).

The *supporting* function of the skeleton is more obvious in large, terrestrial animals that are surrounded by air than in those that live in the more dense medium of water. As well as supporting the body as a whole, the skeleton provides a framework which keeps the major internal organs of the body in constant positions relative to each other.

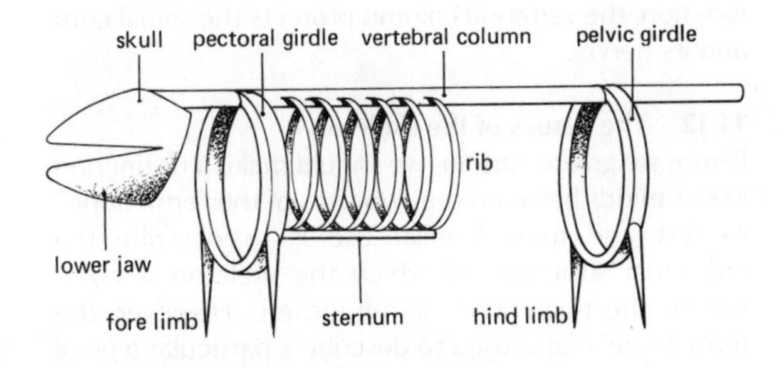

fig 11.1 A diagram showing the general plan of a mammalian skeleton

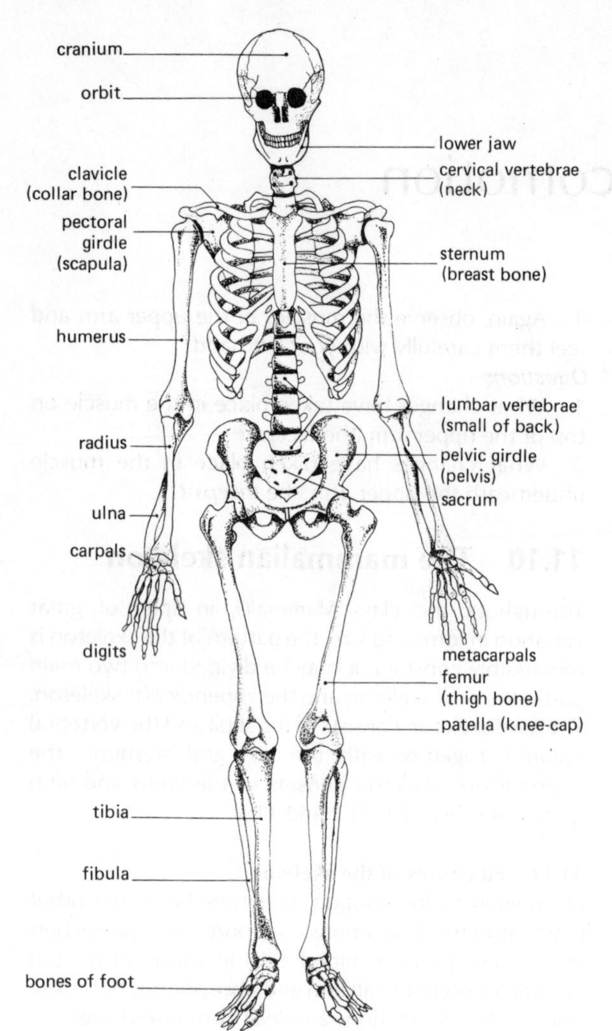

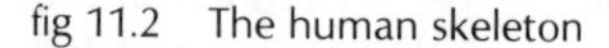

fig 11.2 The human skeleton

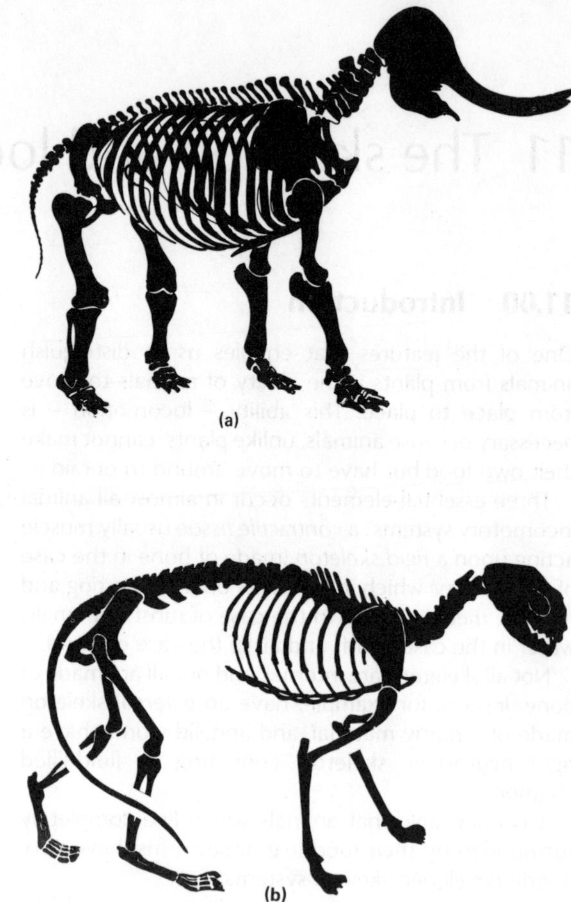

fig 11.3 The skeletons of (a) elephant (b) cat

Certain parts of the skeleton are adapted for the *protection* of delicate structures. The brain is completely surrounded by the skull, and the ribs form an expandable protective framework around the heart and lungs. In addition to its primarily supportive function, the vertebral column protects the spinal cord and its nerves.

11.12 The tissues of the skeleton

Before we go any further, we should make a distinction in our minds between the two uses of the term 'bone'. Its first and most familiar use is to describe the individual structures of which the skeleton is composed: the thigh-bone, shin-bone etc. However, the term 'bone' is also used to describe a particular *type of tissue* which forms only a part, albeit a major part, of each of these structures. The other major type of tissue found in the skeleton is *cartilage* (see fig 11.4).

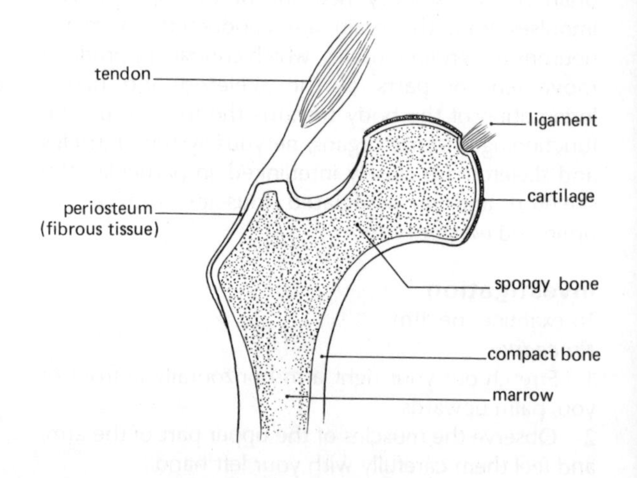

fig 11.4 A section through the head of the femur

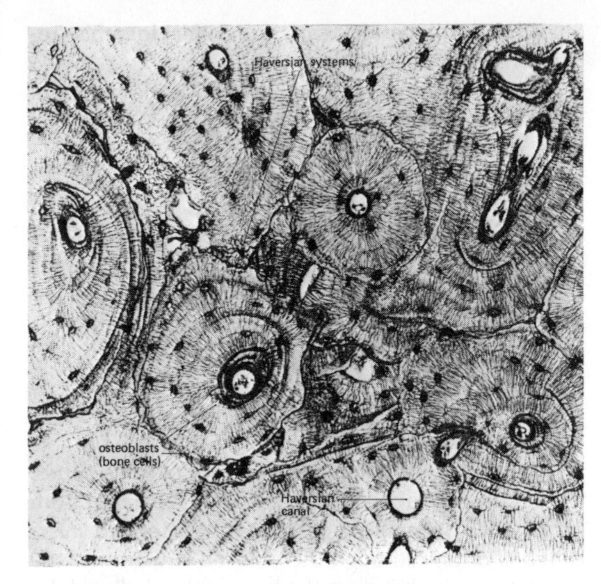

fig 11.5 Microscopic structure of bone tissue

Typically, a bone is covered by a thin layer of dense, hard bone tissue. The hardness is caused by the presence of large quantities of calcium salts (mainly *calcium phosphate*). However, it should not be assumed that bone is non-living. The calcium phosphate forms part of a 'matrix' which makes up only about 70% of bone tissue. This matrix is secreted by bone cells and is perforated by a series of canals (*Haversian canals*) running along the axis of the bone. The canals contain blood vessels, and the mineral-secreting bone cells are arranged around them in concentric circles (see fig 11.5).

In the centre of most limb bones is a soft, fatty substance containing blood vessels. This is the *bone marrow*, where the red blood cells (and certain white cells) are produced. The marrow cavity is continuous with spaces inside the bone tissue at the ends of limb bones. This results in the 'spongy' appearance of the bone in this area.

Where one bone joins another, as in the hip joint, the touching surfaces consist of a smooth layer of clearer, less rigid material known as cartilage or 'gristle'. This tissue owes its elasticity to the protein fibres secreted by its cells and, since it does not contain the hard calcium salts of bone, it acts partly as a *shock absorber* and partly to *reduce friction* between the bones at a joint. In the mammalian foetus, the skeleton is composed of cartilage, which is largely replaced by bone tissue by the time of birth. Some cartilage remains in the shafts of long bones, until growth is completed at the end of puberty.

At the joints, bones are attached to each other by *ligaments* consisting of tough, fibrous and more or less *elastic* tissue. Another type of structure found associated with the skeleton is the *tendon*. Tendons, like

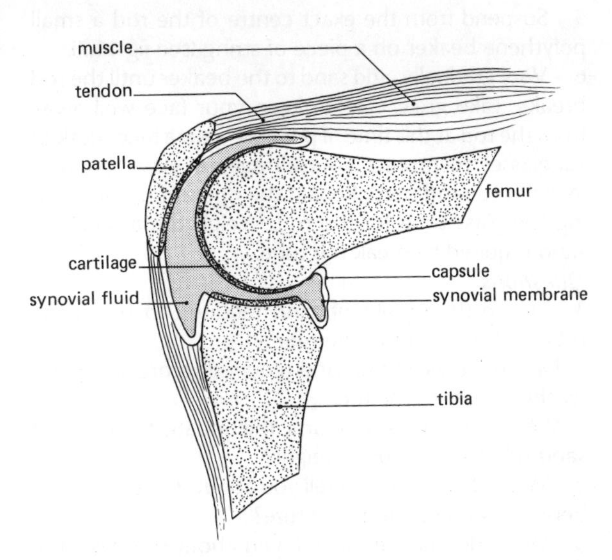

fig 11.6 The hinge joint of the knee

ligaments, consist mainly of fibres, but in their case the fibres are *inelastic*. Tendons occur at the ends of the muscles and serve to join muscles to bones. Their toughness is particularly important where they pass over a bony structure, since they are much less liable than muscles to rupture as a result of vigorous movement.

Experiment

What is the relationship between structure and strength?

Procedure

1 Obtain a glass rod and a piece of glass tubing, each about 20 cm in length.

2 Weigh the rod and the tube.

3 Place two identical chairs (or stools) facing each other, about 18 cm apart.

4 Place the rod across the gap between the chairs, with exactly 1 cm of rod resting on each chair.

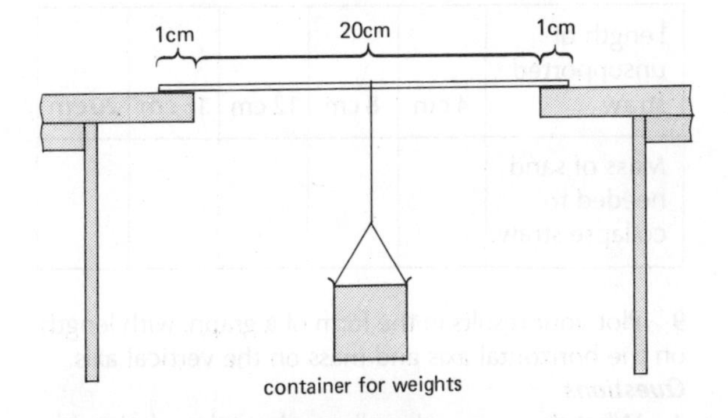

fig 11.7 The experimental apparatus used to investigate the strength of a supporting structure

5 Suspend from the exact centre of the rod a small polythene beaker on a piece of string (see fig 11.7).
6 Very gradually, add sand to the beaker until the rod breaks. Take great care to keep your face well away from the rod at this time: if possible wear a face mask or sunglasses to protect yourself against glass splinters. Weigh the sand used. Repeat the operation, substituting the glass tube for the rod. Weigh the amount of sand required to break the tube.

Questions

1 Divide the weight of sand required to break the tube by the weight of the tube.

Divide the weight of sand required to break the rod by the weight of the rod.

Which structure supports the greatest weight of sand relative to its own weight?

2 What do your results tell you about the relationship between strength and structure?

3 What do your results tell you about the design of limb bones?

Experiment

What is the relationship between length and strength?

Procedure

1 Set up the chairs (or stools) as in the previous experiment.
2 Take a few standard 22 cm drinking straws. Cut one to 18 cm length, one to 14 cm, one to 10 cm and one to 6 cm.
3 Place a 22 cm straw between the two chairs as in the previous experiment, leaving 1 cm lengths resting on each chair.
4 Suspend a polythene beaker (or polythene bag if the beaker is too heavy) from the centre of the straw.
5 Pour sand gradually into the beaker until the straw collapses. Weigh the sand used.
6 Repeat the operation for the other four lengths of straw.
7 Record your results in the form of a table (see below).

Length of unsupported straw	4 cm	8 cm	12 cm	16 cm	20 cm
Mass of sand needed to collapse straw.					

9 Plot your results in the form of a graph, with length on the horizontal axis and mass on the vertical axis.

Questions

4 What do your results tell you about the relationship between length and strength of a uniform tubular structure?

5 What explanation can you offer for this relationship?

6 What do your results tell you about the nature of bones of very tall animals?

11.20 The axial skeleton

The axial skeleton consists principally of the *skull*, together with the *vertebral column* (spine or backbone). For most purposes, the *ribs* and the *sternum* or breastbone are also included under this heading. Its functions are concerned mainly with support and protection.

11.21 The vertebral column

The backbone is made up of a number of small bones, the *vertebrae*, placed end to end and separated from each other by small pads of cartilage termed *intervertebral discs*. Although the detailed structure of the vertebrae varies between the base of the skull and the tip of tail, they all conform to the same basic plan (see fig 11.8).

Each vertebra consists of a central disc of bone, the *centrum*, to the dorsal surface of which is attached the bony *neural arch*. The tunnel formed by placing the neural arches of all the vertebrae end to end provides protection for the spinal cord in the *neural canal*.

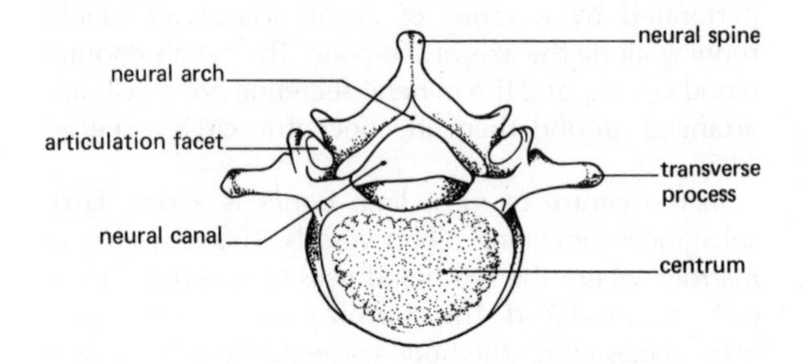

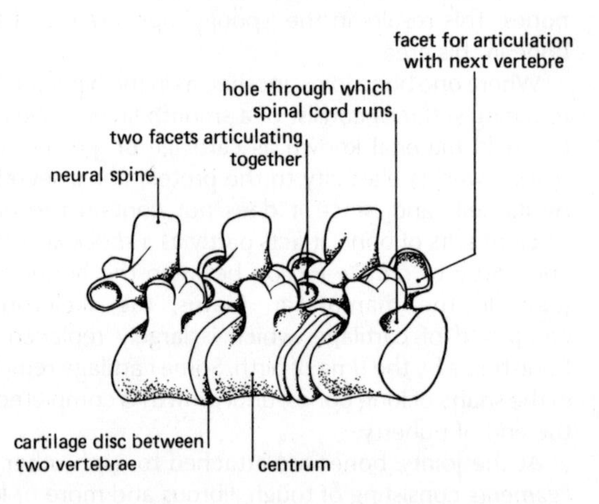

fig 11.8 The structure of a lumbar vertebra and the articulation between several vertebrae *After D. G. Mackean*

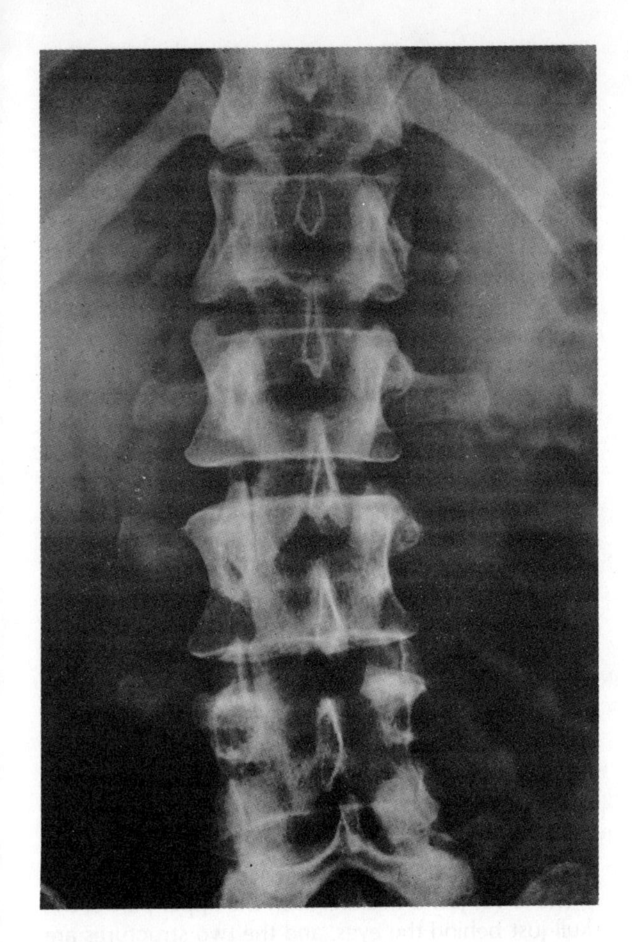
fig 11.9 X-ray of a lumbar vertebra

Typically, seven bony projections arise from the neural arch, these projections serving for the attachment of muscles and/or articulation with other bones. The *neural spine* is situated in the mid-dorsal line of the neural arch, and two *transverse processes* arise from the junction of the arch with the centrum. Four *articulating facets*, two anterior and two posterior, are also present. The downward-facing articulating facets on the posterior surface fit into the upward-facing anterior facets of the vertebra immediately behind (see fig 11.8).

Investigation

To examine the different types of vertebra found in the skeleton of a small mammal

Procedure

1 Study carefully the whole, mounted skeleton of a rabbit (or rat, or dog). (Refer to fig 11.10, if no mounted skeleton is available.)

2 Obtain a large number of loose vertebrae, and arrange them on the bench in a line, with the neck vertebrae at one end and the tail vertebrae at the other end. Use the mounted skeleton for reference.

3 Make large, clear drawings (anterior and lateral views) of the following vertebrae: first neck or *cervical* vertebra (*atlas*), second cervical vertebra (*axis*), one other cervical vertebra, one *thoracic* vertebra, one *lumbar* or middle back vertebra, all the vertebrae from the lower back or hip region (*sacral*), one *caudal* or tail vertebra.

Questions

1 Which vertebra has a very wide neural canal, a small centrum and broad, flat transverse processes?

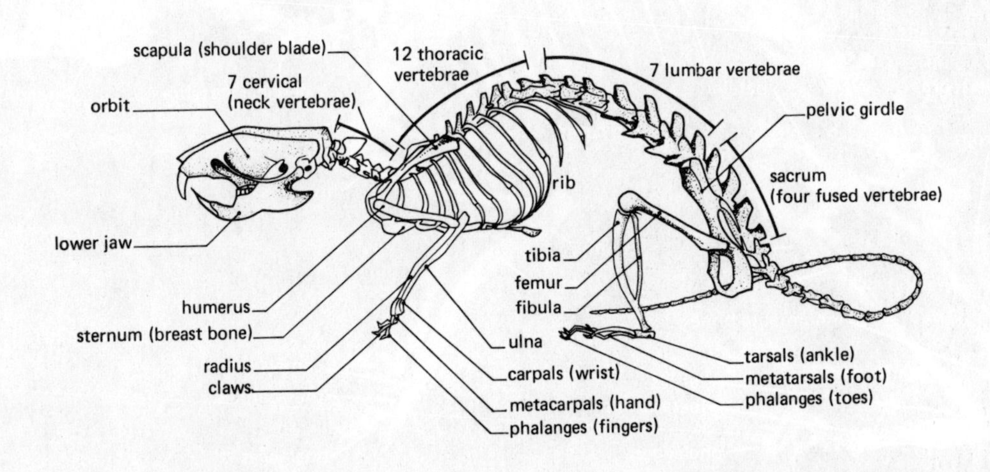

fig 11.10 (a) The skeleton of a rat

2 What kind of movement of the head will this arrangement allow?
3 Note the *odontoid* process of the axis vertebra, which slots into the atlas. What kind of movement of the head does this allow?
4 Examine a thoracic vertebra carefully. Note the large neural spine. What is its function?
5 Note the small depressions at each end of the centrum and on the ends of the transverse processes. What is their function?
6 What differences can you observe between the thoracic and lumbar vertebrae? Present your answer in the form of a table.
7 Count the sacral vertebrae. Are any of them fused?
8 At what points does the sacrum articulate with the pelvic (hip) girdle?
9 How do the more posterior caudal vertebrae compare with the more anterior ones?
10 Examine the mounted skeleton again. How many pairs of ribs can you count?
11 How many of these are connected to the breastbone (sternum)?
12 Note the pieces of cartilage between the first seven pairs of ribs and the sternum. How many pairs of ribs are attached to those in front by cartilage?
13 How many 'floating' ribs, with no ventral attachment are there?

11.22 The skull

Examination of the skull bones of any mammal reveals that the skull is made up of two separate parts. The upper part consists of a brain-box or cranium, and

Vertebra	Human	Rat
cervical	7	7
thoracic	12	13
lumbar	5	6
sacral	5	4
caudal	4	30 (approx)
Total	33	60 (approx)

Table 11.1 Comparison of human and rat vertebrae
(See fig 11.10)

fused upper jaw. Articulating with this is the lower jaw.

The cranium consists of a number of flattened bones, joined to form a rounded case protecting the brain. In newly-born babies, the bones on the top of the cranium may not have joined up completely, resulting in a delicate, unprotected area known as the *fontanelle*.

The cranium also *protects* the organs of special sense. The eyes are partially enclosed in bony sockets and the delicate structures of the inner ear are surrounded by a capsule of bone. The hard palate is a bony structure which runs along the top of the upper jaw, partly enclosing the nasal cavity.

The lower jaw articulates with the upper part of the skull just behind the eyes, and the two structures are connected by powerful muscles.

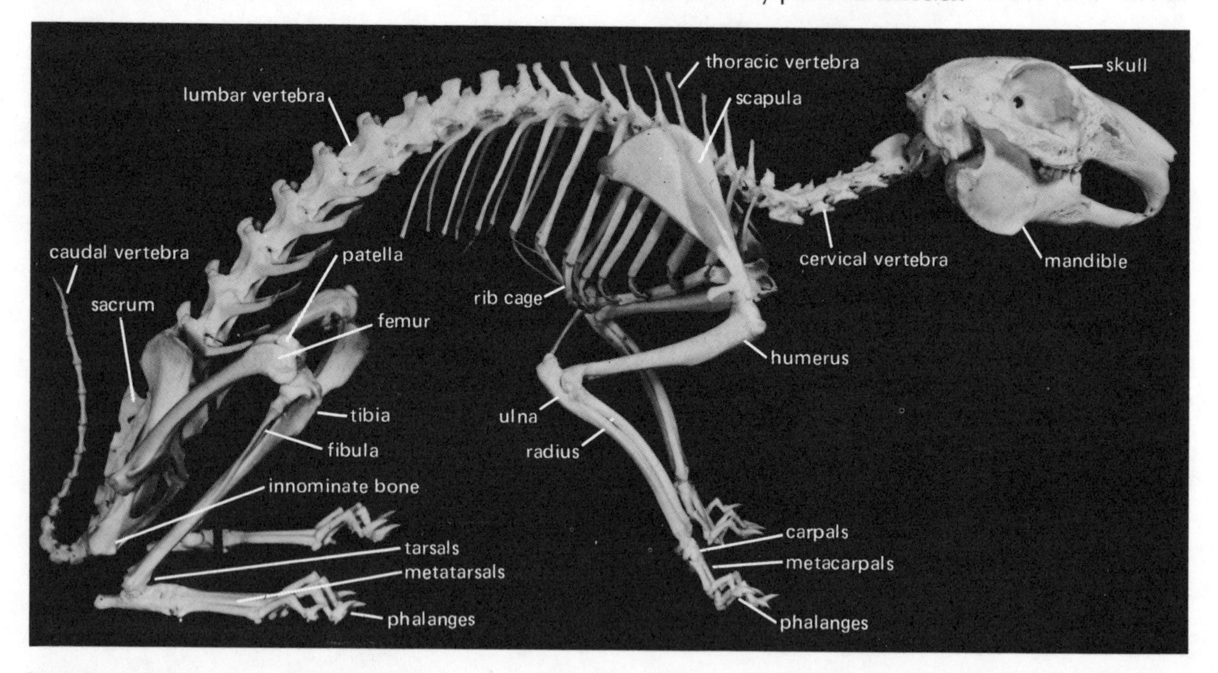

(b) A photograph of the skeleton of a rabbit

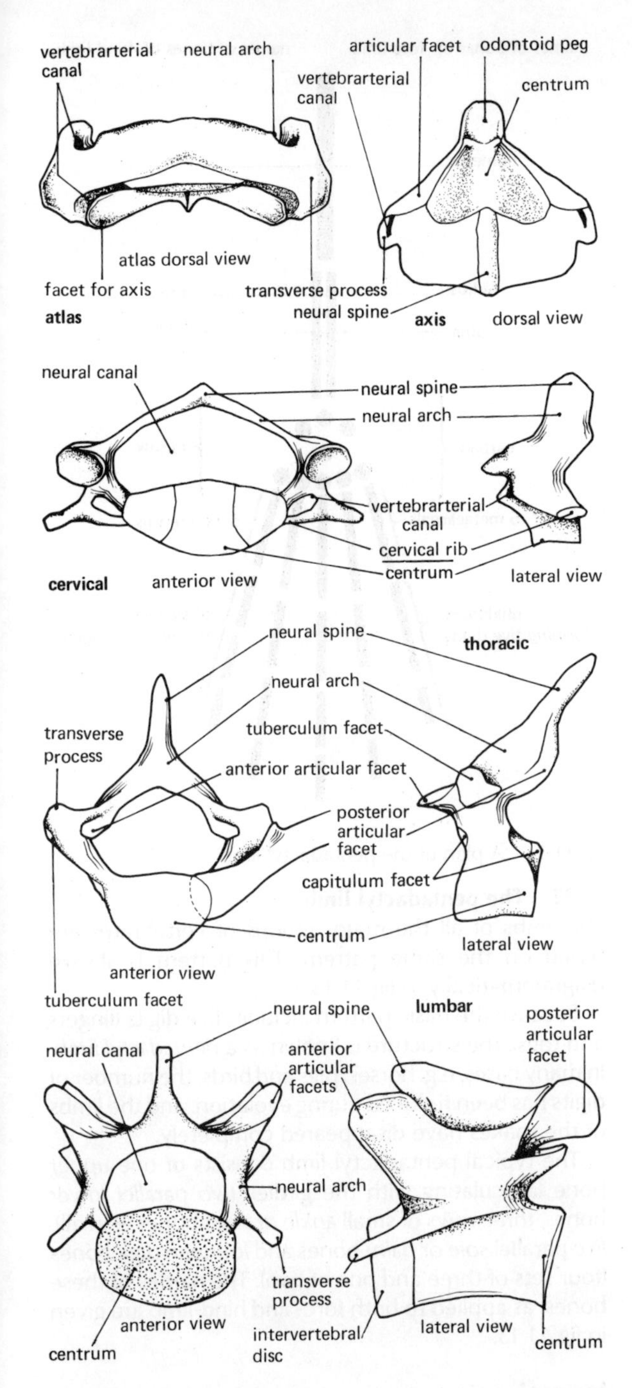

fig 11.11 The different types of vertebrae

11.30 The appendicular skeleton

The appendicular skeleton consists of the *limbs* and the *limb girdles*. Fig 11.12 gives a simplified illustration of these structures, and the details of the arrangements of the bones in Man and in the rat are shown in figs 11.2 and 11.10 respectively.

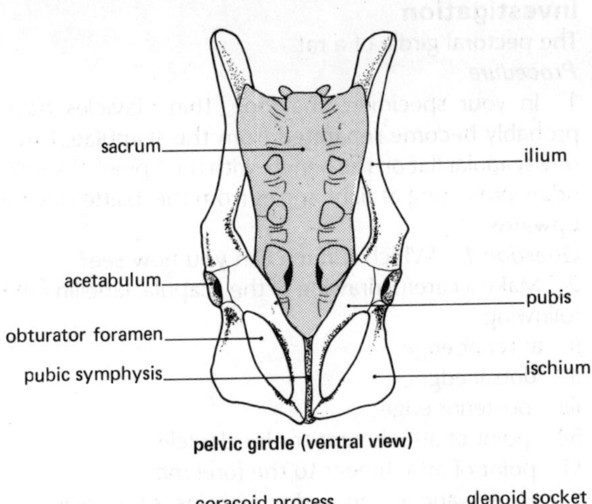

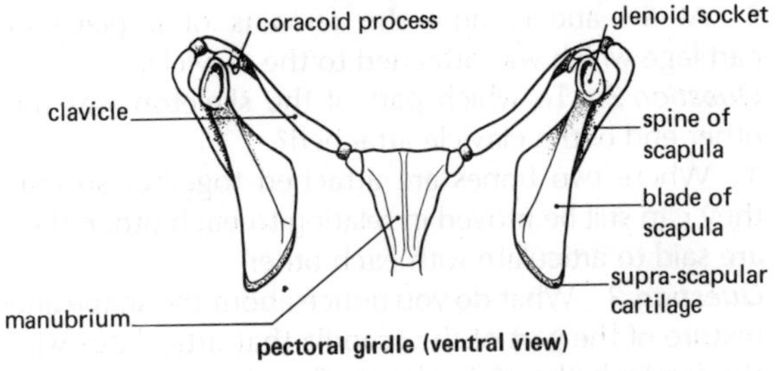

fig 11.12 The pelvic and pectoral girdles of a rat

11.31 The limb girdles

There are two limb girdles; the *pectoral* or shoulder girdle and the *pelvic* or hip girdle. They have the following three main functions:

i) to form a more or less rigid *connection* between the axial skeleton and the limbs;
ii) to provide suitable surfaces for the *attachment* of the *muscles* that move the limbs;
iii) to provide *stability* by separating the limbs.

In mammals, the hind limbs produce most of the power for walking and running. Therefore, the pelvic girdle is much more rigid than the pectoral girdle, and is *fused to the vertebral column* at the sacrum. The three main bones which make up each side of the pelvic girdle (the *ilium*, the *ischium* and the *pubis*) are fused together so tightly that it is usually impossible to detect the joints in an adult skeleton. Thus they are often referred to collectively as the *innominate bone* The pubic bones are fused ventrally in the *pubic symphysis*.

On the other hand, the paired *scapulae* (shoulder blades) and *clavicles* (collar bones), which make up the pectoral girdle, are quite separate and are attached less rigidly to the axial skeleton by means of ligaments and muscles. In many mammals, this arrangement enables the fore limbs to be moved through a wide variety of planes and angles. Unlike the pelvic girdle, the pectoral girdle is not a complete girdle.

Investigation

The pectoral girdle of a rat

Procedure

1 In your specimen, the long, thin clavicles have probably become separated from the scapulae. Place one scapular flat on the bench, with the 'spine' (the long ridge, projecting at right angles from the 'blade') facing upwards.

Question 1 Which surface can you now see?

2 Make a careful drawing of the scapula, labelling the following:

i) anterior edge;
ii) dorsal edge;
iii) posterior edge;
iv) point of attachment to the clavicle;
v) point of attachment to the forelimb.

3 At (iv) above, note the remains of a piece of cartilage which was attached to the clavicle.

Question 2 To which part of the skeleton was the other end of the clavicle attached?

4 Where two bones are attached together so that they can still be moved in relation to each other, they are said to articulate with each other.

Question 3 What do you notice about the shape and texture of the part of the scapula that articulates with the forelimb (the *glenoid* socket)?

Question 4 What is the likely function of the spine of the scapula?

Investigation

The pelvic girdle of a rat

Procedure

1 Use the mounted skeleton of a rabbit to enable you to orientate the rat pelvic girdle correctly.

2 Your specimen will consist of two innominate bones, probably separated because of the breakdown of the cartilage which joins them in life at the pubic symphysis (see fig 11.12).

Question 5 What is the likely function of this cartilage in the female?

3 Make a careful drawing of the lateral view of one innominate bone. Identify and label the following features:

i) the ilium;
ii) the ischium;
iii) the pubis;
iv) the *obturator foramen* (the large 'hole' between the bones);
v) the *acetabulum*, where the pelvic girdle articulates with the hind limb.

Remember that, except in young specimens, you will be unable to see the *sutures* between the three main bones.

4 Examine the inner surface of the ilium.

Question 6 What is the function of the roughened area just above the 'notch'?

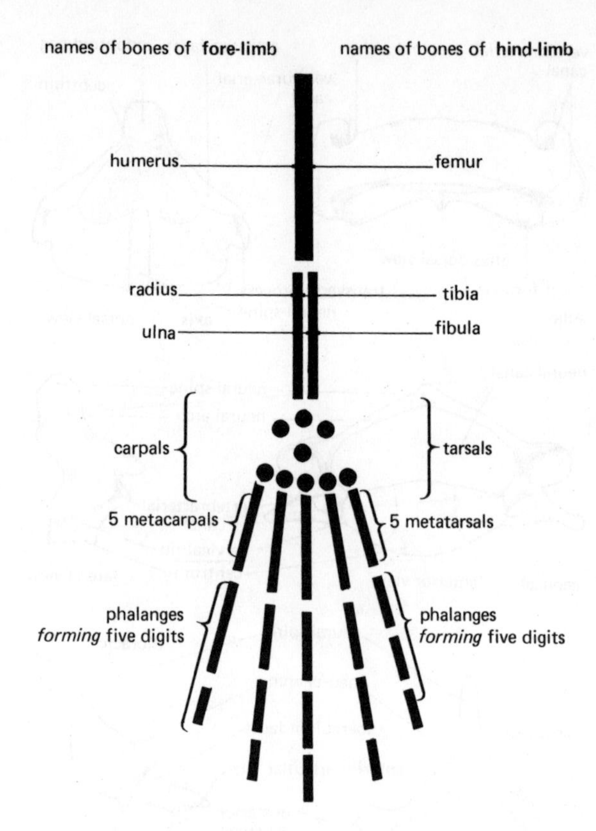

fig 11.13 A plan of the pentadactyl limb

11.32 The pentadactyl limb

The limbs of all the major groups of vertebrates are based on the same pattern. This pattern is shown diagrammatically in fig 11.13.

Because the basic pattern includes five digits (fingers and toes), the structure is known as a *pentadactyl limb*. In many cases, e.g. horses, pigs and birds, the number of digits has been reduced during evolution, and the limbs of the snakes have disappeared completely.

The typical pentadactyl limb consists of *one upper* bone (articulating with the girdle), *two parallel lower* bones, *three rows* of small *ankle or wrist* bones (9 in all), *five* parallel *sole or palm* bones and *fourteen digit bones* (four sets of three and one of two). The names of these bones, as applied to both fore- and hind-limb are given in fig 11.13.

Investigation

To examine the forelimb of a rat

Procedure

1 Obtain the individual bones of a rat forelimb and place them flat on the bench, in correct relationship to each other. Use the mounted skeleton as a model.

2 Make a careful drawing (lateral view) of the assembled bones, labelling the *humerus, ulna, radius, carpals, metacarpals* and *phalanges*. The ulna has a projection (the *olecranon process*), at the elbow end.

Question 1 What is the function of this projection?
3 Hold out your own arm in front of you, with the palm facing downwards (this is known as the *prone* position).
Question 2 What is the position of the olecranon process (use your other hand to locate it)?
4 Now turn your extended arm so that the palm faces upwards (*supine* position).
Question 3 What is the new position of the olecranon process?
Question 4 Is the same mobility of the forelimb possible in the rat or rabbit?
Question 5 What is the position of the rat forelimb?
Question 6 What is the importance of the mobility of the arm in Man?
5 Examine the rat humerus. Note that the head (*proximal*) end is rounded for articulation with the scapula, and that the other (*distal*) end is grooved for articulation with the ulna. Both ends bear a number of processes for muscle attachment.
Question 7 Can you identify another site for muscle attachment?
6 Examine the bones of the forelimb. Note that the ulna and the much more slender radius are not fused but lie in very close contact.
7 Examine the carpals.
Question 8 How many can you count?
8 Note that the first digit is very small compared with the other four; much smaller than the human thumb is in comparison with the fingers.

Investigation
To examine the hind limb of a rat
Procedure
1 Obtain the individual bones of a rat hind limb and place them flat on the bench, in the correct relationship to each other. Use the mounted skeleton as a model.
2 Make a careful drawing (lateral view) of the assembled bones, labelling the *femur, tibia, fibula* (partly fused to the tibia), *tarsals, metatarsals* and *phalanges*. The patella or knee cap may have been lost if the bones have been separated for some time.
3 Examine the tarsals.
Question 9 How many can you see?
4 Note that one of the tarsals projects backwards to form a heel-bone for the attachment of muscles.
5 Examine the metatarsals.
Question 10 What does their length tell you about the way in which the rat moves?
6 Note that the tip of the phalanx at the end of each digit is pointed, for insertion into a *strong claw*.

11.33 Arthropod limbs

Investigation
Structure of an insect leg

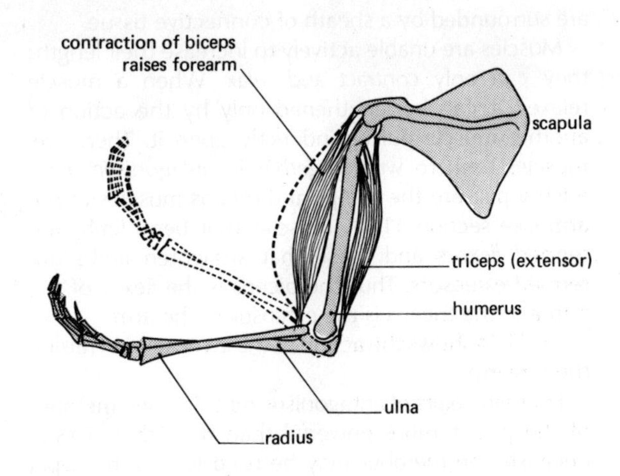

fig 11.14 The antagonistic muscles moving the forearm
After D. G. Mackean

Procedure
1 Remove the hind limb from a locust or grasshopper.
2 Compare it with the skeleton of the hind limb of a rat.
Question 1 What are the main points of similarity between the two limbs?
Question 2 In what respects does the insect limb differ from the rat limb?
Question 3 What major effect are these differences likely to have on the life of the insect?

11.40 How do muscles move the skeleton?

Investigation
The muscles of a rat's hind limb.
Procedure
1 Obtain a rat carcase which has been preserved after a previous dissection. It will probably have been opened ventrally for investigation of the abdominal organs. Pull the skin at the base of the abdomen clear of the body wall and make a cut along the leg to as far as possible below the knee.
The lower you cut, the more difficult it will become to separate the skin from the underlying muscle.
2 Make a second skin cut completely around the top of the leg.
3 Peel back the skin from the top of the leg downwards, using the back of your scalpel to remove the connective tissue between skin and muscle. Expose the heel.
Question 1 What is the general appearance of the calf muscle (*gastrocnemius*)?
Question 2 How does this appearance differ from that near the heel-bone?
4 Use the scalpel (blunt side) to separate the longitudinal bundles of muscle fibres. Note that they

are surrounded by a sheath of connective tissue.

Muscles are unable actively to increase their length: they can only *contract and relax*. When a muscle relaxes, it can be lengthened only by the action of another muscle pulling indirectly upon it. Therefore, muscles tend to *work together* in *antagonistic pairs*. Such a pair are the biceps and triceps muscles of the arm (see section 11.00). Muscles that bend limbs are termed *flexors* and those that straighten limbs are termed *extensors*. Thus the biceps is the flexor of the arm and the triceps is the extensor of the arm.

Fig 11.14 shows the action of these muscles in raising the forearm.

For many pairs of antagonistic muscles, one member of the pair is more powerful than the other. This is because one member may be used for the principal function (e.g. *lifting* of a weight by the arm) and the other member is used simply to return the limb to its original position.

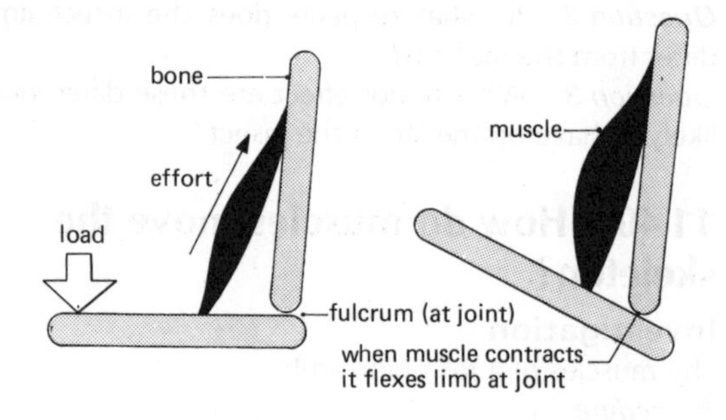

fig 11.15 A diagram showing the human forelimb as a lever

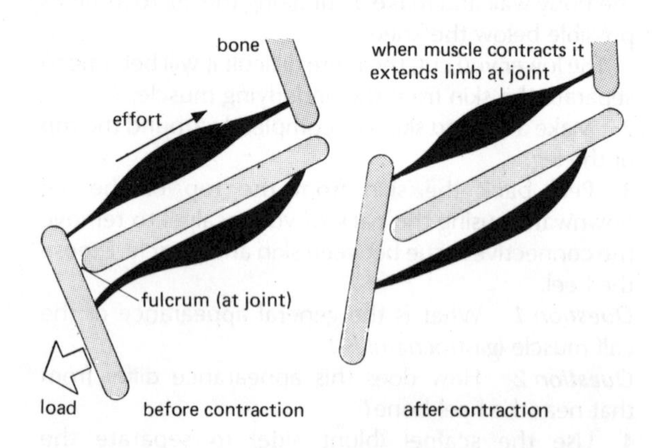

fig 11.16 A diagram showing the rat hind limb as a lever

Limbs work as a series of levers and in the case of the arm, the elbow joint acts as the fulcrum and the effort is produced by the biceps (see section 11.50). Thus the flexor of the arm is more powerful than the extensor.

An example of the opposite situation can be seen in the running action performed by a rat. This is brought about in part by the extension of the lower hind limb, which pushes downwards and backwards on the ground, causes the rat to move forwards and upwards. The gastrocnemius or calf muscle (the extensor) contracts, pulling the heel-bone upwards and bringing the metatarsals into line with tibio-fibula. The muscle on the anterior side of the leg (the flexor) is less powerful, being used only to return the foot to its original position.

11.41 The structure of muscle

As with the term *bone* (see section 11.12), we use the term *muscle* in two different ways. In one sense it is

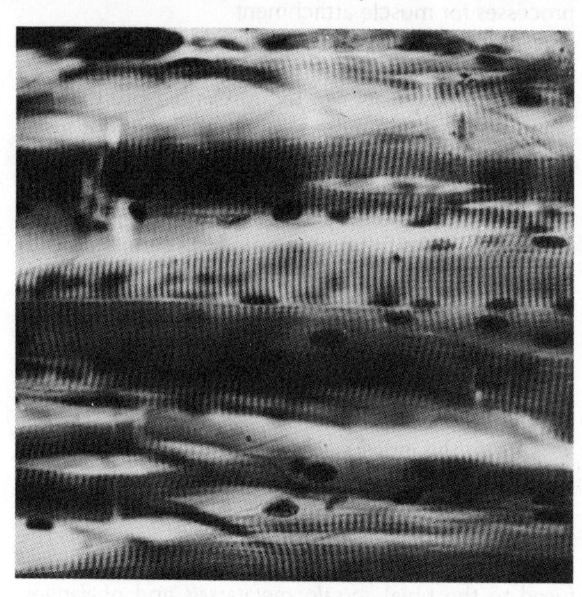

fig 11.17 (a) Microscopic structure of skeletal muscle
(b) Stereograms of muscle structure

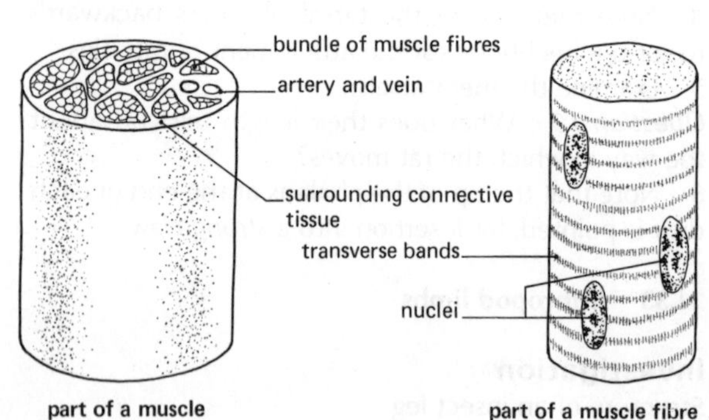

used to refer to *specific organs* (the biceps muscle, the deltoid muscle, etc); in the other sense it refers to one of the *tissues* of which these organs are made up.

The unit of muscle tissue is the muscle cell or fibre. This is an elongated, approximately cylindrical structure, which under the light microscope can be seen to have several nuclei and a series of transverse bands (*striations*) parallel to its short axis. These cells occur in *bundles*, each bundle being surrounded by a sheath of connective tissue. A muscle such as the biceps consists of a *number of such bundles*, together with nerve and blood supplies, all surrounded by a further sheath of connective tissue.

The above description and diagram applies only to the kind of muscle that moves the arms and the legs. This type of muscle is termed *skeletal* muscle, *voluntary* muscle (because it is under the control of the will) or *striated* muscle.

Two other types of muscle tissue are found in mammals, *cardiac* muscle, where the *transversely banded* fibres are *branching*, and *smooth* muscle, where the fibres are *spindle shaped* and *striations are absent*. Cardiac muscle is found in the heart and smooth muscle is found in the bladder and certain other non-skeletal organs.

11.42 Energy for muscle function

In order to contract, muscles require energy. Some of this energy is stored in the muscle in the form of the carbohydrate glycogen (see Chapter 5, section 5.32). This is oxidised to carbon dioxide and water, with the production of ATP for 'contraction energy' during cellular respiration in the muscle fibres.

However, since the amount of glycogen stored in a muscle seldom exceeds one or two per cent of its weight, this is soon used up during vigorous exercise. Energy is then obtained by the oxidation of glucose transported to the muscle through its rich blood supply (the extent of this supply is obvious from the appearance of fresh meat which is skeletal muscle).

If the period of exercise is prolonged and the demand for oxygen exceeds its supply, the muscle cells are able to respire anaerobically, breaking down glucose to lactic acid, and liberating energy for ATP production (see Chapter 5, section 5.41). The build up of lactic acid in this way results in *'muscle fatigue'*, a phenomenon familiar to all athletes.

11.50 The action of muscle through levers and joints

A *lever* is a simple *machine* which enables a certain amount of work to be done. It involves the expenditure of energy in producing an *effort* to move a *load*. All levers have a *fulcrum* or pivot, which is their fixed point of support.

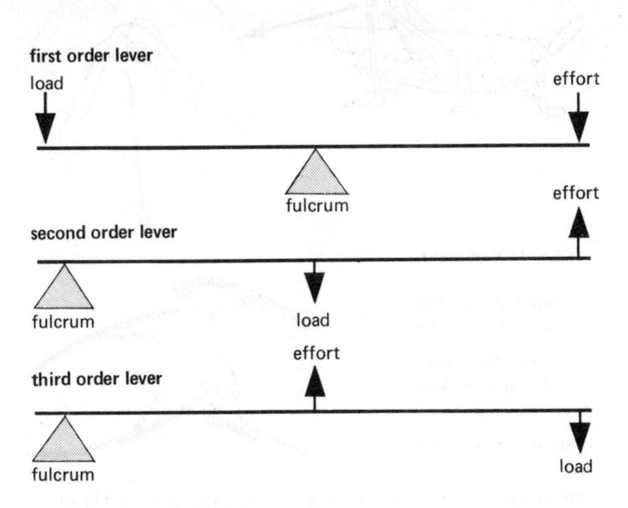

fig 11.18 The three orders of lever

There are three orders of levers, classified according to the relationship between the positions of the load, the fulcrum and the point of application of the effort.

In a *first order lever*, the fulcrum is between the effort and the load. A crowbar and a pair of scissors are two examples of this order.

A wheelbarrow is an example of the *second order of levers*, where the load is between the fulcrum and the effort.

Finally, there are the *third order levers*, where the effort is applied between the fulcrum and the load. This order is used when a ladder is lifted away from a wall.

Examples of all of these orders can be found in the joints of a mammal.

Question 1 Give an example of a joint in the human body for each of the order of levers shown in fig 11.18.

11.51 Joints

In all the above examples, the fulcrum of the lever has been at the position of a joint between two bones. These joints allow varying degrees of movements, and can be classified on this basis into several different groups.

Perhaps the easiest type of joint to understand is the *hinge joint*, such as is found at the *elbow and knee*. This allows movement in one plane only, and so is directly comparable with the simple examples of levers considered in section 11.50.

Ball and socket or *universal joints* are found at the *shoulder* and *hip*. These allow movement in all planes, and thus we can move our arms forward, sideways or in a circle.

Examples of *gliding joints* are found between the carpals and between the tarsals, where one bone moves over the surface of another.

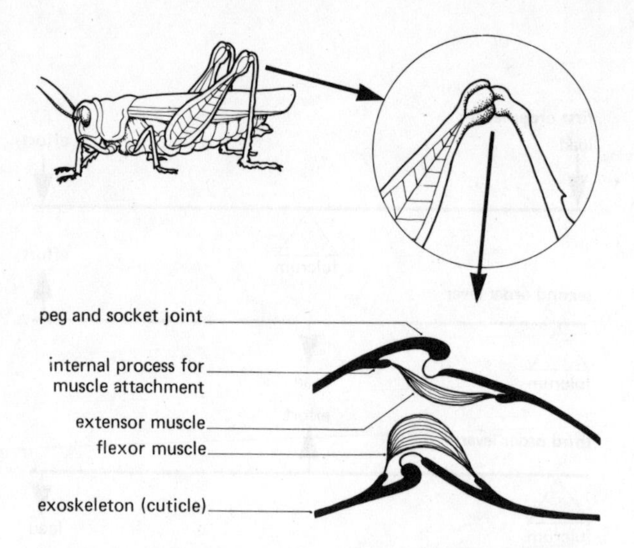

fig 11.19 The peg and socket joint of an arthropod limb

A fourth type of joint is the *pivot*. This allows rotation of one bone around another, and the most obvious example is found between the *atlas* and the *axis vertebrae*, which allows rotation of the skull.

Finally, we should consider the *fixed joint or suture*. We have seen examples of this in the skull and in the innominate bone, where the ilium, ischium and pubis have become fused so that they are not moveable in relation to each other.

The limbs of arthropods contain a particular type of hinge joint, the 'peg and socket' (see fig 11.19). This allows movement in one plane only but, as a single limb contains a number of joints with the pegs and sockets set at different angles, the end of the limb can be placed in almost any position.

Questions requiring an extended essay-type answer

1 a) The vertebral column of a mammal has five regions. *Name* each region and show the basis for distinguishing each one.
b) What similarities are present in the vertebrae from each of the five regions?
c) Describe the ways in which the first two vertebrae of the vertebral column are specialised for the functions they serve.

2 a) Draw a diagram of generalised mammalian vertebra.
b) Name four types of vertebra in the mammalian skeleton and state how each one differs from a generalised vertebra.
c) What are the functions of the sacrum?

3 a) Describe (with the aid of diagrams) the structure of
i) a thoracic vertebra, and
ii) a true rib of a *named* mammal. Show how the rib is attached to the vertebrae and the sternum.
b) Describe the mechanism of breathing in a mammal with particular reference to the structures mentioned in part a).

4 a) Draw a diagram to show the general plan of the pentadactyl limb. Label the bones drawn as though they are part of the forelimb.
b) How is the forelimb of man adapted for use as a tool-holding limb? How are these adaptations different from the forelimb of a rat or rabbit?

5 a) By means of at least two labelled diagrams show how the muscles and bones cause movement at a ball and socket joint in the mammal.
b) How does the arthropod limb *differ* from a mammalian limb in the arrangement of its skeleton and muscles at a hinge joint?

6 a) Make a labelled diagram of a ball and socket joint found in a mammal. Name the joint and state where it is found in the body.
b) By means of a labelled diagram show how either the knee or the elbow joint is moved by muscles. Show in your diagram a typical nerve path by which this voluntary action is brought about.

7 a) What are the functions of the mammalian skeleton?
b) Draw fully labelled diagrams to show the use of three different types of lever in the mammalian body.

12 Disease

12.00 The discovery of micro-organisms

Anton von Leeuwenhoek (1632–1723), a Dutchman, first saw micro-organisms by means of his newly invented microscope.

The magnification of his instrument was no more than 200 times, but he was able to see tiny organisms collected from pools of water. At this time, the appearance of disease in organisms and even the presence of small organisms was attributed to *spontaneous generation*, the continual creation of new life from non-living things. Later Spallanzani, an Italian showed that decay was due to living organisms by demonstrating that substances usually teeming with micro-organisms could be prevented from developing them. He did this by immersing in boiling water experimental flasks containing these substances. *Louis Pasteur* (1822–1895), a Frenchman, continued the work of Spallanzani. He first developed the *germ theory of disease* which postulates that all disease must be caused by micro-organisms. In section 12.10 we shall see the simple experiment performed by Pasteur, which showed decay could only be caused by living material coming from the air. He also developed a method of culturing micro-organisms by putting them into a sterile soup or on a sterile jelly; the organisms being transferred by a sterile needle from the decaying matter.

Robert Koch (1842–1910), a German, designed many of the techniques of handling bacteria still used in the present day. He investigated the disease *Anthrax*, cultured the bacteria and proved convincingly that the disease of cattle was caused by specific micro-organisms. He reasoned that the bacteria spored from the dead animal were released into the soil, and concluded that all dead cattle must be burned to destroy the spores.

Lister, (1827–1912), an Englishman, was a famous surgeon and was the first to try to remove dangerous bacteria from the operating theatre by using a disinfectant, *carbolic acid*. His efforts greatly improved the chances of patients surviving an operation and not dying from bacterial infection of the wound. Antiseptic surgery involved the disinfection of all surgical instruments, surgeons' hands, and the atmosphere.

12.10 Sterile techniques and the investigation of bacterial growth

Micro-organisms of different kinds are found everywhere. In every gramme of soil there are about 100 million bacteria, while the collections of dead cells on the scalp (scurf) contain about 500 million bacteria. The air, bench tops, clothes, skin, finger nail scrapings and gut, as well as many other places, all shelter bacteria.

Fortunately most of these are harmless to human beings, i.e. they are not *pathogenic*. Nevertheless, when experimenting with micro-organisms, we must always take extreme care, for cultures of harmless types can often contain bacteria which cause disease.

When culturing bacteria, even if they are thought to be harmless, always follow the rules below.

i) Wash hands before touching a sterile Petri dish.

ii) Open the Petri dish as little as possible, and replace the lid quickly.

iii) Never cough or sneeze near the dish.

iv) Never touch the infected jelly with your fingers.

v) When cultures are no longer required they should be flooded with strong disinfectant.

vi) After cleaning out the nutrient from Petri dishes, they should be washed and disinfected, and then, if they are glass, heat sterilised.

vii) Wash your hands thoroughly after all operations. Use plenty of soap.

viii) Never put hands near the mouth during experimental work (on no account consume food in the laboratory).

Experiment

Where do micro-organisms come from? Pasteur's experiment.

Procedure

1 Set up four test-tubes as shown in fig 12.1. Tube 2 is not heated, whereas tubes 1, 3 and 4 are heated above the boiling point of water by placing them in a pressure cooker for 15 minutes.

2 Place the tubes in a rack and examine daily.

3 Make a table of results, and record the appearance of the tubes as the days go by.

Questions

1 Which tubes have changed in appearance?

2 What change is noticed in each nutrient broth?

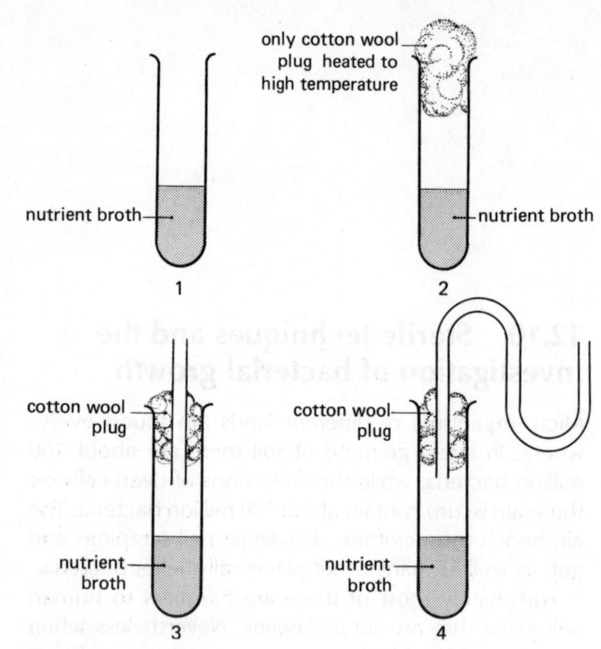

fig 12.1 Pasteur's experiment to investigate the origin of micro-organisms

3 What do you think causes this appearance? Smell the tubes. Is their any relation between appearance and smell? (Do not open tubes 3 and 4).

4 State the factors which have prevented any change taking place in the other tubes.

Experiment

Are there micro-organisms around us in the air and on the body?

Procedure

1 Use sterile glass Petri dishes or plastic Petri dishes which have been pre-sterilised. Pour into five dishes nutrient broth (made from agar jelly – a seaweed extract) containing beef extract. Allow the broth to coagulate at room temperature.

2 Number the dishes 1 to 5. By means of a chinagraph pencil (or felt tip pen), draw lines across the bases of three dishes so that each has two halves A and B.

3 Prepare the Petri dishes as follows:

Dish 1	part A	Press dirty fingers on the jelly.
	part B	Wash the fingers well, using soap, dry and press them on the jelly.
Dish 2	part A	Place nail scrapings onto the jelly.
	part B	Place scrapings from between the teeth onto the jelly.
Dish 3	part A	Place some hairs on the jelly.
	part B	Leave this half of the dish clear.
Dish 4		Open, and cough or sneeze over the jelly.
Dish 5		Open the dish in the laboratory for 30 minutes.

4 Now fix the lids tightly to the bases of the Petri dishes with clear adhesive tape.

5 Place the dishes upside down in an oven at 37°C.

6 Examine the plates in your next lesson (or 2–3 days).

7 Record your results in the form of a table.

Questions

5 Why are the dishes marked on the under surface of the base, and placed upside down in the incubator?

6 Why are the lids tightly fixed with adhesive tape?

7 What appears in the dishes?

8 What is the function of dish 3 part B? Did anything appear in this section? Do you think it worked well?

9 Compare 1A with 1B. What is the significance of your observations?

10 If you were a health officer in a large food factory, what methods would you use to prevent workers from contaminating food?

The micro-organisms grow well in the incubator at 37°C and our experiments have shown that they are present in a number of places in the laboratory and on the body. They can thus be spread easily by the wind

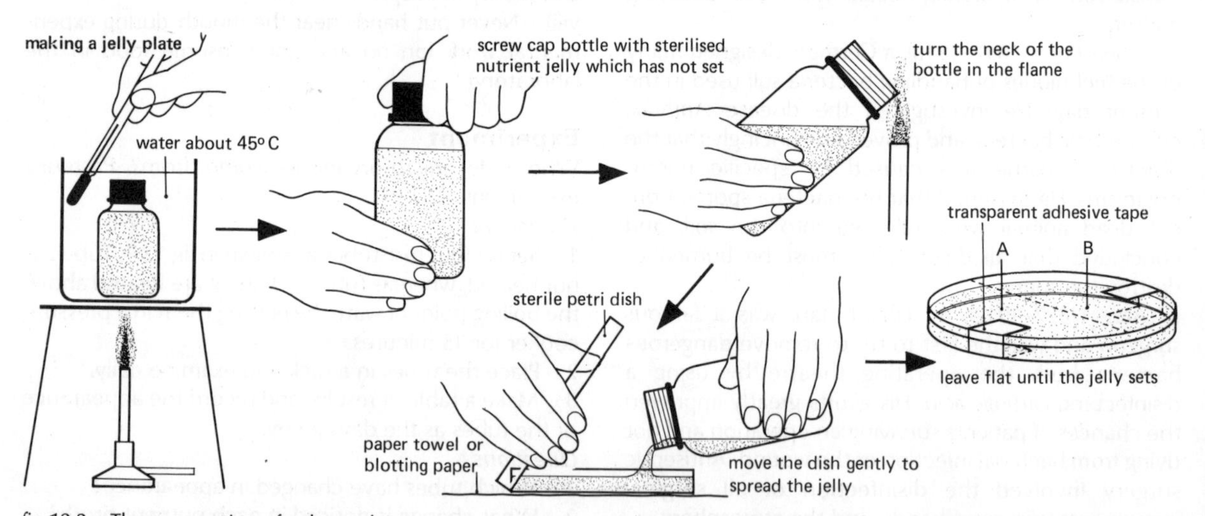

fig 12.2 The preparation of culture plates

and by contact with animals such as insects and Man. We can ask ourselves whether we can do anything about their presence and prevent their multiplication. We know that temperature is a factor in their development, since they grow best at a temperature above that of the laboratory.

Experiment

Does temperature have any effect on the growth of bacterial cultures?

Procedure

1 Prepare five Petri dishes with nutrient broth and when cool open them in the laboratory for 30 minutes.
2 Place the lids on the dishes and then treat them as follows:

Dish 1 Leave at room temperature (20–25°C).
Dish 2 Place in the cool compartment of a refrigerator (3–5°C).
Dish 3 Place in the freezing compartment of a refrigerator (below 0°C).
Dish 4 Place in an incubator (37°C).
Dish 5 Heat in a hot oven or place in a pressure cooker for 15 minutes (130°C+).

3 Examine the plates after 2–3 days.
4 Make a table to show your observations.

Questions

11 At which temperature did the bacteria develop most rapidly? What is the significance of this temperature?
12 From your results, can you suggest the most suitable temperature for storing perishable foods? Give your reasons.

12.20 Disease-causing organisms

12.21 Bacteria

Bacteria are single cells but they do not have a nucleus. The nuclear material (DNA) is spread throughout the cell. The shape of bacteria is their most recognisable character and is used for identification.

i) *Cocci* (singular – coccus) are spherical with a diameter of about 1 μ.
a) *Streptococci* stick together to form a chain (throat infection).
b) *Staphylococci* stick together in irregular bunches (boils).
c) *Diplococci* stick together in pairs (pneumonia).
ii) *Bacilli* (singular – bacillus) are rod-shaped. Some are motile with one or more flagellae (anthrax and typhoid).
iii) *Vibrios* are curved rods which appear comma shaped (cholera).
iv *Spirilla* (singular – spirillum) are rod-like but twisted into a spiral cork-screw shape.
v) *Spirochaetes* are much finer, flexible and also twisted into a spiral (syphilis). They are motile.

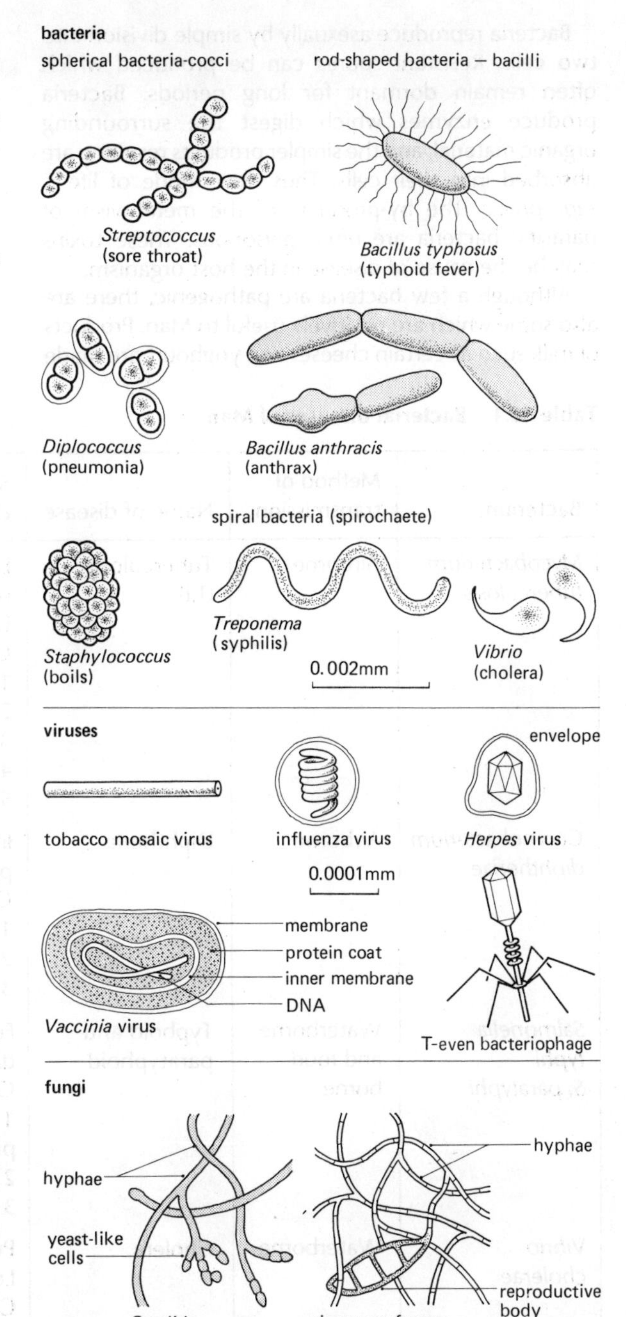

fig 12.3 Bacteria, viruses and fungi

Bacteria differ in their requirements for oxygen during respiratory processes.

i) Aerobic bacteria require oxygen for respiration.
ii) Anaerobic bacteria do not require oxygen and produce their energy by anaerobic respiration (putrefaction).
iii) Facultative bacteria, which constitute the majority of bacteria, can live under aerobic or anaerobic conditions.

Bacteria reproduce asexually by simple division into two cells. Resistant *spores* can be produced which often remain dormant for long periods. Bacteria produce enzymes which digest the surrounding organic material, and the simpler products resulting are absorbed into their cells. Thus their mode of life is *saprophytic*. The by-products of the metabolism of parasitic bacteria are often poisonous. These toxins may be the cause of disease in the host organism.

Although a few bacteria are pathogenic, there are also some which are positively useful to Man. Products of milk such as certain cheeses and yoghourt are made with the help of micro-organisms. The *Acetobacter* group of bacteria are used to produce vinegar from wine. Bacteria are most important as agents of decay. The change of dead organisms to harmless materials which become part of the air and soil has great significance for the further growth of new plants and animals. The work of bacteria is thus part of the great cycle of nature. The decay of organic matter by bacteria is a disadvantage as far as our food is concerned but we must not forget the importance of this work in nature.

Table 12.1 Bacterial diseases of Man

Bacterium	Method of transmission	Name of disease	Symptoms, other characteristics and treatment
Mycobacterium tuberculosis	Airborne	Tuberculosis (T.B.)	Entry of the bacterium is not always followed by serious disease, but it can infect any body organ. Disease of the lung is commonest. Control: 1 Eradication of cattle T.B. 2 Pasteurisation of milk. 3 Drugs e.g. streptomycin, isoniazid 4 Mass radiography to screen cases. 5 Vaccination by BCG vaccine.
Corynebacterium diphtheriae	Airborne	Diphtheria	Infection of nose, throat and larynx releasing a powerful toxin. Control: 1 Injection of antitoxin. 2 Antibiotics e.g.: penicillin and erythromycin 3 Immunisation by diphtheria toxoid.
Salmonella typhi *S. paratyphi*	Waterborne and food borne	Typhoid and paratyphoid	Fever and muscular pains at first which give way to diarrhoea; could lead to death. Control: 1 Severe control of sanitation and water supplies to prevent contamination. 2 Vaccination with killed bacteria. 3 Drugs e.g. ampicillin
Vibrio cholerae	Waterborne	Cholera	Principle feature profuse diarrhoea called 'rice water'. Loss of water from body fluids prime cause of death. Control: 1 Severe control of sanitation and water supplies to prevent contamination. 2 Drugs e.g. tetracycline and chloramphenicol 3 Vaccination gives protection and is useful in the management of an epidemic.
Neisseria gonorrhoeae	Sexual contact	Gonorrhoea	Discomfort on passing urine (particularly in males); discharge of pus from urethra. Control: 1 Sulphonamide drugs. 2 Antibiotics. 3 Avoidance of sexual contact until cured.

Bacterium	Method of transmission	Name of disease	Symptoms, other characteristics and treatment
Clostridium tetani	Contact and skin penetration	Tetanus (lockjaw)	Muscular spasms, particularly of the jaws (hence alternative name of lock-jaw), followed by generalised convulsions. Disease of war but also of infants in India where it is one of the four main causes of death. Control: 1 Immunisation by tetanus toxoid.
Diplococcus pneumoniae	Airborne	Pneumonia	Bacteria often present in the throat but become active if the patient in a weak condition. Lungs are invaded together with respiratory tract causing fever, pain in the chest and shivering. Fluid builds up in the lungs. Control: 1 Drugs e.g. sulphonamide and anti-biotics
Neisseria meningitidis	Airborne	Meningitis	Enters through mucous membrane of the throat and nose, invades the meninges (brain coverings), high fever, severe headache, vomiting and a rash appears during the course of the disease. Control: 1 Prevention of overcrowding in houses, schools and work places. 2 Drugs e.g. sulphonamides and anti-biotics

12.22 Viruses

Viruses are too small to be seen by a light microscope but there is a wide range of size from 200–300 nm to 30 nm. Each virus particle is composed of nuclear material (DNA) enclosed in a coat of protein. The shape is varied (see fig 12.3).

They are all parasites and cannot be cultured outside living cells. They multiply very rapidly and can be transferred through the air or by contact between organisms.

Certain types of virus are known as *bacteriophages* since they cause fatal infections of bacteria. The phages are one of the suitable materials for the study of viruses, and over the last twenty years detailed studies have been made of their behaviour.

Viruses produce disease by attacking particular groups of cells in the animal or plant body. They multiply within these cells and destroy their structure as well as inhibiting their activity. Rabies virus attacks nerve cells in the brain while poliomyelitis virus enters nerve cells of skeletal muscle. Yellow fever virus enters the cells of the mammalian liver while leaf mosaic virus enters and damages the outer cells of the leaf.

Table 12.2 Viral diseases of Man

Method of transmission	Name of disease	Symptoms, other characteristics and treatment
Airborne	Measles	Incubation period 10–14 days; begins with fever, catarrh and harsh coughing. A rash appears inside the mouth (Koplik's spots); a rash appears on the body at the hairline and then spreads downwards. Very dangerous disease in young age groups. Control: 1 Isolation of patient and avoidance of overcrowding in schools 2 Vaccine available
Airborne	Rubella (German measles)	Mild infection compared with measles and complications are rare. Potentially disastrous effects of rubella when it occurs during pregnancy in women. 20% of such women will produce babies with deafness and defects of the heart.

Method of transmission	Name of disease	Symptoms, other characteristics and treatment
Airborne	Influenza	The symptoms arrive with dramatic speed – headache, sore throat, shivering, backache. Fever develops with temperatures to 40°C. The disease is of short duration unless complications develop. Control: 1 Prevention of overcrowding 2 Make patient comfortable and use anti-biotics to prevent secondary infections. The virus produces new strains and spreads rapidly throughout the world in a pandemic. A considerable hazard both in developed and underdeveloped countries.
Food or waterborne	Poliomyelitis (infantile paralysis)	Attacks young and old; infection may produce only a minor illness comparable with influenza. In only a few cases the disease progresses to the nervous system, producing meningitis and attacking the muscle nerves. The muscles are then unable to act. If chest muscles are affected then artificial aids to breathing must be used, the 'iron lung'. Control: 1 Hygienic preparation of food and clean water supplies. 2 Vaccination – latest is oral vaccine. 3 No specific drug treatment.
Airborne	Smallpox	Fever, headache, backache and vomiting are symptoms followed by typical rash on face, forearms and hands spreading to the rest of the body. Control: 1 Drugs e.g. antibiotics to contain secondary infection 2 Vaccination World Health Organisation smallpox eradication programme begun in 1968, by 1975 had achieved extraordinarily successful results.

Table 12.3 Protozoan diseases of Man

Disease	Cause	Transmission	Symptoms, other characteristics and treatment
Malaria	*Plasmodium spp.*	*Anopheles* mosquito bite	Plasmodia injected into the blood multiply rapidly. After 10 days, high fever develops which may be continuous, irregular or occur twice a day. Control: 1 Drainage of the breeding places of mosquitoes. 2 Destruction of larvae with an oil spray. 3 Destruction of adults with insecticide. 4 Destruction of the parasites in Man by drugs e.g. chloroquine and quinine. 5 Preventive drugs e.g. paludrine and daraprim
Amoebiasis (Amoebic dysentery)	*Entamoeba histolytica*	Uncooked food, unhygienic preparation of food.	Causes diarrhoea with loss of blood, fever, nausea and vomiting – can lead to death. Control: 1 Hygienic food handling 2 Prevention of flies that can spread the disease. 3 Drugs – e.g. emetine, antibiotics and sulpha drugs

Disease	Cause	Transmission	Symptoms, other characteristics and treatment
Trypanosomiasis (sleeping sickness)	*Trypanosoma spp.*	Tsetse fly bite	A painless lump develops at the bite, lymph glands become enlarged, fever, enlargement of spleen and liver follow. Later the parasite invades the nervous system resulting in sleepiness and muscular spasms. Control: 1 Control of flies and limitation to certain areas. 2 Fly screens in human dwelling places 3 Drugs e.g. pentamidine in the early stages

Table 12.4 Fungal diseases of Man

Disease	Cause	Symptoms, other characteristics and treatment
Ringworm of the scalp	*Microsporium audouini*	A highly contagious disease by contact, combs hats etc. amongst children. It begins as a small scaly spot which enlarges and older patches are covered with greyish scales. Control: 1 Exclusion of infected children from school. 2 Drugs e.g. antibiotic griseofulvin taken by mouth.
Ringworm of the skin	As for scalp or *M. canis*	Lesions on the skin are seen as pale, scaly discs. There is more inflammation around the edges, causing swelling and blistering. Control: 1 Drugs e.g. griseofulvin
Athlete's foot	*Tinea pedis*	Shows as sodden, peeling skin between the toes that can be subject to secondary bacterial infection. Cure rate is low. Control: 1 Exclude sufferers from swimming pools and changing rooms. 2 Griseofulvin is only used in extreme cases.
Candidiasis (Thrush)	*Candida albicans*	A yeast-like cell, 2–4 microns in diameter. Commonly harmless in the body but infection results from some local reduction in resistance of the tissues. This may occur in the mouth, intestine, vagina, etc. Control: 1 Establish the predisposing factor and change this to clear up the infection. 2 Drug e.g. antibiotic nystatin used as a local cream.

12.23 Protozoa

Protozoa are small animals made up of a single piece of cytoplasm with a nucleus (acellular or unicellular). A living membrane surrounds the cytoplasm. Some protozoa are amoeboid and can change their shape while others possess a whip-like extension of the cytoplasm (flagellum) in order to help them move. They feed *holozoically* ingesting their food and then digesting it, or *saprozoically* by digesting it externally and absorbing the products. Parasitic protozoa can have two hosts, with sexual reproduction occurring in the primary host and the asexual reproduction in the secondary host.

12.24 Fungi

Many bacteria and viruses can parasitise Man and so cause illness, but comparatively few protozoa and fungi disturb Man's normal functions in this way. Fungi are either *saprophytes* or *parasites* and both types secrete enzymes to digest the external organic matter. The fungus produces fine filaments called hyphae which together form a dense mass, the mycelium. The cell wall is made of fungal cellulose and within is a lining of cytoplasm containing many nuclei (see fig 12.3). In addition, food stores such as oil droplets are present.

Fungi are responsible for many *plant diseases*, destroying large areas of crops, if conditions are right

Table 12.5 Fungal diseases of plants

Disease	Cause	Effects and control
Potato blight	*Phytophthora infestans*	Spread by spores in damp conditions. Hyphae spread through tissues, secreting digestive enzymes which destroy both tubers and above-ground parts. Control: 1 Fungicides, used on tubers before planting and sprayed on crops. 2 Burning of affected crops. 3 Storage of tubers in dry, well-ventilated places. 4 Rotation of crops.
Black rust of wheat	*Puccinia graminis*	Black spots on wheat stems are caused by one of many types of fungal spore produced during complicated life cycle, which involves another host plant (the barberry). Control: 1 Removal of barberry bushes, thus interrupting fungal life cycle. 2 Breeding resistant strains. Yellow and brown rusts are caused by related fungi. The fungi destroy leaf tissue and impair photosynthesis. The grains of affected plants are small and shrunken.
Powdery mildew of barley	*Erysiphe graminis*	The mycelium lies on the surface of the host plant and haustoria penetrate the epidermal cells. Control: Breeding resistant strains.

for their reproduction. They reproduce very rapidly by asexual spore formation so that the disease spreads quickly. Parasitic fungi include *smuts, rusts, leaf spot* and *blights* of different plant species and cause huge losses in crops of cereals, potatoes and many other crops.

12.25 Platyhelminthes and Nematodes

A large variety of worms infest Man. They fall into two main groups (i) *Platyhelminthes* (flatworms, tapeworms and flukes). Examples include liver fluke; pork, beef and fish tapeworm. (ii) *Nematoda* (roundworms). Examples include large round worms (*Ascaris*) and threadworms (*Enterobius*).

These worms contribute to a great deal of suffering and disease in the world. They have a complicated life cycle, often including a secondary host as well as the primary host (Man).

Table 12.6 Worms infecting Man

Disease	Transmission	Hosts	Symptoms, other characteristics and control
Tapeworm (*Taenia spp.*)	Through food – under-cooked meat and fish	1 Man 2 Pig, cattle or fish	Encysted embryo in the flesh of secondary host is consumed in undercooked meat or fish. Tapeworm develops in the gut attached to the intestinal wall. Fertilised eggs are passed out with the faeces, eggs are then eaten by animals. The tapeworms cause few symptoms and little damage in Man. Control: 1 Meat and fish to be well cooked at high temperatures. 2 Inspection of meat at slaughterhouses. 3 Proper processing and disposal of sewage. 4 Drugs e.g. similar drugs to those used against malaria: mepacrine and chloroquine, (see fig 12.6).

Disease	Transmission	Hosts	Symptoms, other characteristics and control
Ascariasis (*Ascaris lumbricoides*)	Infected food and water	1 Man (no secondary host)	The worms live in the bowel of Man and they produce vast numbers of eggs that are very resistant when shed with the faeces. When shed by Man, the eggs hatch and the larvae burrow into the lungs and from there reach the gut by way of the pharynx. The worms may obstruct the bowel and the larvae damage the lungs causing malnutrition and death. Control: 1 Proper processing and treatment of sewage. 2 Hygienic food and water supply. 3 Drugs e.g. piperazine is the most effective.
Threadworms or pinworms (*Enterobius vermicularis*)	Eggs swallowed	Man (no secondary host)	Very common, especially in children. Adults live in large intestine. Females migrate to anus to lay eggs, causing itching. Scratching, followed by placing of fingers in mouth, causes reinfection. Control: Washing hands after using toilet or touching anal area.

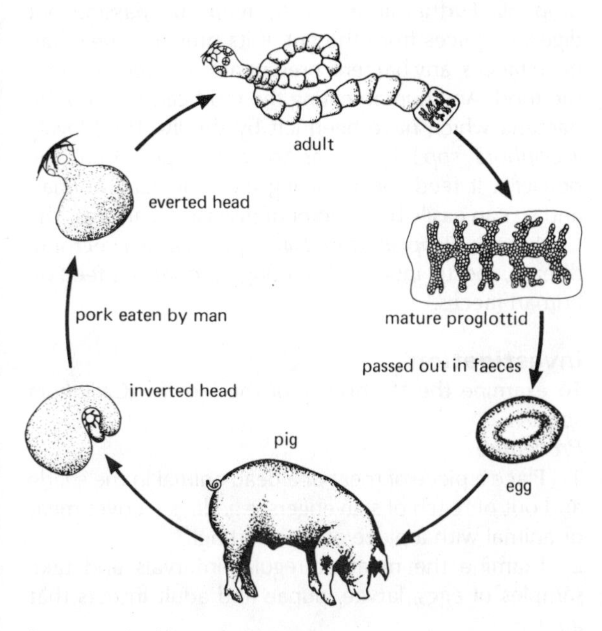

fig 12.4 The life cycle of a tapeworm

12.30 The transmission of disease

Diseases considered in this section are those caused directly by harmful organisms that enter the body of Man. These have been dealt with in sections 12.21 to 12.25, but we can now look at their methods of transmission in greater detail.

12.31 Airborne diseases

Some disease organisms are transmitted from person to person in tiny droplets of moisture. Coughing, sneezing, talking and ordinary breathing project moisture particles into the air. Larger droplets can settle onto food and on other objects in the home or the school. Smaller droplets evaporate quickly leaving bacteria or virus particles suspended in the air so that they can be inhaled. Droplet-borne infection is spread rapidly under conditions of high humidity and overcrowding, such as are encountered in schools, buses, trains and public meetings. Thus head colds and influenza tend to spread rapidly and produce epidemics.

12.32 Waterborne diseases

Drinking water is a source of many diseases, such as dysentery, cholera, typhoid and paratyphoid, which affect the alimentary canal. The spores and active organisms are liberated with faeces and can be spread into water supplies by insanitary conditions. Large numbers of people can become infected quickly in this way, particularly when floods, typhoons and earthquakes have seriously damaged water supplies and sewage disposal systems. Under normal conditions in home and school it is essential that hands should be washed after defaecating or urinating, for infection can be transferred in unclean hands used to prepare food or handle eating pots.

Rivers can spread disease very quickly, so that populations remote from a source of infection can develop cholera and typhoid.

12.33 Food borne diseases

Many organisms transmitted by water can also be carried by some foods. Unwashed hands, exposed septic sores, contaminated water and flies can also spread infection to food during its preparation. Bacterial, viral and worm infections are all possible from contaminated food.

12.34 Contagious diseases

These are diseases spread by direct contact between people or by objects handled by people. Fungal infections, ringworm and athlete's foot, can be transferred from skin to skin, by infected towels, floor surfaces, clothing etc. Smallpox is another disease that can be caught by contact.

The venereal diseases, of which gonorrhoea and syphilis are the most serious, are spread *only* by direct contact. They are passed during sexual intercourse and other types of sexual contact. They can both be cured if treatment is given at an early stage. *Gonorrhoea* is caused by a bacterium and the first symptoms of infection are a burning sensation when a man passes urine. The infection is often not noticed in a woman since the genital duct is separated from the urinary duct. Some 50% of women suffering from the disease do not know that they have it, but nevertheless it can be passed from mother to foetus. Detection in a woman often only occurs because their male partners develop the disease.

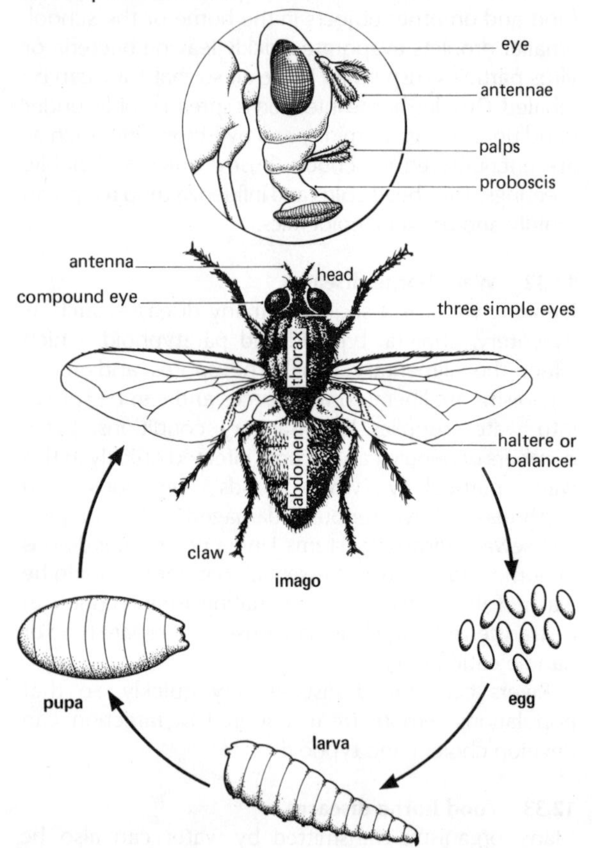

fig 12.5 the life cycle of the housefly

Syphilis, caused by a spirochaete, is much more dangerous because if untreated it can attack the brain and heart to such an extent that death eventually ensues. The first symptom is a sore at the site of the infection, usually on the genital organs. This sore may disappear quickly and no further symptoms develop for up to a year. The secondary stage can then include rashes, swollen lymph glands and disorders of bone, eye and other organs. This is a highly infectious stage. The tertiary stage, non-infective, lasts many years during which different body tissues become damaged.

12.35 Insect borne diseases

Insects can transmit disease organisms in two ways: (i) on the outsides of their bodies, and (ii) on the insides of their bodies.

i) By virtue of its feeding habits, the housefly (*Musca spp.*) is the most important vector of intestinal diseases, for it often feeds on animal dung and human faeces. Micro-organisms present on this material will cling to its legs and body and, as the insect walks across food, they drop off. Furthermore, the fly feeds by passing out digestive juices from the gut. If its previous meal had been faeces, any bacteria present are vomited out onto the food. Any human eating the food can pick up the bacteria which have been left by the fly. The blowfly (*Calliphora spp.*) is similar to, but larger than, the housefly. It feeds on decaying meat or dead animals and so can easily be a vector of disease organisms. The Cockroach (*Periplaneta* or *Blattia* spp.) is also a vector of disease, mostly intestinal, although it does not feed on human faeces.

Investigation

To examine the life history of the blowfly (*Calliphora spp.*)

Procedure

1 Place a piece of meat or a dead animal in the shade and out of reach of scavengers (e.g. dogs – cover meat or animal with a piece of wire netting).

2 Examine the meat at regular intervals and take samples of eggs, larvae, pupae and adult insects that appear.

3 Place the specimens into specimen boxes or petri dishes and observe. Make drawings of the larvae and pupae.

Questions

1 How long do the eggs take to change into larvae?

2 How long do the larvae take to change into pupae?

3 What difference is there between the activity of the larvae and pupae?

4 What happens when the adult fly hatches from the pupa?

ii) Of the diseases carried by insects, malaria is probably the most important. About 500 million people

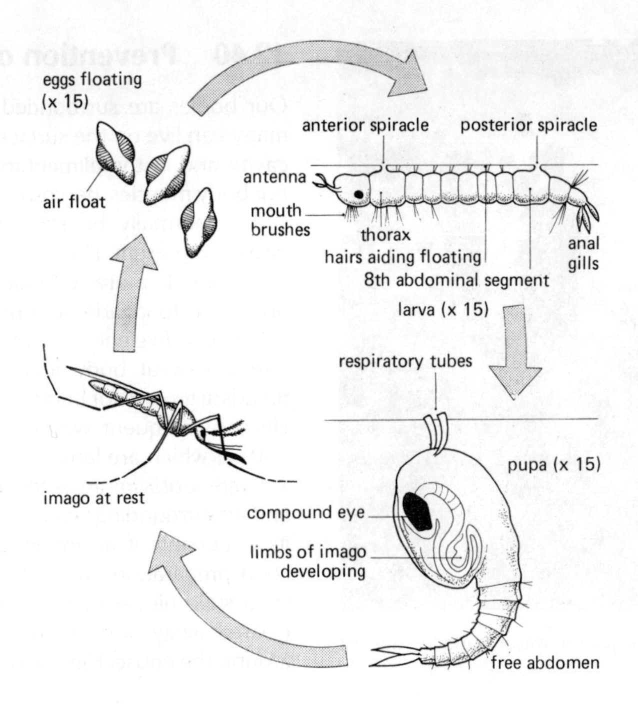

fig 12.6 The life cycle of the *Anopheles* mosquito

are at risk, some 200 million have severe attacks and there is an annual death rate of 2 million people. The disease is caused by the protozoan *Plasmodium* transmitted by the *Anopheles* mosquito. It was in 1895 that a British doctor, *Manson*, showed that mosquitoes transmitted *Filariasis*, and this led *Ross*, an Indian Army doctor, to search for a similar vector for malaria. In 1897 he found a malarial parasite in the stomach of a mosquito, and established the fact that the parasite goes through a developmental stage in this vector. This was the sexual phase of *Plasmodium*; the asexual phase takes place in the liver and blood cells of Man. Prevention of malaria involves attacking both the insect vector and the plasmodial parasite.

1 Control of the insect vector

a) The adult mosquito can be killed by insecticides such as *DDT, gamma B.H.C.* or *Dieldrin*. These are sprayed on the outer walls of buildings, under roof overhangs and inside huts and rooms. The insecticide remains effective for weeks or even months, killing any insect that lands on the sprayed surface.

Cause	Disease	Course of the disease
Plasmodium vivax	Benign tertian malaria	Seldom fatal with attacks of fever every 48 hours
Plasmodium falciparum	Malignant tertian malaria	Severe disease and more often fatal with attacks of fever every 48 hours
Plasmodium ovale	Tertian malaria	Similar attacks of fever every 48 hours
Plasmodium malariae	Quartian malaria	Mild disease compared with the others and with attacks every 72 hours

Table 12.7 Types of Plasmodial parasite

fig 12.7 The spraying of exposed water surfaces against mosquito larvae

b) The larvae are destroyed by spraying *oil* on the surface of stagnant water. The mosquitoes lay their eggs in lakes, ponds, swamps and in any man-made object such as tin cans, pots, barrels, drains and gutters that retain water. The latter should be emptied regularly or sprayed with oil. This substance reduces the surface tension of the water and the larvae sink below the surface, where they are unable to obtain oxygen.

c) The adult mosquito can be prevented from reaching Man. The latter can protect himself at night with mosquito nets over beds, or by fly gauze, carefully fixed over doors and windows.

2 *Control of the Plasmodium parasite*

a) The use of drugs to cure malarial patients, by killing the parasites in the liver and the blood cells, has long been used. The patients are then no longer a source of further infection. *Quinine* was used first but later *Chloroquine* and *Mepacrine* have been developed to kill the asexual stages in the body of Man. *Paludrine* is more effective still and is taken daily, whereas the new drug *Daraprim* is only taken once per week.

b) These latter drugs are also effective as a preventative, killing off the parasites as soon as they are injected into the blood of the human host.

Between 1945 and 1970 three quarters of the population of the world living in malarious areas had been freed from the disease, but even so, 500 million are still liable to contract malaria. Constant vigilance is required, even in areas made free of the disease, and all efforts should be made to prevent it where the mosquito and its parasite are still endemic.

12.40 Prevention of infection

Our bodies are surrounded by micro-organisms and many can live on the surface of the skin, in our buccal cavity and in the alimentary canal. The remainder of our body muscles, nervous system, glands and skeleton should normally be free of micro-organisms unless disease is present. How can we take action to prevent, or make it more difficult, for pathogenic micro-organisms to invade our organs?

1 *Personal hygiene.* Washing with soap and water removes sweat, body oils and dirt which provide a fine breeding ground for bacteria. Clothes must also be kept clean by frequent washing with soap or detergents, both of which are lethal to bacteria.

2 *Hygiene of living areas and social hygiene.* The house and its surroundings can contribute towards disease of its occupants if attention is not paid to cleanliness. Food preparation areas must be kept clean and free from stale pieces of food. Household waste must be cleared away and *not* thrown on the ground in or around the house. Flies can breed in organic waste, and

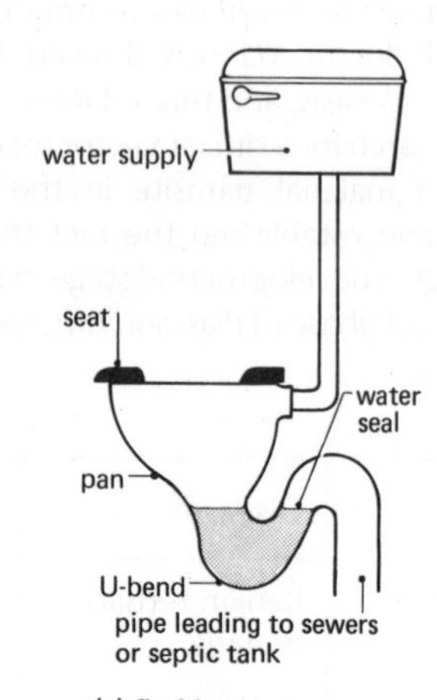

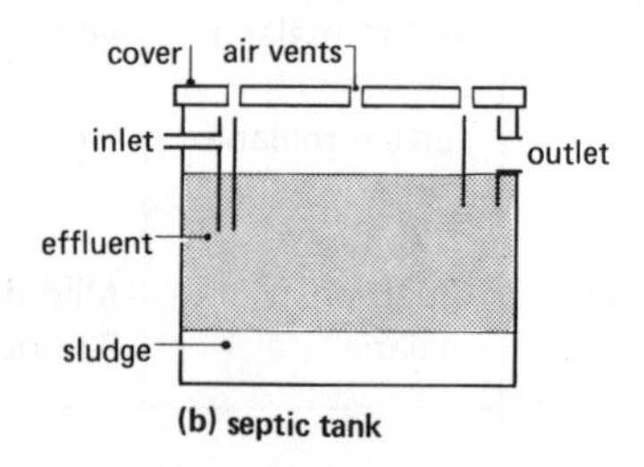

fig 12.8 Sewage disposal

carry bacteria onto food such that *intestinal diseases* are spread e.g. cholera, typhoid and dysentery. If possible, waste food and other matter should be placed in a bin with a tight-fitting lid.

Human faeces and urine are an important link in the life histories of many different parasites. Inadequate disposal of sewage is responsible for the rapid spread of many intestinal diseases and a number of worm infections. To avoid this, fresh human sewage should *not* be used for fertilising crops since it exposes the material to flies and could contaminate the growing plant crop. As far as possible, human excreta should be deposited in flush lavatories and then piped to sewage works where it is treated and made harmless. Flush lavatories need plenty of water to wash away the sewage. The faeces and urine are dropped into a china pan and flushed down an S-bend which effectively acts as a seal separating the excreta from the air and keeping out flies. If the lavatory is not connected to a sewage works, the sewage may be flushed into a large concrete tank from which the water can escape but in which the more solid material remains and gradually decomposes. Such a tank is called a *cesspool* or *septic tank* depending on its design (see fig 12.8).

3 *Community hygiene*. Villages and towns have tended to grow slowly over the years in a rather haphazard fashion, but today local governments exercise control over the type and site of any new building.

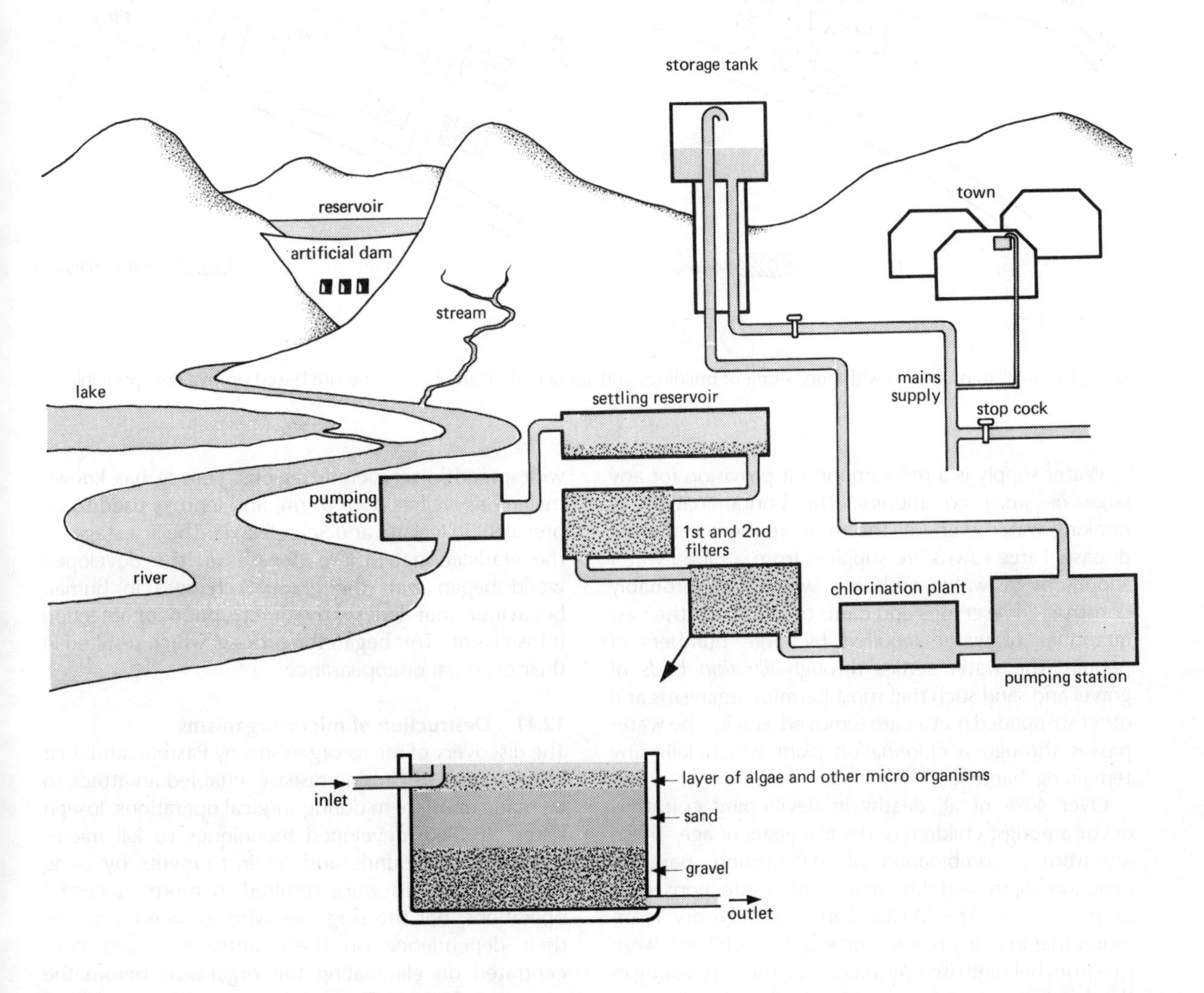

fig 12.9 The water supply of a town, with detail of a filter bed

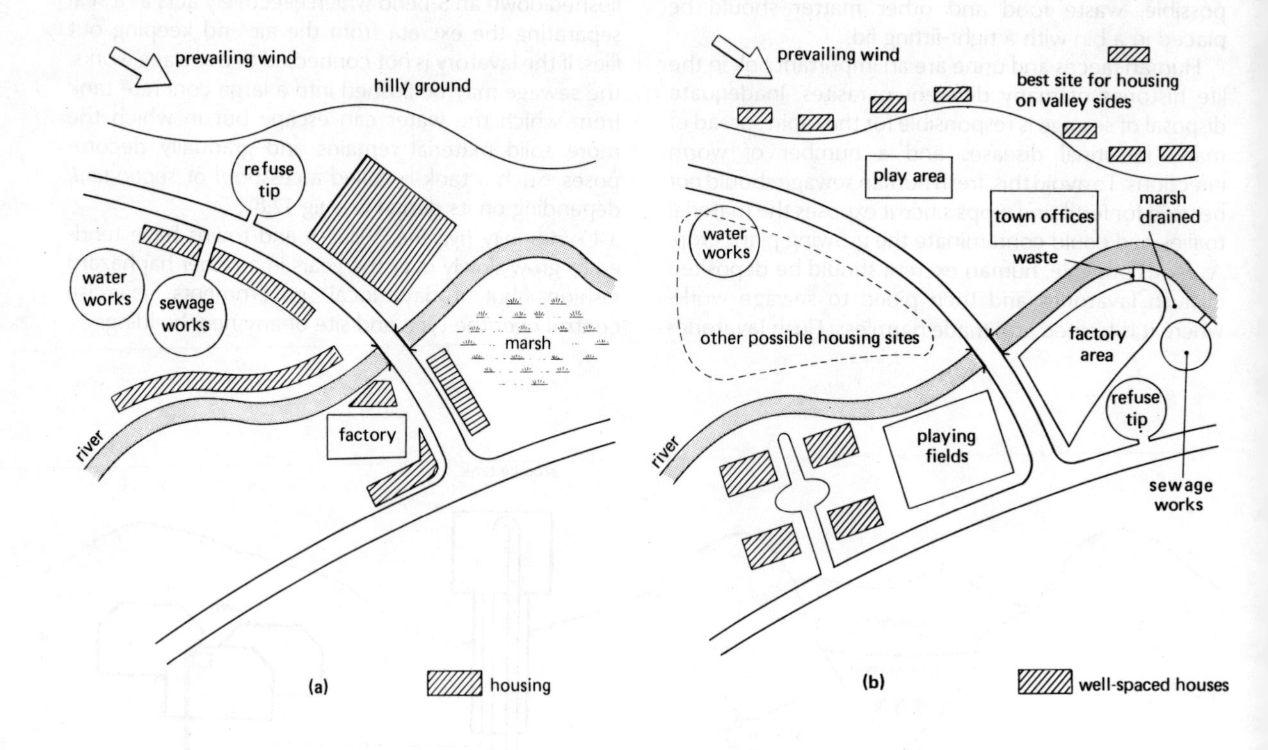

fig 12.10 (a) Plan of a town with poor siting of buildings and services (b) Plan of the same site based on hygienic principles

Water supply is a most important provision for any large or small community. The contamination of drinking water is one of the most frequent causes of disease. Large towns are supplied from a piped water supply by a water authority which has probably constructed reservoirs and dams to hold back the vast quantities of water required by large numbers of people. The water settles through *filtration beds* of gravel and sand such that most harmful organisms and other suspended matter are removed. Finally, the water passes through a *chlorination plant* which kills any remaining bacteria.

Over 40% of all deaths in developing countries occur amongst children under five years of age. These are from a combination of malnutrition, parasitic infection, diarrhoeal disease and other infections such as pneumonia. The WHO diarrhoeal advisory team found that in many regions, only 20% of children were free from helminth (worm) infections. The very young in the developing countries carry such a high incidence of infection that many experience only a few days free of illness in the course of a year. Many diseases now confined to sub-tropical and tropical areas were once widespread in temperate regions. Europe has known malaria as well as hookworm, and leprosy used to be prevalent in Iceland and Scandinavia. The first stages in the eradication of these diseases in the developed world began with the gradual changes in human behaviour that led to the interruption of infection transmission. This began the process which resulted in their eventual disappearance.

12.41 Destruction of micro-organisms

The discovery of micro-organisms by Pasteur, and their identification as causing disease, enabled an attack to be made upon them during surgical operations. Joseph Lister, in 1860, developed techniques to kill micro-organisms in wounds and on instruments by using carbolic acid. His work resulted in more successful operations, but the surgeons who followed, reduced their dependence on these antiseptics and concentrated on eliminating the organisms before the operation began. These *aseptic* practices involved sterilising instruments, gowns, gloves, masks and every other item in contact with a patient.

Sterilisation means the destruction or removal of *all*

micro-organisms. *Disinfectants* and *antiseptics* are chemical substances that destroy micro-organisms, but generally the manner of their use means that they do not destroy all. Most of the harmful organisms are eliminated by these substances. Disinfectants are chemicals used on non-living surfaces such as crockery, cooking utensils, cutlery and drains, whereas antiseptics are used on a living surface such as skin.

Sterilisation is best carried out by *thermal treatment* using an *autoclave* or pressure cooker. These enable temperatures of 115°C to 135°C to be reached which will destroy bacteria and most spores. Radiation sterilisation is now used for medical and surgical instruments. Gamma radiation from radio-isotopes is a powerful penetrating form produced by an isotope of Cobalt (*Cobalt 60*).

Disinfectants are commonest in the liquid form, and their usefulness depends on the resistance of the bacteria and on other factors such as temperature and concentration of the chemical. *Hypochlorites* are readily available as calcium hypochlorite or sodium hypochlorite and their action depends on the formation of hypochlorous acid which liberates oxygen in a highly active state. They were first used to reduce the incidence of childbirth fever, but are now commonly used in the home, laundries, dairies and the food industry in order to clean equipment. *Iodine* (dissolved with potassium iodide in 90% alcohol) is used as an antiseptic for superficial cuts and scratches to the skin. Phenol is rarely used as an antiseptic these days, but a chlorinated phenol (*Hexachlorophene*) is widely used as a skin antiseptic. *Chloroxylenol* (present in 'Dettol') is an antiseptic used to prevent sepsis during childbirth.

12.42 Natural defences of the body

The body is protected externally by the *skin* or *mucous membranes* that act as a barrier against disease organisms. The skin has a dead, horny layer difficult to penetrate, but it also possesses its own population of harmless bacteria together with certain chemical secretions. The respiratory tracts are continually exposed to atmospheric pollution, certain bacteria and viruses. These tubes are covered internally by a layer of mucus which is driven upwards by minute protoplasmic processes called *cilia*. Most of the airborne particles are trapped in the nose or in the mucus of the tracts, and are destroyed by anti-bacterial enzymes present. The mouth, buccal cavity, alimentary canal, vagina and the surfaces of the eye have their own defence mechanisms. The *saliva* has a cleansing action, but, if any bacteria are swallowed, then the *acid* of the stomach will kill them. The acid conditions of the vagina inhibit the growth of pathogenic bacteria. The *tears* secreted across the surface of the eye also destroy bacteria.

Internally, living tissue reacts to injury in a complex way, although this often begins as a simple local reaction to damage. This is called *inflammation* and is recognisable when our skin is burned, cut or infected with pathogenic bacteria. The skin becomes reddened, warm, painful and often swollen. This is due to the dilation of the small blood vessels increasing the supply of blood. As a result, the blood plasma oozes from the capillary walls to accumulate in the tissues. In addition, white blood cells squeeze through the capillary walls and are present in the tissue fluid.

Thus the tissue receives any anti-bacterial substances that are in the plasma while the white blood cells are able to engulf bacteria and cell debris. The *polymorph leucocytes* are the first to begin the scavenging process and are followed by *monocytes*. At the same time, the escaping plasma contains *fibrinogen* which forms blood clots helping to unite the damaged tissues and form a barrier to the movement of bacteria. If the polymorphs can overcome the local invasion of pathogens then the tissue gradually returns to normal. Tissue fluid is absorbed back into the blood stream, fibrin clots break down and the blood flow decreases. The dead tissues are softened by protein-digesting enzymes from the polymorphs and the resulting fluid, called *pus*, accumulates beneath the skin. This is released by bursting through the surface of the skin accompanied by dramatic easing of pain in the tissues, as in the case of a boil or abcess.

The external and internal protection considered above are relatively simple mechanisms. Higher animals also have a much more specific defence against micro-organisms. Their bodies can act against foreign proteins from bacteria, viruses, fungi or cells of another animal of the same species (as in a skin transplant). Any substance that can alter the properties of body cells, when introduced, is called an *antigen*. If the antigen is introduced again at a later date the altered body cells can 'neutralise' the antigen by the *antibodies* which they have produced or, in the case of a skin graft, by surrounding it with cells called *lymphocytes*.

Immunity

It is clear, therefore, that the body's defences can be organised against pathogens causing specific diseases, so that a second infection will not occur in the individual's lifetime. This is called *immunity* and is long lasting against such diseases as smallpox, chicken pox, diphtheria, measles, mumps and poliomyelitis. Unfortunately, permanent immunity does not follow infection from influenza or cold viruses, staphylococci or streptococci.

Artificial immunity to disease can sometimes be induced by introducing into the bloodstream micro-organisms closely related to those that cause the disease. Antibodies to these antigens are then produced,

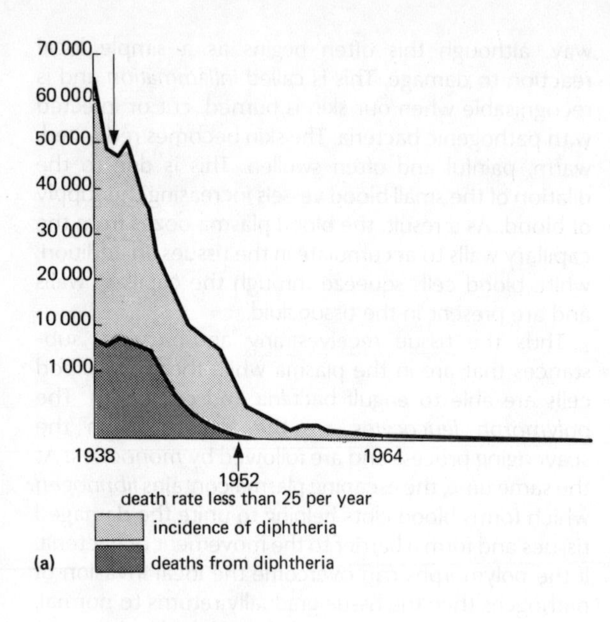

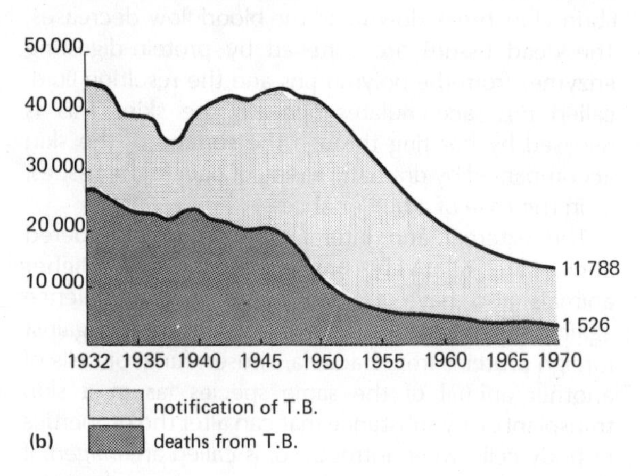

fig 12.11 Graphs showing decrease in (a) diphtheria (b) tuberculosis

and these act against the pathogens without the patient having first been subjected to an attack of the disease. This technique is called *vaccination* or *immunisation* and its action in eliminating many dangerous diseases is a great achievement of modern medicine.

Edward Jenner first attempted vaccination in the early nineteenth century, by using material from a pustule of cowpox which, he discovered, produced immunity from smallpox. Pasteur successfully produced vaccines to protect animals against anthrax, by keeping the bacteria at 42–43°C for a week (instead of at 37°C). The virulence of the bacteria was reduced by this method, and further, they lost their ability to produce resistant spores. This technique is called *attenuation*, and the longer the bacteria are kept at 42°C, the weaker they become. Sir Almroth Wright, researching on vaccines in 1892, was able to produce vaccines from dead micro-organisms such that the dangerous diseases of typhoid and paratyphoid could be contained. Finally, powerful *toxoids* produced by certain bacteria can be filtered out of cultures and used as vaccines to promote the formation of powerful substances in the blood that can neutralise the effects of the toxins.

Where vaccination against disease has been introduced on a large scale, as shown in the table 12.8, there has been a dramatic drop in the incidence of infection. In addition, many governments insist on a *certificate of vaccination* before allowing travel around the world, particularly in respect of entry of people into their countries.

Vaccine production	Antibodies produced against diseases	
	Bacterial disease	Viral disease
1 Living attenuated micro-organisms	Tuberculosis	Yellow fever, Measles, German measles, Poliomyelitis, Rabies
2 Dead micro-organisms	Typhoid, Paratyphoid, Whooping cough, Cholera	Influenza
3 Toxoids	Tetanus, Diphtheria	

Table 12.8 Types of vaccine

12.50 Food preservation

All decay is due to the activity of micro-organisms, particularly bacteria and fungi. Food consumed by Man is susceptible to decay, and furthermore, his diseases are often transmitted by way of food. The proper handling, preparation, storage and cooking of food are all important in preventing wastage, and transmission of disease organisms. In addition to diseases such as cholera, typhoid and dysentery, other mild and sometimes severe illnesses come under the general title of *food poisoning*.

Precautions which serve to reduce the incidence of food poisoning

1 Storage of food in vermin-proof rooms avoids

contamination by faeces of rats and mice.
2 Slaughter and evisceration of animals away from carcase meat or cooked food.
3 Regular disinfection of equipment in slaughter houses and butchers' shops.
4 Exclusion of any human 'carriers' of disease from slaughter houses and food shops.
5 Refrigeration of meat and fish, then thorough thawing before cooking.
6 Thorough roasting or pressure cooking to achieve high temperatures; many spores of bacteria are resistant to boiling for several hours.
7 Avoiding the reheating of food that has been standing in warm conditions.
8 Reducing the handling of food to a minimum and excluding food handlers with septic sores on any part of the body.
9 Education of food handlers in personal hygiene.

Food processing in relation to food hygiene
1 *Killing micro-organisms.* This is possible by thermal processes such as boiling, roasting and pasteurisation. Cans of food are heated in this way and then sealed. The food will remain unharmed unless the tin of food is pierced. Milk contains a wide variety of micro-organisms which cause souring within a short time unless the milk is pasteurised (kept at 72°C for 15 seconds then cooled rapidly to 10°C) and refrigerated.

2 *Prevention of growth of micro-organisms.*
a) *Freezing.* A great variety of food is marketed in frozen form and large quantities are shipped across the world in this state. Spore-bearing organisms are particularly resistant to freezing. Micro-organisms are not killed by freezing; they simply stop growing. Frozen food removed from the freezer will begin to thaw and bacterial growth commences immediately. Thus the food must be eaten as soon as possible.
b) *Dehydration.* Dehydrated foods are useful since they are light in weight and keep for a long time. Water has been removed and this is essential for bacterial growth. The foods can reabsorb water from the

Causative agent	Source	Symptoms	Prevention of transmission
Salmonella spp. causing Salmonellasis	Many animals carry the disease organisms e.g. pigs, calves and poultry. Main source for Man is meat of these animals.	Within 12–24 hours, fever followed by vomiting and diarrhoea. Firm diagnosis needs laboratory tests on faeces. Rarely fatal.	Flies can transmit the bacteria which are excreted in the faeces, therefore environmental control is important. This is difficult to achieve on farms. Fish and shellfish can transmit salmonella especially if in contact with sewage. Refrigeration, complete thawing and thorough cooking will contain and eventually kill bacteria.
Clostridium welchii causing clostridial food poisoning	Widely distributed in nature e.g. soil, sewage and water. Thus many possibilities of food contamination. Spores can survive several hours of boiling water. Many outbreaks traced to meat.	12–24 hours incubation; followed by fever and vomiting abdominal pain and diarrhoea. Infection only lasts about 24 hours; rarely fatal.	Meat should be thoroughly roasted in small quantities, not more than 2–3 kg. Meat should be eaten after cooking, any remaining refrigerated within 1½ hours.
Clostridium botulinum causing Botulism	Anaerobic bacteria living where air is excluded, as in canned, potted or pickled food. The food generally has been treated but spore forms survive.	Rare disease, but a high mortality rate – 50% in reported cases. Takes 24 hours to develop with vomiting, muscle paralysis and constipation.	Adequate heating of food will destroy the spores. Growth of the organism can be inhibited by complete drying, refrigeration, thorough salting or reduction of pH i.e. acid conditions.

Table 12.9 Food poisoning

atmosphere so that they must be packed in air-tight containers (e.g. plastic bags).

3 *Inhibitors of the growth of micro-organisms*.

a) *Acids*. Lactic acid and acetic acid (vinegar) are used to preserve food. They inhibit growth of organisms which cannot stand an acid environment.

b) *Salt*. Meat and fish can be kept for long periods without decay in salt since this stops bacterial growth and enzyme action in the tissues.

c) *Smoke*. Meat products are cured by smoking, which dries the surface and also coats the meat with substances retarding the growth of micro-organisms.

d) *Sugar*. This acts as a preservative when present in high concentrations. Water is not available for the micro-organisms and so growth is inhibited. Both sugar and salt exert an osmotic action on the organisms. Honey and jam are examples which do not spoil readily.

4 *Radiation*. The use of ionising radiations has increased since it has been shown to destroy spoilage and disease-producing organisms. Small doses of radiation can also destroy animal parasites present in food. The *'shelf life'* of raw or cooked foods can be prolonged without changing the taste or appearance.

Questions requiring an extended essay-type answer

1 a) Describe the life history of a *named* insect vector of disease.
b) Name the disease organism transmitted and show how a knowledge of the life history of the insect and the disease organism can be used to implement control methods.

2 a) *Name* a virus disease of man. How is this virus transmitted to its host and how can the disease be prevented or controlled?
b) How would you show that bacteria are present on your skin? (Consider that no microscope is available.)
c) *Name* a disease caused by a protozoan organism transmitted by an invertebrate vector. Describe the life history of the vector and show at which vulnerable stages in its life history it could be controlled.

3 a) Food poisoning can sometimes result from unhygienic handling of food. Give six rules that could be observed in kitchens and shops, and discuss the importance of personal hygiene in all stages of the preparation and handling of food.
b) Describe three methods of food preservation to prevent the growth of micro-organisms. Why are these methods successful?

4 a) Why does damp food (e.g. bread) go mouldy and then become almost liquid?
b) Describe how you would test experimentally your observations in part a).
c) Describe briefly three different methods that you could use to prevent food from going mouldy and decomposing.

5 a) Why is it important to keep an area where food is prepared free from house-flies (*Musca domestica*)?
b) Give a detailed description of an experiment to confirm your statements in part a), and explain the results that you obtain.
c) What methods are available for controlling these pests?

13 Soil

13.00 Introduction

Plants grow in the outermost layers of the earth's crust. These are called *soil*, and were formed, originally, from the breakdown of rock. The exposed rock was broken into smaller and smaller pieces because of the action of heat and cold. The expansion of water as it changes to ice, the expansion of rock under a hot sun and the friction of stones and water are factors which all contribute to this *weathering*. Gradually, small plants (lichens, mosses and ferns) are able to grow amongst the pieces of rock and, after a long period of time, the growth and death of living organisms provide another important part of soil, the *humus*.

Investigation

To obtain samples of topsoil and subsoil and to expose a soil profile

Procedure

1 In an isolated spot, well covered with vegetation, dig a hole in the ground.

2 Clean off the side of the hole so that it is vertical and shows a section through the soil.

3 Measure and record the thickness of any clearly defined layers that you can see. Make a simple diagram to show the thickness of the layers drawn to the correct scale.

4 Note and record in your diagram the details of the appearance of each layer e.g. colour, presence of roots, presence or absence of other living organisms, presence or absence of stones, compactness etc.

5 Remove soil carefully from a grass root system and observe the spread of the roots. Repeat this for a larger herb. Make sketches of these root systems.

6 Take a sample of the soil from every layer and enclose each in a plastic bag. Seal the bags, by tying up the opening. Collect also a sample of the dead material (litter) from the surface of the soil, and seal in a plastic bag. Label each bag. The soil layers or horizons progress upwards from the parent rock (*C horizon*), through the subsoil (*B horizon*) to the organically enriched surface layer (*A horizon*). The total is referred to as the *soil profile*. This profile may be altered by water movement; in the tropics this is mainly due to excessive rainfall washing out (*leaching*) soluble substances from the A horizon and depositing them lower down by drainage of the water through the soil. Leaching is most pronounced when the rainfall exceeds surface evaporation, especially on soil exposed by heavy grazing which has removed the transpiring plants. Thus leaching and consequent acidity are particularly common in sandy soils (see fig 13.1).

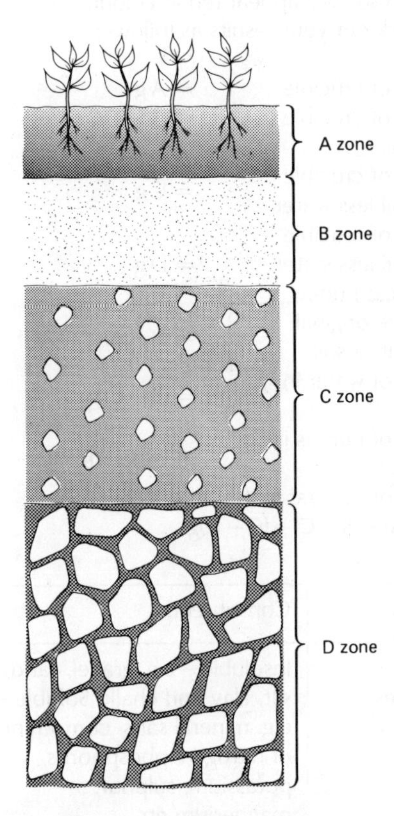

fig 13.1 A diagram of a soil profile

13.10 What is soil?

Experiment

Can the soil be separated into different fractions?

Procedure

1 Take a sample (about 20 g) of the topsoil from the plastic bag and press it into a weighed crucible (metal or porcelain). Weigh the crucible and the soil.

2 Place the crucible and soil into a steam oven or an oven kept at 105°C in order to dry the soil. Heating

should be continued for several days, weighing each day until there is no further loss in weight, (allow the crucible and soil to cool before weighing).

3 Place the crucible and soil on a pipeclay triangle and a tripod and heat strongly with a hot flame. At first the soil will turn black, and there will be a smell of burning organic material, but gradually this will disappear. Allow to cool and then weigh. Continue heating until there is not further loss in weight. By this time the soil will appear red in colour.

4 Work out your results as follows:

Weight of crucible = A g
Weight of crucible + soil = B g
Weight of crucible + soil less water = C g
Weight of crucible + soil less water and less humus = D g
Therefore original weight of soil = (B − A) g
Weight of water in the soil = (B − C) g
Weight of humus in the soil = (C − D) g
Weight of mineral matter in the soil = (B − A) − (B − C) − (C − D) g

Calculate the percentage of each fraction (water, humus and mineral matter) present in the soil sample e.g.:

$$\% \text{ of water in the topsoil} = \frac{(B-C)}{(B-A)} \times 100\%$$

5 Repeat this experiment with topsoil from different sites e.g. forest soil, garden soil, hard-packed soil from a footpath etc.

Questions

1 The amount of water in the soil will vary considerably from sample to sample. What factors can cause this variation?

2 What general differences regarding humus content do you find in soils?

3 What constituent is missing from hard-packed soil such as is found in a footpath?

4 What general differences do you find between topsoil and subsoil?

Experiment

To estimate, quantitatively, the amount of air in a top soil.

Procedure

1 Take a small tin (volume about 200 cm³) and make one or two holes in its base (with a hammer and nail). Press its open end into some topsoil having first removed the surface litter.

2 Scrape the soil away from the sides of the tin and

Fraction	Constituents	Function
Rock particles	Insoluble – e.g. gravel, sand, silt, clay and chalk. Soluble – e.g. mineral salts, compounds of nitrogen, phosphorus, potassium, sulphur, magnesium etc.	Provides the framework of the soil and is mainly derived from the underlying rock.
Humus	Decaying plant and animal matter.	This gives the soil a darker colour. It absorbs and retains large amounts of water. When breakdown of humus is complete, mineral salts are available for use by the plant.
Air	Principally oxygen and nitrogen.	The air provides oxygen for the respiration of soil organisms and plant roots. It provides nitrogen for fixation by the nitrogen-fixing bacteria of the soil.
Water	Rain water seeping downwards or capillary water from underground sources.	Dissolves the mineral salts and makes them readily available to plants. Carbon dioxide produced by living organisms dissolves in the soil water.
Living organisms	Bacteria and many other small organisms (see table 13.2).	Bacteria assist in the breakdown of humus and play an important part in the nitrogen cycle. Other organisms feed on plant and animal material, and thus form complex food webs. The aeration and drainage of soil is improved by the burrowing of soil animals.

Table 13.1 Soil fractions and their functions

Phylum	Class	Name of animal	Structure	Habitat
Nematoda		Roundworms		Water film. Occasionally trapped by fungi.
Platyhelminthes		*Turbellaria* (flatworms)	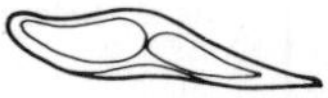	Under logs, decaying vegetation, damp moss. Water film.
Annelida	*Oligochaeta*	*Lumbricidae* (earthworms)		Burrowing; strongly affect soil fauna. Rare in acid, waterlogged or very dry soils.
		Enchytraeidae (potworms)		Soil cracks, rarely burrowing. Small and white.
Arthropoda	*Arachnida*	Mites		Pore spaces.
		Pseudoscorpions		Pore spaces.
		Spiders		Leaf litter, soil crevices, under stones.
	Crustacea	*Isopoda* (woodlice)	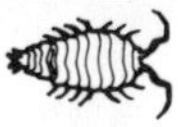	Leaf litter, soil crevices, under stones.
	Insecta	*Collembola* (springtails)		Pore spaces.
		Orthoptera (grasshoppers)		Active burrower.
		Coleoptera (beetles)		Active burrowers and channellers.
		Diptera (flies)		Pore spaces.
		Hymenoptera (Ants)	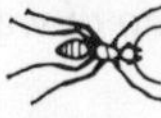	Active burrowers.
		Lepidoptera (Butterflies and moths)	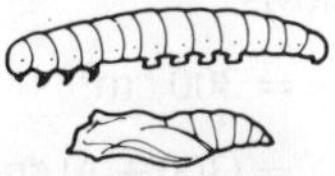	Crevices and pore spaces.
	Myriapoda	*Diplopoda* (millipedes)	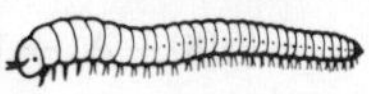	Leaf litter, pore spaces, crevices.
		Chilopoda (centipedes)		As above.
Mollusca	*Gastropoda*	*Pulmonata* (slugs and snails)	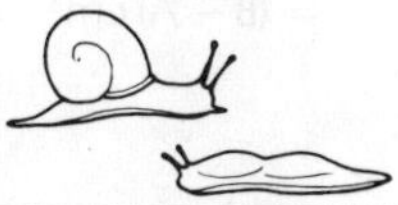	Leaf litter, larger crevices of upper soil, under stones.

Table 13.2 Key to soil organisms (see Chapter 15 for details of classification)

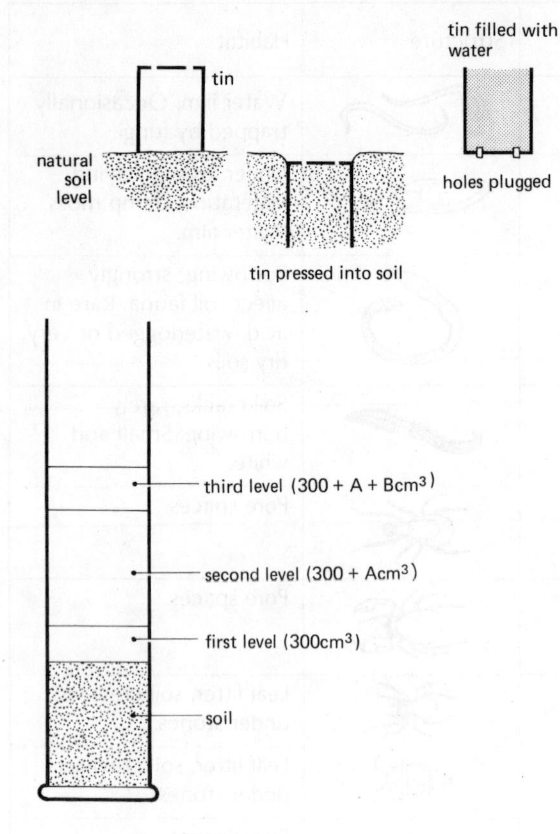

fig 13.2 The experimental apparatus to determine the quantitiy of air in soil

then remove the tin carefully so that it is completely filled with soil.

3 Pour or scrape all of the soil into a large measuring cylinder (e.g. 1 000 cm^3) containing 300 cm^3 of water. Shake or stir the soil until no more air bubbles appear.

4 Measure the new volume of the water.

5 Plug the holes in the tin with clay or gum. Fill the empty tin with water, then pour this into the measuring cylinder and measure the new volume of water (see fig 13.2).

6 Work out your results as follows:

Initial volume of water in the measuring cylinder	$= 300\ cm^3$
Volume of water plus soil	$= (300 + A)\ cm^3$
Volume of water plus soil plus water from the tin	$= (300 + A) + B\ cm^3$
Therefore volume of the soil	$= A\ cm^3$
Therefore volume of the tin	$= B\ cm^3$
Therefore volume of air in the soil sample	$= (B - A)\ cm^3$

Questions

5 What further calculation would give a better indication of the air content?

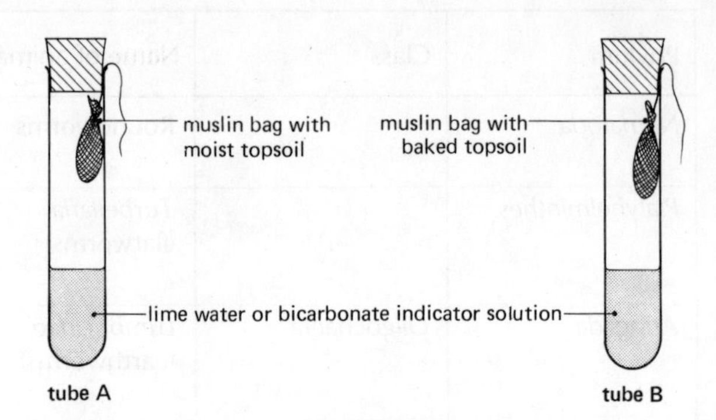

fig 13.3 The experimental apparatus to investigate the presence of living organisms in the soil

6 Where are the likely sources of error in this experiment?

7 What do you consider to be the importance of air in the soil?

Experiment

Does soil contain a population of living organisms?

Procedure

1 Set up the test-tubes as shown in fig 13.3.

2 Tube A contains moist topsoil and tube B contains a similar amount of sterilised soil. Soil can be sterilised by placing it in a steam oven, or any oven at 105°C, for about thirty minutes.

3 Leave the tubes for several days and observe the appearance of the lime water or bicarbonate indicator. You may shake the tubes gently.

Questions

8 What change takes place in the indicator (or limewater)?

9 What do you conclude from your observations?

Investigation

Can the living organisms in soil be extracted and identified? Do topsoil and subsoil differ in their populations of living organisms?

Procedure

1 Assemble two sets of apparatus as shown in fig 13.4. This is known as a *Tullgren funnel* and can be constructed fairly simply.

2 In one funnel, place a sample of topsoil and in the other, a sample of subsoil.

3 Switch on the lamp of each funnel and leave for 24 to 48 hours.

4 Soil organisms will move from the soil, through the grid and into the preservative.

5 Identify the organisms as broadly as possible from the key shown in table 13.2.

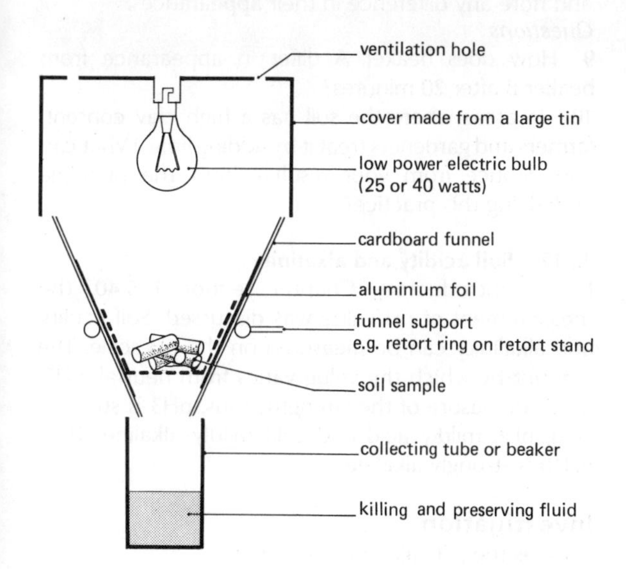

fig 13.4 The experimental apparatus to separate living organisms from the soil

6 Make a comparative table of the organisms found in the topsoil and subsoil, indicating type and numbers of each organism. (See table 13.1.)

7 Use the same apparatus to compare different types of soil.

Questions

10 What environmental factors cause the soil organisms to move out of the soil?

11 Why is the aluminium foil placed inside the funnel?

12 What general differences do you observe between the topsoil and subsoil in respect of the soil organisms collected? Account for these differences.

13.11 Characteristics of different types of soil

Experiment

To investigate the rise of water (capillarity) through different types of soil

Procedure

1 Set up the apparatus as shown in fig 13.5. The glass wool helps to retain the soil in the tubes.

2 Note the time and date.

3 At hourly and then daily intervals, note the rise of water up through the soil, and measure the height above the free water surface. Note the time and date.

4 Record your observations in the form of a table and draw a graph for each tube.

Questions

1 Which soil shows the greatest rise of water in the early stages of the experiment?

2 Explain this result.

3 Which soil shows the highest rise of water at the

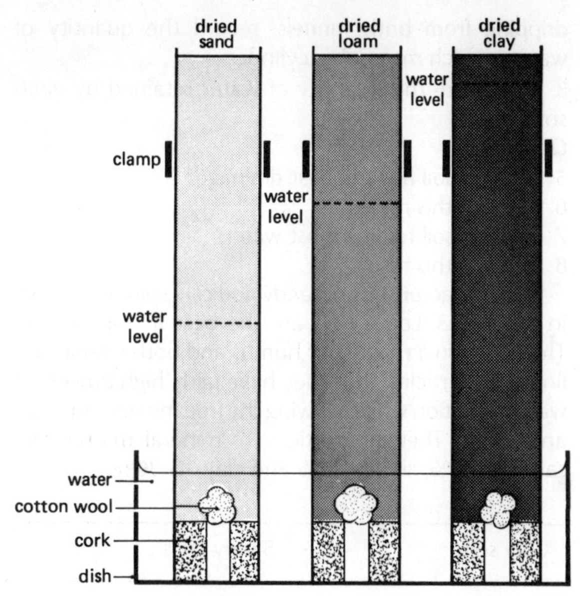

fig 13.5 The experimental apparatus to compare the rates at which water rises in sand and clay soils

end of the experiment?

4 Explain this result.

Capillarity is an important factor in soil fertility. In times of drought, water can rise from the water table to a high level in some soils.

Experiment

To investigate the drainage of water in different types of soil

Procedure

1 Set up the apparatus as shown in fig 13.6, with the filter papers containing dried samples of sandy and clay soils.

2 Pour 50 cm^3 of water onto each soil sample and leave to drain through. When the water has stopped

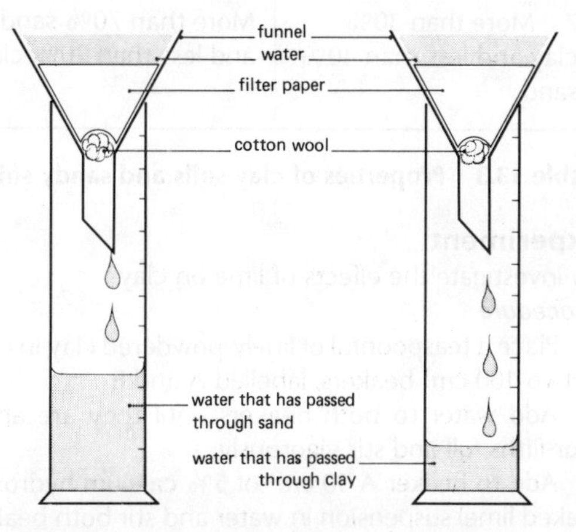

fig 13.6 The experimental apparatus to compare the drainage of sand and clay soils

dripping from both funnels, record the quantity of water in each measuring cylinder.
3 Calculate the quantity of water retained by each soil.
Questions
5 Which soil has the best drainage?
6 Explain this result.
7 Which soil retains most water?
8 Explain this result.

The characteristics of *sandy* and *clay* soils are shown in table 13.3. Loam soils are the best all-round soils. They contain a mixture of humus and both coarse and fine rock particles, thus they have fairly high powers of water retention while allowing the free movement of air and water. Their proportions of mineral matter are: sand 40–70%, silt 20–30% and clay 10–30%.

Clay soils	Sandy soils
1 Very small air spaces between the particles.	Larger air spaces between the particles.
2 Water does not drain well and is easily absorbed.	Water drains well and is not so easily absorbed.
3 The soil remains wet – retains water.	The soil dries out easily – cannot retain water.
4 Water can rise to a high level by capillarity.	Water cannot rise to a high level by capillarity.
5 Rich in dissolved salts.	Poor in dissolved salts.
6 The soil requires a long time to warm up after a rainy season and is heavy to cultivate.	The soil warms up quickly after a rainy season and is easy to cultivate.
7 More than 30% clay and less than 40% sand.	More than 70% sand and less than 20% clay.

Table 13.3 Properties of clay soils and sandy soils

Experiment
To investigate the effects of lime on clay
Procedure
1 Place a teaspoonful of finely-powdered clay in each of two 100 cm³ beakers, labelled A and B.
2 Add water to both beakers until they are about four-fifths full and stir vigorously.
3 Add to beaker A 10 cm³ of 5% calcium hydroxide (slaked lime) suspension in water and stir both beakers again.
4 Leave for 20 minutes: then examine both beakers and note any difference in their appearances.
Questions
9 How does beaker A differ in appearance from beaker B after 20 minutes?
10 In areas where the soil has a high clay content, farmers and gardeners treat it by adding lime. What can you deduce from your results about the principle underlying this practice?

13.12 Soil acidity and alkalinity
In the Introductory Chapter section I 5.40, the measurement of pH value was discussed. Soil acidity and alkalinity can be measured on the pH scale. The amount by which the value varies from neutral (pH7) gives a measure of the strength; thus, pH3 is strongly acid, pH6 mildly acid and pH8 mildly alkaline. Thus pH10 is strongly alkaline.

Investigation
What is the pH value of a soil sample?
Procedure
1 Place about 2 g of soil on a Petri dish and soak it with BDH Universal Indicator. Leave for at least 2 to 3 minutes.
2 Tilt the Petri dish so that the indicator drains out of the soil.
3 Compare the colour of the indicator with the chart supplied with the indicator solution.
4 (Alternative to steps 1, 2 and 3 above). Soak the soil sample with distilled water, drain off and test with wide range Universal Indicator Test Papers. Obtain a more accurate result using a second test paper of more limited range. Compare with the colour chart supplied with the papers.

13.20 Loss of fertility

Experiment
What effects do slope and plant cover have on the soil?
Procedure
1 Set up three seed boxes (about 40 cm × 20 cm × 5 cm). Put equal amounts of soil in two of the boxes and for the third cut a turf from a well grassed area and place it in the box so that it fits exactly.
2 Each box should have a V-shaped opening in one end.
3 Set up the two boxes of soil inclined at 20° and 40° to the horizontal and the box of turf at 20°.
4 The V-shaped opening should be at the lower end of the box and directly below, a container should be placed to catch water.
5 Obtain a tin can, and pierce some holes in the bottom. Hold the can over one of the boxes and pour in a measured amount of water. Move the can around sprinkling water evenly over the soil. Repeat for the other two boxes, using equal amounts of water. Collect

the water that runs off in the containers placed under the boxes.

6 Examine the water in the containers and notice its appearance. Stir the water vigorously and pour each one in turn into a measuring cylinder.

7 Allow the soil to settle and measure the volume of the soil and the volume of the water.

8 Enter your results in the form of a table, as shown below.

Condition	Volume of soil	Volume of water
Box of soil inclined at 20°		
Box of soil inclined at 40°		
Box of soil with grass turf		

Questions

1 Which box lost most soil and water? Explain your findings.

2 Which box lost least soil and water? Explain your findings.

3 What name is given to this loss of soil from the surface of the earth?

When natural vegetation covers the soil, heavy rain is deflected by leaves and thus drops lightly onto the soil. The soil is also *bound together* by the root systems and humus, so that it is able to absorb vast quantities of water. When natural vegetation is removed, by the felling of trees and shrubs, by over-cultivating or by overgrazing with sheep or cattle, then the sun's rays fall directly on the soil and dry out the humus. When the rain comes, erosion results.

In tropical regions, rainfall is often heavy over a short period of time with the result that several centimetres of water fall on exposed ground during a single storm. The rain falls faster than it can be absorbed into the ground and so water builds up on the surface of the soil. A very shallow slope results in a *large mass of water* moving downhill and *carrying soil with it*. This type of erosion is called *sheet erosion*.

The practice of growing crops in round heaps or on ridges does not stop erosion. The water flows around and between the ridges, forming rivulets which often have a greater force than a single sheet of water. As a result much more damage is done; the soil is washed away and a steep-sided channel is formed. This type of erosion is called *gully erosion* (see fig 13.7(a)). The storm water has an increasingly powerful action as the gully deepens, through the washing away of subsoil. In the dry season the gully hardens and, since there is no topsoil, new vegetation cannot become established. This process, if continued, can cause large areas of land to become devastated (see fig 13.7). This is not usually a great problem in the United Kingdom, since rainfall is moderate and distributed throughout the year, and the rate of evaporation is low.

Another effect of heavy rainfall is to dissolve out the soluble salts from the soil and wash them away. This process, known as *leaching*, poses particularly difficult problems for hill farmers.

Depletion of salt content also results, of course, from over-cultivation. If crop plants are grown in a field year after year, the salts are used to build plant tissues and are not replaced naturally.

fig 13.7 (a) Gully erosion

(b) Sheet erosion

Any *trampling* of soil which *destroys the plant cover,* provides a ready made situation for gully erosion to occur. Footpaths, animal tracks and dirt roads become the starting point of erosion, especially where they wind up a hill or mountain. This is becoming a particularily severe problem in some of our areas of greatest natural beauty, such as the Pennine Way and Snowdonia, where thousands of tourists tread the same paths every year.

Any vegetation removed from slopes must be replaced as quickly as possible. Logging operations removing timber should be quickly followed by a reafforestation programme. The growing of crops in hilly country is often carried out on terraces, or with planting along strips that follow the contour lines and that alternate with strips of grass or 'cover crops'.

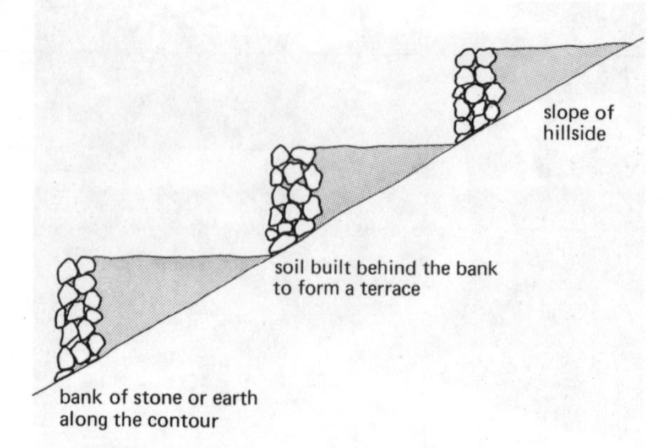

fig 13.8 A diagram of a section through a terraced hillside

fig 13.9 Strip cropping

fig 13.10 A shelter belt of trees

Another type of erosion occurs in flat areas of country, where the topsoil has been impoverished by overcropping or overgrazing. The humus no longer binds the soil, and in arid conditions, the wind blows the topsoil away as dust. The Harmattan wind blowing from the Sahara in West Africa carries large quantities of soil and dust. The United States of America has also experienced this *wind erosion* which has resulted in the formation of dust bowls. Again the problem is relatively mild in Britain, but some wind erosion of light, sandy soils and of East Anglian 'black fen' does occur.

13.21 Conserving and renewing soil fertility

Rotation of crops

The method of crop rotation allows the maximum possible use of land because fields are kept under constant cultivation. The method needs careful planning, bearing in mind the following points.

1 The length of time through which the crops may be rotated is variable (1–5 years).

2 Crops included in the rotation:

i) make different demands on the soil, and

ii) have varying root depths to draw on different levels for mineral salts.

3 Rotation does not increase fertility and therefore regular addition of animal manure, green manure, compost or artificial fertilisers must be made.

4 Leguminous crops should be included since they make little demand on the soil nitrogen and can also be ploughed in as a green manure.

5 An adequate cover of vegetation maintained at all times helps to prevent excessive leaching and soil erosion.

6 Crop rotation prevents the establishment of a crop disease or insect pest, which might flourish on a single crop repeated year after year.

Selection of crops should include one or two *main crops* such as wheat or barley, alternating with legumes and *root* crops such as beet. *Deep-rooted* crops should alternate with *shallow-rooted* crops. Deep-rooted crops help to increase soil depth by breaking up the subsoil.

Continuous crop production in a crop rotational cycle is only possible with adequate manuring replacing lost minerals, salts and humus. If this is not possible, then, for some period of the cycle, the ground must be allowed to recover, that is, lie *fallow*. This can be achieved in various ways, by putting down to grass and feeding cattle or by growing a short-term legume which can be ploughed in as a green manure. In the former example the dung and urine of the animals replace the nitrogenous matter.

Reafforestation or shelter belts

The accelerating effect of wind over flat areas of country can be stopped by planting shelter belts of trees, between farms or between large areas of cultivation. These trees reduce the wind speed and thus prevent erosion of topsoil, but in addition, they provide shade for cattle.

Renewal of humus

The function of humus in binding soil is most important in preventing erosion, and furthermore, its breakdown in the soil provides valuable mineral salts. It is essential, therefore, to maintain the humus fraction of the soil. This is usually achieved by one of the following two methods.

a) *Manure.* Animal manure is seldom available, but another source is human manure. Sludge from sewage farms can also be used as manure, but with this source, great care must be taken that careless disposal does not spread intestinal infections.

b) *Compost.* All waste plant and animal materials can be collected into a heap and allowed to rot down as compost. There is a large variety of these materials which can be used e.g. weeds, kitchen waste, feathers, animals intestines, urine and plant stalks. The heap should be prepared in layers with some shallow layers of soil added at intervals. Moisture and air are essential in the first stages of decomposition, and as a result, high temperatures are produced within the heap. This first stage of decay, induced by bacteria is followed by fungal decay as the temperature falls. The heap can be turned over at this stage to let air in for a further bacterial stage which also needs to be moist. The final product should be dug into the soil and not left exposed to the sun so that it dries out.

Artificial fertilisers

Adding salts directly to the soil leads to increased growth and a bigger yield of crops. Salts added in this way are called artificial fertilisers and are produced in factories or formed as a waste product of certain industrial processes. The most common elements lacking in highly cultivated soil are nitrogen, phosphorus and potassium, known by their symbols NPK. They are supplied in the form of their salts: potassium sulphate; ammonium sulphate and calcium phosphate. Great care should be taken to *avoid continual application* of artificial fertilisers, for they do not replace humus, and this alone can bind the soil together and prevent erosion. Over-use of fertilisers may also result in excessive nitrate concentrations in the ponds and lakes that receive run-off water from the fields. In some instances this may have an adverse effect on drinking-water supplies.

Questions requiring an extended essay-type answer

1 a) How would you make an analysis of soil to determine the water, humus and mineral content?
  b) What is the importance of air in soil? How could you describe the percentage of air in a given soil?

2 a) Sand and clay are important constituents of many soils. What are their properties? How do the proportions in which they are present affect soils?
  b) What other components are essential in a fertile garden soil? Describe their importance.

3 a) For each of the following constituents of a fertile soil, describe their importance for the growth of a rooted plant
    i) soil air,
    ii) soil water,
    iii) clay particles, and
    iv) micro-organisms.
  b) How is the fertility of soil maintained in a natural habitat?

4 a) Describe an experiment to determine whether a soil sample contains bacteria.
  b) What other micro-organisms are present in soil? How would you recognise these under the microscope?
  c) How would you analyse a soil sample to determine its degree of acidity or alkalinity? Why is such information important to the farmer?

14 Organisms and their environment – Ecology

14.00 Introduction

Ecology is the study of plants and animals in relation to their surroundings. Plants and animals depend upon each other. This has already been discovered to some extent, when the topic of Nutrition was studied in Chapters 1 and 3.

During these studies of nutrition, it was found that green plants manufacture their own food materials in the process of photosynthesis; they are, therefore, termed *producers*. Animals are dependent upon plants for their own food, and for this reason they are called *consumers*. Photosynthesis enables green plants to utilise solar energy from the sun, and thus there is a flow of energy from the sun, to green plants, animals and eventually to decomposers.

It is now possible to see the *interdependence* that exists between green plants and animals. In Chapter 3, it was discovered that interdependence exists in other ways; saprophytes, parasites and symbionts all have modes of life which show a high degree of interdependence between themselves and other organisms in their surroundings. From the studies of the nitrogen and carbon cycles, it will have been realised that these cycles are based on interdependence between living organisms and their physical surroundings.

14.10 Habitat, population, community, ecosystem

The surroundings in which animals and plants live is their *environment*. A *habitat* is a particular area of the environment, and some examples of different types of habitat are shown in fig 14.2. Any one habitat may have a large number of different plant and animal species living in it. If one species is more numerous than the others it is then said to be the *dominant* species. The term given to all the members of a *single species* of plant or animal, living in a particular habitat, is a *population*. Seabirds living on a cliff is an example of a population.

Populations tend to change in size over a period of time. Such a change is called a *fluctuation*. In some years, the populations of migratory locusts are small, and little damage is done to agricultural crops. However, in some years, the populations are very large and enormous swarms of locusts destroy thousands of hectares of agricultural land. The reasons for these fluctuations are not always known, and much research is being carried out to determine the causes.

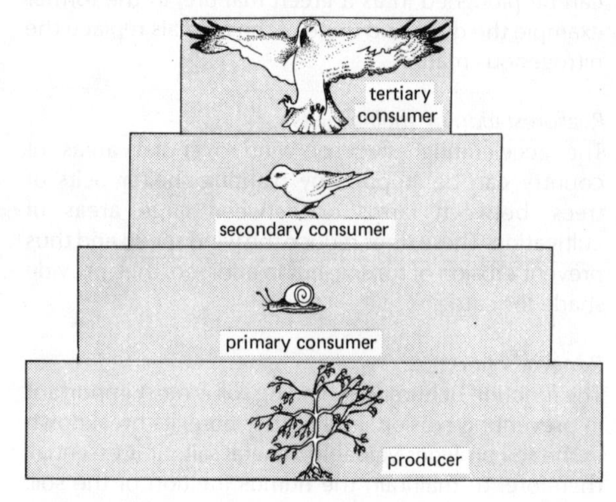

fig 14.1 The energy flow between the sun, producers, and consumers

All plants and animals living in a particular habitat, make up a *community*. A community consists of populations of *different* species, living together and showing varying degrees of dependence on each other.

An *ecosystem* is a 'natural' unit consisting of *all* the plants, animals and micro-organisms in an area functioning together with the non-living (physical) factors of the environment.

14.20 Distribution

If a part of the environment is examined, it will be discovered that populations of plants and animals are not distributed evenly, but rather their *distribution* shows a certain pattern.

Investigation

What is the world distribution of lungfish?

Procedure

Look at the map (fig 14.3). It shows the world

(a) Freshwater

(b) Temperate forest

(c) Savanna

(d) Desert

fig 14.2 Different types of habitat

distribution of different forms of lungfish.

Questions

1 How many types of lungfish are considered in this distribution study?

2 Are the types distributed evenly throughout the world?

3 What can be said about the distribution of lungfish in general?

4 What can be said about the distribution of the different types of lungfish in particular?

The distribution of animals and plants displays a world wide pattern largely determined by *climate*. Thus, there are tropical rain forests, deciduous forests in temperate regions, tropical grasslands and temperate grasslands etc and all of these have their characteristic animals. Within these large areas of the world, where the type of organism is determined by climate, smaller areas will be conditioned by other factors such as the *soil type* and the *amount of rainfall*.

Distribution has been considered in terms of a world-wide pattern. Let us now narrow this down to a smaller area of land.

Investigation

Is there a relationship between the distribution of the Golden Eagle and mountains of Europe?

Procedure

Look at the maps fig 14.4

Questions

5 Do the two distribution patterns display a similarity?

6 What can be deduced by comparing the two distribution patterns?

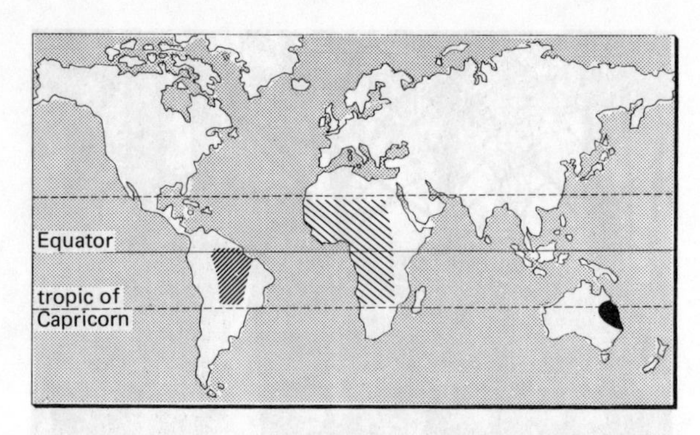

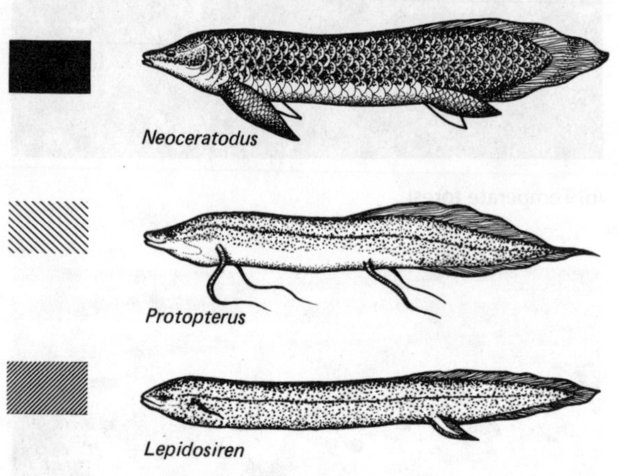

fig 14.3 A world distribution of some different types of lungfish

7 Is it helpful, when considering the reasons for the distribution of one particular species, to know something about its habitat?
8 Name another example already studied, where a comparison between two different distribution patterns led to a greater understanding of the problem being investigated.

The examples discussed so far, have been concerned with distribution on a large scale. However, the same studies can be made on a much smaller scale. Later in this chapter, exercises will be carried out which will allow an investigation to be made of the distribution patterns in a much smaller area within or near the school grounds.

14.21 Factors affecting distribution

There are a number of factors which affect the distribution of plants and animals. These can be grouped into two types; *abiotic* (physical) factors and *biotic* factors.

Abiotic factors
These are factors such as temperature, rainfall, light intensity, humidity, oxygen concentration, wind direction and velocity, salinity, topography, pH, carbon dioxide concentrations and concentration of pollutants, all of which have a significant bearing on plant and animal distributions. Temperature, rainfall and light intensity are all part of the overall climatic effect of vegetation and its accompanying animals. On a much smaller scale, however, temperature may determine the distribution of organisms in a small area e.g. animals present in a small patch of stones. Rainfall, in terms of humidity, may determine the presence of small animals under the bark of a tree, while light determines the types of climbing plants or epiphytes present on trees.

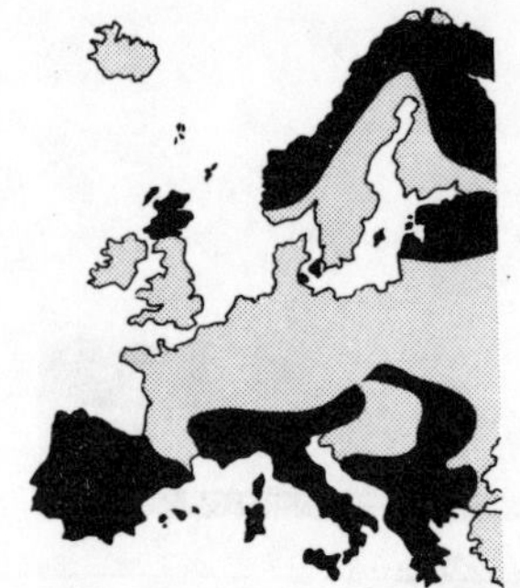

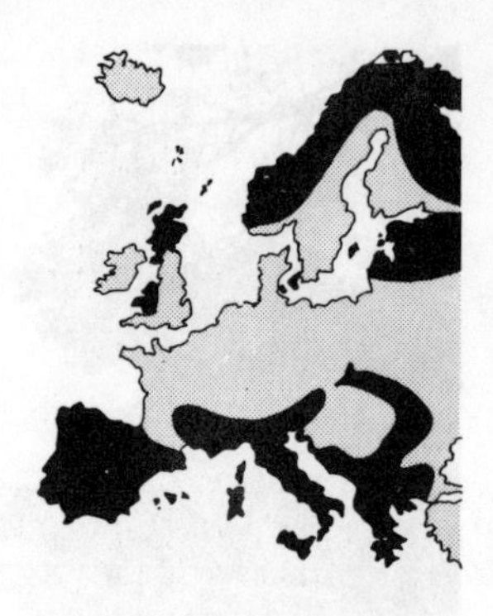

fig 14.4 *Left* Distribution of the Golden Eagle (*Aquila chrysaetos*) in Europe. *Right* Distribution of mountainous regions in Europe

14.22 Temperature

Temperature is measured by thermometers, and there are different types of thermometer for measuring different ranges of temperature. The temperature of a natural habitat does not remain constant for very long, and if a series of temperature measurements are made it is easy to see the fluctuations.

14.23 Rainfall

Rainfall is measured by means of a rain gauge. This has a metal cylinder which is embedded in the ground. A conical funnel is placed in the top of the cylinder, and this funnel serves to collect the falling raindrops, directing them into the collecting cylinder. The contents of the cylinder should be measured every day and the amount of rain recorded. The apparatus is then set up again in order to catch any rain that may fall over the next 24-hour period.

Ideally, recordings should be made every day, and records for the whole year should be compiled in the form of a bar chart (see Introductory Chapter). It may be that the school operates its own rain gauge. If not, a rain gauge can be easily made (fig 14.5).

Investigation

What is the rainfall over a given period in a certain habitat?

Procedure

1 Make a series of recordings (once a day) of the rainfall in the school grounds. It is best if the results are recorded over a period of a year, but it may be more convenient to carry out this activity over a shorter period of time. It will be necessary to decide the best position in which to place the rain gauge, and it will probably be necessary to discuss this with other people, before a decision is made. Use the results to draw a bar chart for the time period measured.

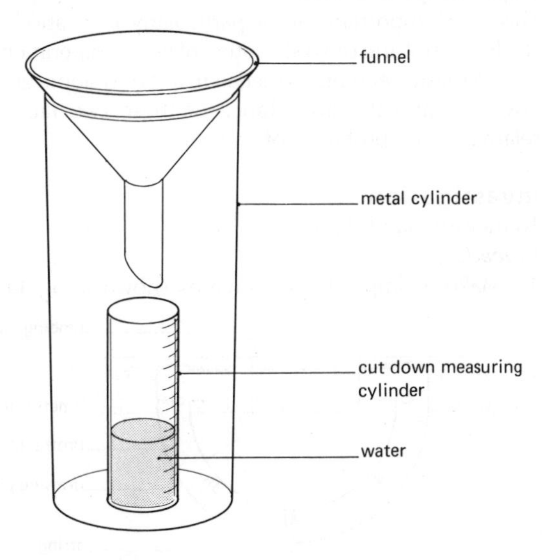

fig 14.5 A diagramatic representation of a rain gauge

14.24 Light

In order to measure light, it is necessary to use fairly expensive and complex equipment. If it is possible to obtain a light meter such as the type associated with a camera, this can be used to detect different light intensities. It has already been discovered how some organisms respond to changes in light intensity. In Chapter 7, (Response), experiments were carried out with blow fly larvae to demonstrate their preference for conditions of very low light intensity. In Chapter 10, (Growth), it was shown that plants react to differing light intensities in different ways.

The flowers of Morning Glory (*Ipomoea*) open soon after dawn in most tropical countries. This is probably in response to increasing light intensity. The flowers remain fully open for only a short time, and by mid-day, most of them have already started to close up and die.

14.25 Humidity

Relative humidity is a measure of the amount of water vapour in the atmosphere. Many organisms distribute themselves in relation to the relative humidity. In Chapter 7, experiments were carried out with blow fly larvae to find out how they distributed themselves in relation to different humidities. Humidity is an important abiotic factor which greatly influences the distribution of organisms, particularly small invertebrates and higher plants.

14.26 Other factors

Other abiotic factors such as oxygen concentration, degree of salinity (salt concentration) and pH (acidity or alkalinity) can also be measured, and the last of these factors, pH, has been considered in the work on soil. Measurements of oxygen concentration and salinity are complex exercises and it is not possible to consider them here. However, it must be stressed that they are extremely important factors affecting the distribution of many organisms.

14.30 Special measurements

14.31 Wind

The direction and strength of the winds are two important factors affecting many habitats. In particular, plants that produce wind-dispersed fruits and seeds are dependent on the prevailing wind for effective dispersal of their future offspring. Many flying animals, particularly the smaller insects, are influenced by velocity and direction of prevailing winds. In addition, it has already been discovered that increase in movement of air currents causes an increase in the rate of evaporation of water. This may be important in removing surface waters such as small puddles, thus eliminating possible water supplies for other organisms. Transpiration is also speeded up by an increase in conditions of wind velocity and thus, many plants are directly affected in their water relations by changes in wind velocity.

Investigation

To make a simple directional wind gauge

Procedure

Make a simple wind gauge as shown in fig 14.6(a). This can be used to determine the direction in which the wind blows, relative to the school grounds. In order to do this, it will be necessary to place the wind gauge in a reasonably elevated position such as an easily accessible roof top, or on top of a long pole. It will be necessary to establish the main points of the compass relative to the position of the wind gauge when the latter instrument is positioned. Take several readings every day (morning, noon and evening) and record the results over a period of one year. It will then be possible to determine if the prevailing wind always blows from the same direction, or whether its direction changes at different times of the year.

In addition, it is possible to make a simple piece of apparatus in order to estimate the velocity of the wind; the method of construction can be seen in fig 14.6(b). It

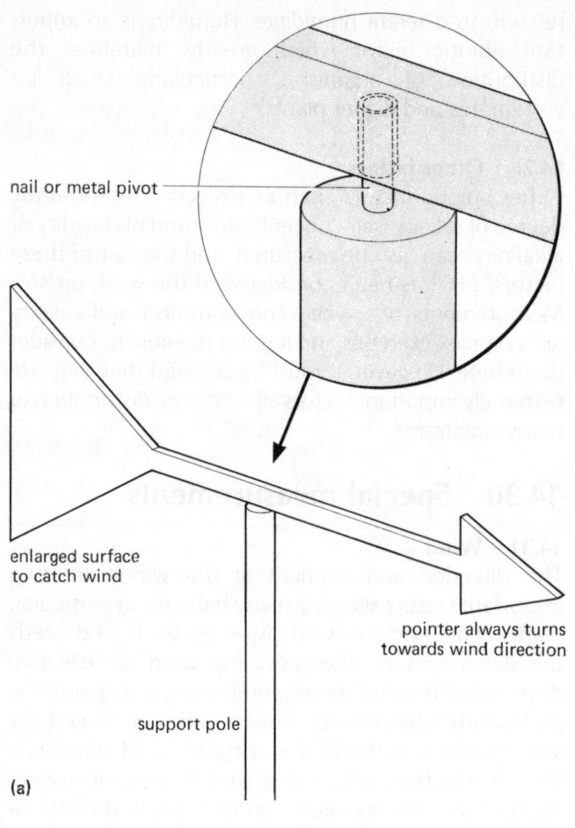

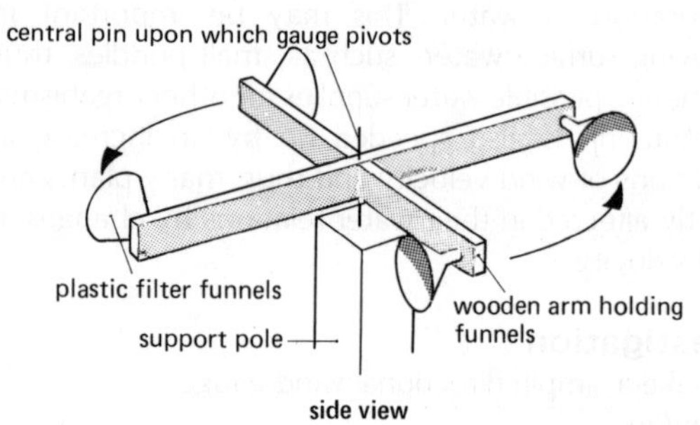

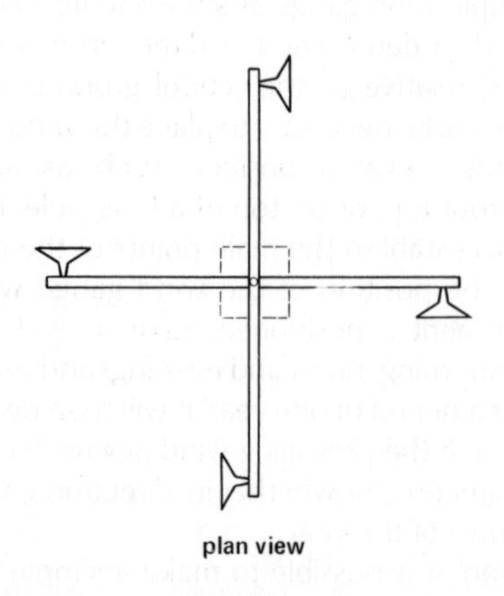

fig 14.6 Simple wind gauges to detect (a) wind direction and (b) velocity

is probably helpful to paint one arm of this apparatus a bright colour in order to help in counting the number of revolutions per unit time. It will also be necessary to work out some kind of scale so that the number of revolutions per unit time indicates a specific wind speed. In order to keep the results simple, it is best to have four major categories of wind; no wind (zero velocity), low velocity, medium velocity and high velocity.

14.32 Slope of the land

This is an important factor particularly in relation to studies of soil and the distribution of water. Experiments have already been performed in relation to slope, grass coverage and the importance of these two factors, relative to the problems of erosion.

Investigation

To measure land slope.

Procedure

1 Make a simple slope gauge as shown in fig 14.7.

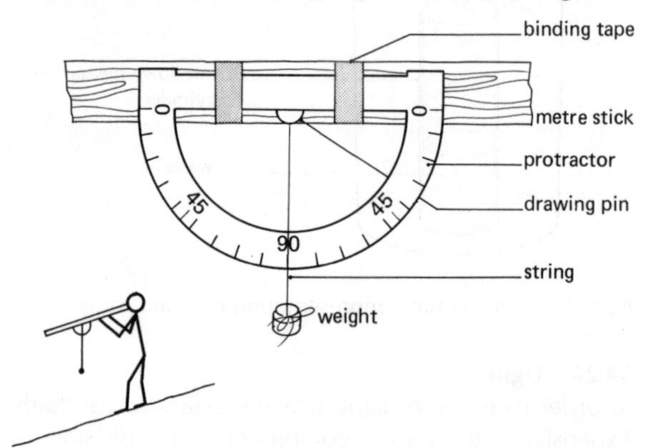

fig 14.7 A simple slope gauge, and the method of operation

2 Find some ground, in or near the school grounds which has a definite slope to it.

3 Use the slope gauge to measure the degree of slope. Practice this measurement technique a number of times on different slopes.

14.33 Height

Sometimes, ecologists are required to measure vertical heights. This can be done by using a very simple technique as shown in fig 14.8(a).

Investigation

Can the height of a tree be measured? It will be necessary to work with a classmate. A 2 metre pole and a long piece of string which has knots at 1 metre intervals tied along its length are necessary for this activity.

1 Ask the classmate to hold one end of the string at the base of the tree selected.

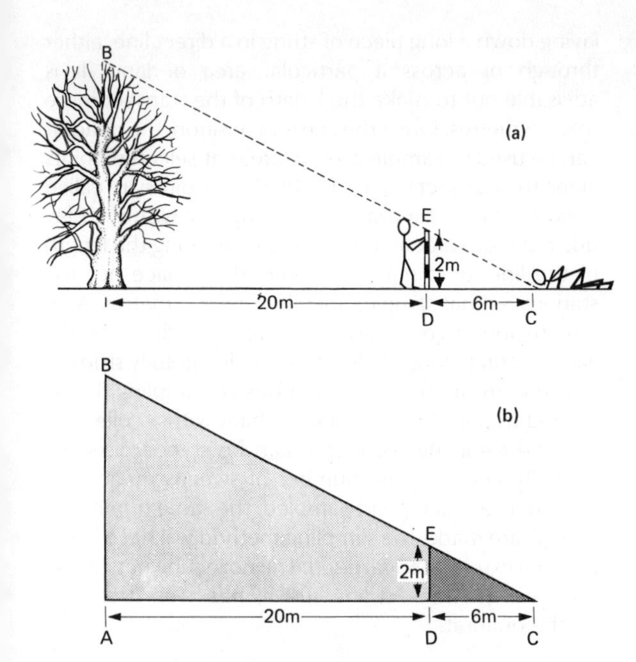

fig 14.8 (a) A diagram showing the method used to estimate the height of a tree (b) A principle of similar triangles

2 Measure a straight line distance of about 20 metres from the base of the tree, outwards. It may be necessary to measure a greater distance than this, depending on the relative height of the tree.
3 Mark the point measured, and record the measurement of the horizontal length from the tree base.
4 Ask the classmate to stand the 2 metre pole vertically at the marked point.
5 Walk back from the pole, in a straight line, as though it were a continuation of the horizontal line from the base of the tree. Kneel down and place the head close to the ground so that the eyes are as close to the ground as possible.
6 Look, from this position close to the ground, at the top of the 2 metre pole and, by moving the position of the body forwards or backwards, (keeping the eyes at ground level), line up the top of the 2 metre pole as closely as possible, with the top of the tree. When this has been done, record exactly the position of the eyes relative to the ground.
7 Measure the distance from the 2 metre pole to where the eyes were at ground level.
8 Now apply the principle of similar triangles as shown in fig 14.8(b) in order to calculate the height of the tree. Follow the example given and then substitute the experimental measurements.

Example

Triangle ABC has one side 26 m long and a height x. The similar triangle DEC has one side 6 m long and a height 2 m. The two known sides of DEC are in the ratio of 6:2 or 3:1, i.e. the vertical side is ⅓ the length of the horizontal side DC.

The triangle ABC has a horizontal side AC of 26 m. It is similar to DEC, and therefore its vertical side AB must also be ⅓ the length of the horizontal side AC. The height x (the height of the tree) is $26/3 = 8.7$ m.

14.40 Biotic factors

So far in this chapter the studies have been concerned mainly with abiotic factors, i.e. the physical aspects of the environment operating within the habitat. However, as stated at the beginning of the chapter, plants and animals are dependent upon each other. It is, therefore, impossible to study the ecology of a particular species or population without considering the effects on that species or population, of the other living forms in the habitat. These other living forms comprise the biotic factors. Biotic factors can be classified under the three headings:

1 effects of other animals (grazers, browsers, insect pests, pollinating insects etc);
2 effects of other plants (shading, climbing, parasites etc);
3 effects of Man (agriculture, grazing, forestry, urbanisation etc).

Although a species can be studied in isolation, a truer picture of its ecology is obtained only when it is examined relative to other living things around it.

A little later in this chapter some specific ecological studies will be carried out and the effect of other plants and animals on a particular species will become more obvious. However, before beginning these ecological surveys, it is necessary to introduce the various methods of investigation which may be required and also to demonstrate the various pieces of equipment available and useful in these studies.

14.50 Methods of investigation

The method adopted depends on the type of investigation to be carried out. A simple exercise might consist of collecting the different types of plants and animals living in a particular habitat, identifying them and counting their relative numbers. The exercise could be taken a stage further by recording the position in the habitat from which the different organisms were collected. Even in this simple type of investigation, a number of necessary steps exist. These are:

1 choosing the habitat or part of the habitat to be studied;
2 deciding how best to sample the habitat to be studied;
3 collecting and counting the different organisms found there;
4 identifying the different organisms collected;

5 measuring important abiotic factors in the habitat;
6 recording the results of the study;
7 drawing conclusions from results obtained.

These seven major steps are carried out in all ecological studies, and it will be necessary to use them in the three ecological studies to be made in this chapter.

14.51 Choosing the habitat

It is important to remember that a habitat containing too many different species is often more difficult to study, and it is, therefore, better to choose a habitat with only a few different types of organism.

14.52 Sampling

This is particularly important. Usually, the area to be studied is too big to enable all the organisms living there to be studied. Therefore, only certain areas or positions can be investigated. This involves the technique of *sampling*. This is usually carried out by means of a *quadrat* (see fig 14.9). This is a square, frequently made of wood, which may be further subdivided into smaller squares by lengths of string or wire. The standard quadrat is 1 m², but smaller quadrats can be used if only a small area is being studied.

The quadrat can be thrown at random in the habitat area, and the place where it lands, within the boundaries of its sides, is carefully studied. This normally involves collecting all the living organisms from within this area and counting and identifying them. Several throws of the quadrat can be made and the same procedure adopted each time.

A more precise use of the quadrat is to use it in conjunction with a *line transect*. A line transect involves laying down a long piece of string in a direct line, either through or across a particular area of land. It is advisable not to make the length of the transect more than 25 metres. Once the line is in position, the quadrat can be used to sample specific areas at selected points along the transect (see fig 14.10). These points are called *stations*. It may be that the investigator wishes to lay down the quadrat at 1 metre intervals along the length of the line, or it may be preferred to space out the stations, so that samples are taken every 3 metres. As in the previous investigation, the habitat enclosed within each quadrat along the line transect is carefully studied.

There are many other methods of sampling. If it is desired to sample a particular habitat with a collecting net, the sampling technique can be standardised by carefully counting the number of sweeps made with the net. If, at each point sampled, the same number of sweeps are made, the sampling method will have been, to some extent, standardised. The collecting net can be used along a transect line, just as has been described for the quadrat.

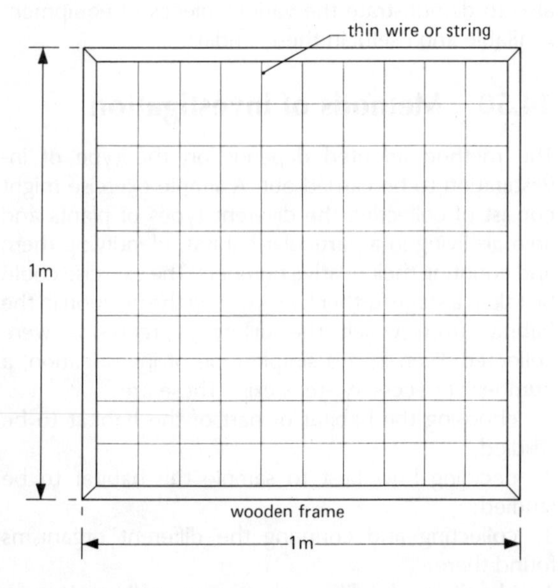

fig 14.9 A standard one metre quadrat

14.53 Methods of collecting

The methods used depend upon the type of habitat being investigated. Generally speaking, collecting requires something to catch the organisms with (if they are animals) and something in which to put the captured animals. The various types of equipment used for collecting and securing captured organisms are shown in fig 14.11. It is important that any containers used to put plant and animal specimens in should be clearly labelled. This information will be required when on return to the laboratory the collection is examined.

14.54 Methods of estimating frequency, density and percentage cover

Density can be measured in a number of ways. The most obvious way is to count the number of organisms concerned in a given area. If a quadrat is being used along a line transect, a count may be required of the number of animals of a particular species at different stations. Having done this, an estimate of the number of

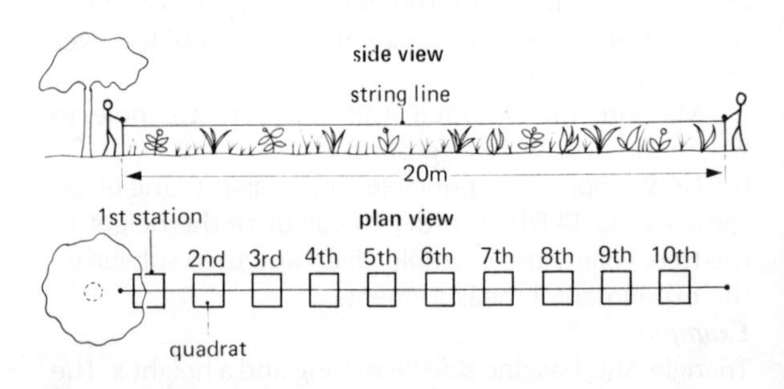

fig 14.10 A line transect showing the positions of the sampling stations.

animals of the species concerned per m² can be made. This provides a measure of density.

If a quadrat 1 m², divided up into smaller squares is used, the number of smaller squares in which the particular species occurred can be counted. This gives a measurement of the *frequency of* the *distribution* of the organism.

When investigating plants, it is often difficult to estimate the number of individuals of a particular species. It is easier to estimate the *percentage cover*, – the area of ground covered by the particular species being investigated.

Investigation

The technique of using the metre quadrat.

Procedure

Examine fig 14.12. It is a diagram of a quadrat (divided into smaller squares) placed in position in a habitat. Five different species of plant are found within the quadrat, and the amount of ground cover they give is indicated by the various forms of shading.

Questions

1 What is the dominant species?

2 Estimate the percentage cover of each of the five species.

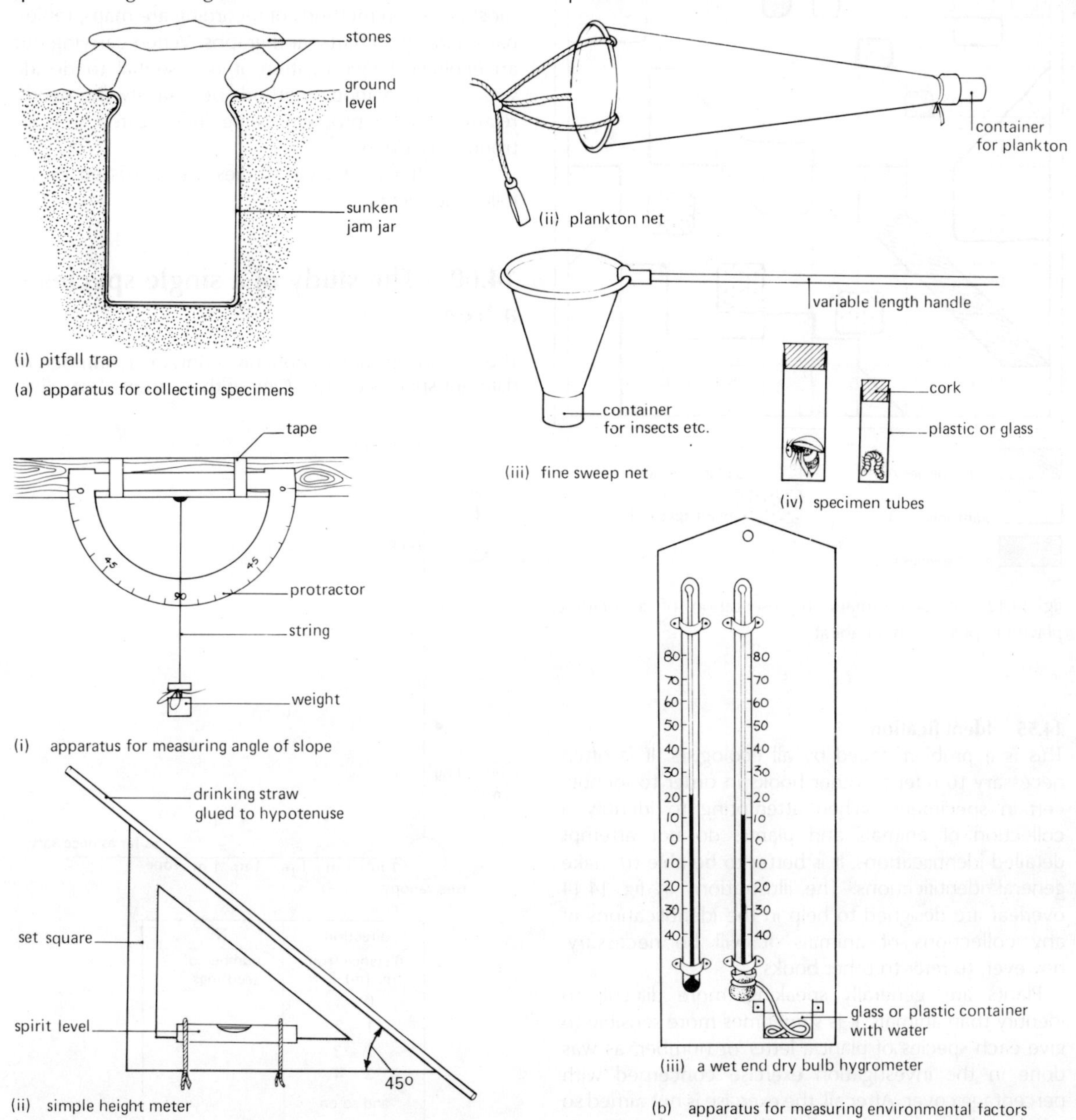

fig 14.11 The different types of apparatus and equipment used in carrying out an ecological survey

It will now be realised that there are many methods and techniques which can be used in making simple ecological investigations. It may be necessary to discuss the important points of an investigation with a teacher, before beginning a study.

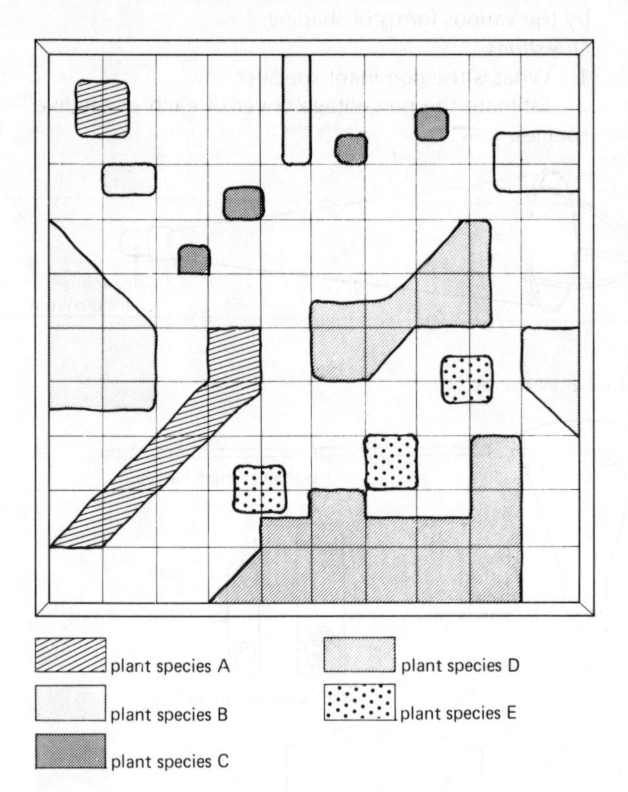

fig 14.12 A diagrammatic representation of a quadrat placed in position in a habitat

14.55 Identification

This is a problem faced by all ecologists. It is often necessary to refer to other books in order to identify certain specimens. When attempting to identify a collection of animals and plants, do not attempt detailed identifications. It is better to be able to make general identifications. The illustrations in fig 14.14 overleaf are designed to help in the identifications of any collections of animals. It will be necessary, however, to refer to other books.

Plants are, generally speaking, more difficult to identify than animals. It is sometimes more sensible to give each species of plant a letter or number, as was done in the investigation exercise concerned with percentage cover. After all, the exercise is not aimed so much at the names of the individual plants rather as in how they live in the habitat being investigated.

14.56 Measuring abiotic factors

Some of these factors have already been studied at the beginning of this chapter. When considering the types of ecological study to be engaged in, the most important abiotic factors to consider are those such as light, wind, rainfall, humidity and structure and pH of the soil. Apart from light, these are all easy to measure, and the materials required to detect and measure their presence have already been discussed.

14.57 Recording results

Results can be recorded in a number of ways. The five most common methods of recording are maps, tables, bar charts, pie charts, and graphs. When carrying out an ecological investigation, it is essential to decide which of the various methods best satisfy the needs, relative to the problem being investigated (see Introductory Chapter).

Two major ecological studies are discussed in the following section.

14.60 The study of a single species – a tree

The school grounds probably contain a number of different species of tree. One of these trees, if carefully

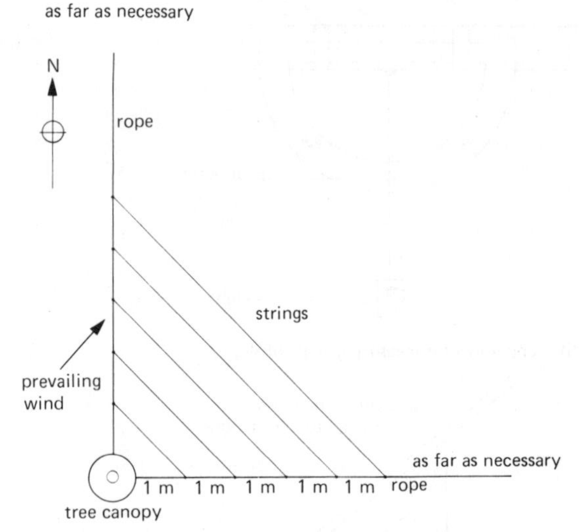

direction	NE
distance from tree (m)	number of seedlings
0 – 1	
1 – 2	
2 – 3	
3 – 4	
and so on	

fig 14.13 Dispersal of fruits and seeds from a tree

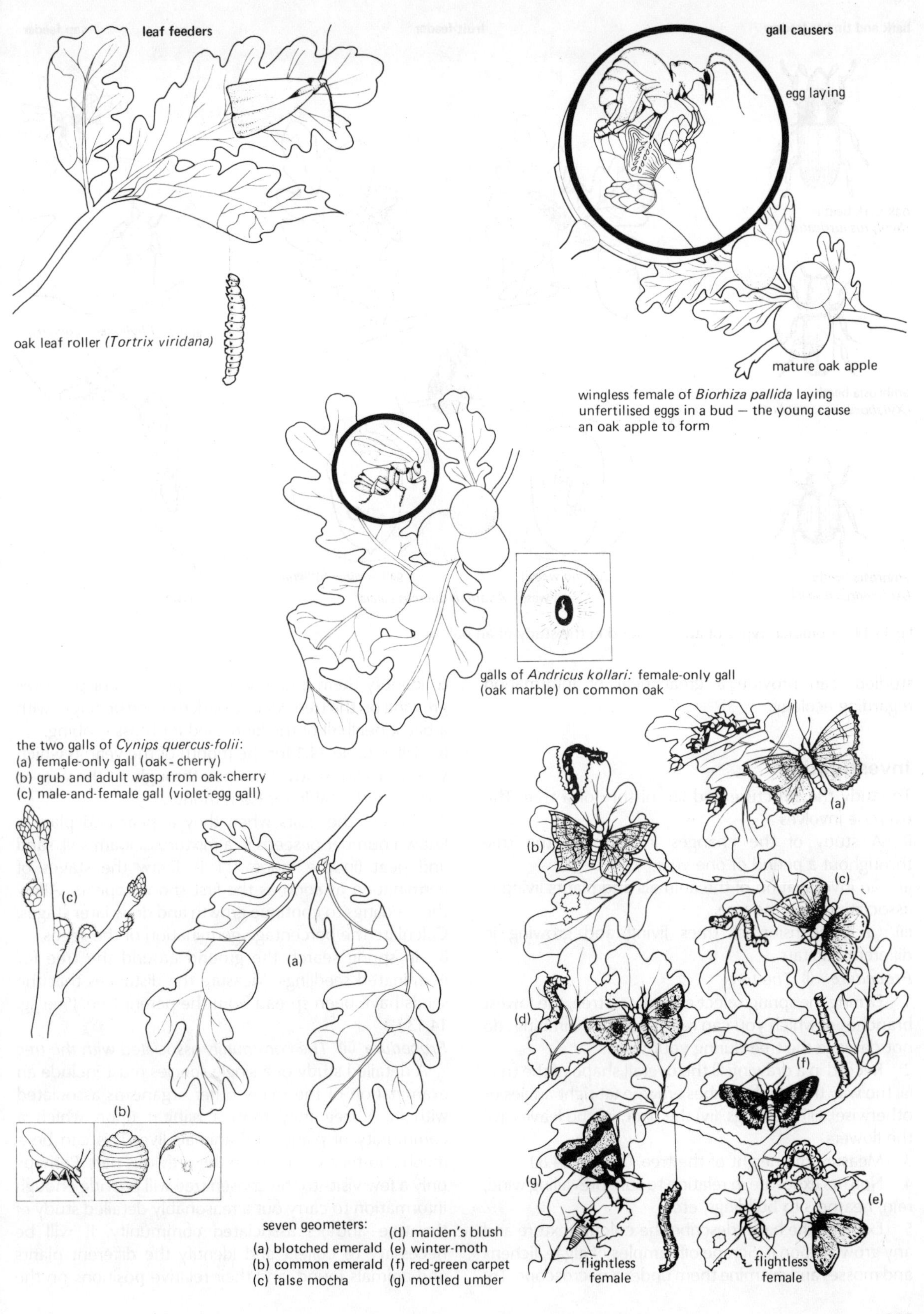

oak leaf roller *(Tortrix viridana)*

wingless female of *Biorhiza pallida* laying unfertilised eggs in a bud – the young cause an oak apple to form

galls of *Andricus kollari:* female-only gall (oak marble) on common oak

the two galls of *Cynips quercus-folii*:
(a) female-only gall (oak-cherry)
(b) grub and adult wasp from oak-cherry
(c) male-and-female gall (violet-egg gall)

seven geometers:
(a) blotched emerald
(b) common emerald
(c) false mocha
(d) maiden's blush
(e) winter-moth
(f) red-green carpet
(g) mottled umber

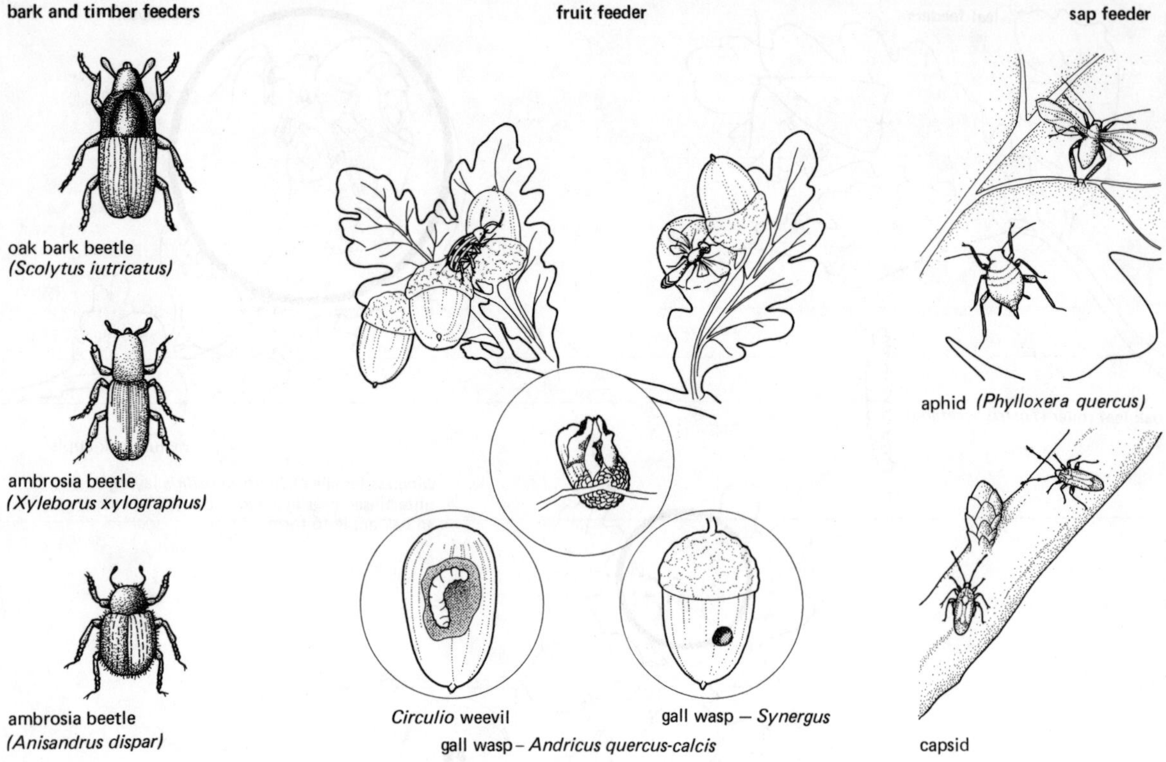

bark and timber feeders

fruit feeder

sap feeder

oak bark beetle
(Scolytus iutricatus)

ambrosia beetle
(Xyleborus xylographus)

ambrosia beetle
(Anisandrus dispar)

aphid *(Phylloxera quercus)*

Circulio weevil

gall wasp – *Synergus*

gall wasp – *Andricus quercus-calcis*

capsid

fig 14.14 Common types of animal found in the study of an oak tree

studied, can provide a great deal of information regarding ecology.

Investigation

To study the structure and life history of a tree. This exercise involves

i) A study of the changes in form of the tree throughout a period of one year,

ii) an investigation of the animals and plants living in association with the tree,

iii) a comparision of trees living and growing in different habitats.

Procedure (i) *The tree*

1 During the spring select an isolated tree, the lowest branches of which you can reach without climbing; do not damage the tree during your work.

2 Record in a drawing (i) the overall shape of the tree, (ii) the way that the branches emerge (at right angles or otherwise), (iii) the twigs, (iv) the bark, (v) the leaves, (vi) the flowers.

3 Measure the height of the tree. (See fig 14.8.)

4 Note its exposure in relation to sun, prevailing wind, rain, nearness to buildings, etc.

5 Examine the bark; describe the colour, texture and any growth upon it. Scrape off samples of algae, lichens and mosses and examine them under a microscope. Try to identify them and any other epiphytes or parasites growing on the tree. Make a bark rubbing on paper with a black heelball of the kind used for brass rubbing.

6 Fill in table 14.1 for the month of the year in which you start your study. Continue throughout the year to complete the table for each month.

7 Collect the fruits when they appear and plant a known number of seeds in a mixture of loam soil, sand and peat (in the ratios 2:1:1). Draw the stages of germination as soon as the first shoot appears. Allow the seedlings to continue growth and draw later stages. Calculate the percentage germination of the seeds.

8 In spring search the ground around the tree for germinated seedlings. Measure the distances that the seeds have been spread from the parent tree. (See fig. 14.13.)

Procedure (ii) *The community associated with the tree*

A detailed study of a single species must include an examination of the many other organisms associated with it. A tree may form a habitat upon which a community of plants and animals live. This can be a much shorter exercise than the previous part. Perhaps only a few visits to the chosen tree will provide enough information to carry out a reasonably detailed study of the tree and its associated community. It will be necessary to collect and identify the different plants and animals found, and their relative positions on the

Structure	Observation	Months of the year											
		J	F	M	A	M	J	J	A	S	O	N	D
Leaves	No leaves present												
	Many young leaves												
	Leaves change colour												
	Leaves fall												
Shoots	Many new shoots												
	Few new shoots												
	No new shoots												
Flowers	No flowers												
	Flower buds												
	Few open flowers												
	Many open flowers												
Fruits and Seeds	No fruits												
	Young fruits												
	Ripe fruits												
	Old fruits fallen												
	Seed dispersal												
Climatic Conditions	Temperature												
	Rainfall												
	Humidity												

Table 14.1 Study of a single species

tree must be recorded.
1 Collect any animals from the following sites: (i) the litter at the base of the tree, (ii) holes and hollows on the trunk of the tree, (iii) under loose bark, (iv) in the forks of branches.
2 Draw a distribution map for the types of organism found at the different sites listed in 1 above.
3 Collect any animals on the tree at the following sites: (i) on or in the buds, (ii) on or in the leaves, (iii) on or in the flowers.
4 Collect samples of any plants growing on the tree. These may be: (i) algae, lichens and mosses growing on the bark, (ii) large epiphytes such as ferns growing on the trunk or branches, (iii) occasional higher plants, including parasites, growing in the forks of branches.
5 For sections 1, 3 and 4 above complete a table similar to table 14.2. See fig 14.14 for some of the common organisms found on oak trees.

Procedure *(iii) Comparison of trees living in different habitats.*

1 Locate a group of trees of the same species as the one studied. Examine these to discover whether any differences from your previous observations can be attributed to their living in close proximity.
2 If possible find other specimens of this species growing on different soils. Examine these also to discover whether this new factor has any obvious effect on the tree and its inhabitants.

This study will give an opportunity of investigating some of the abiotic factors that might affect the distribution of the members of the community. It might be interesting to measure the humidity of different parts of the tree. Holes in the bark may be more humid and cooler than the surface bark directly exposed to the atmosphere. It is certainly true that the regions underneath the bark and in holes in the trunk will have a much lower light intensity compared to the more

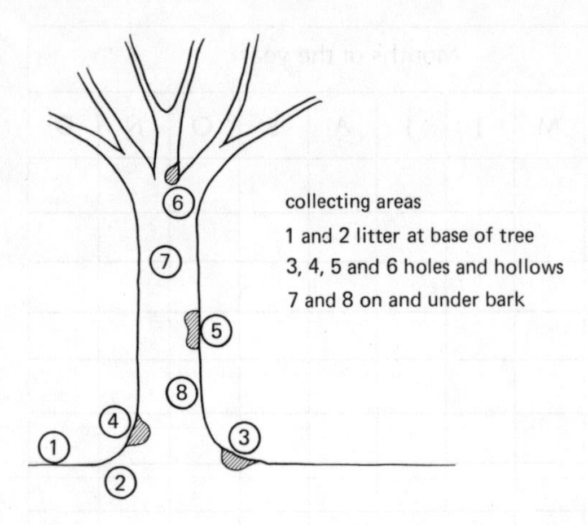

fig 14.15 A map of a tree showing areas for sample collection

exposed areas. There are many abiotic factors that might be considered in this work, but probably those most relevant are the factors of temperature, humidity and light. When looking for plants and animals, do not destroy the areas in which they are found. If a stone or a detached branch at the base of the tree, are turned over, always return them to their normal positions when the search has been completed.

In addition to drawing a distribution map of the different organisms found on the tree, a summary table of the results should be drawn up (see table 14.2).

Organisms collected		
Number on container	Position collected	Name of plants or animals

Table 14.2 Information for distribution map

14.62 The study of a population of earthworms

Earthworms are common to varying degrees throughout the British Isles. Choose a small wood or a piece of fairly open grassland which also contains some trees. Make a line transect (about 30 m in length) so that the line stretches from open grassland to a piece of land under the cover of the trees. Using a 1 m^2 quadrat, sample at frequent intervals along the transect. The worms may be brought to the surface by spraying on the soil a .55% solution of formalin from a watering can. This is made up by mixing 25 cm^3 of 40% formalin in 4.5 dm^3 (4 500 cm^2) of water, and this amount should be sprayed on ⅓ of a 1 m^2 quadrat. Repeat for the whole transect. All the earthworms that rise to the surface should be collected and the population can be expressed in terms both of numbers (no/m^2) and of weight (g/m^2) (biomass). Expression of population size in numbers only may be misleading because of variation in size of individual specimens.

Questions

1 Are the earthworm populations evenly distributed along the transect? If not what is the cause of the uneven distribution?

(Further data that may help the answers to question 1 are analyses of soils from each quadrat. Use the technique detailed in Chapter 13 to find out the amounts of water, humus and mineral matter in the topsoil and subsoil.)

2 Is there any relationship between the earthworm population and the available food supply in terms of vegetation cover and humus content of the soil?

3 Is there any biotic factor that changed along the transect? Perhaps some parts of the transect were drier than others or perhaps some parts received more light than others (this could be affected by plant cover).

14.70 Food chains

From the studies so far, it will have been realised that there is interdependence between species. In Chapters 2 and 3, it was discovered that the various forms of nutrition are closely associated, and thus, generally speaking, animals are dependent upon plants. The situation is very complicated, and within a single community there are many cross links and associations. However, this can be simplified, to some extent, in order to show the important features.

In general terms, green plants are termed *producers*, and animals are termed *consumers*. Thus, in terms of energy, green plants can utilise solar energy in order to make their own food, while animals are dependent upon green plants for obtaining food, and therefore, energy. It has been discovered, already, that no organism lives in total isolation. This is true with regard to *energy flow* within the community.

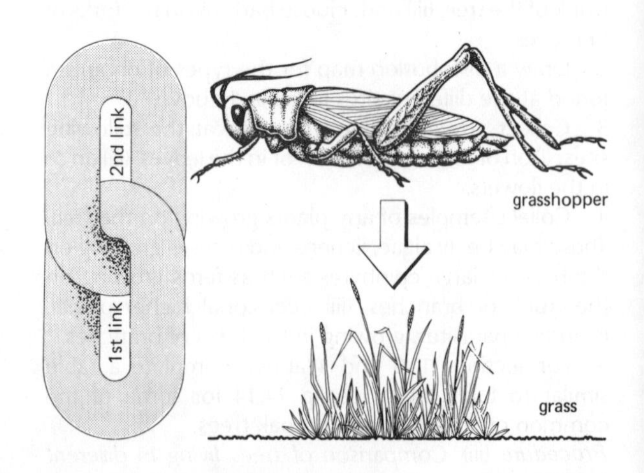

fig 14.16 A simple 2-link food chain

From the study of the tree (section 14.60) it is clear that the leaves produce food for many different organisms. In one case the leaf may be eaten by a caterpillar. This is an example of a two-link food chain involving one plant and one animal similar to the two-link chain shown in fig 14.16. Most chains consist of more than two links, and some contain many links. An example of a three-link chain is shown in fig 14.17. The leaf (producer) forms the first link, the caterpillar (primary consumer) forms the second link and the woodpecker (secondary consumer) forms the third link. Food chains can thus be generalised as: plant → herbivore → carnivore. In larger chains, there are tertiary consumers and even quaternary consumers.

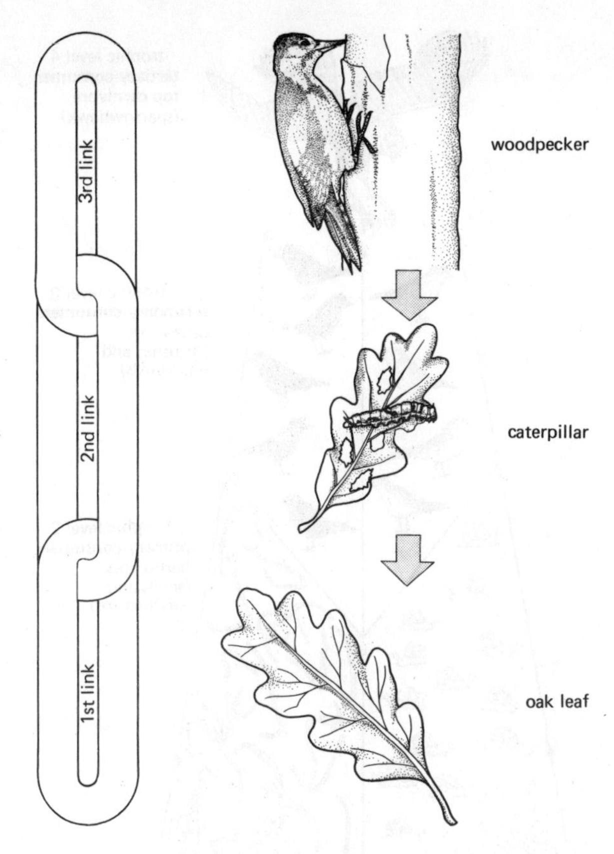

fig 14.17 A simple 3-link food chain

14.71 Food webs

Within a community, associations are such that simple food chains rarely exist. Even an elementary study of a community, such as the one carried out on the tree, indicates that most associations concerned with *energy levels* consist of food webs, rather than food chains. A simple food web is shown in fig 14.18. A food web is thus a number of interconnected food chains with a number of organisms at each feeding level.

14.72 Pyramid of numbers

A careful examination shows that numbers of individuals present in any one link of a food chain is usually greatest as the chain proceeds from the top

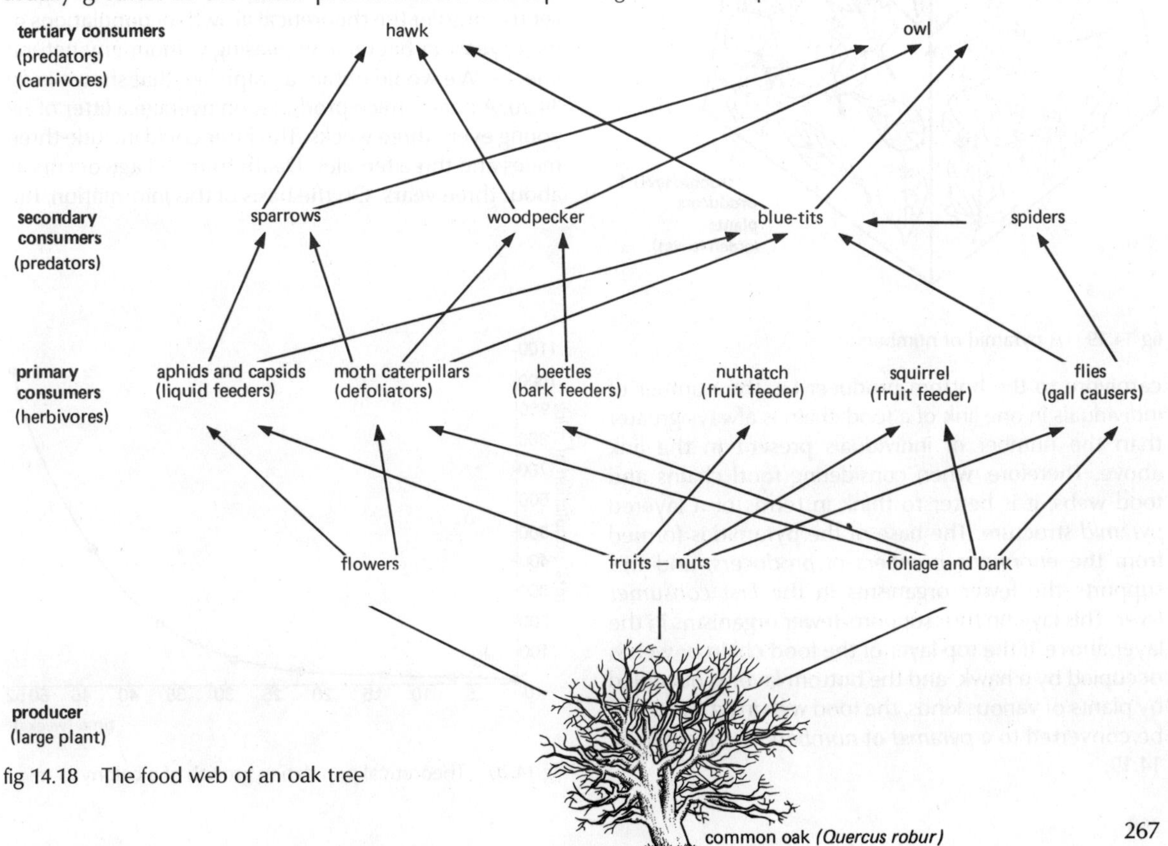

fig 14.18 The food web of an oak tree

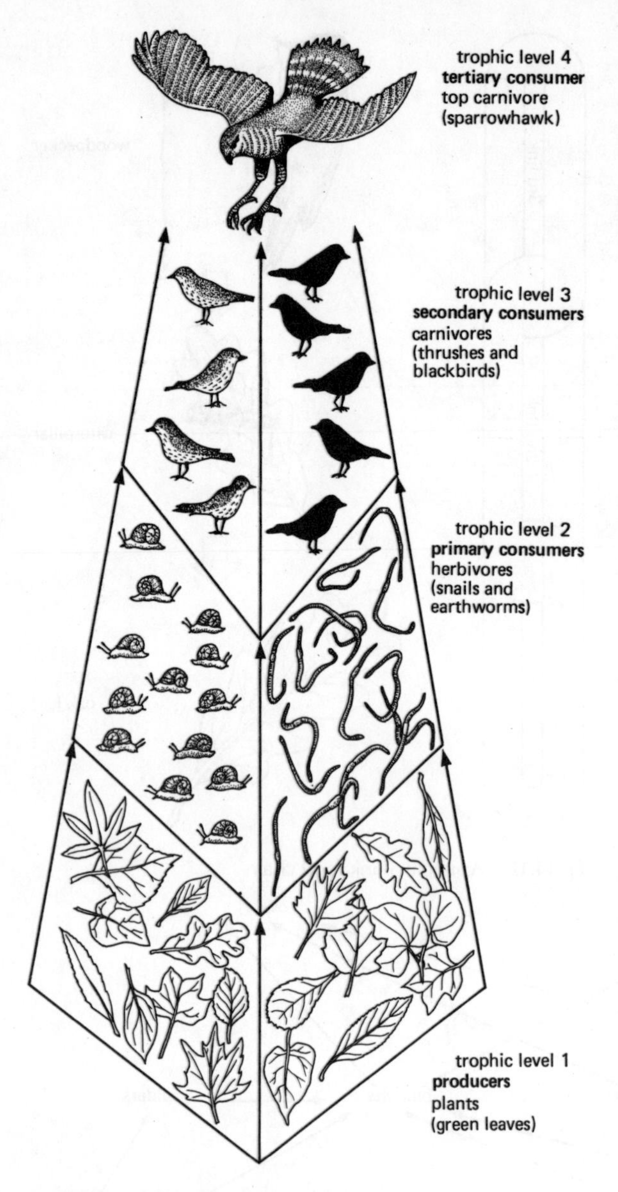

fig 14.19 A pyramid of numbers

carnivore to the bottom producer i.e. the number of individuals in one link of a food chain is always greater than the number of individuals present in the link above. Therefore, when considering food chains and food webs, it is better to think in terms of a layered *pyramid* structure. The base of the pyramid is formed from the *enormous numbers of producers*, and this supports the fewer organisms in the *first consumer layer*. This layer in turn supports fewer organisms in the layer above. If the top layer of the food chain or web is occupied by a hawk, and the bottom layer is occupied by plants of various kinds, the food web in fig 14.18 can be converted to a *pyramid* of *numbers* as shown in fig 14.19.

Two obvious facts can be seen when examining a pyramid of numbers. Firstly, the number of organisms in each layer *decreases* from the bottom to the top of the pyramid. Secondly, in general terms, the physical size of the organisms in each layer *increases* from the base to the top of the pyramid.

14.73 Energy within food chains and food webs

Very little has been discussed in relation to the flow of energy within a food chain or food web. However, it will have been realised that all living organisms are dependent, either directly as in the case of green plants or indirectly as in the case of animals, on solar energy. As energy is transferred up through a food chain, or within a food web, much of it is lost. For example, there is a loss of available energy in Man's agricultural food chains. In a food chain involving the sun, a producer (maize), a first consumer (cattle) and a second consumer (Man), of the incident energy, about ½% in the cow or bullock is available to Man. If man were to eat maize, 15% of the incident energy would be made available for his body functions.

14.80 Population

Animal populations are kept in check by three factors:
1 disease
2 predators
3 limitations of the food supply.

Let us consider the theoretical growth of populations of mice, yeast or bacteria, increasing without any natural checks. We would obtain a graph like that shown in fig 14.20. A pair of mice produces on average a litter of six young every three weeks. This litter could include three males and three females. Death from old age occurs at about three years. On the basis of this information, the

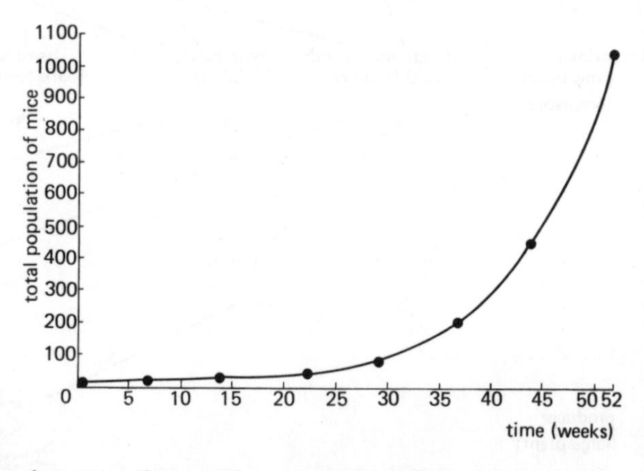

fig 14.20 Theoretical population growth of a colony of mice

theoretical growth rate over a period of one year would be as follows:

Time (weeks)	0	7	14	21	28	35	42	49
Population of mice	2	8	14	38	80	194	434	1016

A graph of these figures would resemble that in fig 14.20 (see also Chapter 11). It is clear from these graphs that unhindered population growth results in a huge and continuous increase in numbers. In real populations, however, the natural checks result in changes in the curve such that it does not go on rising. Figure 14.21 represents the population growth of sheep introduced into the island of Tasmania in 1814. The population increased rapidly over a period of fifty years, but then the curve levelled out as the limiting factors took effect. Figure 14.22 shows a population of deer which increased rapidly when *predators* (lions, coyotes and wolves) were dramatically reduced in the habitat. The population of deer increased very rapidly until it overshot the estimated carrying capacity based on food supply.

Questions

1 What explanation can you give of the rapid increase in the deer population from 1905–1924?

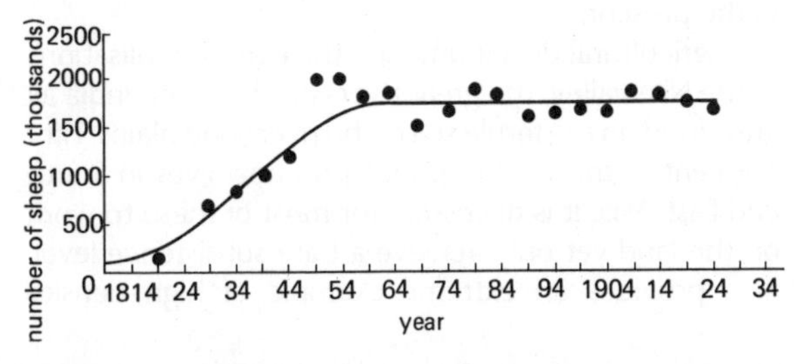

fig 14.21 Population growth of sheep introduced into Tasmania. The dots represent average numbers over five-year periods

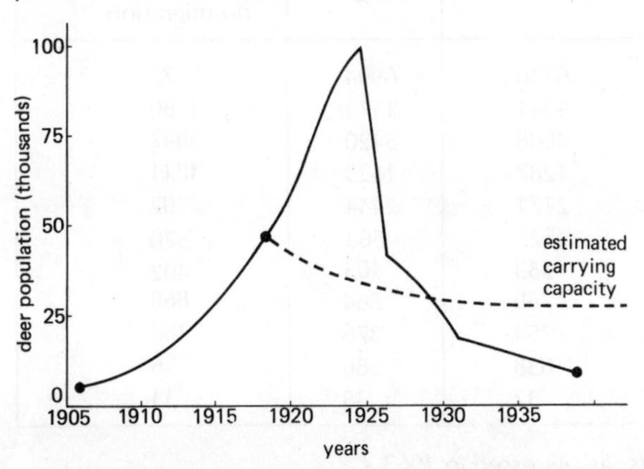

fig 14.22 Population explosion in Kaibab deer following removal of predators in an isolated area in America

2 If the estimated carrying capacity of the habitat was about 25000 to 30000 deer, what happened to the vegetation between 1918 and 1924?

3 What explanation can you give for the sudden downward curve of the population between 1924 and 1930?

4 What would you expect to have happened to the deer population and the predators after 1935?

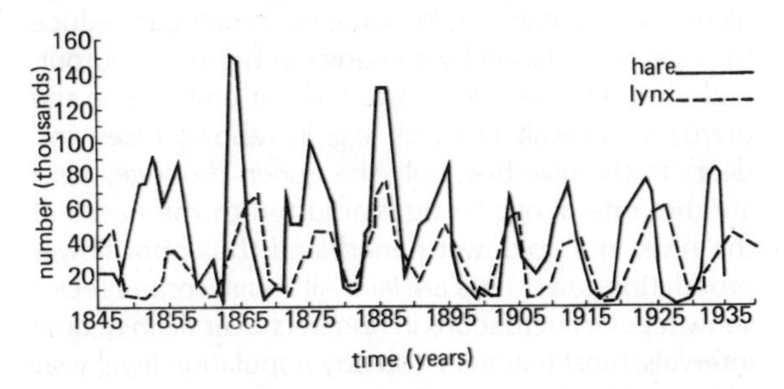

fig 14.23 Changes in the abundance of lynx and snowshoe hare

Figure 14.23 shows this aspect of prey-predator relationship over a long period of time, indicating that these fluctuations in numbers are closely related. Examine fig 14.23 and answer the following questions.

Questions

5 Which animal population increased first? Why should this be so?

6 Why is this then followed by a peak in the numbers of the second animal?

7 Why is this second peak always below (less numerous than) the first peak?

Once a population has reached an equilibrium point due to natural checks then the population remains fairly stable (see fig 14.21 Tasmanian sheep). Despite the enormous reproductive potential of plants and animals, the controls acting on a population are quite effective. The parasites and predators feed on those less fitted to survive. Early death results from these factors and thus the population is kept in check.

In addition to the three checks already mentioned, other mechanisms operate in keeping the numbers of each generation below its theoretical reproductive potential. These are the density-dependent factors.

i) *Territorial behaviour* (see Chapter 7) The number of parents are reduced by establishing breeding territories such that each pair occupies an area of sufficient size to supply the needs of itself and its offspring. The males defend the area against other members of the same species. An intruder is attacked and usually withdraws, leaving the occupying pair in sole possession. This territorial behaviour ensures that the food available in the habitat is shared among the breeding pairs, and furthermore it can prevent breeding amongst the surplus members of the population. When food is plentiful for a species the territories become smaller, allowing more breeding pairs.

ii) *Reduction of offspring produced* Some animals reduce the numbers of their offspring and thus safeguard against over depletion of food supplies. Fruit flies living under crowded conditions produce fewer eggs. Laboratory rats and mice living in a crowded environment reach a population limit, even though plenty of food is available. Some mammals can reduce the rate of ovulation by a change in hormone output, and in others there is a resorption of embryos in the uterus as a result of stress (e.g. in rabbits, foxes and deer). In the hive-bee, only the queen produces eggs and her rate of egg-laying is adjusted to the needs of the hive. In a cold, wet period (and thus poor flower production) fewer eggs are laid. Although some species show regular fluctuations in numbers over definite time intervals, most maintain a steady population level year after year and thus they seem to be regulated in a homeostatic manner (see Chapter 6).

14.81 Human populations

In human populations, the checks that apply to other animals, namely disease, predators and food supply are not all equally relevant. The problems of infectious and contagious diseases have been greatly reduced in recent years. Man's population growth is not conditioned therefore by this first factor, and furthermore he has no predators. War is often thought to eliminate large numbers of people but up to the second world war (1939–1945) more people died of disease during wars than from injury. Food supply has always been a problem with Man and, as the population has increased rapidly in the twentieth century, more than half of the people in the world suffer from malnutrition. Even so, it can be seen that natural population checks are not really operating to any extent on humans, and therefore the graph of population increase follows that of a theoretical curve (see fig 14.24), rather than a curve of a natural animal population which levels out or fluctuates as checks come into operation. Table 14.2 shows the projected figures of world population and it is apparent that the world population will double between 1969 and the year 2000. Fig 14.25 shows these figures in the form of a histogram indicating the proportionate increase in the major areas of the world. The first million was reached about 2000 years ago and 500 million was reached in the seventeenth century. The population doubled again in only 200 years by the mid-nineteenth century, and the next doubling was in 100 years, whereas currently doubling takes only about 30 years. Great fears are expressed about the effects of over population. A group of Nobel prize-winners has stated that:

'Unless a favourable balance of population and resources is achieved with a minimum of delay, there is in prospect a dark age of human misery, famine, under-education and unrest which would generate a growing panic, exploding in wars fought to appropriate the dwindling means of survival.'

There are three factors which have brought about an increase in population in the past and continue to do so in the present.

i) Agricultural development – the earliest civilisations of the Nile valley, the great rivers of China and India all grew food on the fertile soil of the river flood plains. Fifty per cent of the world's population now lives in South and East Asia. It is necessary for most of these to work on the land yet only achieve a bare subsistence level. Java provides an extreme example of high density

Region	1969 population	Projection			
		Low	Medium	High	Constant fertility, no migration
World total	3551	5449	6130	6994	7522
Developed regions	1078	1293	1441	1574	1580
Underdeveloped regions	2473	4155	4688	5420	5942
East Asia	1182	1118	1287	1623	1811
South Asia	809	1984	2171	2444	2702
Europe	456	491	527	563	570
Soviet Union	241	316	353	403	402
Africa	344	684	768	864	860
Northern America	225	294	354	376	388
Latin America	276	532	638	686	756
Oceania	19	28	32	35	33

SOURCE: United Nations, *World Population Prospects as Assessed in 1963.*

Table 14.2 Estimate of population in the year 2000 (in millions)

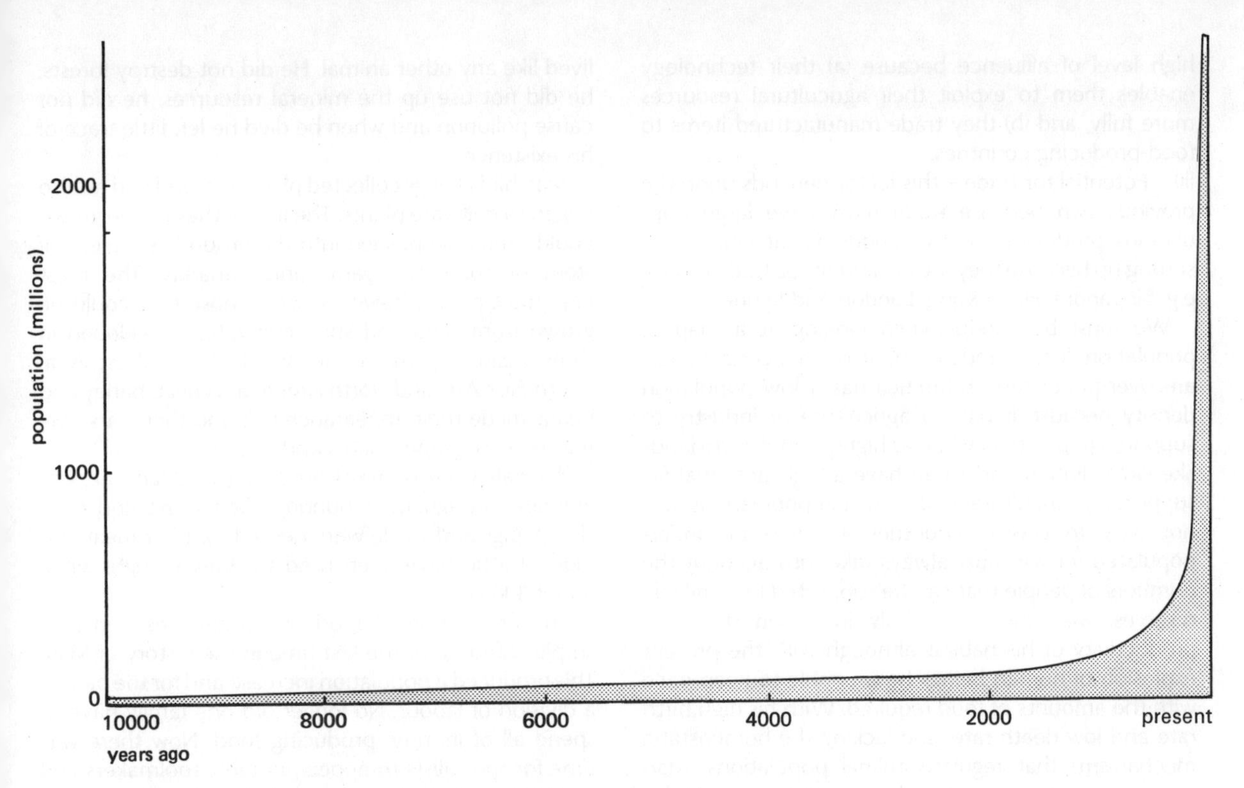

fig 14.24 Growth of the human population from 10000 years ago to the present time

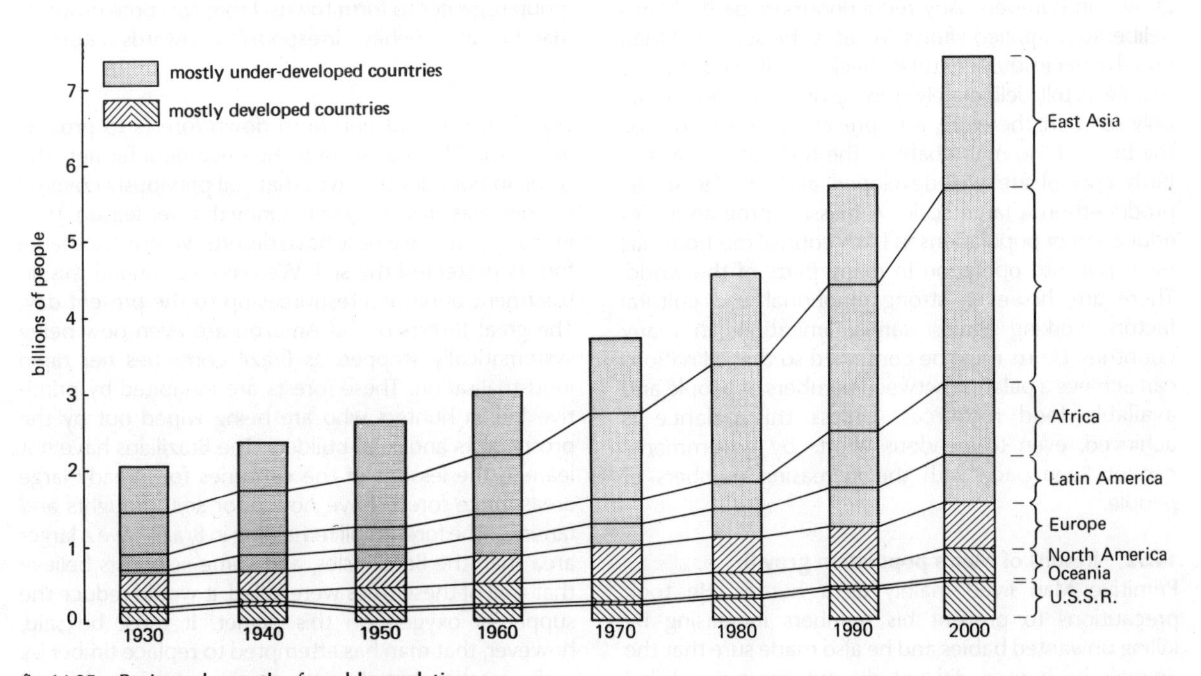

fig 14.25 Projected growth of world population

population supporting more than 1000 to the square mile (2.29 km²).

ii) Industrial development – the great centres of population in Europe are correlated with areas in which coal is mined. Associated with these areas are large industry-supported populations, from Scotland eastwards through Europe to Poland. Industrial development leads to population build-up such that Holland has 960 to the square mile, England 900 and Belgium 790. These populations are of course supported at a high level of affluence because (a) their technology enables them to exploit their agricultural resources more fully, and (b) they trade manufactured items to food-producing countries.

iii) Potential for trade – this factor depends upon the previous two because trade must have large populations producing food or goods. Great cities have sprung up because they are on land or sea trade routes, e.g. Singapore, Hong Kong, London and Sydney.

We must be careful when looking at a map of population density and pronouncing that certain areas are over-populated. Antarctica has a low population density because it has no agriculture or industry to support a population, whereas highly populated islands like Great Britain and Japan have a high potential for supporting agricultural and industrial populations. It is not easy to classify countries as over- or under-populated for we must always take into account the numbers of people that can be supported by available reserves. Man has continually increased the food productivity of his habitat although with the present explosion of human beings he is unable to keep pace with the amounts of food required. With his high birth rate and low death rate, and lacking the homeostatic mechanisms that regulate animal populations, Man cannot look to any natural process to restrain his growth in numbers. Any reduction must be by Man's deliberately applied efforts. What is the solution? Man could never institute natural checks by allowing disease to take its toll, deliberately starving or killing people. The only solution, therefore, is to prevent, by birth control, the birth of too many babies. The technical means of birth control are well-developed and can be mass-produced on a large scale. A massive programme of education of populations in birth control methods has been put into operation in many parts of the world. There are, however, strong emotional and cultural factors working against family limitations in many countries. Efforts must be continued so that all nations can achieve a balance between numbers of people and available food resources. Unless this balance is achieved, even tremendous efforts by governments cannot keep pace with the increasing numbers of people.

14.82 Results of Man's population growth

Primitive Man lived mainly as a hunter. He took precautions to prevent his numbers increasing by killing unwanted babies and he also made sure that the animals he hunted did not die out. He never killed pregnant females and hunting usually stopped during the breeding season. Primitive hunting peoples had small families because such small groups could survive more easily and the children of these families probably inherited their parent's low fertility. Thus stone-age Man made little impact on his environment and he lived like any other animal. He did not destroy forests, he did not use up the mineral resources, he did not cause pollution and when he died he left little trace of his existence.

Man had always collected plants for food and he also began to cultivate plants. The first of these to be grown could simply be pushed into the ground as a piece of stem or root, e.g. yams and bananas. The most important plants, however, were those that could be grown from seed and such cultivation developed in three main regions of the world: South-West Asia, South-East Asia and North America. Wheat, barley and beans made their appearance first and then oats, rye, maize, peas, grapes, dates and apples.

Animals were domesticated by those familiar with animals through their hunting. Goats first and then sheep, pigs and cattle were herded for their meat and hides. Cattle have been used to draw ploughs since about 3000 B.C.

The production of food by farming resulted in a surplus of food for the first time in the history of Man. This produced a population increase and for the people a division of labour. No longer did one family have to spend all of its time producing food. Now there was time for specialists to appear, notably toolmakers and potters, and at the same time the populations began to group together to form towns. From this time onwards Man began to behave irresponsibly towards nature.

Forests

Man began to cut and burn down forests to provide agricultural land as soon as he became a farmer. The result in hot countries was that soil previously covered by trees was eroded by rain during the wet season. Thus in many places we now have deserts, where once large forests protected the soil. We have continued this ill-treatment of natural resources up to the present day. The great forests of the Amazon are even now being systematically stripped as Brazil continues her rapid industrialisation. These forests are inhabited by primitive Indian hunters who are being wiped out by the prospectors and road builders. The Brazilians have not learned the lessons of the centuries for already large areas, once forest, have now poor soil, droughts and famines. The forests still remaining in Brazil have a larger area than the British Isles, and some scientists believe that if all of these trees were felled, it would reduce the supply of oxygen on this planet. It must be said, however, that man has attempted to replace timber by reafforestation but the rate of consumption of wood still outstrips the rate of replacement. The manufacture of newsprint alone consumes vast quantities of woodland every year.

Fossil fuels

The industrial revolution of the nineteenth century was

essentially brought about by a change in the main source of energy used. Up to this time the muscles of man, his horses and his oxen had supplied the necessary energy. The Pyramids were built by thousands of slaves over many years. Rich people and land-owners employed vast numbers of servants in their homes and on their farms. It was the increase in the production of coal and the invention of the steam engine that provided a much greater source of energy from 1800 onwards. Industry increased by means of this energy and industrial nations became much richer, although this wealth was not widely shared. In 1750 most people in Britain worked on the land, but by 1900 only one tenth of the population worked on farms, and today only one person in fifty is employed in agriculture. These figures indicate the great change brought about by the industrial revolution.

The spread of industry has been speeded over the last seventy years by the discovery and use of oil as an energy producer. This fossil fuel is being used at a prodigious rate and, although Britain now has her own oil supply, it is not expected to last beyond the year 2000. The supply of coal could last longer but both fuels are being expended rapidly and they can never be replaced. Natural gas is the most recent discovery and by 1977 the whole of England was being supplied with North Sea gas. These natural fuels were formed during a period of 100 million years but, with present consumption, all will have gone by the year 2100.

Other types of mining

Iron ore is used in great quantities and turned into sheet steels. Whether it is made into cars, beer cans, washing machines or bicycles it is eventually abandoned to rust away into iron oxide. This is dispersed and the iron cannot be recovered so that it is a net loss to this planet. Other metals likewise are mined, processed and eventually discarded. Farming relies more and more on artificial fertilisers containing phosphates and nitrates. These occur in natural deposits which have been mined and distributed around the world, spread on the land and finally washed into seas or lakes.

14.83 Pollution

Air pollution

Air is a mixture of gases, 78% nitrogen, 21% oxygen, 0.03% carbon dioxide and minute amounts of argon, neon and helium. The pollution of the atmosphere in industrial countries is causing increasing concern. The following are the main pollutants.

i) *Dust* This consists of small particles produced by industrial processes such as brickworks, cement works and power stations.

ii) *Smoke* Industrial furnaces and domestic fires produce about one thousand million kilogrammes of smoke every year. In certain atmospheric conditions this smoke combines with fog to cause smog, which at its worst can be lethal. After the great London smog of 1952 that killed 4000 people in five days, the Clean Air Act of 1956 enabled authorities in towns and cities to set up smokeless zones. Since that time the appearance of London in winter has changed completely. December sunshine has increased by 70% in the last fifteen years, and the estimated smoke emission is down by 50%. This has been achieved by the use of electricity and smokeless fuels in the clean air zones.

iii) *Sulphur dioxide* The combustion of fuels such as coal and oil produces sulphur dioxide. This is an irritant, choking gas. In moist air it forms sulphurous acid and this, falling in rain, can cause corrosion of stone and metal. The use of smokeless fuels has reduced the concentration of this pollutant gas by half.

iv) *Carbon monoxide* This highly poisonous gas is produced by petrol engines. In a concentration of 1000 p.p.m. it kills rapidly, but at a concentration of 100 p.p.m. it causes headaches and stomach cramps. Measurements have been made during London rush hour traffic and these show concentrations as high as 200 p.p.m. The most heavily polluted air containing this gas is found in Tokyo where traffic policemen have to wear gas masks.

v) *Lead* This is a very dangerous atmospheric pollutant and is mainly produced by the internal combustion engines of cars and lorries. Lead is added to petrol to make the engine run more smoothly. It is expelled in car exhausts and settles on vegetation at the roadside. Thus blackberries should never be picked from hedgerows near busy roads.

vi) *Smoke from cigarettes* This is far more dangerous than any other form of pollutant. This is true for smokers and non-smokers alike, but in particular inhaling the smoke directly by the smoker is the principal cause of many deaths from lung cancer.

vii) *Nitrogen oxide* This substance is also produced by the motor vehicle and is another possible cause of lung cancer.

We cannot afford a society totally free of pollution, but we must aim at a cost-effective effort to improve the atmosphere incorporating an acceptable small level of risk.

Experiment

To investigate quantitatively the effects of pollution

Procedure

1 Obtain nine beakers (or jam jars) labelled A to I and two-thirds fill them with tap-water. Fill a tenth beaker (J) with distilled water.

2 Add to the beakers the following substances:

A 5 cm^3 of garden soil
B 5 cm^3 of common salt (sodium chloride)
C 5 cm^3 of detergent
D 5 g of sodium nitrate

E 5 g of superphosphate of lime
F 5 cm^3 of car oil
G 5 cm^3 of concentrated sulphuric acid
H 5 cm^3 of common weed killer (e.g. paraquat, weedol)
I no addition (control) – tap water
J no addition (control) – distilled water.

3 Add to each container 10 fronds (lobes) of the common duckweed (*Lemna*).

4 Stand the cultures on a well-lit bench.

5 Count the number of living fronds present in each container after the following intervals:
(i) 2 days (ii) 4 days (iii) 7 days (iv) 2 weeks (v) 3 weeks (vi) 4 weeks. Record also brown or dying fronds.

Freshwater pollution

Probably 50% of the total water in the rivers of the British Isles is polluted to an unacceptably high level, but River Authorities are taking more and more action to improve this situation. The river Thames has been polluted for many years and as late as 1957 there were no fish present in its lower reaches. By 1968, however, forty-one species had been recorded, this in spite of the fact that the Thames receives 2300 million dm^3 (litres) of treated sewage effluent every day.

The pollution of rivers, streams and lakes can be caused by the following.

i) *Detergents* River weirs and locks used to produce considerable amounts of foam because of unused detergents flushed into rivers in sewage. This problem has been largely overcome by producing detergents that are easily decomposed by bacteria in sewage works. Detergents in household waste that pass into sewage works still contain phosphates, even after treatment, and so these are discharged unchanged into rivers.

ii) *Sewage* Heavily-polluted rivers may have raw sewage pumped into the water, but local authorities are now endeavouring to treat all sewage before discharge. Unfortunately the increase in size of populations of towns often outruns the capacities of sewage works. Bacteria in rivers can break down small quantities of sewage, but when it is discharged in large quantities the bacteria increase very rapidly, using up all of the oxygen in the water. In these deoxygenated waters all other living organisms will die. Even treated sewage can damage the freshwater environment because the high amounts of phosphate cause an abnormal growth of algae (called a 'bloom'). This effect is most noticeable in lakes and reservoirs and it is called eutrophication. The algae eventually die and their decomposition removes oxygen from the water.

iii) *Farming* The modern practice of factory-farming where stock is kept in covered accommodation all the time also contributes to water pollution. When stock (cattle, sheep, fowls) lives in open fields the manure is spread around and helps to keep the soil fertile. In factory farms, the manure is washed away with large amounts of water and this *slurry*, as it is called, is often discharged into rivers.

Farmers also use chemical fertilisers in their fields instead of, and to supplement, manure. This also causes problems, because rainfall dissolves the unused fertilisers and washes them into rivers and lakes. Much nitrate finds its way into freshwater from this source and can cause eutrophication in the same way as phosphate. Furthermore it can still be present in drinking water that has passed through a filtration plant. A continual check on nitrate levels must be maintained by water authorities and in some areas special bottled water has been supplied for babies where the nitrate content of water is too high.

iv) *Industrial waste* Factory waste can contain many different chemicals which are generally poisonous. These pollutants may include lead, mercury, zinc, cyanide and copper which are cumulative poisons, building up in the living food chains as they pass from organism to organism. Oil, detergents, sulphur compounds and suspended particles are also dangerous constituents of factory effluent into freshwater.

Sea pollution

Seventy per cent of the Earth's surface is covered by the salt water of the seas. Into these flow the rivers of every country, carrying their own pollutants and adding to pollutants of all types from the towns and factories that border the seas. The following are the major pollutants.

i) *Sewage* Large centres of population bordering the seas have always considered it their right to pump sewage into the sea. It is true that the sewage emerges from large pipes often a mile out to sea but unfortunately it can be swept back by tides on to the beaches. The bacterial content of the sea water off bathing beaches can rise alarmingly and bathers may suffer infections of throat and ear as a result. In 1973 Italy had an outbreak of cholera caused by people eating shellfish contaminated by the disease organisms in sewage. The countries bordering the Mediterranean are largely to blame for the polluted state of its waters. The Spanish government has allocated large amounts of money in an effort to treat sewage and so protect the tourist industry.

ii) *Oil* It is considered that 5 to 10 billion kilogrammes of oil are released every year into the seas of the world. This includes illegal discharge by tankers as well as accidental losses from damaged tankers. More and more oil drills are in the sea and oil can escape from these in a variety of ways. The tanker *Torrey Canyon* wrecked off the British Isles released 100 million kilogrammes of crude oil, killing some 100 000 seabirds and necessitating the expenditure of some £1 500 000 in order to clean up the beaches of the West Country. The ultimate catastrophe, the collision of two 300000 ton super-tankers at sea, occurred off South Africa's southern Cape Coast in December 1977. Only one ship was fully laden with 250 million kilogrammes of crude oil. It was two weeks before a change of wind blew a large slick of oil onto the beaches of the South Africa coast. In March 1978, the *Amoco Cadiz* the largest oil tanker ever wrecked was grounded off the French Atlantic Coast. It lay there in two weeks of continuous gale and eventually lost all of its millions of gallons of crude oil. The fishing and the tourist industries of this part of France were ruined and it cost £10000 a day for many weeks to clear the oil from miles of beaches.

Oil, being a natural product, can be broken down by bacteria in time but the efforts of man to break down very large quantities of oil spillage can be dangerous to wild life. Detergents used to clear the oil are poisonous to many small sea animals.

iii) *Metals and chemicals* Rivers carry large quantities of these substances into the seas and they become concentrated in organisms as they move up through the food chains. Minimata in Japan is a small town in which forty-six people died and many remain permanently injured from eating shellfish containing mercury released into the sea by a plastics company. The dangerous substances dumped in the seas are too numerous to mention but they include canisters of nerve gas, radioactive waste and pesticides.

14.84 Conservation

The wildlife of the world has suffered a severe blow at the hand of Man. Many species have become extinct and many others such as the leopard and certain whales are approaching the stage when they will also disappear. There are many demands to protect wildlife, but should we do so? Let us consider some of the reasons.

1 The useful aspects of wild life and wild places

a) The study of wild plants has provided us with many of our medicines. New crops have been developed by the incorporation of wild stock into breeding programmes. Wild animals originally provided the source of our domestic animals and Man still makes use of wild animals for food by the herding of reindeer, and the domestication of red deer and antelope (Eland). Wild plants and animals must not be allowed to disappear or their breeding and genetic potential will be lost for ever.

b) The biological control of pests can be continued only if wild populations of plants and animals are maintained. Only from these sources can we find the control organisms to safeguard our food supplies from pests. This method must be used more and more to avoid the use of chemical pesticides, particularly the persistent poisons that become incorporated into the natural food chains.

c) The countryside must be maintained, not only for agricultural purposes but as a place where the populations of large cities can breathe freely and use it for their recreation and enjoyment. With more freedom from a shorter working week in the future, more people will wish to pursue outdoor activities ranging from mountain-climbing to fishing. it is most important that these varied interests should be satisfied and for this reason National Parks have been set up all over the world. The U.S.A. has very large National Parks attracting millions of visitors every year. Here wildlife is protected with regulations for control of shooting and erection of buildings. The British Isles has ten National Parks with a total area of 13500 km². These often include farms and houses but the government and Parks authorities endeavour to satisfy all the conflicting interests while protecting the wildlife of the Parks. The game parks of Kenya and Tanzania attract vast numbers of tourists from all over the world who come to see the variety of wildlife enclosed within them. These parks contain a small fraction of the enormous herds of game that used to cover East and Central Africa.

2 Aesthetic appreciation of wild life

By no means the least important aspect of wildlife in its natural surroundings is its beauty. Birds, butterflies, moths, flowers, fungi and fish all provide a source of study and hobby for countless enthusiastic specialist students of their variety of form, lifecycle and behaviour.

Let us therefore consider ways in which we can avoid the future extinction of animal and plant species. The answer lies in conserving what we still have and trying to revitalise old and damaged habitats. In 1977 the President of the U.S.A. exhorted his country to conserve energy, and at the same time an oil drilling platform in the North Sea lost thousands of kilogrammes of oil through human error in the fitting of a valve. 1977 also saw the tenth anniversary of the Council of Europe's Centre for Nature Conservation. This centre has achieved a great deal in this short space of time and has shown that only Man can preserve Nature in our densely populated countries, and that it is the duty and responsibility of all of us to safeguard our natural resources. In order to do this there must be public understanding and assistance. Each country in the Council of Europe has set up a National Council for Conservation and in 1970 the European Council launched a campaign, known as European Conservation Year, to influence public opinion.

To conserve and revive our environment nationally and in the continent of Europe we must look towards the following solutions:

1 The protection of wildlife.

a) National Parks already established in the British Isles and Europe must be maintained and new areas of outstanding beauty must be added to the list of those already in existence.

b) Hunting animals for their skins (e.g. leopards) and their products (e.g. whales), and other rare animals must be forbidden. More species must be protected by law, whether it be the taking of bird's eggs (Protection of Birds Act 1954) or the importation of skins for making coats.

c) Rare animals must be preserved and bred in zoos and parks.

d) Exploitation of the remaining wild places of the earth for minerals or timber must be stopped or strictly controlled.

2 Prevention of habitat destruction

a) Pollution must be tackled at all levels. The discharge of chemical waste and sewage into lakes and rivers, the emission of smoke and chemicals into the atmosphere and the fouling of the seas must be stopped by international action.

b) The misuse of land must be halted. The growth of large scale farming resulting in loss of habitat for countless organisms must be controlled. The decline of 278 of the rarest British plants has been measured between the years 1963 to 1970 and has been calculated at 30%. This was due to increased agriculture and the elimination of hedges and copses (see fig 14.27). The decline of habitats has not been compensated for by the production of new habitats. Old-fashioned mixed farming resulted in a by-product of varied wildlife and we must try to achieve some compromise between increased food production and efficient wildlife conservation.

c) Wherever the waste of deep or open cast mining is left on the surface of the ground this must eventually be landscaped and replanted so that the natural habitat can renew itself.

3 Reduction in the rate of loss of resources

a) The forests of Europe and the world must no longer be cut down to provide enormous quantities of newsprint which has a life of only one day. Very little of this mass of paper is used again although many conservation groups in society do make an effort to use it again by collecting and repulping. Each nation must introduce or extend its reafforestation programme to include not only quick-growing conifers (softwoods) but also the longer maturing hardwoods.

b) The burning of coal, oil and gas must be slowed down so that the vital energy resources are dissipated for ever. All nations with the technology available must continue to research actively for new methods of producing energy from tidal power, tidal barrages, windpower and hydroelectric schemes.

c) There must be recycling where possible of the items used in modern industrial societies. An important first step would be the reuse of the so-called 'disposable' glass bottles.

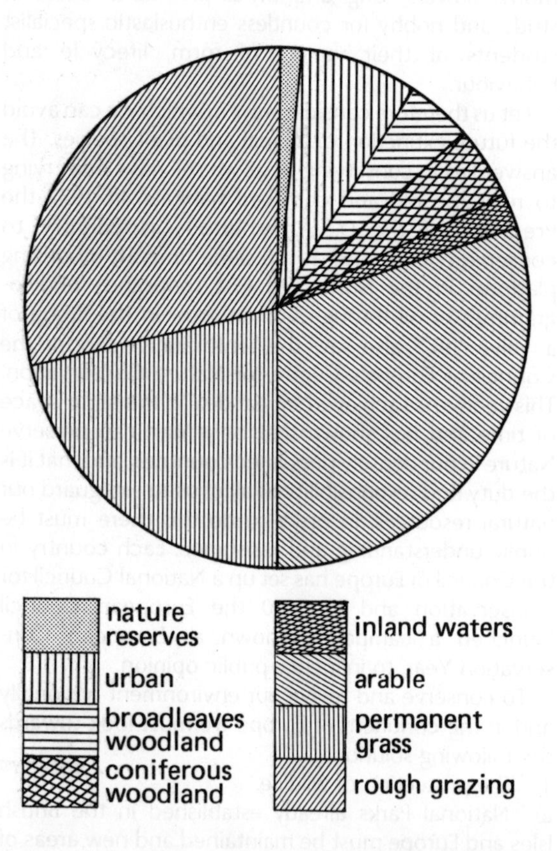

fig 14.26 The pattern of land use in the British Isles – nature reserves take up only 0.8% of the land surface

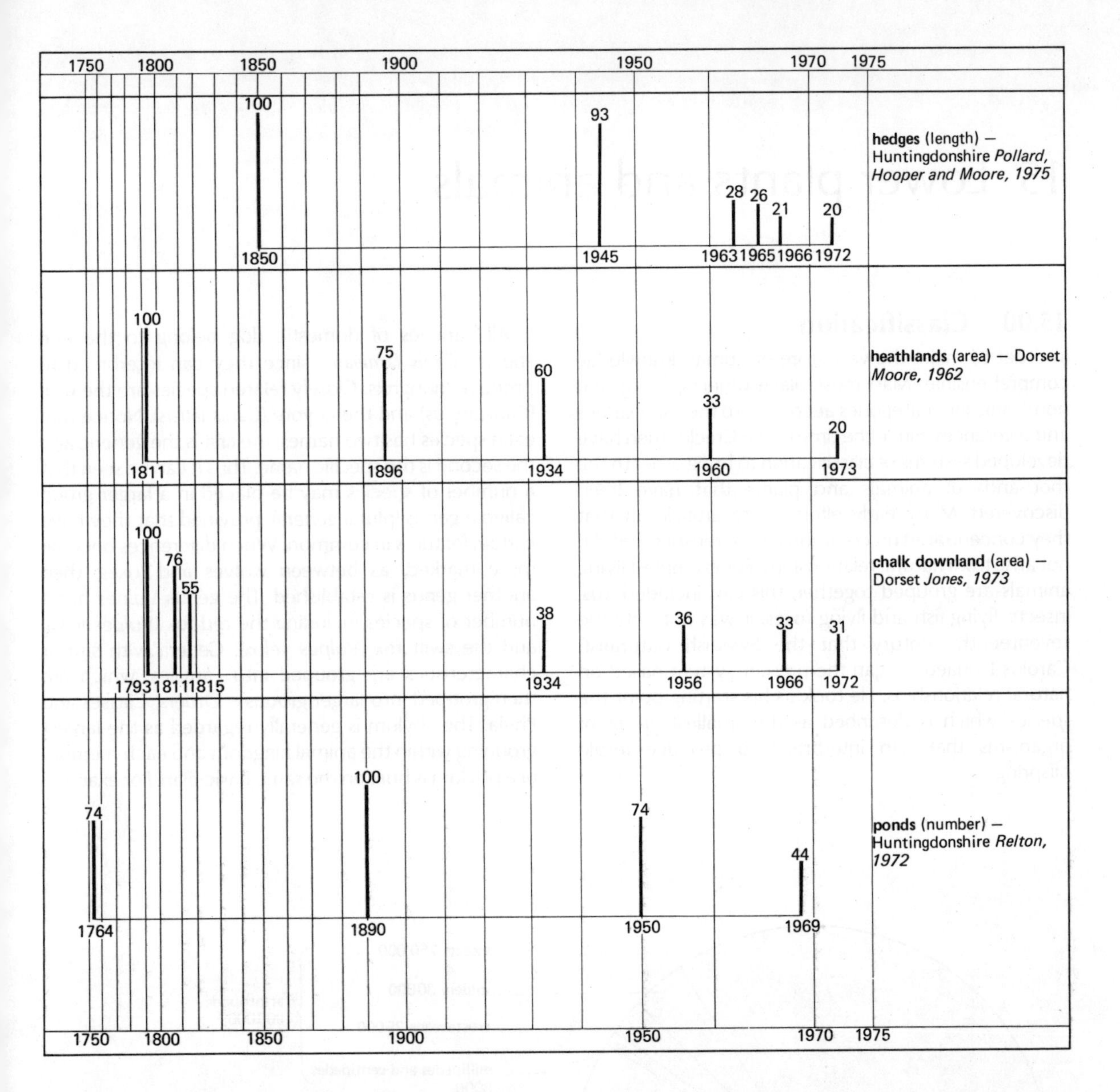

fig 14.27 Four types of land loss in England (All values are expressed as percentages of the largest value recorded in each case)

Questions requiring an extended essay-type answer

1 Describe the appearance of a *named* tree, and give an account of the changes that you have seen throughout a year in the life of the tree.

2 a) What is meant by a habitat as applied to animals and plants?
b) Describe the seasonal changes in a *named* habitat that you have observed throughout the year. Refer to **three** *named* organisms that live in the habitat.

3 From your observations of a *named* habitat show how **three** *named* animals are, in each case, adapted to the habitat in their locomotion, feeding and reproduction.

4 a) Describe the methods you would use to carry out a study of the distribution of a plant species in a habitat.
b) By means of an outline map, describe the distribution pattern of a *named* plant that you have studied.
c) Discuss the biological and physical factors of the environment that could influence the distribution of this organism.

15 Lower plants and animals

15.00 Classification

In order to make the vast store of human knowledge comprehensible, Man must place objects, living and non-living, into categories according to their similarities and differences. Since the time of the Greeks, men have developed systems of classification to bring order to the thousands of animals and plants that have been discovered. Many early efforts were artificial' in that they concentrated on common characteristics that did not arise from natural relationships. For example if flying animals are grouped together, this can include birds, insects, flying fish and flying foxes. It was not until the seventeenth century that the Swedish naturalist, Carolus Linnaeus began the present system based on natural relationships. He took as his starting point the *species* which is described as the smallest group of organisms that can interbreed to produce fertile offspring.

All varieties of domestic dog belong to the one species *Canis familiaris,* since they can interbreed to produce mongrels. Closely related species are the wolf (*Canis lupus*) and the coyote (*Canis latrus*). Notice that each species has two names, the first is the generic and the second is the specific name. Thus it can be seen that a number of species may be placed in a larger group called a *genus* (plural *genera*), provided that they have certain features in common. When differences become more marked, as between wolves and foxes, then another genus is established. The genus *Vulpes* has a number of species including the red fox (*Vulpes fulva*) and the swift fox (*Vulpes velox*). Genera with similar characteristics are grouped into Families, which are then grouped into larger groups – Orders, Classes and Phyla. The Phylum is generally regarded as the largest grouping within the animal kingdom and each member of a phylum is built on the same basic plan. For example

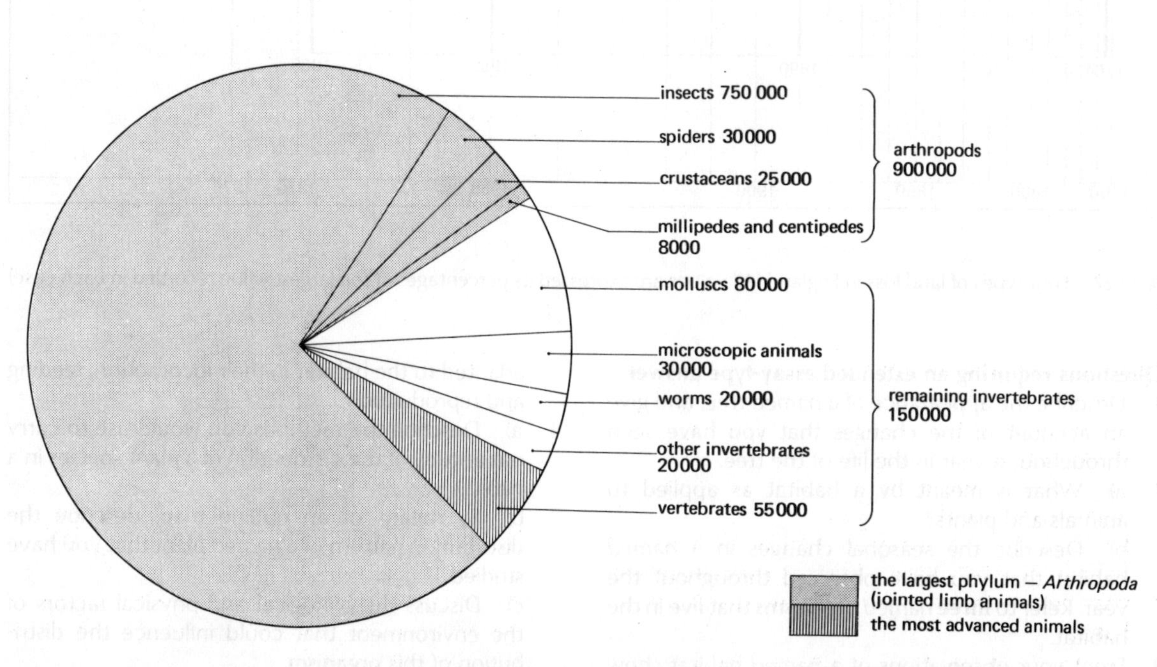

fig 15.1 A pie chart showing the number of species in major animal groups

the phylum Arthropoda includes all animals with an exoskeleton and jointed legs.

The wolves and dogs mentioned above would therefore be completely classified as follows:

Kingdom	Animalia	
Phylum	Chordata	
Sub-phylum	Vertebrata (Craniata)	
Class	Mammalia	Possess milk glands and diaphragm
Order	Carnivora	meat-eating, large canine teeth
Family	Canidae	long slender limbs, four toes on the front feet, non-retractile claws
Genus	*Canis*	dog-like
Species	*C. familiaris*	(domestic dog)
	C. mesomelas	(black-haired jackal)
	C. lupus	(wolf)

Once a classification scheme has been formalised, a method must be devised whereby other scientists can determine the identity of a particular organism. This is done by constructing an identification key based on the classification. To illustrate the relationship between classification and identification keys consider how you would classify and make a key for a collection of the present British coinage (50p, 10p, 5p, 2p, 1p, and ½p). Obtain a set of these coins and then sort them as shown.

The coins can be classified by selecting two different characteristics several times in succession to produce a dichotomous key.

This type of classification as applied to organisms, is more commonly presented in the form below.

1 metal	—	bronze	see 3
		cupro-nickel	see 2
2 shape	—	7-sided	50p
		round	see 4
3 design	—	feathers	2p
		no feathers	see 5
4 design	—	rose or lion	10p
		thistle	5p
5 design	—	portcullis	1p
		crown	½p

Investigation

To construct dichotomous keys and to identify the following:

1 Modern methods of transport – car, bus, cycle, three-wheeler car, motor-cycle, motor-cycle and sidecar, six-wheeled lorry, etc.

2 Staff cars in the school car-park – colour, make, four seats, two seats, four door, two door, saloon, sports car, etc.

3 Members of a biology class – hair colour, eye colour, male, female, skin colour, etc.

4 Trees and shrubs in the school grounds – leaf size, leaf shape, leaf texture, colour of twigs, presence of leaf stalk, etc.

Examine the two keys following for the flowering plant Family Ranunculaceae, Genus Ranunculus (which includes the common buttercup) and the Order Hymenoptera (which includes the wasps, bees and ants). These two keys are typical of many keys available

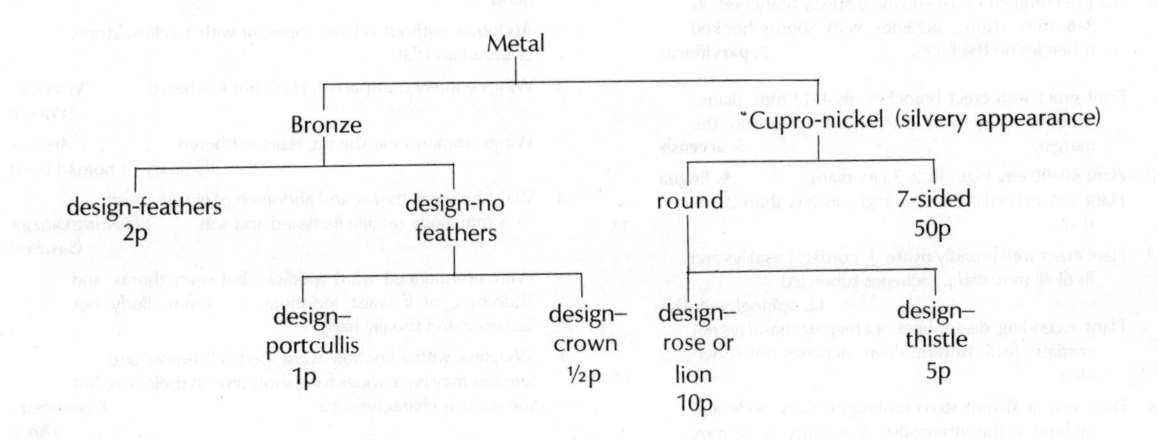

for identification of every possible group of organisms. Once you understand the principle, they are not difficult to use. Note that they are not strictly dichotomous in all respects: at some stages the organisms may be placed in three or more groups, all but one of which give a final identification (see stage 1 of the Hymenoptera key).

Part of a dichotomous key from Clapham, Tutin and Warburg for the species of the genus Ranunculus (the buttercup genus).

---

6 Stem tuberous at base, ± appressed-hairy above; achenes minutely pitted, with hooked beak. **3. bulbosus**

Stem not tuberous, with spreading hairs throughout; achenes usually with a few tubercles just within the conspicuous border and with almost straight beak. **6. sardous**

7 Lvs glabrous or nearly so. 8
Lvs hairy. 9

8 Fls 0.5–1 cm. diam.; achenes in an oblong head; in wet places, and on bare mud. **13. sceleratus**

Fls 1.5–2.5 cm. diam. but less when, as often, petals are reduced or 0; achenes in a spherical cluster; woods and shady hedgebanks. **8. auricomus**

9 Plant with fleshy root-tubers at the base of a short erect stock; fls 2.5–3 cm. diam. with very glossy petals; stem-lvs 1(–2), small **4. paludosus**

Plant without root-tubers; stem lvs usually more than 2, the lower ones large. 10

10 Plant with long runners; fl.-stalk furrowed. **2. repens**
Plant without runners; fl.-stalk not furrowed. **1. acris**

11 Plant decumbent or ascending, diffusely branched; fls 3–6 mm. diam.; achenes with shortly-hooked tubercles on the faces. **7. parviflorus**

Plant erect with erect branches; fls 4–12 mm. diam.; achenes spiny with the largest spines on the margin. **5. arvensis**

12 Plant 60–90 cm. high; fls 2–3 cm. diam. **9. lingua**
Plant not exceeding 60 cm. high; fls less than 2 cm. diam. 13

13 Plant erect with broadly ovate ± cordate basal lvs and fls 6(–9) mm. diam.; achenes tubercled. **12. ophioglossifolius**

Plant ascending, decumbent or creeping; basal lvs not cordate; fls 5–18 mm. diam; achenes not tubercled. 14

14 Plant with a filiform stem rooting at every node and arching in the internodes; fls solitary, 5–10 mm. diam.; achenes 1 mm. **11. reptans**

Plant, if creeping, not rooting at every node; fls 1–several; achenes 1.5–1.8 mm. **10. flammula**

15 Plant with no finely dissected submerged lvs. 16
Finely dissected submerged lvs present. 18

16 Lvs deeply 3(–5)-lobed, the lobes cuneate, distant; receptacle pubescent. **16. tripartitus**

Lvs shallowly lobed; receptacle glabrous. 17

17 Lf-lobes broadest at their base; fls 3–6 mm. diam. **14. hederaceus**

Lf-lobes narrowest at their base; fls 8–12 mm. diam. **15. omiophyllus**

18 Petals usually less than 6 mm.; stamens 5–15. 19
Petals usually more than 6 mm.; stamens 12–numerous. 20

19 Floating lvs never present; achenes pubescent to hispid (at least while immature). **19. trichophyllus**

Floating lvs always present, usually less than 1.5 cm. across, deeply 3(–5)-lobed; submerged lvs few; achenes glabrous. **16. tripartitus**

20 Submerged lvs very long (8–30 cm.) and parallel; floating lvs 0; receptacle glabrous. **17. fluitans**

Submerged lvs very rarely exceeding 8 cm.; floating lvs present or 0; receptacle pubescent. 21

KEY TO THE ORDER HYMENOPTERA

(Bees, Wasps, Ants and Sawflies)

1. Densely furry, with none of the underlying cuticle visible. Black or brown, or with 1 to 3 yellow, buff or orange stripes. Hind tibiae with spurs . . . . APOIDEA (Bumble bees)

— Moderately furry, with some of the underlying cuticle visible. Uniformly brown. Hind legs with tibiae flattened and broad, without spurs . . . . . APOIDEA (Honey bees)

— Slightly furry or bare. Either uniformly black brown, metallic blue or green, or with at least 4 yellow stripes or black spots on the abdomen 2

2. Abdomen cylindrical with yellow stripes or a broad brown band 3

— Abdomen without yellow stripes, or with 1 yellow stripe, or broad and flat 4

3. Wings entirely transparent. Hairs not feathered . VESPOIDEA (Wasps)

— Wings smokey near the tip. Hairs feathered . . APOIDEA (Solitary or nomad bees)

4. Waist between thorax and abdomen, slight or absent. > 6 mm. Body usually flattened and soft TENTHREDINOIDEA (Sawflies)

— With pronounced waist (pedicel) between thorax and abdomen, or if waist indistinct, < 4 mm. Body not flattened and usually hard 5

5. Wingless, with a knobbly waist (pedicel). (Males and females may have wings for a short time in their lives, but the waist is characteristic.) . . . . FORMICOIDEA (Ants)

— Winged, with smooth waist . . . . PARASITICA* (*For further separation—6*)

6. Wings without veins, or with a vestigial vein along the leading edge. Antennae with angular bend or wing margins hairy . . . . . . CHALCIDOIDEA or PROCTOTRUPOIDEA

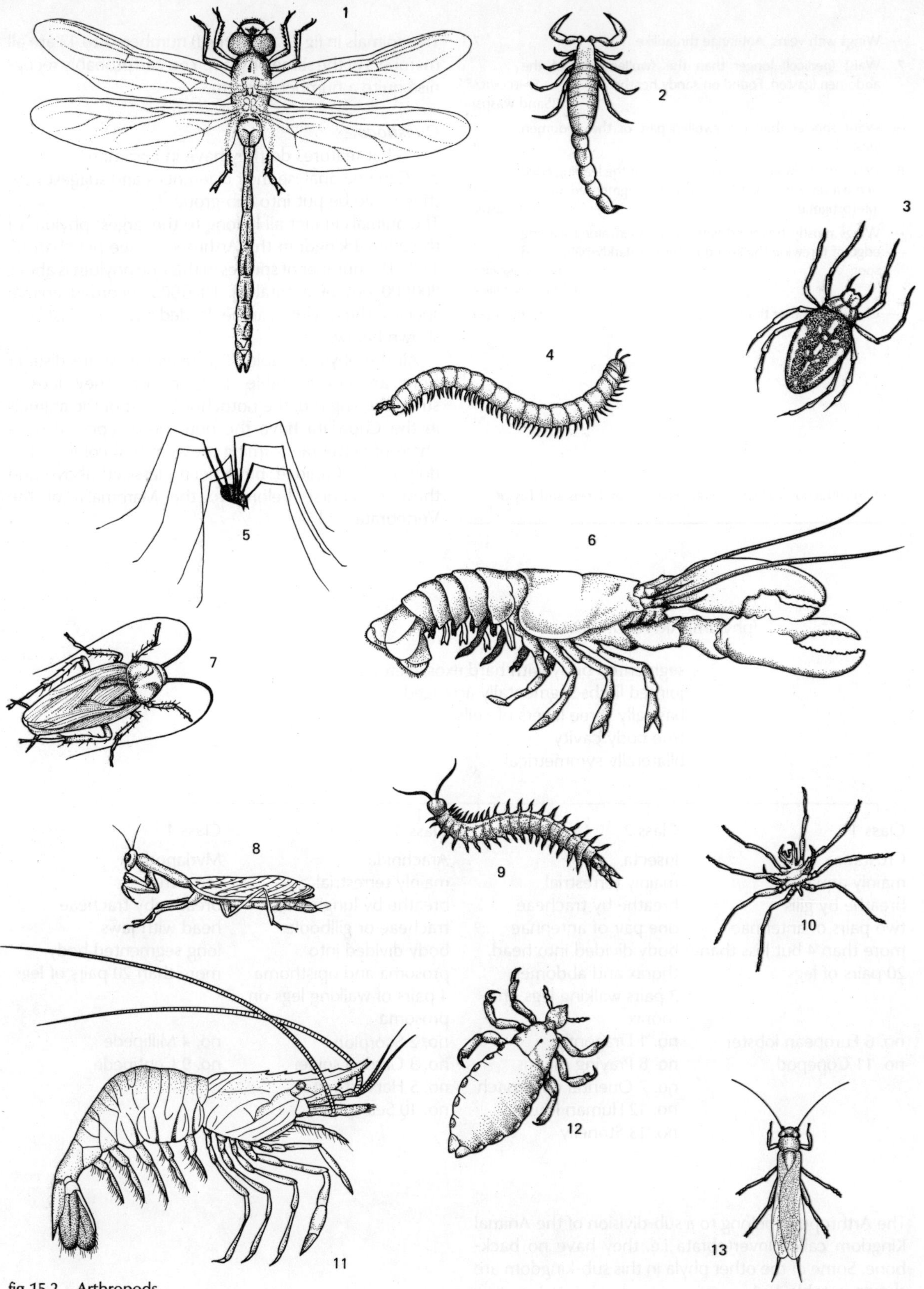

fig 15.2 Arthropods

— Wings with veins. Antennae threadlike 7

7. Waist (pedicel) longer than the swollen part of the abdomen (gaster). Found on sandy heaths . SPHECOIDEA (Sand wasps)

— Waist shorter than the swollen part of the abdomen (gaster) 8

8. Wings with few veins concentrated near the leading edge of the forewings, which have *no* darkly-pigmented spot (pterostigma) . . . . CYNIPOIDEA (Gall wasps)

— Wings mostly traversed with veins. Vein along leading edge of forewing thickened to form a darkly-pigmented spot . . . . ICHNEUMONOIDEA (Ichneumon flies)

— Characters other than above . . . . . Other taxa

From 'Introduction to Experimental Ecology' by Lewis and Taylor.

The animals in fig 15.2 (overleaf) numbers 1 to 13 are all members of the same phylum. You can probably recognise quite a number of them.

Questions

1 What features do they have in common?

2 Can you analyse their differences and suggest how they could be put into sub-groups?

The animals in fact all belong to the largest phylum of the animal kingdom the Arthropoda (see pie chart fig 15.1). The number of species in this one phylum is about 800000 out of a total of 1000000 recorded animal species. The phylum can be divided into four classes as shown below.

All the phyla in table 15.1 Invertebrata are distinct from animals in table 15.2, in that they lack a strengthening rod, the notochord. Most of the animals in the Chordata have the notochord replaced by a chain of vertebrae forming the vertebral column. The dog family, Canidae, have been classified above and these of course belong to the Mammalia of the Vertebrata.

(phylum) ARTHROPODA

segmented body with hard exoskeleton
jointed limbs segmentally arranged
basically three layers of cells
true body cavity
bilaterally symmetrical

Class 1	Class 2	Class 3	Class 4
Crustacea mainly aquatic breathe by gills two pairs of antennae more than 4 but less than 20 pairs of legs	Insecta mainly terrestrial breathe by tracheae one pair of antennae body divided into head, thorax and abdomen 3 pairs walking legs on thorax	Arachnida mainly terrestrial breathe by lungbooks, tracheae or gillbooks body divided into prosoma and opisthoma 4 pairs of walking legs on prosoma	Myriapoda terrestrial breathe by tracheae head with jaws long segmented body more than 20 pairs of legs
no. 6 European lobster no. 11 Copepod	no. 1 Dragonfly no. 8 Praying mantis no. 7 Oriental cockroach no. 12 Human louse no. 13 Stonefly	no. 2 Scorpion no. 3 Garden spider no. 5 Harvest mite no. 10 Sea spider	no. 4 Millipede no. 9 Centipede

The Arthropoda belong to a sub-division of the Animal Kingdom called Invertebrata i.e. they have no backbone. Some of the other phyla in this sub-kingdom are shown in table 15.1.

Table 15.1 The Animal Kingdom – Invertebrata

Phylum	Characteristics	Examples
Protozoa	Mainly aquatic; single celled; generally one nucleus; move by pseudopodia, cilia or flagella; some parasitic;	*Amoeba, Paramecium, Euglena, Plasmodium*
Porifera	Aquatic; pores through which water circulates;	Sponges
Coelenterata	Mostly marine; body composed of two layers of cells; single digestive cavity with one opening; mouth surrounded by tentacles; possess stinging cells (nematocysts);	*Hydra* Sea-anemones, corals, jellyfish
Platyhelminthes	Flattened segmented worms; alimentary canal but no mouth; body composed of three layers of cells; no body cavity; two classes are parasitic;	planarians, flukes, tapeworms
Nematodes	Unsegmented worms, pointed both ends; gut with mouth and anus; three layers of cells; many parasitic;	Hookworms, threadworms, roundworms
Annelida	Segmented worms; body composed of three layers of cells; gut with mouth and anus; bristles (chaetae) in two classes; some are parasitic;	earthworms, lugworms, ragworms, leeches
Arthropoda	See page 282	
Molluscs	Unsegmented soft body often with a shell; large single muscular 'foot';	Snails, slugs, mussels, octopus
Echinodermata	Unsegmented, radially symmetrical; five arms; spiny skin with hard plates; possess tube feet; all marine;	Starfish, sea urchin, bristle star, sea cucumber

Table 15.2 The Animal Kingdom – Chordata

Phylum	Characteristics	Examples
Chordata	Notochord present in adult; tubular, dorsal, hollow nerve cord; closed blood system; post-anal tail;	
Sub-phylum		
Acrania	No true brain or skull, heart or kidneys	*Amphioxous*
Craniata or vertebrata	Well developed head and brain; muscular ventral heart; kidneys, organs of excretion; two pairs of limbs; endoskeleton;	
Class		
Pisces (fish)	Paired fins; gills for gaseous exchange; external scales; lateral line system;	Cod, plaice, carp
Amphibia	Paired pentadactyl limbs; gills present in the tadpole larva; lungs in the adult; soft skin, no scales; middle and inner ear, no outer ear;	Frog, newt, toad, salamander
Reptilia	Paired pentadactyl limbs; lungs for gaseous exchange; dry scaly skin; large heavily yolked eggs with shell are laid; no larval stage;	Grass snake, turtle, lizard, crocodile
Aves (birds)	Paired pentadactyl limbs, forelimbs are wings; lungs for gaseous exchange; feathers over the body, legs have scales; heavily yolked eggs with a calcareous shell are laid, no larval stage, warm-blooded;	Chaffinch, eagle, heron, stork
Mammalia	Paired pentadactyl limbs, lungs for gaseous exchange; skin bears hair and two types of gland, sebaceous and sweat; primitive mammals lay eggs (oviparous) but the majority are viviparous; warm-blooded; possess milk glands to feed the young; external, middle and inner ear; diaphragm separates the thorax from abdomen;	Mole, mouse, dog, cow, elephant, whale, bat, monkeys, apes, man

The Plant Kingdom is classified into groups in the same way as the Animal Kingdom. The features by which they are classified however are very different. Table 15.3 shows the main features of the groups and figs 15.7 – 15.15 show some examples of representative plants.

We can classify flowering plants (Angiospermae) into two further broad groups, the monocotyledons, which have one seed leaf, and the dicotyledons, which have two seed leaves. One example of each group is shown fully classified below:

Kingdom	Plantae	Plantae
Division	Spermatophyta	Spermatophyta
Class	Angiospermae	Angiospermae
Sub-class	Dicotyledones	Monocotyledones
Order	Ranales	Liliflorae
Family	Ranunculaceae	Liliaceae
Genus	*Ranunculus*	*Endymion*
Species	*repens*	*non-scripta*
	(Creeping Buttercup)	(Bluebell)

The common name buttercup is given generally to all members of the genus *Ranunculus* with bright yellow flowers and divided leaves. The importance of the specific name then becomes apparent if we look at the following

Ranunculus acris (meadow buttercup)
R. bulbosus (bulbous buttercup)
R. sardous (hairy buttercup)
R. parviflorus (small-flowered buttercup)
R. arvensis (corn buttercup)

The genus *Endymion* has only two species. The other one is *E. hispanicus*, an introduction from Spain that has become naturalised as a result of escaping from gardens. The common name is again confusing here because the 'bluebell of Scotland' is the Harebell, *Campanula rotundifolia*, not *Endymion non-scripta*.

Table 15.3 The Plant Kingdom

Group	Characteristics	Examples
Thallophyta Algae	Aquatic; possess chlorophyll, with or without other pigments; body not differentiated into root, stem or leaf; sexual reproduction by swimming gametes.	Green, red and brown seaweeds, *Spirogyra*
Fungi	Mainly terrestrial; no chlorophyl, thus parasitic or saprophytic; mycelium consisting of filaments called hyphae; cell wall of fungal cellulose; sexual reproduction.	*Rhizopus*, Mushroom, *Penicillium*
Bryophyta	Chiefly terrestrial; may show stem and leaf-like structures; chlorophyll; well-defined sexual reproduction with swimming male gametes; alternation of generations with dominant haploid gametophyte; diploid sporophyte parasitic on the gametophyte; spores of one type.	
Hepaticae (liverworts)	Flattened gametophyte branching continuously into two; rhizoids unicellular; simple sporophyte with no chlorophyll.	*Pellia*
Musci (mosses)	Erect gametophyte with axis and leaf-like structures; Rhizoids multicellular; sporophyte differentiated possessing chlorophyll and stomata.	*Funaria*
Pteridophyta	Mainly terrestrial; highly differentiated body with stem, leaf and root structures; high degree of internal tissue differentiation; chlorophyll; well defined sexual reproduction with swimming gametes; alternation of generations with dominant diploid sporophyte independent of haploid gametophyte; spores of one type or two types.	Male fern (*Dryopteris*)

Group	Characteristics	Examples
Spermatophyta	Chiefly terrestrial; body differentiated into root, stem and leaves; high degree of internal tissue differentiation; xylem and phloem commonly with secondary thickening common;chlorophyll photosynthetic pigment; well defined sexual reproduction with fertilisation by means of pollen tubes; alternation of generations – male gametophyte shed as pollen grain; both gametophytes considerably reduced; fertilisation resulting in a seed which may be borne naked or enclosed in a fruit.	
Gymnospermae	Trees and shrubs of world wide distribution; xylem elements are tracheids; flowers are cones; seeds are naked.	Pine, yew, redwood tree and other conifers
Angiospermae	Wide variety of forms, including trees, shrubs and herbaceous plants; xylem elements are vessels and tracheids; flowers typical, usually hermaphrodite with seeds enclosed in a fruit.	Bluebell, oak, dandelion, ash, beech, buttercup, chestnut, hazel, grass, onion, maize

Investigation

Classify the organisms shown in figs 15.3 to 15.6

Procedure

Use the characteristics labelled on each specimen to determine the kingdom, phylum and class of each plant.

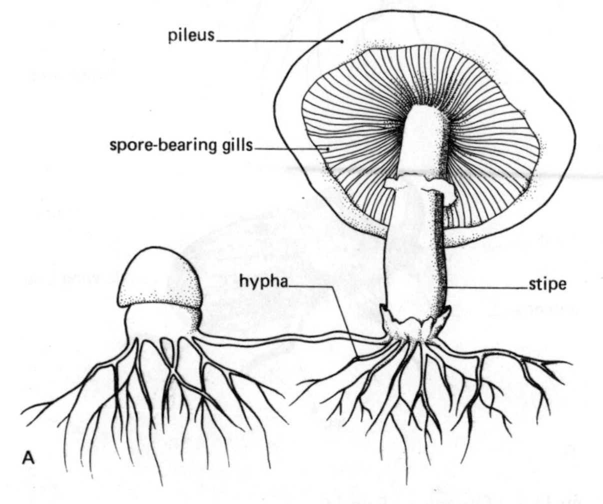

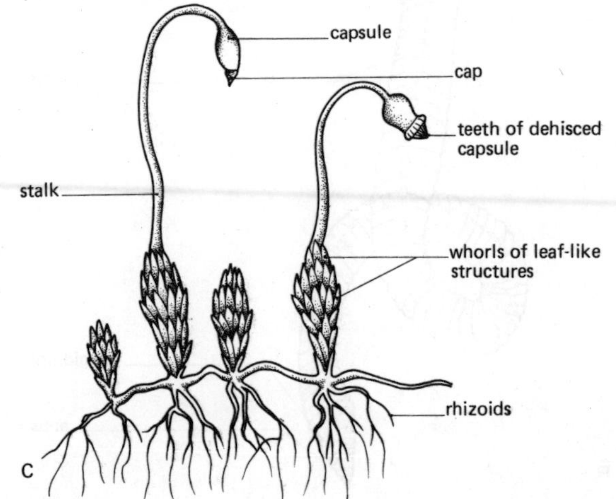

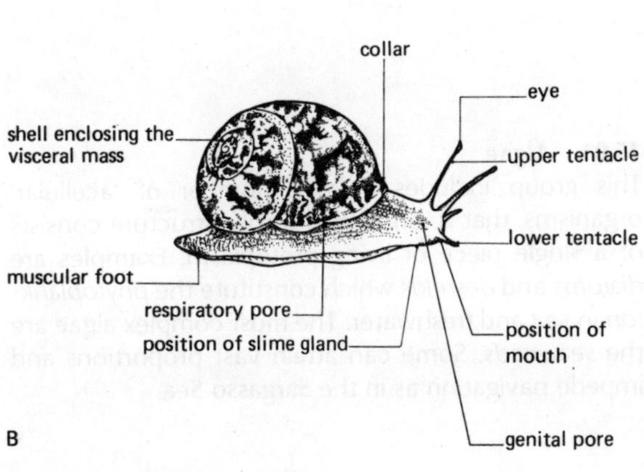

fig 15.3 Organisms A and B

fig 15.4 Organisms C and D

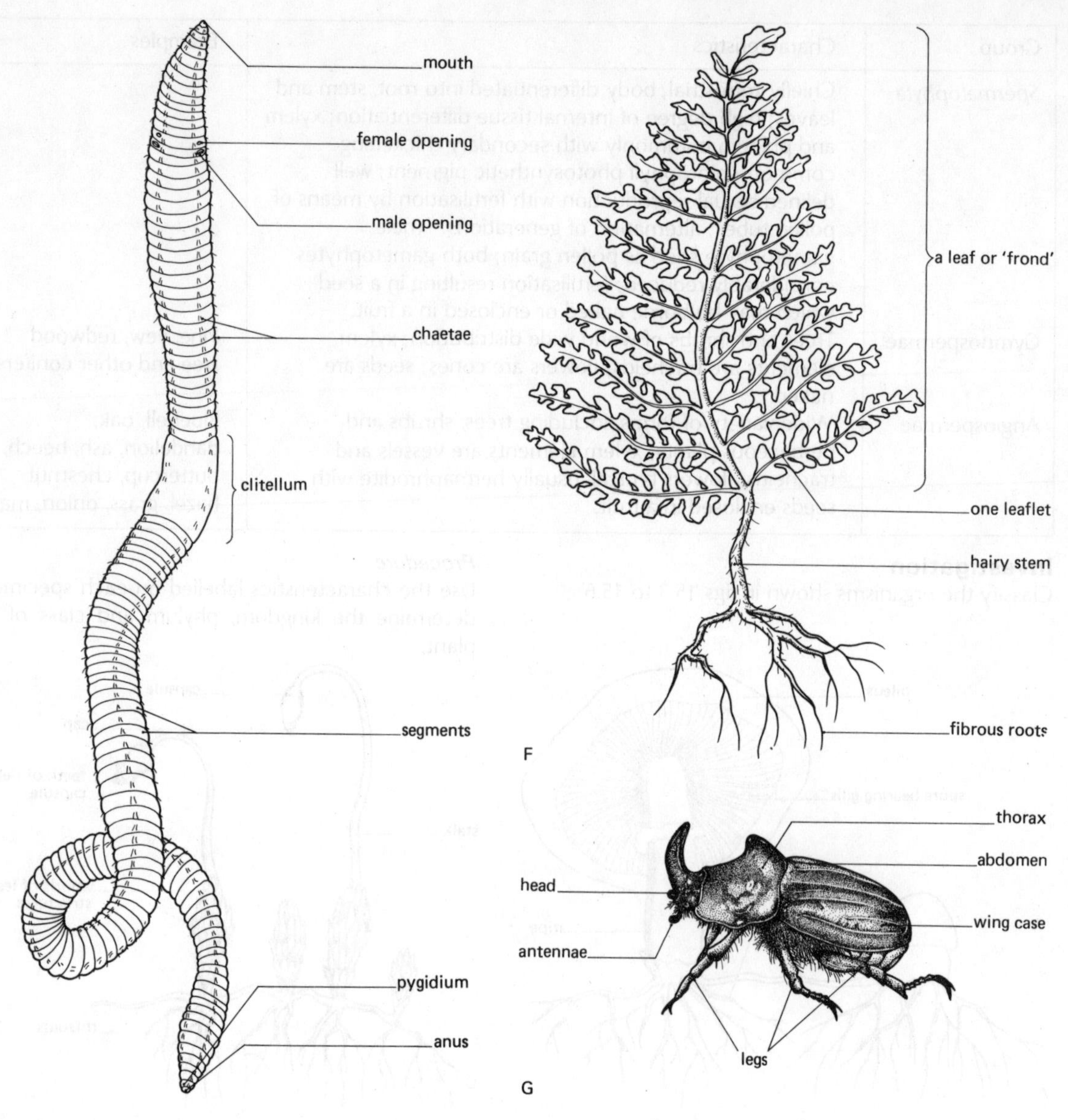

fig 15.5 Organism E

fig 15.6 Organisms F and G

Lower plants

This term is reserved for those plants which do not produce flowers during their life cycle. They include a very wide range of types, from the microscopic bacteria and diatoms, to the more complex ferns. One group, the algae, constitute a dominant group of marine plants. Algae and fungi do not possess the stem, root and leaves characteristic of many plants. Their structure is far simpler. Mosses have stem, and leaf-like structures, but they are lacking roots and the conducting tissues found in higher plants. Ferns have true stems, roots and leaves and conducting tissue.

15.01 Algae

This group includes a large number of 'acellular' organisms, that is, organisms whose structure consists of a single piece of living protoplasm. Examples are *diatoms* and *desmids* which constitute the *phytoplankton* in sea and freshwater. The most complex algae are the *seaweeds*. Some can attain vast proportions and impede navigation as in the Sargasso Sea.

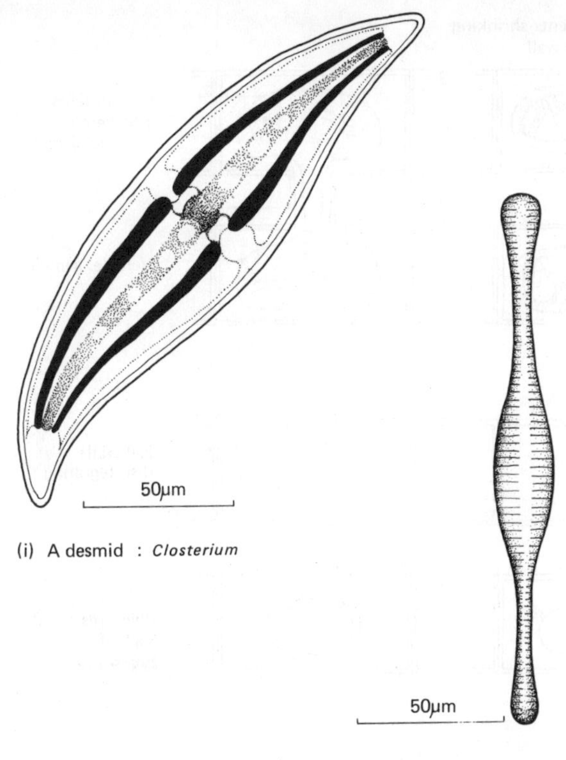

fig 15.7 Phytoplankton

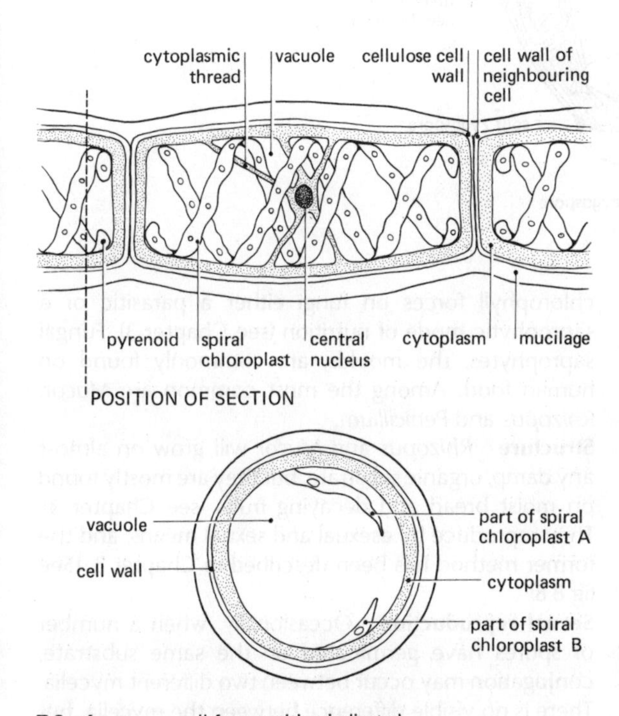

fig 15.8 (a) Individual cells of *Spirogyra* (two views)

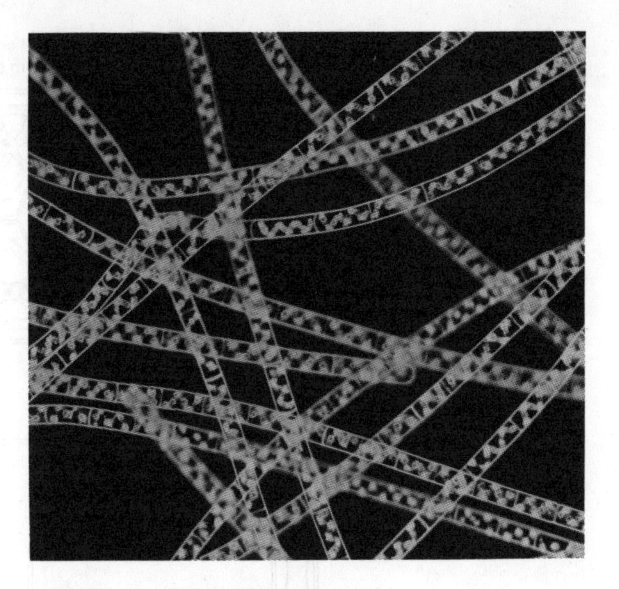

fig 15.8 (b) *Spirogyra* filaments

Spirogyra

Structure *Spirogyra* is one of the algae that forms a slimy green tangle of threads on the surface of ponds and ditches. When examined microscopically, each thread or *filament* is seen to consist of a number of cylindrical cells joined end to end and covered by a thick layer of slime or mucilage (see fig 15.8).

Each cell is enclosed in a firm cellulose cell wall lined by *cytoplasm*, threads of which extend into the centre of the cell and support the *nucleus*. Running through the *central vacuole* and embedded in the cytoplasm is a green spiral band. This is the *chloroplast*. Certain areas in the chloroplast do not contain chlorophyll and appear colourless. These are the *pyrenoids*, thought to be concerned with starch synthesis.

The filament grows in length by the transverse division of each mature cell.

Life history Reproduction is mainly *asexual* by a process called fragmentation (see Chapter 8).

Occasionally, *sexual* reproduction or *conjugation* occurs (see fig 15.9). Two filaments lying parallel may be seen to produce a protuberance from each cell which grows to meet a protuberance from a cell in the opposite filament. When they touch, their end walls break down and a series of *conjugation tubes* are formed between opposite cells. During this time, the cell contents have separated from the cell walls to form a rounded mass in which individual structures are not visible. This is called the *gamete*. Though there is no visible difference between the gametes of different filaments, it is usual for all the gametes of *one* filament only to migrate through the conjugation tubes. Gametes of the other filament remain immobile. The *active* gamete is referred to as the *male*, and the *passive* gamete as the *female*.

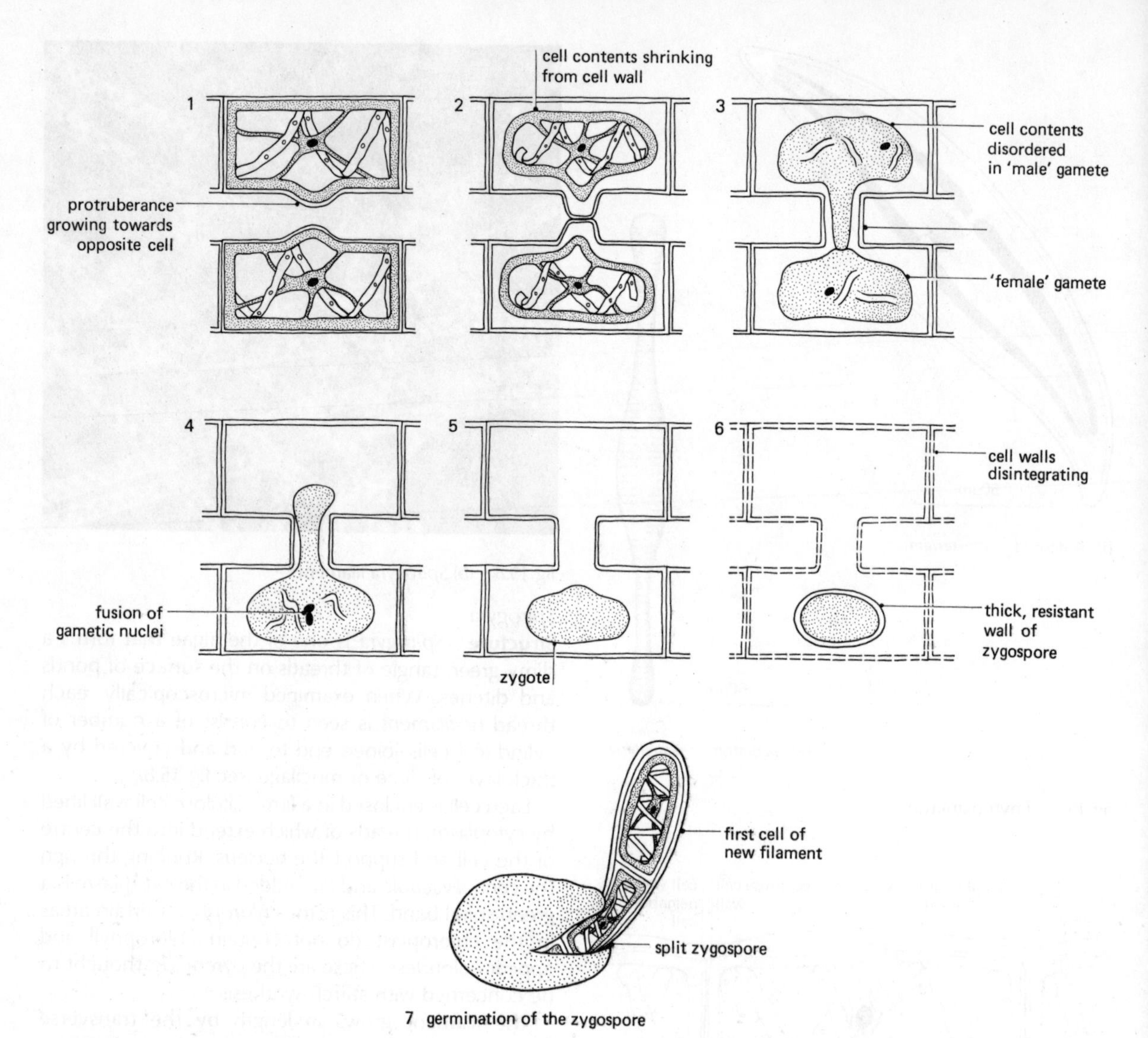

fig 15.9 Conjugation in *Spirogyra*

Each male gamete joins with a female gamete and their nuclei fuse to form the *zygote*. This becomes oval, and enclosed in a thick wall to form the *zygospore*. The zygospore contains oil stores and is important for the survival of *Spirogyra* in adverse conditions such as the drying up of the pond. Germination of the zygospore occurs when conditions are favourable; the wall bursts open and a new filament grows out. Note that each cell behaves independently during conjugation. *Spirogyra* is sometimes regarded as a colony of individual cells held together by mucilage.

15.02 Fungi

This group differs from the algae in that its members do not possess chlorophyll. It includes acellular organisms such as yeast, and large plants such as mushrooms and toadstools and bracket fungi on trees. Lack of chlorophyll forces on fungi either a parasitic or a saprophytic mode of nutrition (see Chapter 3). Fungal saprophytes, the moulds, are commonly found on human food. Among the most common are *Mucor, Rhizopus* and *Penicillium*.

Structure *Rhizopus* and *Mucor* will grow on almost any damp, organic substrate, but they are mostly found on moist bread and decaying fruits (see Chapter 3). They reproduce by asexual and sexual means, and the former method has been described in Chapter 8. (See fig 8.8).

Sexual reproduction Occasionally, when a number of spores have germinated on the same substrate, conjugation may occur between two different mycelia. There is no visible difference between the mycelia, but they are thought to differ physiologically, and are referred to as + and − strains (see fig 15.11). Hyphae

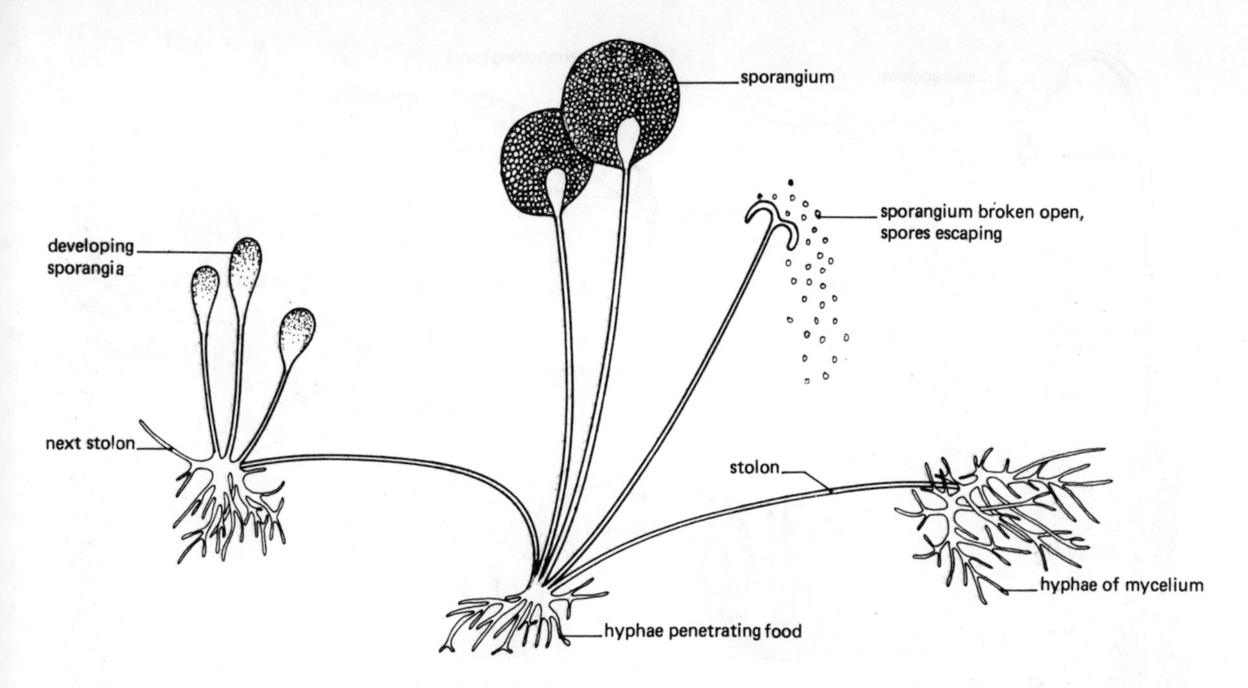

fig 15.10 A mycelium of *Mucor*

from the opposing strains meet and their contents migrate to the tips which become swollen. Cross walls are formed which separate the tips from the remainder of the hyphae. The separated tips are called *gametangia*. The walls between the − and + gametangia dissolve, and nuclei from the − strain fuse with nuclei from the + strain. A thick outer covering of calcium oxalate grows around the resulting structure which is now called the *zygospore*. This is a resistant body which can withstand drought. It can be airborne, and if it lands on a suitably moist substrate, it will germinate to form a sporangium.

15.03 Bryophyta

Mosses form this group together with other shade-dwelling types of plant, called *liverworts*. Liverworts have a flat, wavy form, unlike mosses which are more specialised. Moisture is essential for the life of these plants.

Structure of mosses Mosses grow in clumps on trees, rocks, or moist soil. At first glance, the moss plant appears to have roots, leaves, and a stem, but, microscopically, these structures appear totally different from the true roots, leaves and stems of higher plants. The stalk has no conducting tissue; the leaves are only one cell thick except in the centre, and the *rhizoids*, unlike roots, consist of only a few individual cells. The principal role of the rhizoids is anchorage. They are not the only part of the plant which can absorb water and mineral salts. A typical 'tufted' moss bears a cluster of rhizoids at its base, attached to short

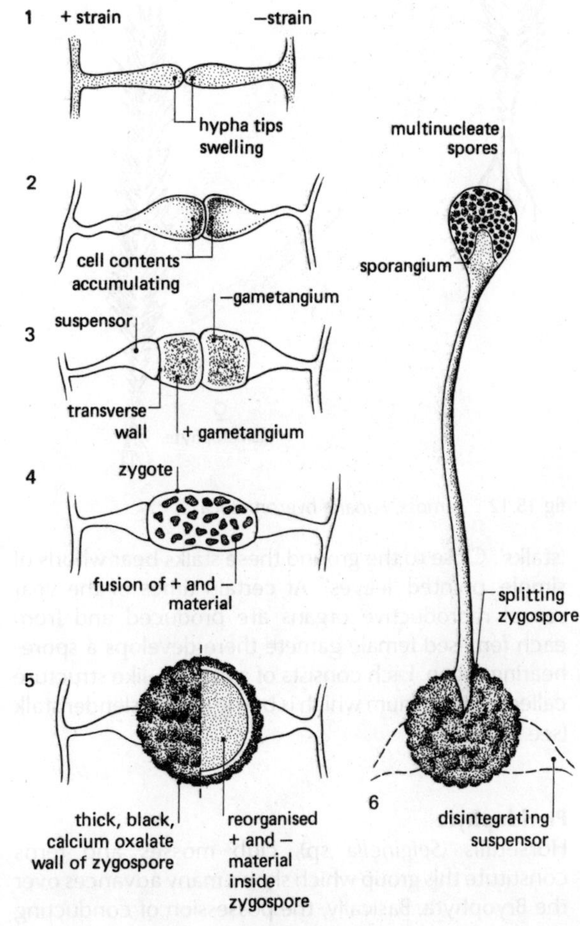

fig 15.11 Conjugation in *Mucor*

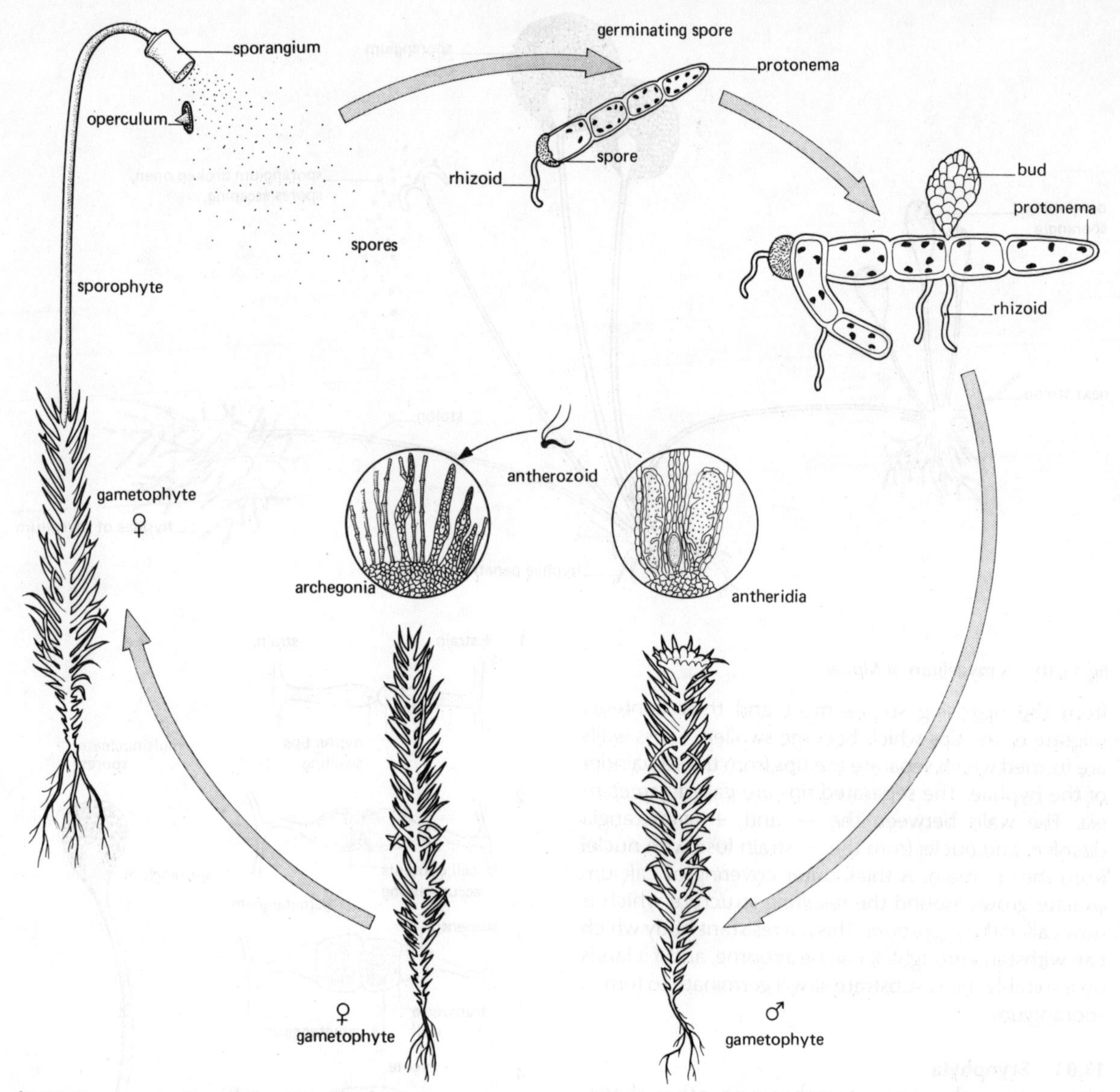

fig 15.12 A moss, *Funaria hygrometrica*

'stalks'. Close to the ground these stalks bear whorls of simple, pointed 'leaves'. At certain times of the year sexual reproductive organs are produced and from each fertilised female gamete there develops a spore-bearing organ. Each consists of a capsule-like structure called a sporangium which is borne on tall, slender stalk (see fig 15.12).

Pteridophyta

Horsetails (*Selginella* sp), club mosses and ferns constitute this group which shows many advances over the Bryophyta. Basically, the possession of conducting tissue has enabled this group to colonise drier habitats where liverworts and mosses cannot survive. True ferns may be found in water, in arid country, or even growing on trees as epiphytes (see Chapter 16).

Ferns

Structure Ferns possess true stems, roots and leaves. Microscopically, they differ very little from those of flowering plants. In most ferns, such as *Dryopteris* the stem is in the form of an underground rhizome (see Chapter 4). Arising from the underside of the rhizome are much-branched *adventitious* roots. These are able to absorb water and mineral salts. The leaves or fronds develop from buds on the rhizome and are the only visible part (see fig 15.13) of the plant. All young fronds and the bases of old ones are covered in brown scales called *ramenta*.

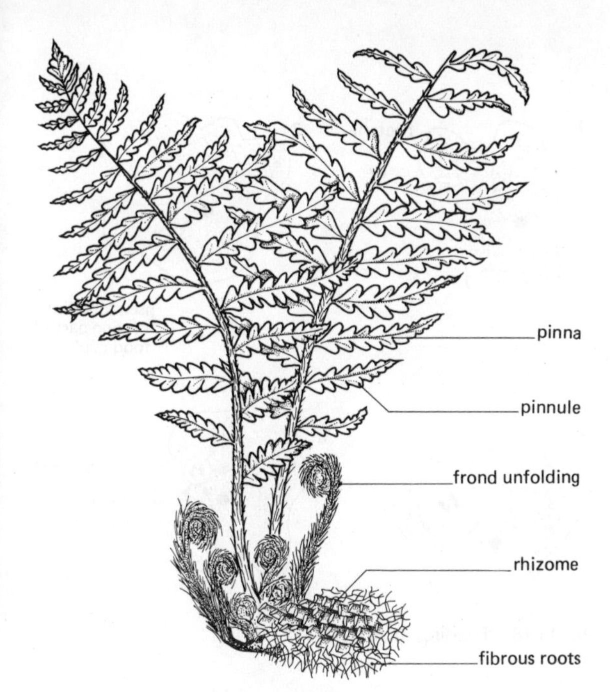

fig 15.13 A fern, *Dryopteris felix-mas*

Mature fronds at a certain season will be found to bear reproductive bodies (sporangia) on their underside (see fig 15.14). In *Dryopteris* these are arranged in clusters called *sori* on the under-surface of the frond leaflets. Each sorus is covered by a protective layer of cells called the *indusium*.

Released spores do not germinate directly to form a new fern plant. They produce a minute, heart-shaped structure called the *prothallus* (see fig 15.15). Sexual reproductive organs are borne on the underside of the prothallus. The male organs produce mobile gametes which swim to the female organs and fertilise the female gamete. The resulting zygote germinates to produce a new fern plant. This alternation of asexual reproduction in the fern plant and sexual reproduction in the fern prothallus is called *alternation of generations*.

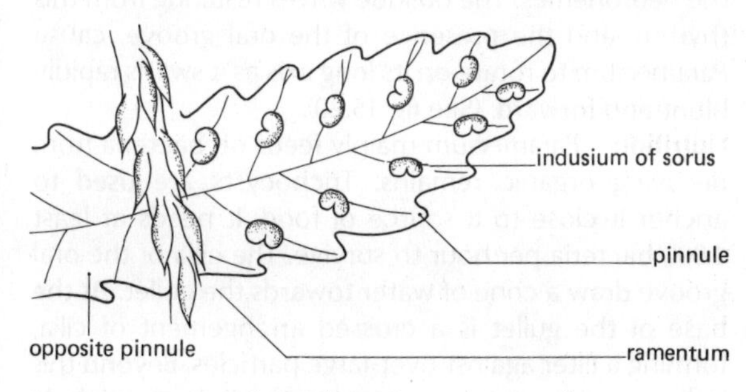

fig 15.14 Sori on the undersurface of a pinnule

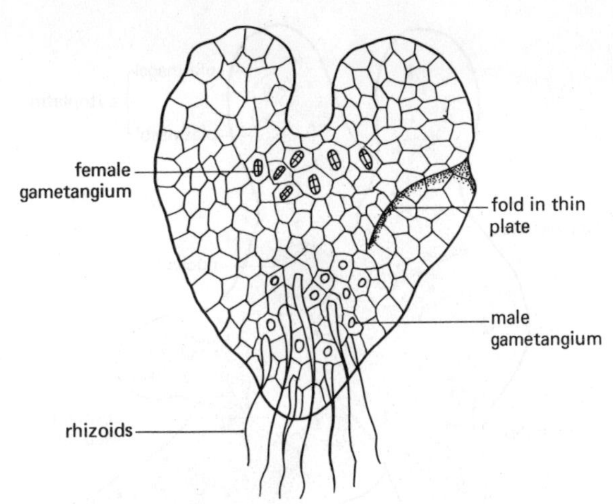

fig 15.15 A fern prothallus

15.10 Lower animals

The term 'lower animals' is generally equated with 'invertebrate animals', that is animals without a vertebral column or backbone. Lack of an endo-skeleton has limited the size of the vast majority of invertebrates. The largest invertebrates are octopuses, squids, cuttlefishes and large crustaceans (e.g. lobsters). Their size is modest when compared with that of a whale or an elephant. Lack of size should not be equated with lack of success. More than 75 per cent of all the animal species in the world belong to the Class *Insecta*.

The simplest lower animals are the *Protozoa* or acellular animals. Much of the zooplankton is composed of *Protozoa*.

15.11 Protozoa

Amoeba proteus

This protozoan is found in mud at the bottom of ponds and ditches.

Structure It is one of the larger examples of the group, reaching a maximum length of 0.25 mm. It can just be seen with the naked eye. Microscopically, it is seen as a jelly-like form which is constantly changing its shape. A dense spherical area, the nucleus, is surrounded by greyish cytoplasm in which granules are constantly streaming. This is the *plasmasol* (endoplasm). At the margins of the cell, surrounded by the plasma (cell) membrane, is a clear outer layer of cytoplasm, the *plasmagel* (ectoplasm). (See fig 15.16). Contained within the plasmasol are clear regions containing debris of bacteria and diatoms. These are the *food vacuoles*. A larger clear region, whose size increases until a maximum is reached is the *contractile vacuole*. This suddenly disappears on reaching maximum size only to grow again from a small vacuole of water. It is the means by which *Amoeba* gets rid of excess water.

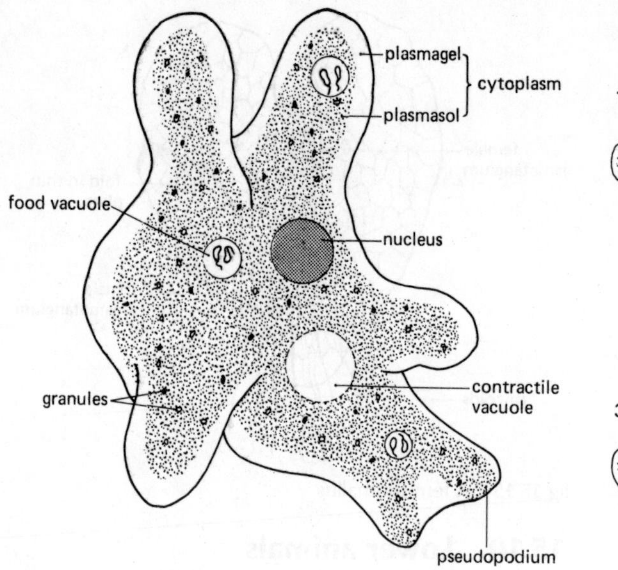

fig 15.16 *Amoeba proteus*, a Protozoan

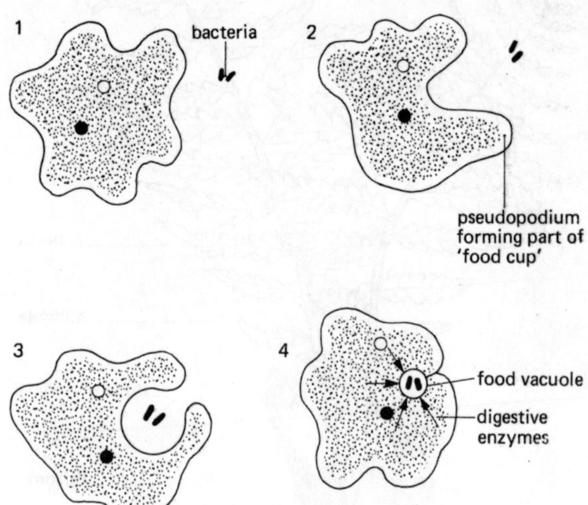

fig 15.18 Feeding in *Amoeba*

Movement Movement is a slow, streaming motion. Plasmasol streams over plasmagel to produce a projection called a *pseudopodium*. As one pseudopodium is formed in a particular direction, so another is withdrawn. (See fig 15.17.)

Nutrition *Amoeba* feeds on bacteria, diatoms, desmids and decayed remains of larger plants. Ingestion is by a process of *engulfing*. On encountering suitable food, pseudopodia encircle it together with a drop of water and a food vacuole is formed. (See fig 15.18.) Enzymes are then passed from the plasmasol into the vacuole and digestion occurs. Soluble products diffuse into the plasmasol and any undigested remains are passed out through the plasmagel.

Paramecium caudatum

Paramecium caudatum is found in freshwater ponds and ditches where there is a high concentration of decaying organic matter. Its distribution essentially coincides with that of bacteria.

Structure It is about 0.24 mm (240 μ) long and again is just visible as a speck to the naked eye. It has a definite shape and is referred to as the 'slipper animalcule' because of its slipper shape (see fig 15.19). The anterior end is rounded, the posterior pointed. The shape is maintained by a well-developed *pellicle*. This is virtually an elastic exoskeleton the outer layer of which is the cell membrane. It is constructed in hexagonal ridges from which rod-like organs, the *trichocysts* emerge from pits. Pairs of *cilia* arise from depressions between the ridges. The *basal granules* of the cilia are connected by conducting fibres called *neuronemes*.

Lining the pellicle is the clear plasmagel, inside which lies the granular plasmasol. Intruding on one side of the animal into the plasmasol is the *oral groove* which leads into the *gullet*. *Food vacuoles* form at the base of the gullet and enter the plasmasol. Also contained in the plasmasol are an anterior and a posterior contractile vacuole. These are fed by *formative canals*. Two nuclei are centrally placed in the plasmasol, the larger is called the *meganucleus* and the smaller, close to it, the micronucleus.

Movement The cilia beat rhythmically producing an effect like wind passing over a field of young corn. This is *metachronal* rhythm and it is effected by the neuronemes. The oblique waves resulting from this rhythm, and the presence of the oral groove, cause *Paramecium* to rotate on its long axis as it swims rapidly blunt end forward. (See fig 15.20).

Nutrition *Paramecium* mainly feeds on bacteria from decaying organic remains. Trichocysts are used to anchor it close to a source of food. It needs at least 1 000 bacteria per hour to survive. The cilia of the oral groove draw a cone of water towards the gullet. At the base of the gullet is a crossed arrangement of cilia, forming a filter against over-large particles. Beyond this is the *cytostome*, a soft area of plasmasol where a globule

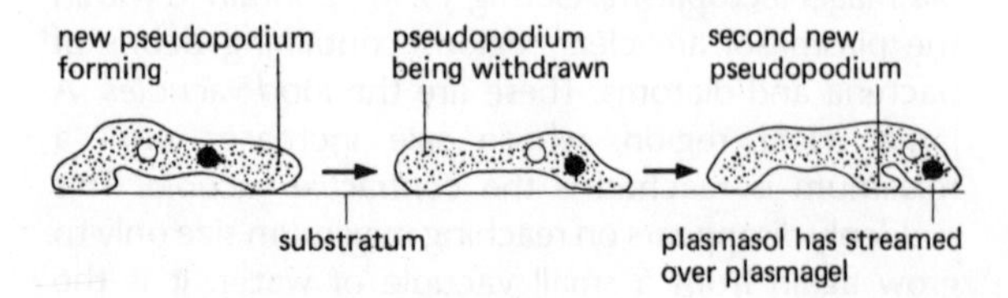

fig 15.17 Movement in *Amoeba*

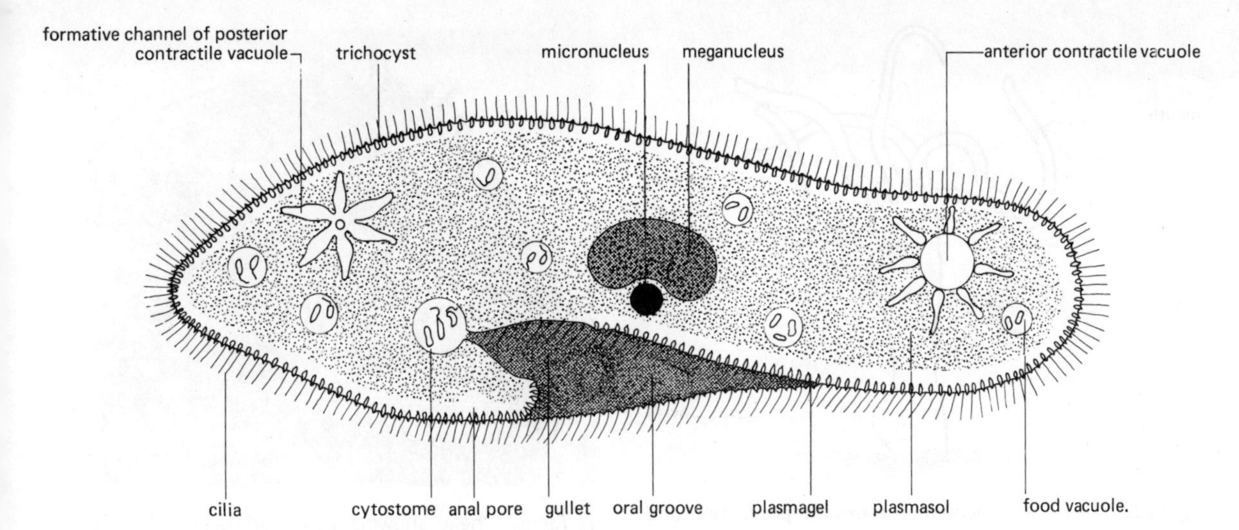

fig 15.19 *Paramecium caudatum*, a Protozoan

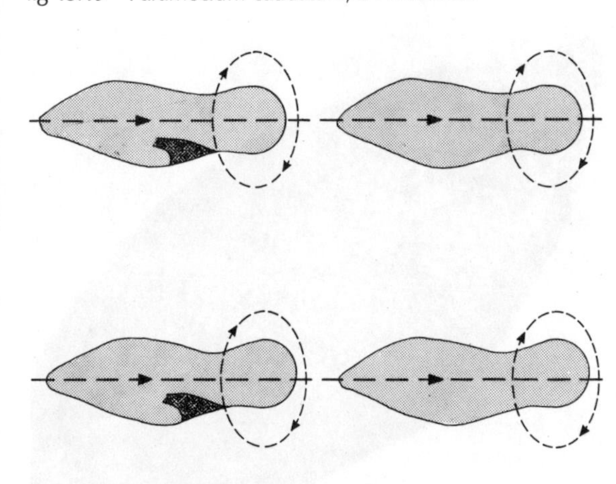

fig 15.20 Movement in *Paramecium*

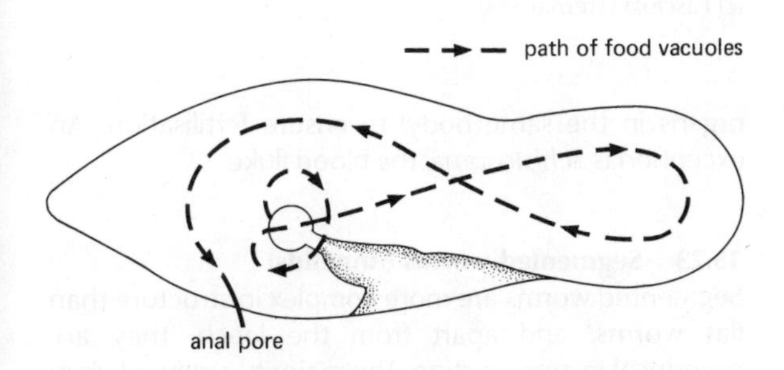

fig 15.21 Route of food vacuoles in *Paramecium*

of water containing food particles increases in volume and forms a food vacuole. Food vacuoles are formed continuously and circulate in a set route in the plasmasol. (See fig 15.21.) During circulation, digestion occurs as in *Amoeba*. Undigested waste is discharged at the temporary *anal pore* behind the cytostome. The pellicle is less well formed at this point.

15.20 Metazoa

The bodies of the majority of the lower animals consist of many cells and together with the vertebrates belong to the group known as the *Metazoa*.

15.21 Coelenterata

Sea anemones and corals

These organisms, although multicellular, do not have their cells organised into tissues. The body consists of only two layers of cells (ectoderm and endoderm), hence they are described as diploblastic. The phylum includes sea anemones, jellyfish and corals.

Hydra

This is a small coelenterate common in freshwater ponds. It has a hollow cylindrical body. At the top is a small cone, in the centre of which is the mouth. Around the base of the cone are six to eight hollow tentacles. At the opposite end of the body there is an adhesive disc by which the organism can attach itself to a stone, a water plant or even to the underside of the water surface. The mouth opens into a single digestive cavity.

Hydra feeds on small water crustaceans such as the water flea (*Daphnia*) which it captures by means of its tentacles. These contain small, stinging threads discharged by special cells, the nematoblasts. The water flea is then pushed through the mouth and into the digestive cavity. Once in the enteron it is digested and its products absorbed by the endoderm cells. The indigested remains are egested through the mouth.

Hydra can reproduce asexually by means of buds (see fig 15.22). It is also hermaphrodite, producing testes in the upper half of the body and ovaries in the lower half. The testes burst and liberate sperms into the water. These swim and fertilise the ova of another *Hydra*. The fertilised egg divides, producing a hollow ball of cells

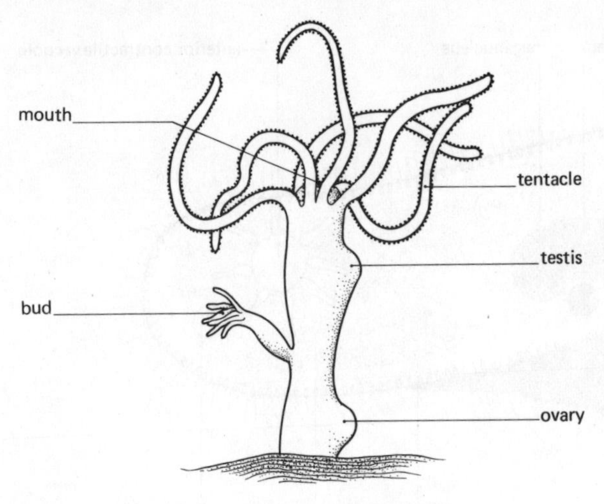

fig 15.22 *Hydra*, a coelenterate (asexual reproduction and sexual reproduction)

(blastula) around which is secreted a horny case. The embryo in its case falls to the bottom of the pond and develops into a new *Hydra* in the following spring.

15.22 Flatworms (Platyhelminths)

These are the most primitive of worms. A few members of the group, the Turbellaria, are free-living and are found in fresh water, but the majority of the group are very specialised parasites, some of them in association with Man. All the worms are flat-bodied dorso-ventrally, but their general shape varies enormously. Turbellarians are flat and ciliated for free swimming; liver flukes are shaped like a leaf and tapeworms are very long, consisting of numerous egg-laden segments, the proglottides. A tapeworm may be as long as 2½ m. Blood flukes are thread-like and very small. Many of the parasitic species have complex life cycles involving larval forms in a secondary host. Most parasites are hermaphrodite (have male and female reproductive organs in the same body) to ensure fertilisation. An exception is *Schistosoma*, the blood fluke.

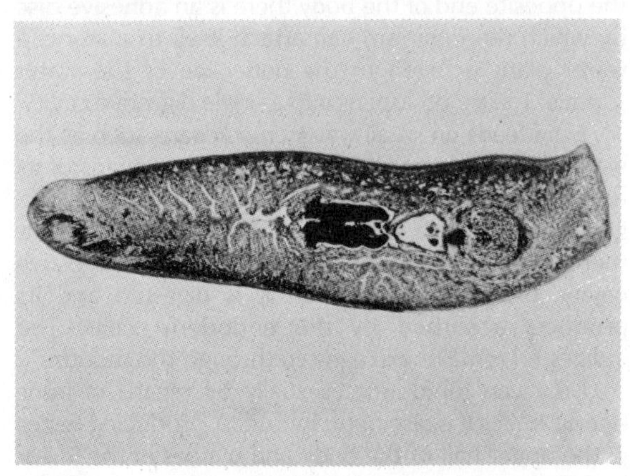

(a) *Planaria* (Turbellaria)

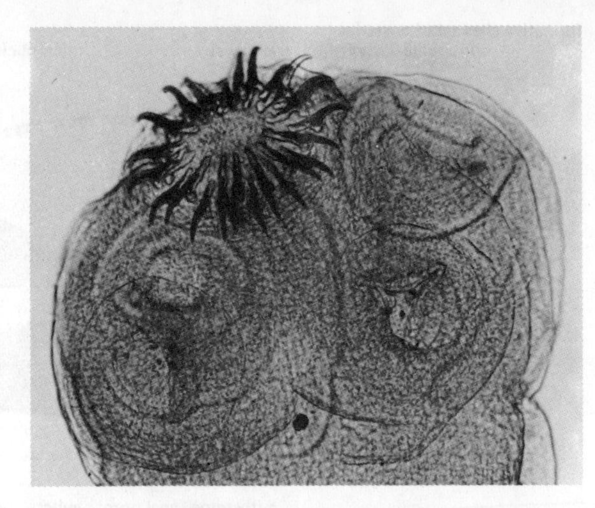

(b) *Taenia* – 'head' showing hooks and suckers (Cestoda)

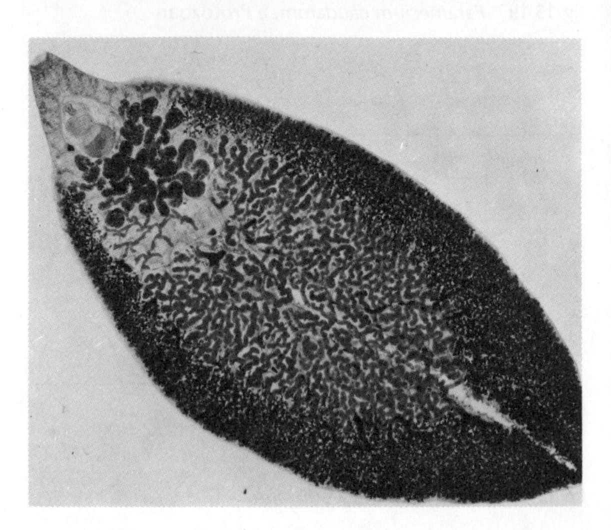

(c) *Fasciola* (Trematoda)

fig 15.23 Examples of flatworms (*Platyhelminthia*)

15.23 Segmented worms (Annelids)

Segmented worms are more complex in structure than flat worms, and apart from the leech, they are cylindrical in cross section. The majority are free-living; some are marine e.g. bristleworms, others are fresh-water e.g. *Tubifex* worm and leeches, but the commonest live in the soil e.g. earthworm. In all cases, the body is divided into similar sections called segments (see fig 15.24).

Earthworm

There are several species of earthworm, but the differences between them are small. They are found

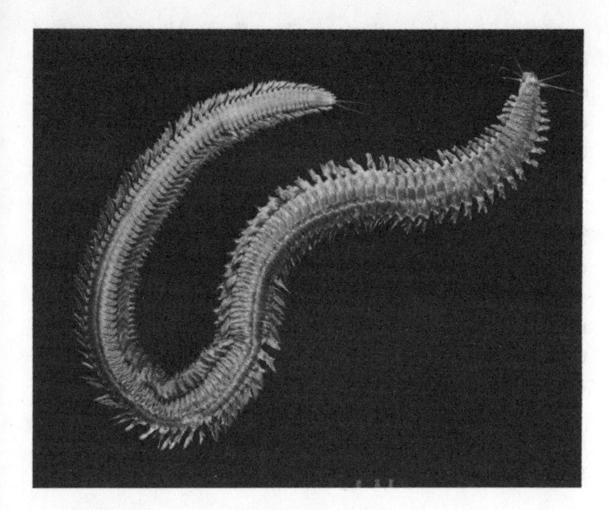

(a) bristleworm

(b) lugworm

(c) leech

fig 15.24 Examples of Annelids

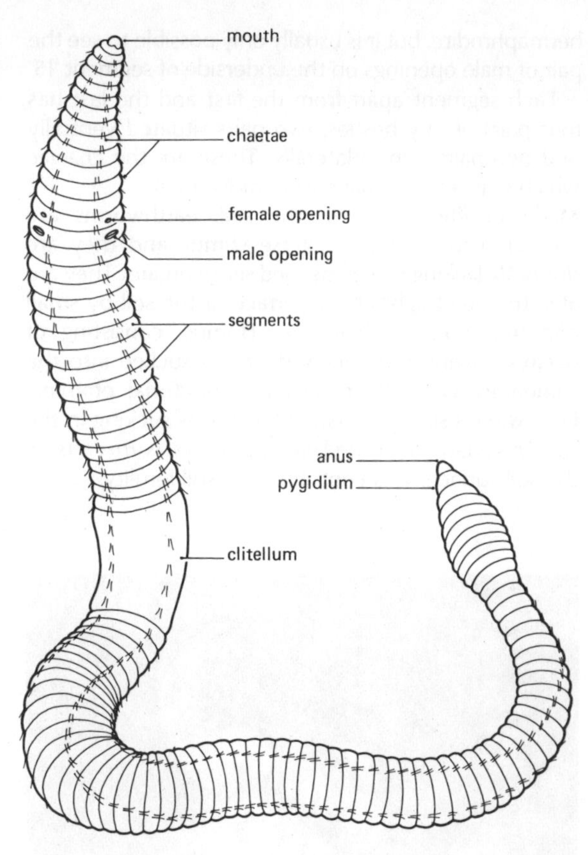

fig 15.25 *Lumbricus sp*, an earthworm

commonly in damp soils rich in humus, but are rarely to be found in dry, sandy soil. They make tunnels underground and usually only emerge at night, or after heavy rainfall.

Structure An adult earthworm is up to 15 cm long. It has a streamlined shape, tapering at anterior and posterior ends. The posterior end is more flattened than the rest of the worm. There are usually about 150 segments from mouth to anus. The dorsal surface is a brown-reddish colour, the ventral surface being paler for camouflage. The earthworm feels soft and moist to the touch because of the secretion of a lubricant from small glands in the skin. In addition to helping movement, this secretion is a bactericide and fungicide preventing the skin from becoming diseased. Gaseous exchange occurs through the skin, aided by this moist secretion. A blood vessel is usually visible beneath the skin in a median dorsal position.

Anterior to the first segment is the flap-like *prostomium* which overhangs the mouth. The *peristomium,* which is the first segment, surrounds the mouth. About 30–36 segments from the anterior in a sexually mature worm is the saddle or *clitellum*. This is a pale glandular structure which produces the cocoon into which the worms lay their eggs. Earthworms are hermaphrodite, but it is usually only possible to see the pair of male openings on the underside of segment 15.

Each segment apart from the first and the last has four pairs of tiny bristles, two pairs situated ventrally and two pairs ventro-laterally. These are the *chaetae* which play an essential part in movement.

Mode of life Living in the soil, earthworms are protected from many adverse stimuli and they are distinctly lacking in well-defined sense organs. They are able to detect light at the surface of the soil by small *photoreceptors* in their skin. Humus, consisting of decaying plant and animal debris, is sucked into the alimentary canal by means of a suctorial pharynx. Burrowing is similarly helped by worms sucking in the soil. This is later deposited via the anus as *worm casts* on the soil surface. This contributes to soil fertility.

Movement through the soil is effected by two sets of muscles below the skin, and by the soil-gripping chaetae. The anterior extends by contraction of the circular muscles. Chaetae at the anterior anchor it into the ground and the posterior segments are pulled forwards by contraction of the longitudinal muscles.

When worms reproduce, they copulate, becoming attached in the anterior regions from head to clitellum, and exchange sperm. They then separate. The sperms are stored in sperm sacs and are later passed into the cocoon together with the eggs. Fertilisation occurs inside the cocoon. The incubation period is about a month.

Worms have some powers of regeneration. If a worm is cut in two, the head end only will regrow the missing segments.

(a) squid

(b) file shell

(c) whelk

(d) octopus

fig 15.26 Examples of Molluscs

15.24 Molluscs

This group of animals includes all those invertebrates possessing a shell. Normally the shell affords external protection, but in slugs, squids, cuttlefishes and octopuses the shell is reduced and its remains have become enclosed inside the body. Movement is brought about in most cases by a large muscular *foot*. Tentacles, bearing sense organs, are a common feature of the group. Many molluscs are found on the seashore, including bivalves with hinged shells such as scallops mussels, and limpets (see fig 15.26).

Roman snail (Helix pomatia)
This snail lives amongst wet vegetation in shady places. It is particularly plentiful during the rainy season. (See fig 15.27.)

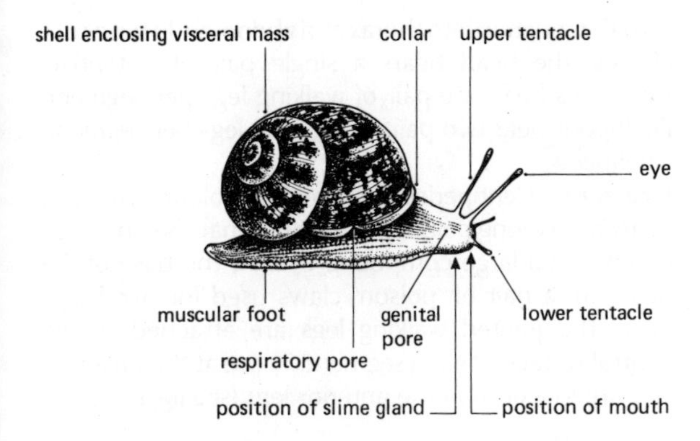

fig 15.27 *Helix pomatia*, the Roman snail

Structure The body is extremely soft and can be completely retracted into the hard calcareous shell.

The body consists of the head, foot and hump or *visceral mass*. The visceral mass is permanently inside the shell and is coiled in a similar way to the shell. It contains all the main organs. The head bears two pairs of tentacles, the upper, longer pair being tipped by small black eyes. The mouth is on the underside of the head and has three lips. The head passes imperceptibly into the foot which is moist, wrinkled and muscular. The shell is coiled and yellow, with dark markings. At the edge of the shell is the thick fleshy *collar* which is a part of the visceral mass.

Mode of life This mollusc feeds on plant material. It grips the food with its lips and rasps the pieces off with its sandpaper-like tongue or *radula*.

Movement is effected by contractions of the muscular foot, and progression is aided by the secretion of slime from a slime gland below the mouth. Snail slime trails may be seen on vegetation. In adverse conditions, the snail retracts into its shell and secretes a thin membrane over the aperature. When the weather is dry, the snail remains dormant inside its shell.

Snails are hermaphrodite, but mating occurs and sperms are exchanged. The sperms are stored and used later to fertilise the ova in each snail. Eggs are laid in batches which hatch in a few days.

15.25 Arthropods

This is the largest group of animals. It includes all those invertebrates which have a jointed exoskeleton. The exoskeleton is made of a tough substance called *chitin*. In order to grow, all arthropods must shed their exoskeleton from time to time. This is called moulting or *ecdysis* (see Chapter 10). Muscles are attached to the inside of the exoskeleton, so it plays a vital part in movement as well as protection. Segmentation is evident in all arthropods at some stage in their life history.

Arthropods are dominant invertebrates in the sea (e.g. crabs, lobsters), in freshwater (freshwater shrimps, waterboatmen, pondskaters), in the soil (centipedes, millipedes, termites, ants, scorpions) and in the air (mosquitoes, dragonflies, etc). Figs 15.23 to 15.27 show a variety of these organisms.

Arachnids
Spiders, scorpions and ticks belong to this class of arthropods. They have a body divided into two parts: the cephalothorax and the abdomen. Four pairs of walking legs are attached to the underside of the cephalothorax. The head bears swollen palps and no antennae.

Structure The cephalothorax consists of the fused head and thorax. The head area bears a pair of compound eyes on the dorsal side. Ventrally, the mouthparts are jointed and end in *pincers*. *Palps* or feelers lie close to the mouthparts. The pincers of some spiders produce a poisonous venom for paralysing prey.

Four pairs of long, jointed, hairy legs arise from the ventral surface of the thoracic region of the cephalothorax.

The abdomen is broad, compact, and segmented; *spinnerets* are situated on the posterior segment. These are used to spin the thread for web construction.

Mode of life Spiders are basically carnivorous, and insects are their main diet. Some species hunt their prey, but many construct webs in which insects can become trapped in the sticky silk. Silk may be used to transport young spiders for long distances and as a 'life-line' if a spider falls from its web. Some spiders may use it for wrapping eggs.

Breathing in spiders is partly by tracheae as in insects (see Chapter 5) and partly by *'lungbooks'*. These are like blood-filled plates, suspended in a small chamber which is open to the air. Air is actively pumped into the chamber.

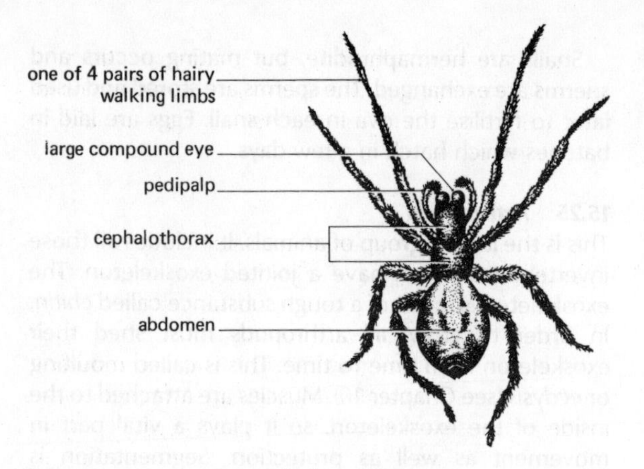

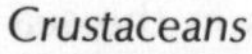

fig 15.28 A common spider (*Arachnida*)

Crustaceans
Crustaceans are mainly aquatic arthropods which breathe with gills found on their thoracic or abdominal segments. They all bear two pairs of antennae on their head and all possess more jointed limbs than arachnids. Examples include crabs, crayfish, barnacles, shrimps, waterfleas and woodlice. Woodlice live in dark, moist places on land.

Prawn
These belong to the same order as crabs, lobsters and crayfish, though they are considerably smaller. All these animals, the *Decapods* are characterised by possessing ten walking legs.
Structure The exoskeleton is particularly hard in decapods, as the chitin is reinforced with chalk. The head and thorax are fused and are covered by the *carapace*. The head bears one very long pair of jointed antennae, and a second, shorter pair. There are a pair of compound eyes at the end of moveable stalks. The mouthparts are complex, and adapted for scavenging.

The first pair of the the five pairs of walking legs arising from the ventral surface of the thorax are modified to form long limbs with pincers for grasping prey. At the base of the walking legs are feathery gills for gaseous exchange.

Six individual segments of the abdomen are visible. On the ventral surface of the first five segments are paired flattish plates called *swimmerets* which aid swimming. The last segment bears a pair of tail plates which help propulsion backwards through the water.

Myriapods
This class includes both Chilopods (centipedes) and Diplopods (millipedes). All species of Myriapods are terrestrial. Chilopods have a flattened body of 15–20 segments, Diplopods have more segments and a more cylindrical body. In both types, the segments are all similar except for those of the head regions. There is

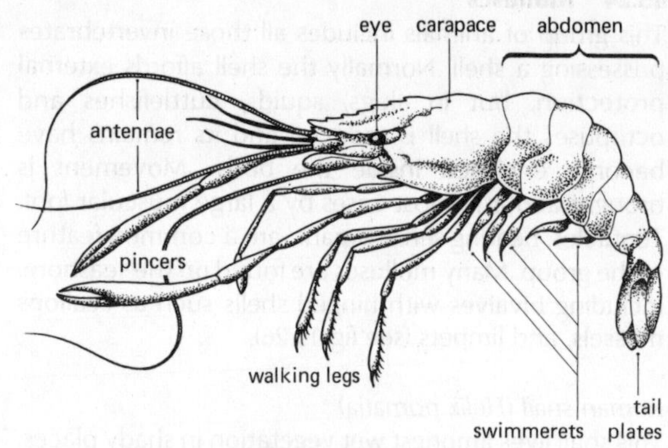

fig 15.29 A prawn (*Crustacea*)

no differentiation into thorax and abdomen. In both subclasses, the head bears a single pair of antennae. Chilopods bear one pair of walking legs per segment, Diplopods bear two pairs of walking legs per segment.

Centipede
Structure Centipedes are brown in colour. The head bears a very long pair of jointed antennae. Behind each antenna is a large compound eye. At the base of the head are a pair of 'poison' claws used for paralysing prey. The jointed walking legs are attached to the ventral surface of each segment. Those at the posterior end are longer than the anterior legs (see fig 15.30).

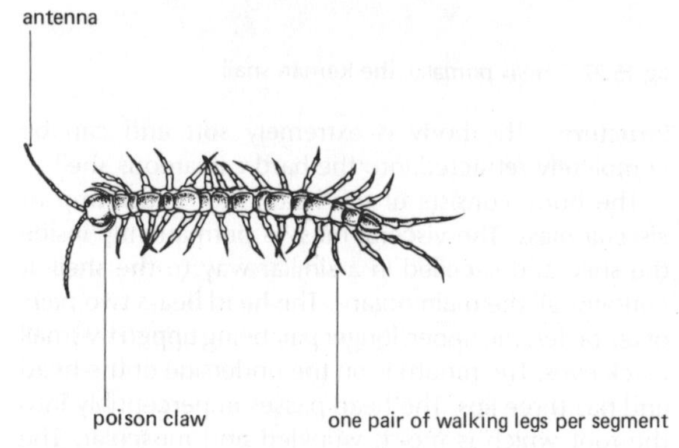

fig 15.30 A centipede (*Myriapoda*)

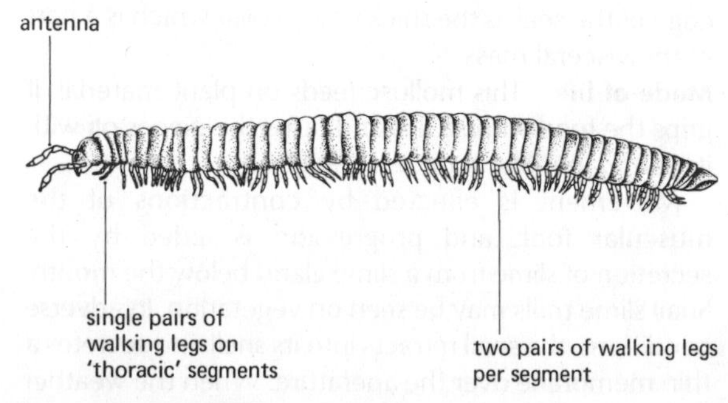

fig 15.31 A millipede (*Myriapoda*)

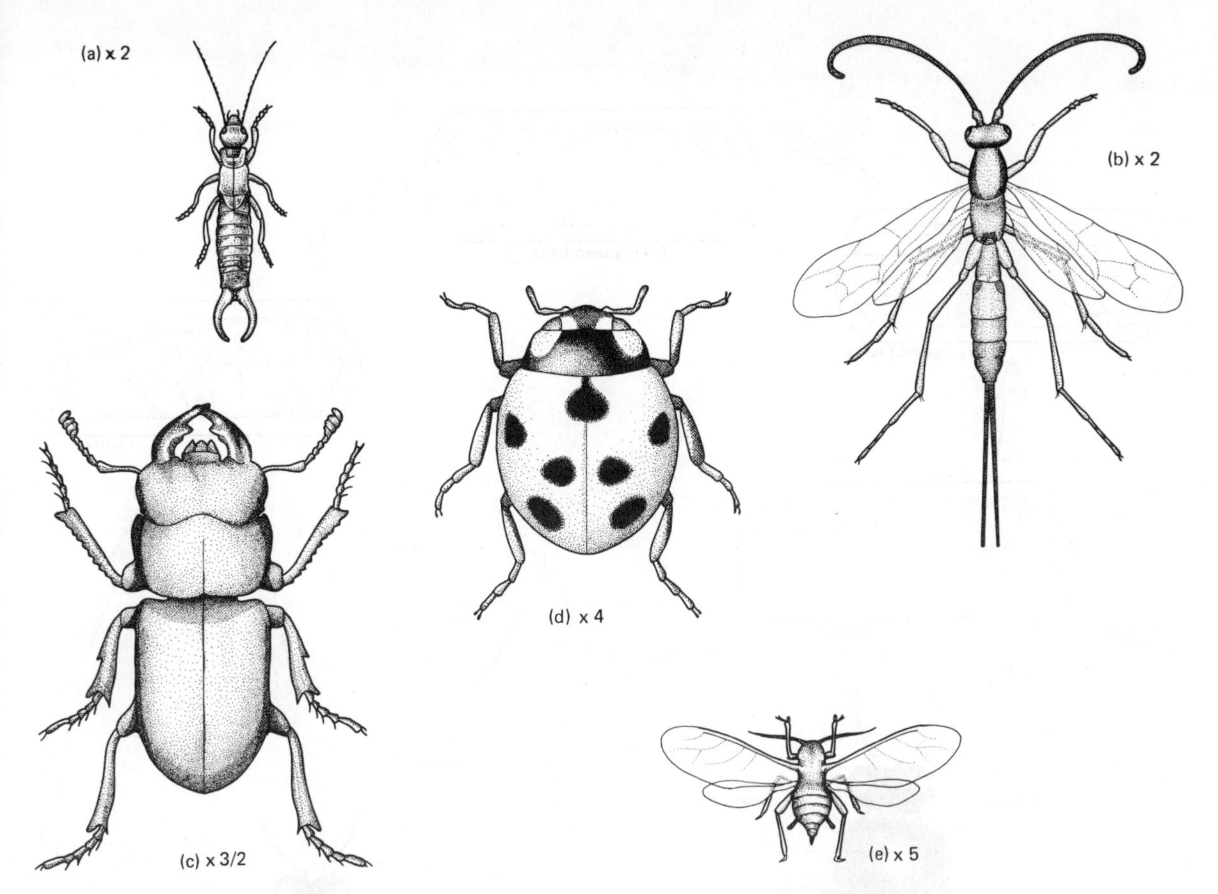

fig 15.32 Examples of insect (a) Earwig (b) Ichneumon fly (c) Stag beetle (d) Ladybird (e) Aphid

Mode of life Centipedes are carnivorous. Their prey includes insects, spiders and worms which they actively chase and paralyse.

Millipedes

Millipedes are dark in colour and their segments appear narrower than those of centipedes. The head is less conspicuous than that of a centipede and the antennae much shorter and the eyes smaller. The mouthparts are palp-like. The dozens of jointed walking legs are very much smaller in relation to total body size than those of centipedes.

Mode of life The less well developed sense organs and smaller legs of the millipede are related to its mode of life. It does not have to pursue prey since it is herbivorous and lives on leaf litter. It burrows into the soil and if touched, curls up into a flat coil for protection.

Insects

Numerically, insects are the most biologically successful group of animals. They have colonised most types of habitat and the variety of their adaptations is very great (see fig 15.32). Despite this variety, certain features are common to all insects. Every insect has a body divided into a distinct head, thorax and abdomen in its adult form. All insects possess large compound eyes for efficient vision. They all possess paired antennae, and three pairs of walking legs attached to the underside of the three segments of the thorax. Most of them possess one or more pairs of wings which arise from the dorsal surface of the thorax.

Insects are classified into two groups depending partly on their life history. The first group includes cockroaches, locusts and grasshoppers. Their life cycles include *incomplete* metamorphosis (see Chapter 10).

The second group includes the majority of insects, all beetles, butterflies, moths, bees, wasps, true flies etc. Their life cycles exhibit *complete* metamorphosis (see Chapter 10).

Cabbage White Butterfly (Pieris brassicae)

Fertilisation is internal. The eggs are small, conical, ribbed structures, yellow in colour and are laid on the undersurfaces of leaves of members of the cabbage family (Cruciferae) particularly those of *Nasturtium*.

The larval stage is called a caterpillar and it hatches from the egg about a week after laying. They are hairy, yellow-green in colour with black or brown papillae. The body has a head, three thoracic segments and ten

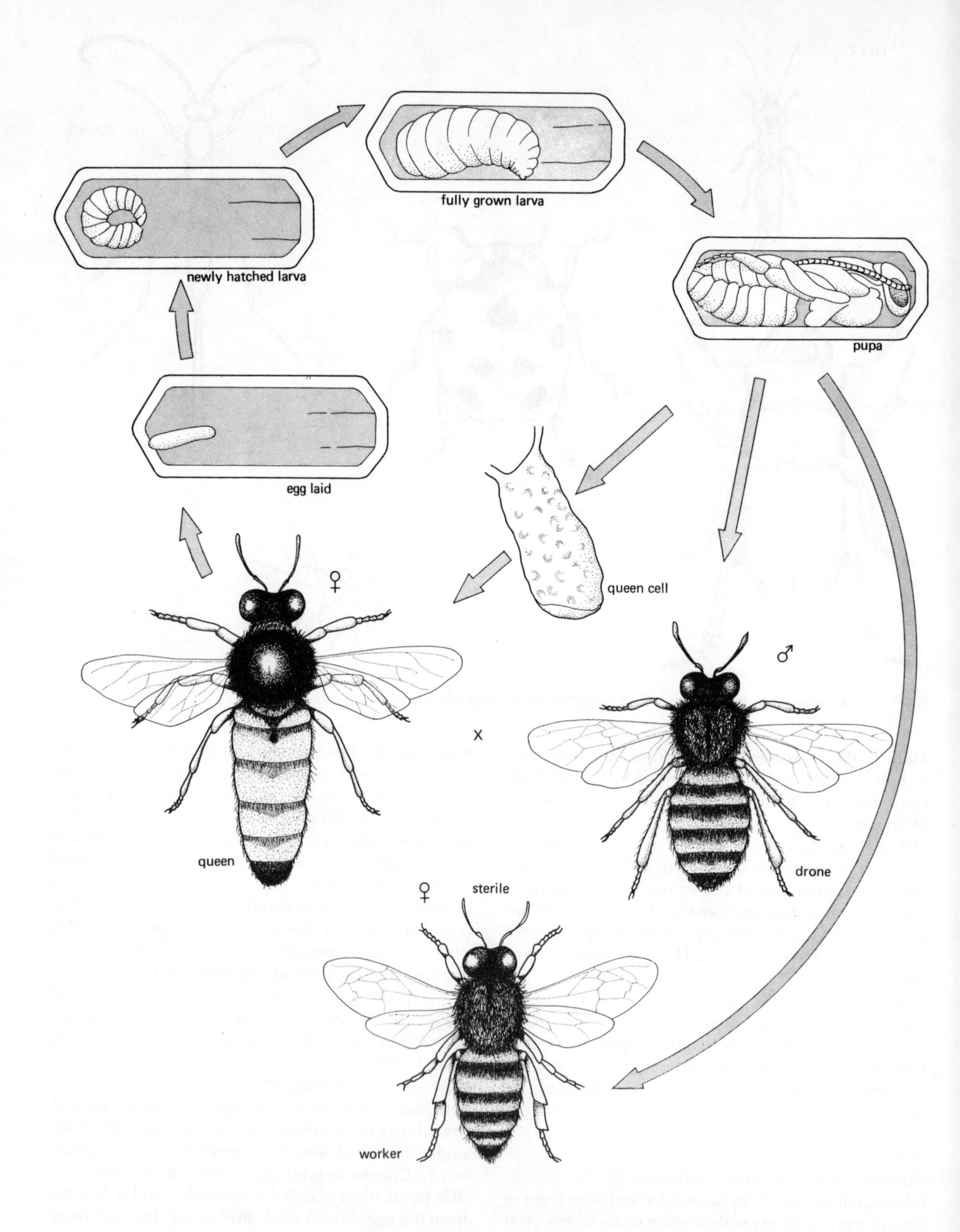

fig 15.33 Life history of the honey bee (*Apis mellifera*)

abdominal segments. Each thoracic segment has a pair of five jointed walking legs and segments three to six of the abdomen each have a pair of pro-legs. The latter are fleshy with a semi-circular series of hooks by which the animal clings to its silk thread. The last segment also has pro-legs which form a pair of claspers. The head has three pairs of simple eyes or ocelli and large biting mouthparts (mandibles) with which the larva feeds continuously. Projecting from the labium is a spinneret which produces a silken thread secreted by the paired silk glands. During this process the head moves from side to side so that a sticky silk ladder is formed to which the larva clings with its pro-legs. The caterpillar sheds the exoskeleton five times, growing between moults until it reaches full size and finally, before the last moult, it climbs upwards to pupate.

It now spins a silk girdle by which it is suspended. It moults and then the exoskeleton hardens to form the pupa. It remains in the pupal stage for three weeks or for the whole winter if pupation takes place at the end of the summer. During this time the internal structure undergoes complete reorganisation to form the body of the adult butterfly. The sex of the adult that emerges is distinguishable by the spots on the wings (see fig 10.18).

Each imago consists of head, thorax and abdomen, together with two pairs of wings and three pairs of legs on the thorax. The head has two large compound eyes and a long coiled proboscis which is used for sucking up nectar from flowers. The head also bears a pair of antennae. The male is attracted to the female by her scent which he picks up by his antennae. Mating takes place during flight.

The Cabbage White Butterfly produces larvae which attack food crops, particularly cabbages, cauliflowers and Brussels sprouts. These crops can be sprayed with insecticide, but biological controls also operate with many birds feeding on the larvae and certain ichneumon flies laying their eggs in the bodies of the caterpillars: the ichneumon eggs hatch into larvae which feed on the caterpillar tissues.

The hive bee (Apis mellifera)

Fertilisation is internal. The young female (queen) leaves the hive and on her nuptial flight is followed by numerous males (drones). One drone succeeds in mating with her and then dies immediately. The queen now has a store of spermatozoa sufficient to ensure egg fertilisation for the next three years. Each egg is laid singly in a cell of the honeycomb. The queen lays two types of egg: fertilised eggs, which develop into sterile females (workers) and fertile females (queens), and unfertilised eggs, which develop into males (drones). The fertilised egg develops into a worker or a queen, depending upon the type of food provided for the larvae by the workers. The cells of the queen larvae are enlarged and altered compared with those of the worker larvae. The cells of the drone larvae are longer and thicker-walled than those of the workers.

Worker The worker egg hatches after three days and after five days the larva is fully grown, with a head and thirteen segments. It has moulted six times during this period. The honeycomb cell is now capped and within two days the larva pupates. After eight days the adult worker emerges. The worker bees go through a series of different stages as follows:

i) they clean the wax cells and feed the older worker larvae with pollen and honey;

ii) after ten days they begin to secrete a nutritive fluid on which they feed the young larvae;

iii) after fourteen days, they undertake the general cleaning duties in the hive, distributing and storing food. They ventilate the hive by beating their wings at the entrance of the hive;

iv) after twenty one days, the workers take trial flights from the hive, and also act as guards preventing intruders getting into the hive;

v) finally they complete their lives by several weeks of foraging for food.

Queen The queen has a longer abdomen than the worker, although both are females. She is fed throughout her life by the workers upon partially digested food. She spends her life laying eggs at the rate of about 1000 per day, although this can rise to a peak of 3000 in June. Egg-laying commences in February and continues until September. The reigning queen in a hive departs in early summer together with a number of workers in a 'swarm'. This collection of queen and workers can be enclosed in a container by a bee-keeper and set up in a new hive. In the old hive a young queen is allowed to emerge by the workers and she then proceeds on her nuptial flight.

Drone The drone develops from an unfertilised egg by the process of parthenogenesis. It is distinguishable from the worker by its broader and larger body. The drones are outnumbered in the hive by workers by about 10 to 1. They perform no tasks in the colony except to follow the queen on her nuptial flight. They are completely dependent on the workers for their predigested food. At the end of the summer most of the drones are driven from the hive by the workers and those that remain are bitten to death so that none survive the winter.

The Anopheles mosquito

Structure Mosquitoes are much smaller than bees. They rarely exceed 2 cm in length. The male and female of *Anopheles* can be distinguished by their heads. The male has inconspicuous jointed antennae and mouthparts for sucking plant juices, the female has antennae with feathery branches and mouthparts adapted for sucking blood as well as plant juices.

The head of the female is dominated by a very large pair of compound eyes and the proboscis. The proboscis is formed from highly modified basic mouthparts

Table 15.4 Structure of the worker, queen and drone

	Worker	Queen	Drone
Head	Eyes do not meet in the midline of the head Mouthparts adapted for i) sucking – labium and maxillae (proboscis) ii) cell construction – mandibles	Eyes do not meet in the midline of the head Proboscis shorter and mandibles larger than in the worker	Eyes much larger and meet in the midline of the head Proboscis short and feeble with small mandibles
Thorax			Large thorax to accommodate powerful flight muscles
	Three pairs of legs First pair – has 'comb' for cleaning antennae and proboscis Second pair – 'prong' for digging out pollen from the pollen basket Third pair – 'pollen basket' carries the pollen collected by the bee	Three pairs of legs Comb and prong present No pollen-collecting apparatus	Three pairs of legs Comb and prong present No pollen collecting apparatus
	Two pairs of wings equal in length to the abdomen	Two pairs of wings shorter than the abdomen	Two pairs of wings equal in length to the abdomen
Abdomen	Equal in length to the wings A barbed sting present at the end of the abdomen – it cannot be withdrawn and thus stinging results in the death of the worker Wax glands present	Broader and longer than that of the worker Sting is unbarbed thus can be withdrawn without causing injury Wax glands present	Abdomen larger than that of worker Sting is absent (the sting is a modified ovipositor (for egg-laying) and thus cannot be present in males). Wax glands absent

contrasting with those of the cockroach. The labium forms a tube, split on its upper surface, which surrounds and protects the other mouthparts. These include 4 sharp stylets a sucking tube, and a salivary tube. The skin is pierced with the stylets and the sucking and salivary tubes are inserted. The labium folds back.

The hairy thorax bears a single pair of long delicate wings on the dorsal surface of the second segment. The second pair of wings are reduced to *halteres*, used in balancing. A pair of slender, long walking limbs are attached to the underside of each thoracic segment.

The abdomen, consisting of nine visible segments, points obliquely upwards when *Anopheles* is at rest, unlike that of *Culex* which droops.

Mode of life Male mosquitoes are rarely found near human habitation, since they live off plant juices. Female mosquitoes also feed on plant juices, but after fertilisation, they must feed on blood. Sucking up blood from infected humans they are the unwitting vectors of the malarial parasite, *Plasmodium*. (See Chapter 12.) Sporozoites of *Plasmodium* eventually invade the salivary glands of the mosquito. Saliva is then poured into another human blood vessel down the salivary tube. It contains an anticoagulant to stop the blood from clotting. *Plasmodium* parasites are injected into the blood at the same time.

The life-cycle of *Anopheles* is basically the same as that of the Cabbage White Butterfly (see Chapter 10). The mosquitoes mate in the long grass near stagnant water and the female lays about 100 eggs on the surface of the water. They are cigar-shaped eggs with floats at each side (see fig 15.34). They hatch after 2–3 days into 'wrigglers' – actively moving larvae well adapted for aquatic life. They breathe using an air tube which obtains atmospheric air, and anal gills which utilise dissolved oxygen. The former are more efficient. They feed on plankton for about a week and then change into an active pupa (see fig 15.34). The pupa also breathes air through a pair of respiratory tubes, but can

Class	Habitat	Body form	Antennae	Movement	Breathing
Arachnida	Mainly terrestrial, some aquatic	Cephalothorax and abdomen	No antennae	4 pairs of thoracic walking limbs	Tracheae and lungbooks
Crustacea	Mainly aquatic	Segmented, or head; thorax; abdomen with head and thorax frequently covered by carapace	2 pairs of antennae	Variable number of walking limbs usually in excess of 5 pairs	Gills
Myriapoda i) Chilopoda	Terrestrial	Segmented	One pair	One pair of walking limbs/segment	Tracheae
ii) Diplopoda	Terrestrial	Segmented	One pair	Two pairs of walking limbs/segment	Tracheae
Insecta	Terrestrial, aquatic, aerial	Head; thorax and abdomen distinct in adult	One pair	Three pairs of thoracic walking limbs, one or two pairs of wings attached to thorax	Tracheae

Table 15.5 Characteristics of different arthropod classes

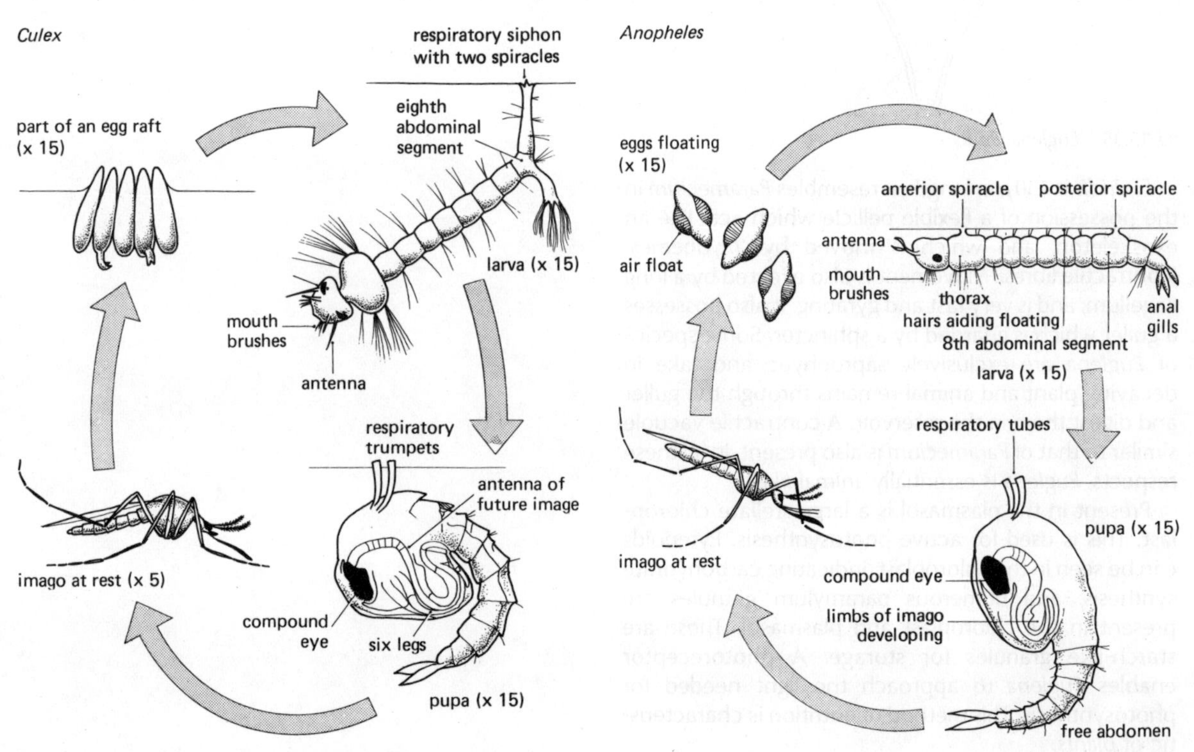

fig 15.34 Life history of *Culex* and *Anopheles* mosquitoes

extract oxygen from the water with its anal gills if disturbed. Inside the pupal case, which only covers head and thorax regions, the drastic reorganisation of the larval tissues to form the structures of the imago can be seen. When metamorphosis is complete, the pupa case splits open, and the imago emerges.

15.30 Plant or Animal?

Euglena viridis

This non-cellular organism is classified by botanists as an Alga, and by zoologists as a Protozoan. It can be seen to possess characteristics of both (see fig 15.35). *Euglena* species are smaller than *Paramecium*, varying from 30 μ to 200 μ.

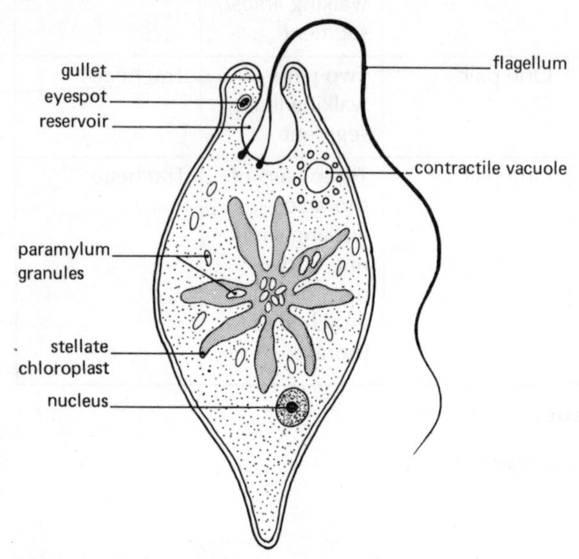

fig 15.35 *Euglena viridis*

E. viridis is 130 μ in length. It resembles *Paramecium* in the possession of a flexible pellicle which acts like an exoskeleton, and which is moved by *'myonemes'* (contractile fibres). Movement is also effected by a long flagellum, and is very fast and gyrating. It also possesses a gullet which is guarded by a sphincter. Some species of *Euglena* are exclusively saprophytic and take in decaying plant and animal remains through the gullet and digest them in the reservoir. A contractile vacuole similar to that of *Paramecium* is also present. In all these respects, *Euglena* is essentially *animal-like*.

Present in the plasmasol is a large stellate chloroplast. This is used for active photosynthesis. Pyrenoids can be seen in the chloroplast, indicating carbohydrate synthesis, and numerous paramylum granules are present in the chloroplast and plasmasol. These are starch-like granules for storage. A photoreceptor enables *Euglena* to approach the light needed for photosynthesis. This method of nutrition is characteristic of *plants*.

Questions requiring an extended essay-type answer

1 Compare and contrast the structure, nutrition and life cycle of a *named* mould and a *named* filamentous alga.

2 *'Mucor* is a very efficient saprophyte'. Discuss this statement with reference to the feeding and life cycle of this mould.

3 Describe the structure, feeding and movement of a *named* protozoan. Draw diagrams to illustrate each part of your answer.

4 a) Explain why mosses and ferns are sometimes termed 'lower' plants.

b) Write a brief survey of the main groups of 'lower' animals. (invertebrates) State the main characteristics of any **two** of these groups and *name* two members of each.

5 a) What are the main characteristics of insects?

b) Choose one *named* example of an insect which undergoes complete metamorphosis and describe its life history.

6 Choose a *named* alga and a *named* protozoan and show how the first is classified as a plant, and the last as an animal. Is *Euglena* a plant or an animal?

16 Higher plants

16.00 Introduction

Higher plants are those plant organisms possessing a well developed, complex vascular system, and reproducing by means of a specialised structure, the flower. All the plants in this group contain the pigment chlorophyll, a green colouring material essential in the process of photosynthesis.

Much of the structure of the higher plants, together with the associated physiology, has been studied in earlier chapters. The internal structure of roots, stems and leaves has been covered in detail and, those aspects relating to growth and reproduction have also been dealt with in depth.

This chapter will summarise much of the relevant information integrated throughout earlier parts of the book, and will also serve to complete the study of the higher plants by discussing the differences between herbs, shrubs and trees, and by comparing external and internal features of monocotyledons and dicotyledons. Information will also be given relating to structure and life history of three types of tree.

16.10 Herbs, shrubs and trees

All higher plants are grouped into one of these three categories, and the classification is based upon size, structure and complexity.

16.11

Herbs These are short, delicate plants the stems of which rarely become woody. They are divided into three main types according to their life cycle: annuals, biennials and perennials.

Annuals grow, flower and complete the life cycle within one year; they then die and the species is continued through germination of the seeds. Examples include wheat, peas and tomatoes.

Biennials grow throughout the first year and accumulate food reserves in storage organs. In the second year they flower and then, having completed the life cycle, they die. An example is the carrot, where the storage organ is a large tap root.

Perennials grow and flower over a number of years; examples are daffodil, *Crocus* and dandelion. Perennials often have well developed storage organs. (See Chapter 4.)

Shrubs – *Woody Perennials* These always grow larger than herbs, and often form woody tissues. They do not die off in the autumn. Examples include rose and privet. These types of plant are often used in a decorative manner and shrubs are common features of many gardens.

Trees – *Woody Perennials* These are large, woody plants. It is sometimes difficult to distinguish a tree from a shrub but, generally speaking, trees grow much taller and produce much thicker stems or trunks.

16.20 The main parts of an herbaceous, dicotyledonous plant

These comprise root, stem leaves and flowers. The example shown in fig 16.5(b) illustrates the relative positions of these main parts. The major functions of the different parts have already been discussed at length in previous chapters. Table 16.1 on p. 306 summarises this information.

The leaf – *external structure* The leaf consists of three main parts; the lamina (leaf blade), the petiole (leaf stalk) and the leaf base. The main parts of a leaf of lime are shown in fig 16.1. Leaves can be grouped according to their shapes, and some examples of the different forms are illustrated in figs 16.2 and 16.3.

Leaf arrangement Leaves are arranged on plants in a number of different ways. The most usual arrangement is in a *spiral* pattern where a leaf develops at each node of the stem, and the attachment of the leaves, to the stem, follows a spiral or twisted path. A common plant which displays this type of leaf arrangement is balsam.

In some plants the leaves occur singly at the nodes but, instead of being arranged in a spiral pattern, they arise alternately on opposite sides of the stem. Leaves arranged in this way are said to be *alternate*.

Sometimes, the leaves arise in pairs directly opposite each other. Leaves arranged in this way are said to be *opposite* and, if they are arranged so that one pair is at right angles to the pairs directly above and below, the arrangement is said to be *decussate*. A number of plants exhibit this type of leaf arrangement and the lilac is an example.

The internal structure of the leaf has been dealt with in detail in Chapter 1 (Nutrition).

The stem The stem has three main functions. It acts as a supporting structure, it contains a transport system and it acts as a storage organ. These functions have been covered in depth in different parts of this book. The internal structure of the stem has also been covered when the relevant physiological processes have been studied. However this latter information will be summarised a little later in this chapter, by means of detailed micrographs of stems and roots of

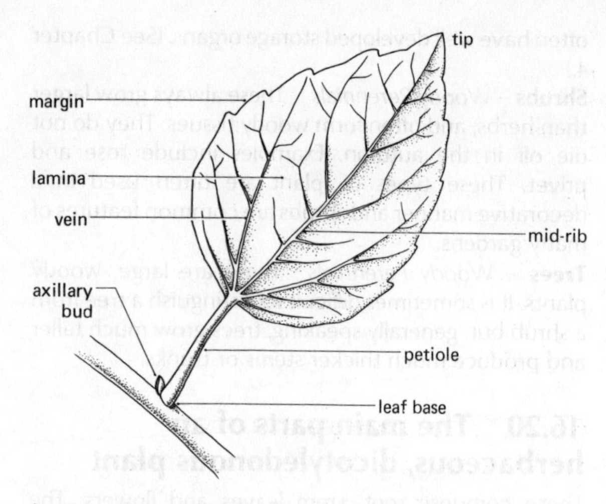

fig 16.1 The external features of a leaf of lime

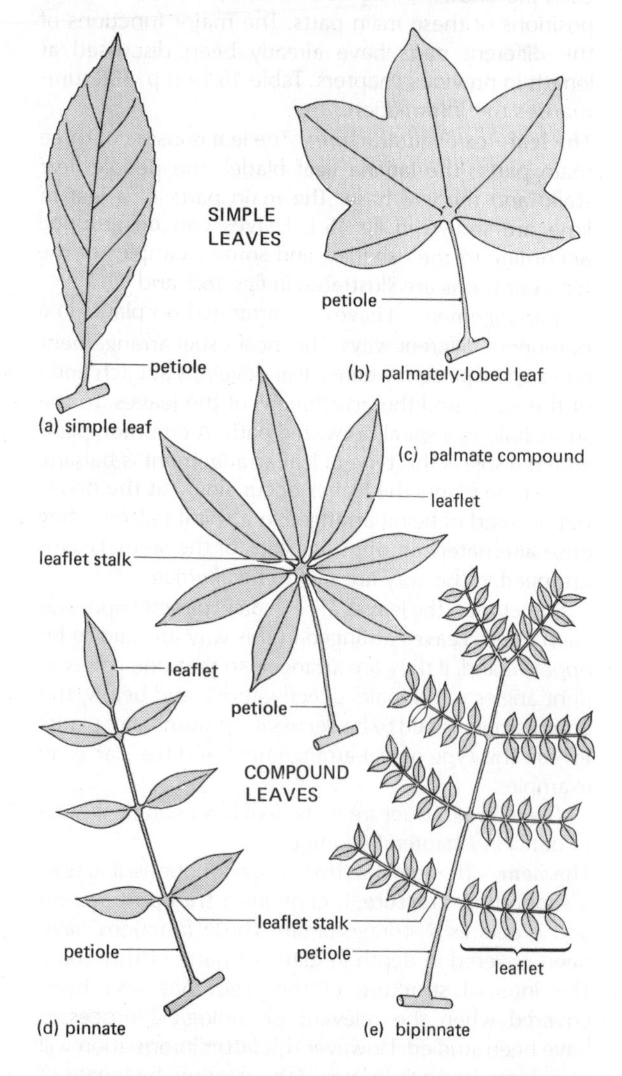

fig 16.2 The main types of leaf shape

monocotyledons and dicotyledons.

The root The root is concerned mainly with anchorage, and with absorption and transport of water and mineral salts. However, as seen in Chapter 4, it can also function as a storage organ in certain types of plant.

Some roots become specially modified for support and, in very exceptional situations, roots can also function in 'breathing'. Some examples of these specialised root structures can be seen in fig 16.4.

Plant part	Function
Root	Anchors the plant in the soil, absorbs water and mineral salts and transports them upwards to the stem; in some cases it serves as a storage organ (see Chapter 4). Some roots become specially modified for support.
Stem	Continuous conduction of water and mineral salts upwards to the leaves; supports leaves and holds them out to receive sunlight; supports flowers; also serves as a storage organ in some cases (see Chapter 4).
Leaf	Absorbs carbon dioxide to help in photosynthesis, also contains chlorophyll to help in this process; stomata in the leaf allow loss of water vapour (transpiration). Some leaves become modified for special functions e.g. climbing, storage, etc.
Flower	The organ of reproduction; the fertilised ovary forms the fruit, which contains seeds; after dispersal, some seeds will germinate to produce new offspring.

Table 16.1 The main parts of an herbaceous dicotyledon and their functions

16.30 Monocotyledons and dicotyledons

Flowering plants are divided into these two major groups. The classification is based on the structure of the seed. In monocotyledons, there is only one embryo leaf (cotyledon) present in the seed, but in dicotyledons two embryonic leaves are present (see Chapter 10, Growth). The monocotyledons include the grasses and cereal crop plants. Monocotyledons are usually fairly

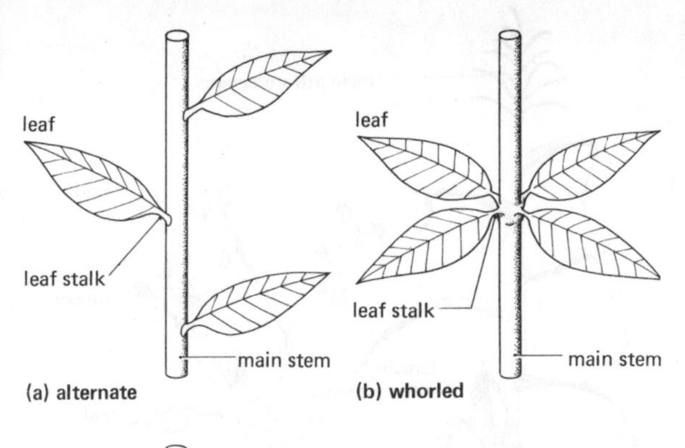

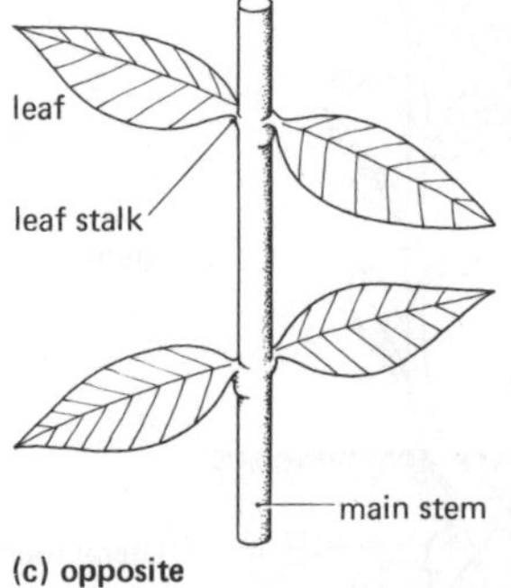

fig 16.3 The main types of leaf arrangement

small plants with little secondary thickening. However, one type of grass in particular, bamboo, does produce very thick, strong stems which, in some parts of the world are used in building.

Another group of monocotyledons become 'tree-like' and may grow to a very great height and girth. These are the palms, some of which are commercially important in the production of coconuts, dates and palm oil.

16.31 External features of monocotyledons and dicotyledons

An example of each of these groups is illustrated in fig 16.5. If a comparison is made of the two plants, some of the main external characteristics of the two divisions are easily seen. This information is summarised in table 16.2.

16.32 The stem, a comparison of internal structure in Monocotyledons and Dicotyledons

A detailed look at the internal structure of the stems illustrates additional differences between the two types of plant. The micrograph in fig 16.6(a) shows a transverse section (low power) of the stem of a sunflower, *Helianthus* (a dicotyledon). Fig 16.6(b) shows a high power view of one of the vascular bundles of this plant.

The micrograph in fig 16.7(a) shows a transverse section (low power) of the stem of maize (*Zea*) (a

fig 16.4 (a) Buttress roots

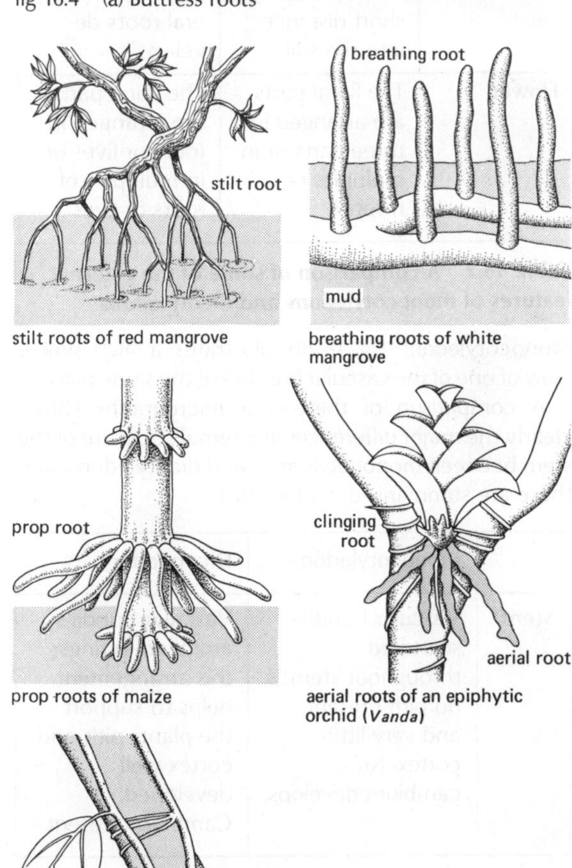

fig 16.4 (b) Some additional modified root structures

Plant structure	Monocotyledons	Dicotyledons
Leaf	Leaves often long and 'ribbon-like'; the veins are arranged in a pattern parallel to the long axis of the leaf	Leaves of different shapes (dependent upon species). The veins are arranged in a net or reticulate pattern
Root	The system is usually fibrous, and the individual roots penetrate only a short distance into the soil	The system usually comprises a main root (tap root) from which lateral roots develop
Flower	The floral parts are arranged in three parts or in multiples of threes	The floral parts are arranged in fours or fives or in multiples of fours or fives

Table 16.2 A comparison of some of the external features of monocotyledons and dicotyledons

monocotyledon). Fig 16.7(b) illustrates a high power view of one of the vascular bundles of the same plant.

A comparison of these four micrographs shows clearly the major differences in internal structure of the stem between monocotyledons and dicotyledons, and these are summarised in table 16.3.

	Monocotyledon	Dicotyledon
Stem	Vascular bundles scattered throughout stem; no pith present and very little cortex. No cambium develops.	Vascular bundles arranged in rings; this arrangement helps to support the plant; pith and cortex well developed. Cambium present.

Table 16.3 A comparison between the internal structure of the stem of a monocotyledon and a dicotyledon

16.33 The root, a comparison of internal structure in Monocotyledons and Dicotyledons

The micrographs in fig 16.8 and fig 16.9 are of transverse sections of the roots of maize and sunflower. A study of these shows the major differences between the root structure of monocotyledons and dicotyledons, summarised in table 16.4.

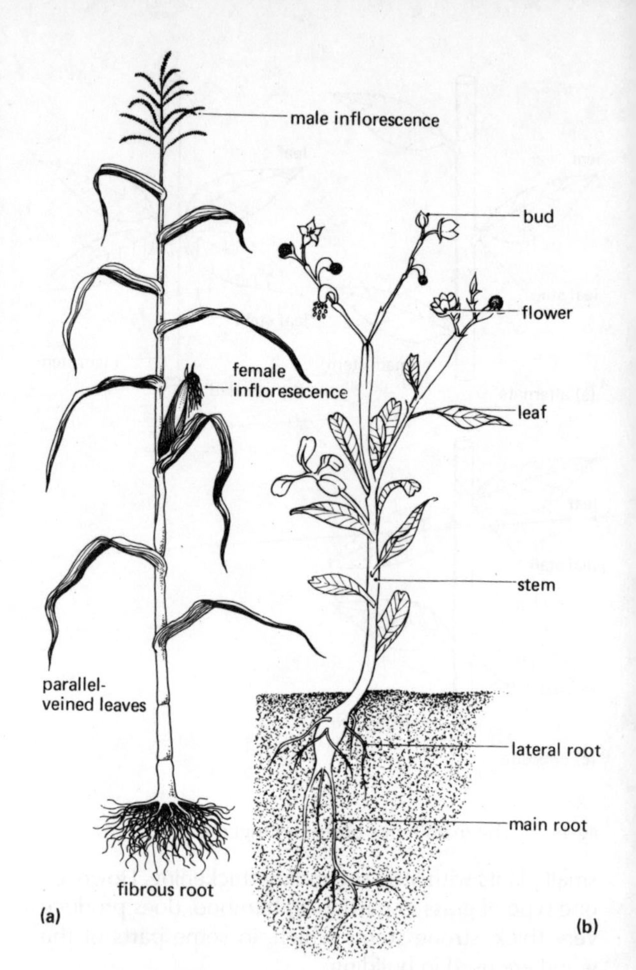

fig 16.5 The external features of (a) a monocotyledon and (b) an herbaceous dicotyledon

	Monocotyledon	Dicotyledon
Root	There is a wide pith surrounded by rings of conducting tissue; the numerous xylem and phloem bundles are arranged alternately.	There is very little pith; the xylem is arranged centrally in the form of radiating arms. The phloem is positioned between the xylem. There are few bundles. Later development of cambium.

Table 16.4 A comparison between the internal structure of the root of a monocotyledon and a dicotyledon

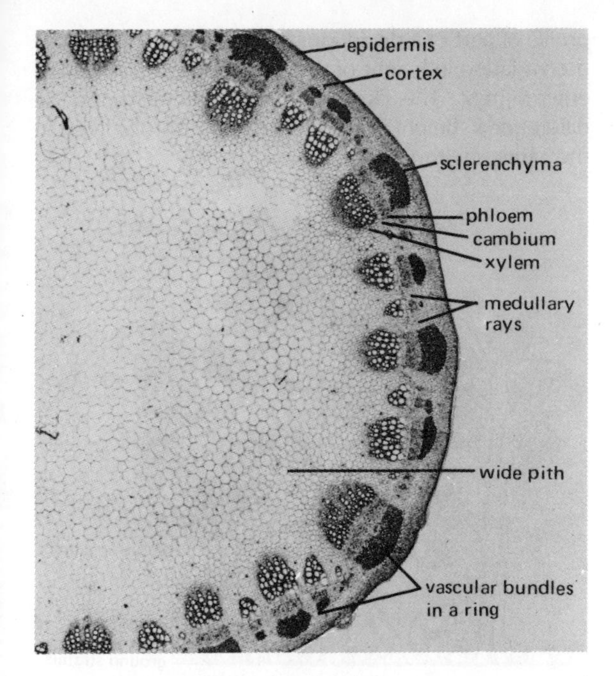

fig 16.6 (a) T.S. of sunflower, *Helianthus* stem as seen under low power

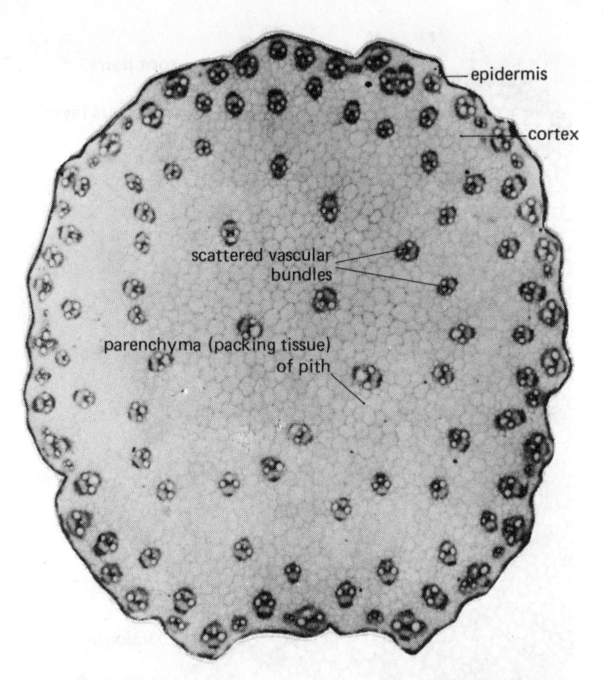

fig 16.7 (a) T.S. of maize, *Zea* stem as seen under low power

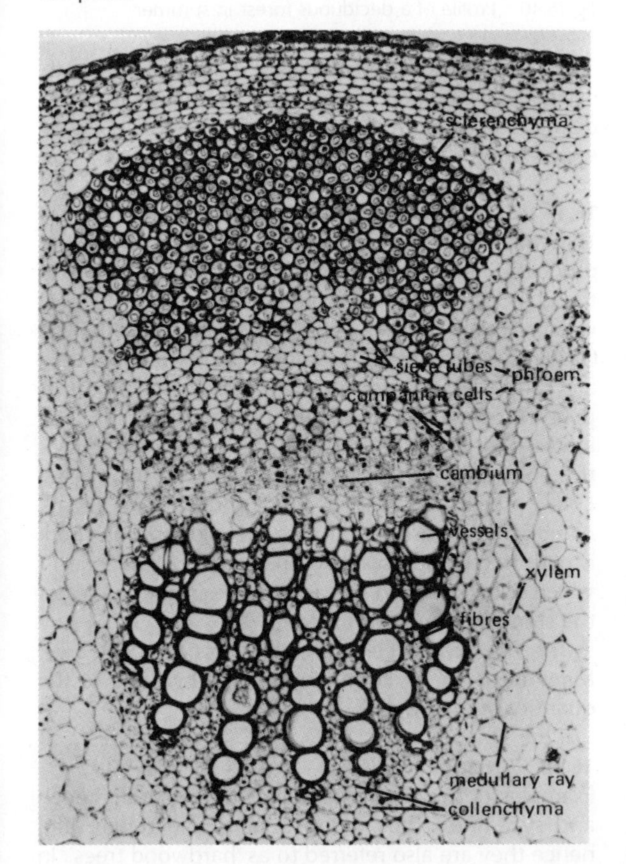

fig 16.6 (b) T.S. of a *Helianthus* vascular bundle as seen under high power

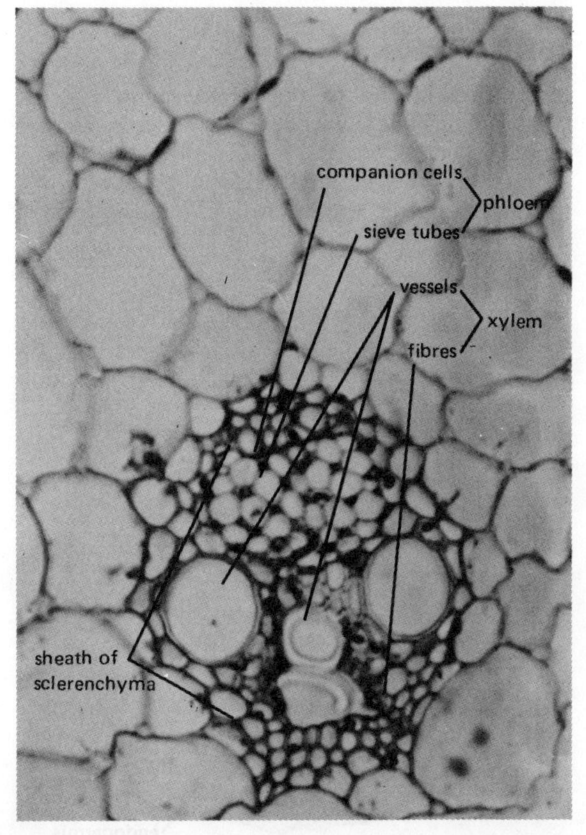

fig 16.7 (b) T.S. of a *Zea* vascular bundle as seen under high power

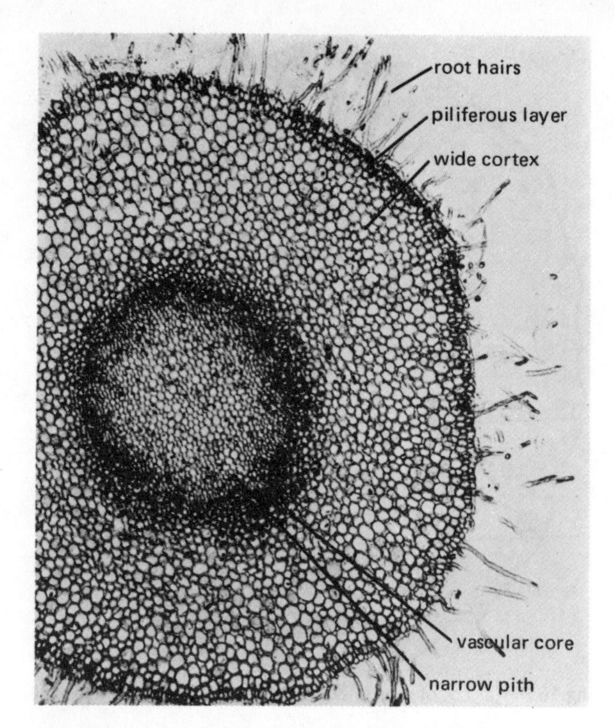

fig 16.8 T.S. of a *Helianthus* root as seen under low power

16.40 Trees

Most trees belong to the gymnosperms or the dicotyledons, but a few examples such as the palms are monocotyledons.

Trees vary in height, depending upon the species and the environment in which they live. The diagram of a profile of part of a deciduous forest (fig 16.10) illustrates the variation in height of the trees living in this kind of environment. The illustration also demonstrates the difference in height between herbs and shrubs living in the same environment.

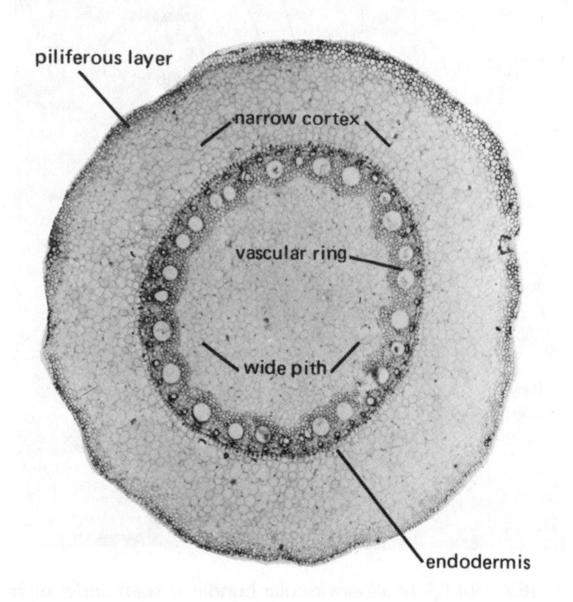

fig 16.9 T.S. of a *Zea* root as seen under low power

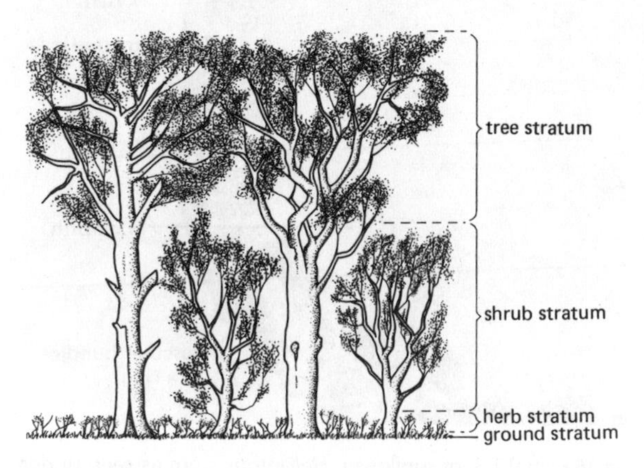

fig 16.10 Profile of a deciduous forest in summer

Apart from increasing in height by means of growth in a vertical direction, trees also grow by increasing the diameter and circumference of the trunk. This is called secondary thickening.

16.41 Secondary thickening

This was discussed in Chapter 10. It involves the production of new cells from the cambium layer. This tissue is capable of cell division in which the new xylem cells are produced on the inside of the cambium (towards the centre of the stem) and new phloem cells are produced on the outside of the cambium. This process can be followed by studying fig 16.11. The increase in girth due to secondary thickening causes the epidermis to rupture. This is replaced by a dead, corky protective layer of *bark* produced by a cork cambium lying just beneath the old epidermis. The bark is perforated at intervals by small gaps called *lenticels*, which allow diffusion of gases between the tissues of the stem and the atmosphere.

The new xylem cells form the 'wood' of the tree. Many trees are of commercial value because of the quantity of their wood.

16.42 Broad-leaved trees

Examples of this type of tree include elm, oak, beech and sycamore. Their wood is hard and tough, and hence they are also referred to as 'hardwood trees'. In most species, the trees are bare of leaves during the winter: the trees are thus said to be *deciduous*. In late

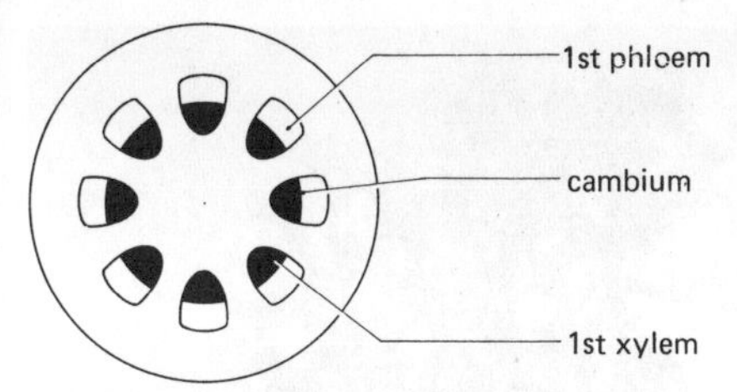

1 only primary tissues are formed

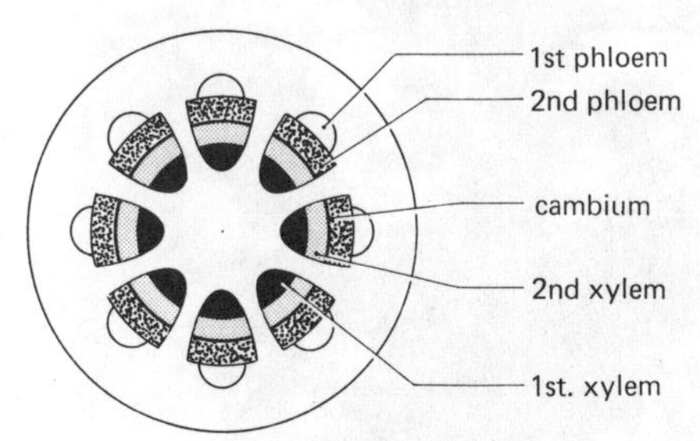

2 some secondary xylem and phloem produced

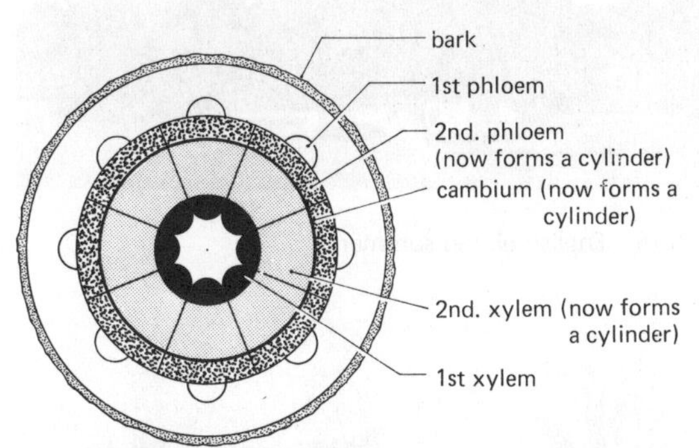

3 enough secondary xylem and phloem has been produced to form a complete ring of each tissue

fig 16.11 The process of secondary thickening

fig 16.12 English oak in summer

fig 16.13 English oak in winter

autumn, the cell contents of the leaves break down and the soluble products pass back into the tree. Then cells in a layer (the *absciss layer*) at the base of each leaf become detached from each other and round off. A layer of cork is formed under the absciss layer and the vascular bundles become blocked, thus cutting off the water supply from the leaf, which withers and drops off. This is known as *abscission*. All the above features (hard wood, broad leaves and winter leaf fall) contrast strongly with those of the other major group of trees, the conifers.

The oak (Quercus)

The English oak (*Quercus robur*) is, at least historically, one of the most important timber trees. Indeed it is reasonable to speak in historical terms, since the oak does not reach timber size until it is about 120 years old. It may live as long as six- or eight hundred years. Oak timber has traditionally been used for boat building and is still used for furniture making, parquet floors and heavy constructional work. The bark of the cork oak (*Q. suber*) is used in the commercial production of cork.

The oak is essentially a woodland tree, although it is commonly seen isolated in British meadows, and the fruits (acorns) are often planted in the shelter of young larches. These are later removed when the oak sapling is sufficiently well established to survive alone.

Separate male and female flowers appear on the same tree in April or May. The male flowers are yellowish-green catkins about 5 cm long, and the female flowers are tiny, spherical and pink, occurring in clusters of two or three (see fig 16.14). The ecology of an oak tree is investigated in Chapter 14.

The elm (Ulmus)

The English elm, *Ulmus procera*, is an important feature of many landscapes in southern England. Its strong, hard wood has been used traditionally for the

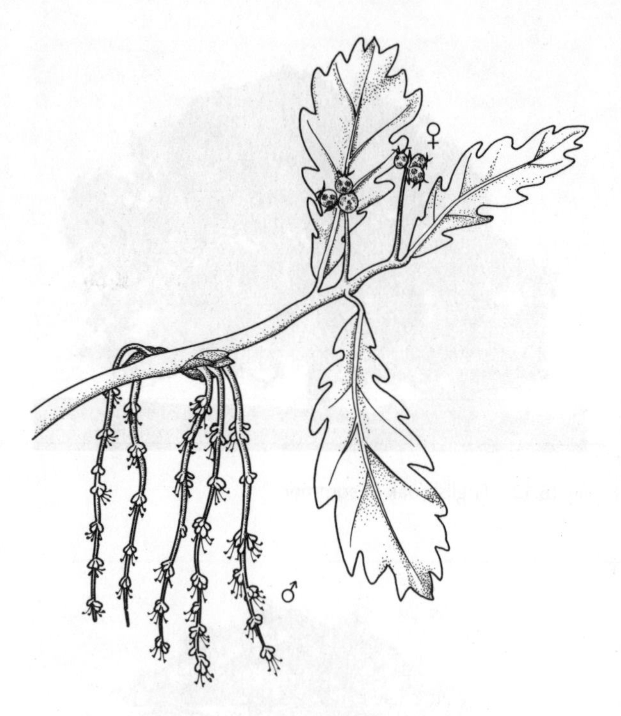

fig 16.14 Male and female flowers of oak

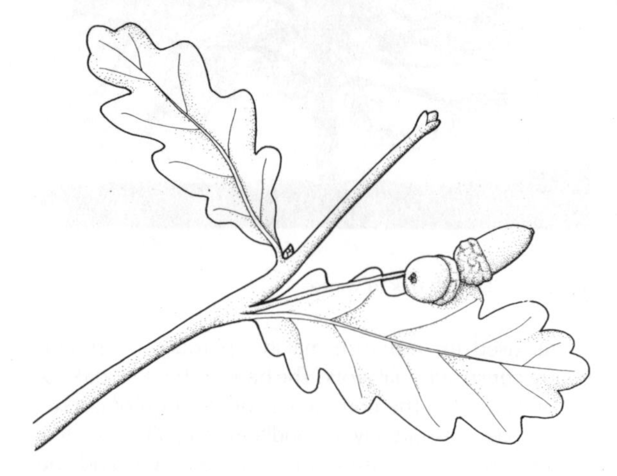

fig 16.15 Oak twig with leaves and fruits (acorns)

manufacture of furniture and coffins.

The flowers, which appear in February, are small and rust-coloured. Clusters of flat, oval, pale-green fruits may be seen in March. In fact, the English elm seeds have a low fertility rate and most trees are propagated from suckers. (See figs 16.19 and 16.20.)

Sadly, thousands of these most beautiful trees have been killed in recent years by Dutch Elm disease. This is caused by the fungus *Ceratocystis ulmi,* which blocks the conducting tissues of the trees. This fungus is carried from elm to elm by a small beetle *Scolytus scolytus*. (See fig 16.17.)

fig 16.16 English elm in summer

fig 16.17 English elms with Dutch Elm disease

fig 16.18 English elm in winter

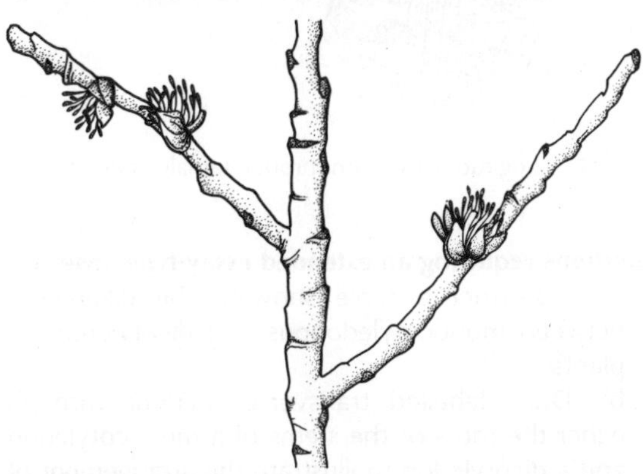

fig 16.19 Elm flowers

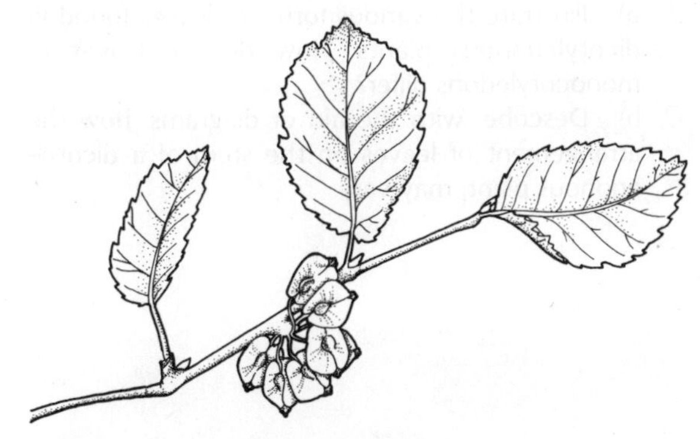

fig 16.20 Elm twig with leaves and fruits

16.43 Coniferous trees

About 90% of the wood used in Britain comes from conifers grown in Canada, Scandinavia and (to a lesser extent) Wales and Scotland. These trees are so-called because their seeds are borne, not enclosed in flowers like those of the monocotyledons and dicotyledons, but open on the scales of cones. Their leaves, which are narrow and needle-like, are not shed regularly every winter: the trees are 'evergreen'. The wood of conifers is much softer and more easily workable than that of broad-leaved trees. Hence they are referred to as 'softwoods'. Examples include the pines, firs, cedars and spruces.

The pine (Pinus)

The Scots pine, *Pinus sylvestris*, is one of the few conifers native to Britain. It grows wild in the Scottish highlands and will survive in very poor soil. Other varieties have been imported into the United Kingdom from North America and Corsica. The timber of pines, which have tall, straight trunks, is used for telegraph poles, pit props and building.

fig 16.21 Scots pine

fig 16.22 Male and female cones of Scots pine

The trees bear both male and female cones. The small male cones produce pollen one year after they first appear at the bases of terminal buds. The much larger female cones, however, take three years to develop. Fertilisation does not take place until a full year after pollination.

The female pine cone has long been used in forecasting weather. More correctly, it reflects changes in humidity, the scales moving apart in dry conditions. This allows the winged seeds to be released in conditions which enable them to be blown well away from the parent tree before germination begins. (See figs 16.23 and 16.24.)

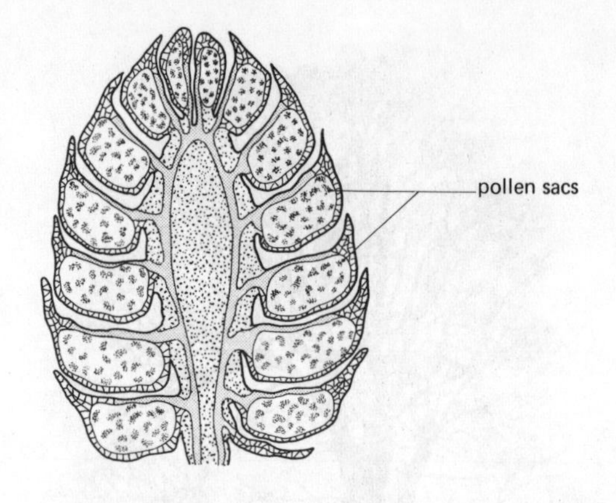

fig 16.23 Longitudinal section through male cone of *Pinus*

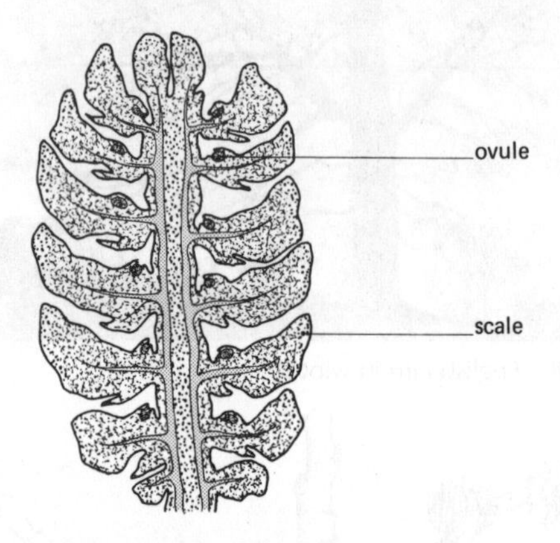

fig 16.24 Longitudinal section through female cone of *Pinus*

Questions requiring an extended essay-type answer

1 a) Construct a table showing the differences between monocotyledonous and dicotyledonous plants.
b) Draw labelled transverse sections through *either* the roots *or* the stems of a monocotyledon and a dicotyledon to illustrate the arrangement of the plant tissues in each.

2 a) Illustrate the various forms of leaves found in dicotyledonous plants. How do the leaves of monocotyledons differ?
b) Describe, with the aid of diagrams, how the arrangement of leaves on the stem of a dicotyledonous plant, may vary.

17 Vertebrate animals

17.00 Higher Animals

The term 'higher animals' is usually reserved for those animals which possess an endoskeleton, based on a backbone or vertebral column. Possession of an endoskeleton has enabled many of these vertebrates to achieve a large size, although the largest, the whale, must be supported by water.

These animals appeared later in geological time than the invertebrates. It can be deduced from this that the vertebrates evolved from the invertebrates and are consequently 'higher' in the evolutionary tree. They tend to be larger and more complex than invertebrates.

Many of the main habitats of the world are dominated by the vertebrates as a result of 'adaptive radiation'. Vertebrates living in the sea evolved specialised structures to succeed in that environment, e.g. fishes evolved fins for swimming. Those living on land evolved a variety of features equipping them for survival in terrestrial conditions. Reptiles, birds and mammals became adapted for land living, and amphibians for functioning in water and on land. Birds evolved a large number of characteristics which enabled them to dominate the air and land.

The study of a representative animal from each of the five classes of vertebrates will illustrate the wide variety of vertebrate characteristics.

17.10 Fishes

The stickleback is a freshwater fish found in ponds and lakes.

External features The general form of the fish is 'torpedo-shaped' or streamlined when seen in lateral view as in fig 17.1. There is no external division between head, body and tail. When viewed from the front, the stickleback appears to be very narrow. Such a shape offers the least possible resistance to the water and is an aid to rapid swimming. The fish is covered in overlapping waterproof scales, the free margins of which are posterior. The colouring of the scales provides camouflage against predators. The lower surface and sides of the fish are often lighter than the upper surface. This conceals the fish when light is shining on it from above. In addition, darker stripes are present on the sides of the fish and these enable it to blend with the water weeds among which it lives, thus camouflaging the animal.

The head has a horny, toothless mouth through which water enters and passes over the gills. The gills are covered on each side by an operculum, which is a flap through which water leaves the gills. There are four gills beneath each operculum (see fig 5.11). In Chapter 5 (Respiration) there is an account of the role of these structures in breathing. The eyes are lidless and the cornea is flat and horny to avoid damage by objects in the water.

Along each side of the body can be seen the lateral line. This indicates the presence of special cells which can detect vibrations and pressure changes in the water. The tail of the stickleback is relatively short when compared with that of other fishes, but the *caudal (tail) fin* is large and powerful. Like all the fins, it is supported by horny fin rays. Along the dorsal mid-line is a *dorsal fin*

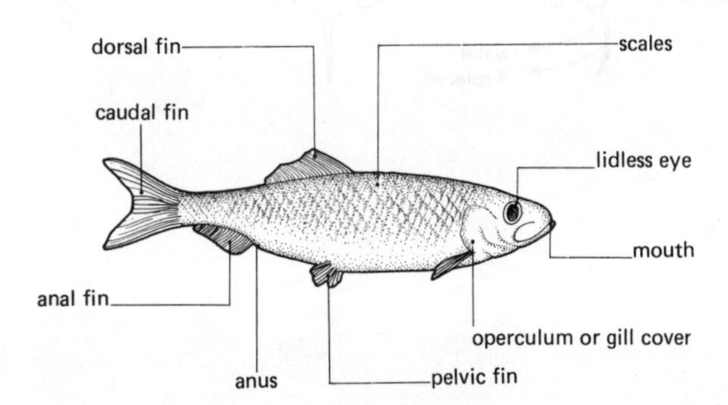

fig 17.1 (a) Lateral view of the herring

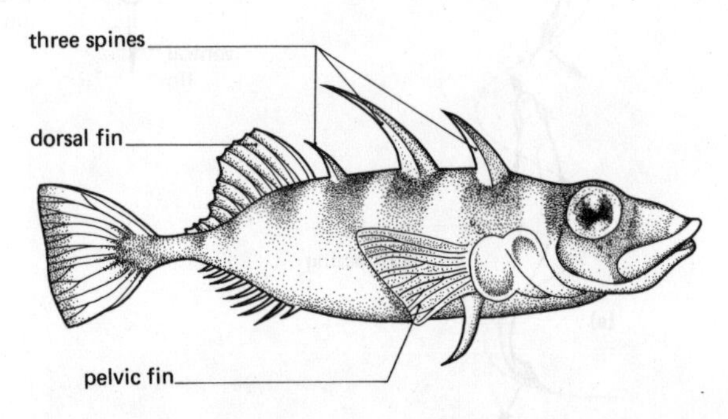

fig 17.1 (b) Lateral view of the stickleback

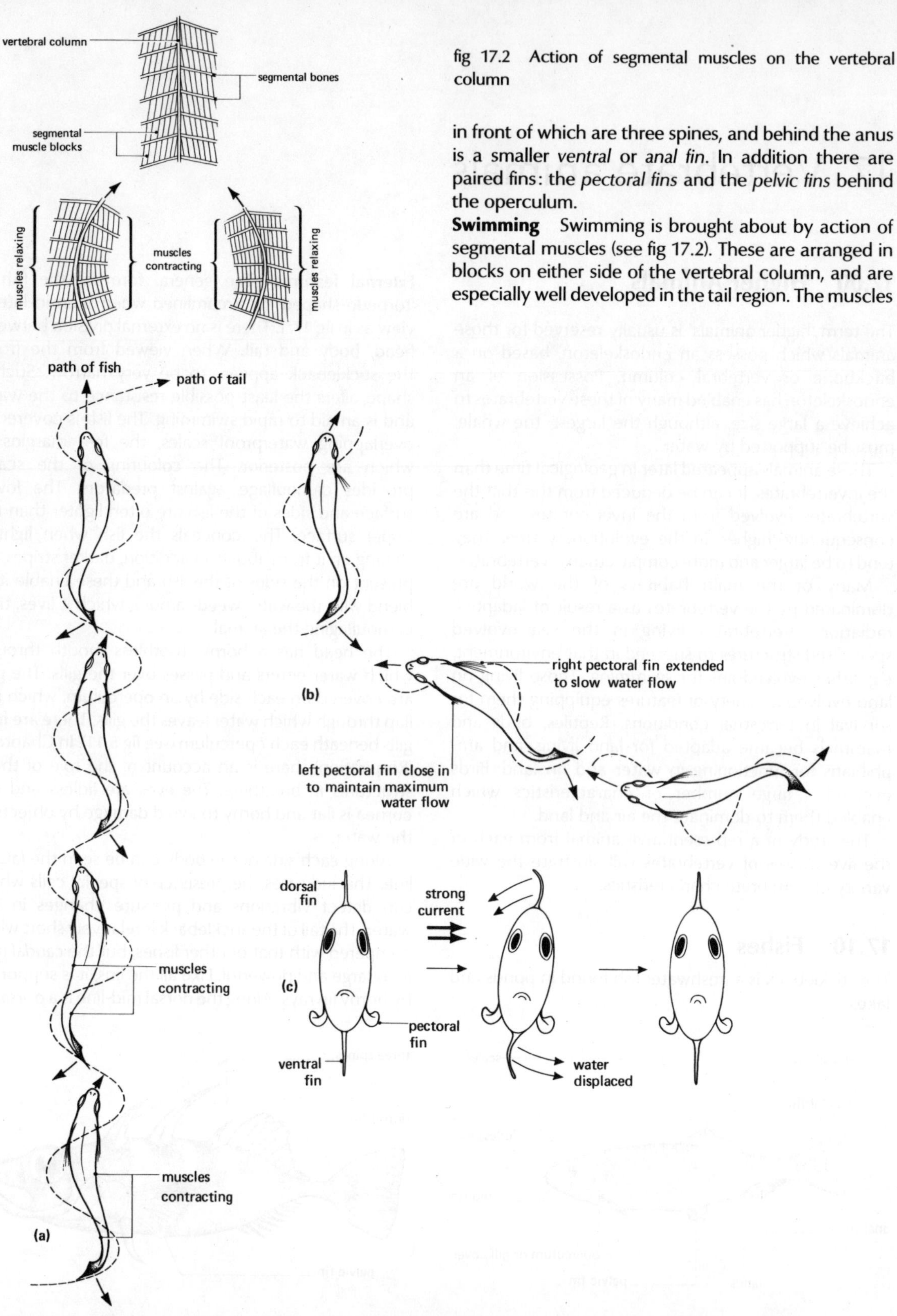

fig 17.2 Action of segmental muscles on the vertebral column

in front of which are three spines, and behind the anus is a smaller *ventral* or *anal fin*. In addition there are paired fins: the *pectoral fins* and the *pelvic fins* behind the operculum.

Swimming Swimming is brought about by action of segmental muscles (see fig 17.2). These are arranged in blocks on either side of the vertebral column, and are especially well developed in the tail region. The muscles

fig 17.3 Swimming movements of fish (a) Forward progression (b) Change of direction (c) Maintenance of stability

contract alternately on either side, causing the tail to bend and thus push the water aside. The caudal fin increases this effect and steers the fish through the water (see fig 17.3(a)). Speed, braking, and, to a lesser extent, control of the depth of swimming, are controlled by the paired fins. The further these are pushed out into the water, the greater the resistance to the water and the slower movement becomes in the water. In some fishes, the paired fins also aid steering. In this case only one of the pair is extended, slowing movement of water on that side and causing the fish to turn towards that direction (see fig 17.3(b)). The median dorsal and ventral fins stabilise the fish by preventing rolling or 'wobble' (see fig 17.3(c)). They are particularly well developed and important in angel fishes.

Depth of swimming is aided by the internal air or swim bladder which can be inflated by gases diffusing from the blood. Inflation of the swim bladder makes the fish more buoyant and thus its body rises easily in the water without fin movement. Deflation has the opposite effect (see fig 17.3(d)). Sharks have no swim-bladder and must keep swimming to stop their bodies slowly sinking.

Life history The life history of the stickleback is unusual in that it involves nest-building, courtship and parental care of the young (see fig 17.4). During early spring in small streams and ponds the male stickleback establishes his territory in a small area of shallow water where he will build his nest. The male stickleback digs a shallow hole by fanning movements of the tail and by scooping up mud with his mouth. With strips of waterweed he builds a roof over this hollow area so that a small tunnel is formed. Female sticklebacks entering the territory are approached and courted by the male. The body of the female, swollen with eggs, is ready for mating and she is attracted to the male by his red belly. The male zig-zags in front of the female, performing a kind of dance. During this display he leads her towards the nest, into which he pushes his own head as he arrives at the entrance. The female then pushes past into the nest so that her head emerges from one end and her tail from the other. The male nudges the tail of the female and this seems to be a signal for egg-laying. This process is completed in a few seconds and the female leaves the nest to be replaced by the male. He sheds sperms over the eggs and fertilisation occurs. The female swims away and takes no further part in the development of the fertilised eggs or the rearing of the young. The male remains on guard and eventually, when the young fish hatch, continues to defend them. The eggs and the young fish in the nest are supplied with a stream of fresh water by the male 'fanning' the nest with his tail. Hatching takes place about ten days after fertilisation and the young fish are protected for another twenty one days by the male, after which time they swim away from the nest.

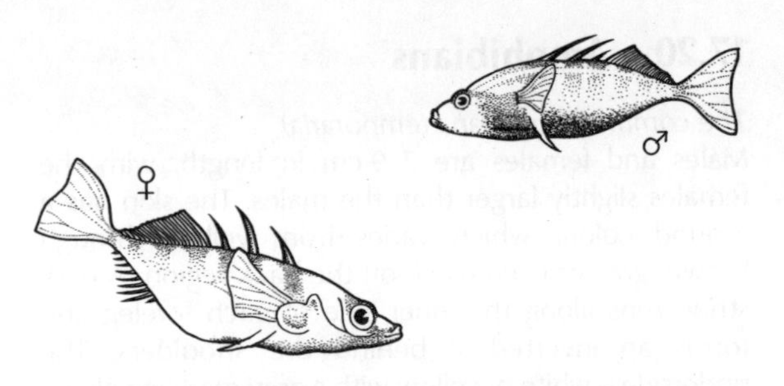

(i) red belly of male attracts female – swollen belly of female attracts male

(ii) male zig-zags towards the nest

(iii) male directs female into nest

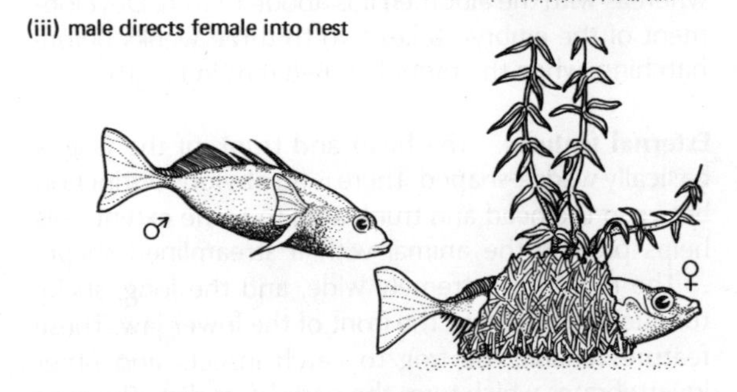

(v) female lays eggs in the nest – male encourages egg-laying

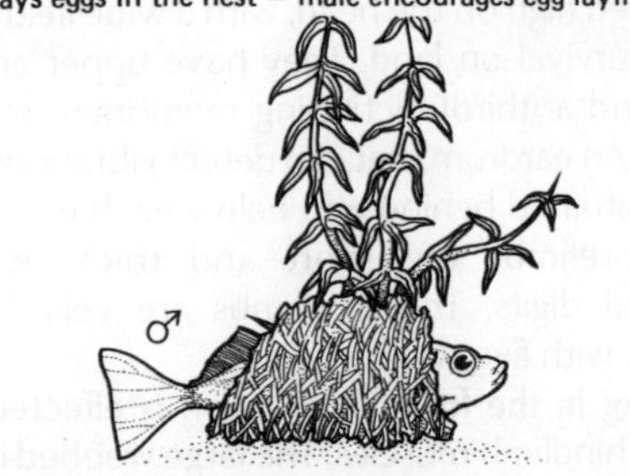

(vi) male follows female into nest to fertilise eggs

fig 17.4 Courtship and mating of the three-spined stickleback

17.20 Amphibians

The common frog (Rana temporaria)

Males and females are 7–9 cm in length, with the females slightly larger than the males. The skin has a ground colour which varies from yellow through brown, grey or even black on the back region. A dark stripe runs along the inner side of each foreleg and forms an inverted V behind the shoulders. The underside is white or yellow with a grey marbling effect. The skin is smooth or with little flat warts. The male has a pair of thumb pads on the forelimbs and the sides of these pads are covered with small black horny outgrowths. The webbed hind limbs are well-developed in both sexes, and the forelimbs are very powerful in the male, especially in the breeding season. The frog can jump a distance of six to seven times its own length by means of its hindlimbs. Both sexes emerge from hibernation in March and they then migrate into the pond in which they originally developed as tadpoles. Once in the water the male clasps the female with his forelimbs such that his fingers meet in front of the female. The male and female both croak loudly during their copulating period. The female lays about 1000 eggs in an hour, and in a complete spawning she produces about five batches of eggs. The male sheds sperms over the eggs as they enter the water and fertilisation occurs immediately. The fertilised eggs sink to the bottom of the pond. Their coating of albumen swells and the clump of spawn rises to the surface of the water where it remains floating. The egg, which has a high yolk content, is about 1.5 mm in diameter, whereas with the albumen it is about 10 mm. Development of the embryo takes two to three weeks before hatching, when the tadpole is 6–8 mm in length.

External features The head and trunk of the frog is basically wedge-shaped. There is no obvious distinction between the head and trunk and, to some extent, this helps provide the animal with a streamlined shape.

The mouth is extremely wide, and the long, sticky tongue is attached to the front of the lower jaw. These features enable the frog to catch insects and other invertebrates which form the basis of its diet. The eyes are situated high on the head, with a wide field of vision to help survival on land. They have upper and lower eyelids and a (third) nictitating membrane to protect the eyes. An eardrum that can detect vibrations in air or water is situated behind and below each eye.

The forelimbs are short and thick, with four unwebbed digits. The hindlimbs are very long and powerful, with five webbed digits.

Swimming in the frog Propulsion is effected by the powerful hindlimb muscles. The large webbed digits are able to displace large quantities of water. This makes frogs efficient swimmers. When the hindlimbs are

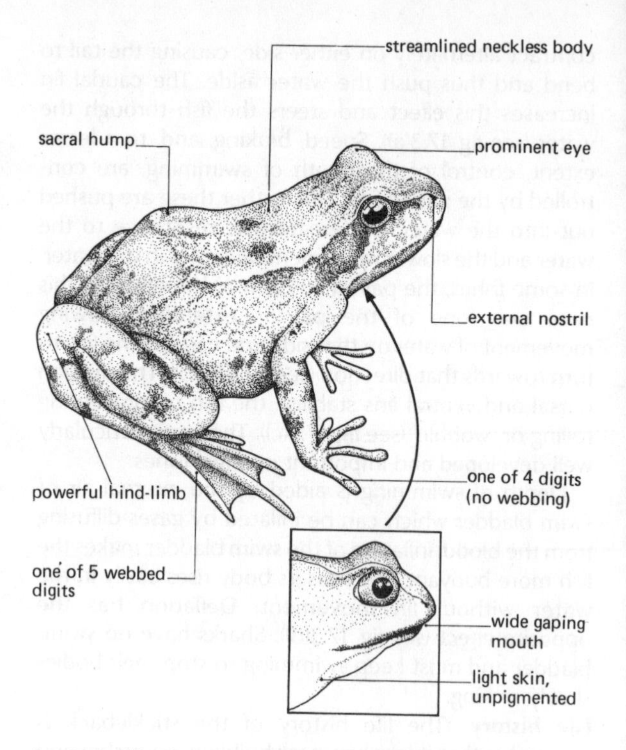

fig 17.5 (a) The common frog (*Rana temporaria*)

(b) The common frog (*Rana temporaria*)

extended, the general shape of the frog is not unlike the streamlined shape of a fish (see fig 17.6(a)). Steering is partly controlled by the webbed digits on the hind limbs, but the forelimbs play an important part (see fig 17.6(b)).

On land, frogs move by leaping or hopping (see fig 17.7). This is brought about by the hindlimb muscles. The webbed digits are pushed against the ground and this provides the necessary thrust for take off. Landing shock is absorbed by the stout forelimbs and the moveable pectoral girdle (not fused to the vertebrae).

Life history Details of the life history of the common frog are given in Chapter 10. All amphibians share certain basic features in their reproduction, being dependent on water for fertilisation. Even tree frogs need the water accumulated in leaves in order to lay their eggs. Male toads possess 'nuptial pads' on their forelimb digits. These are used to grasp the female in mating. The male mounts on the back of the female and remains in that position for a long time, squeezing her eggs out. Sperms are poured over the eggs, effecting external fertilisation. Not all the eggs will be fertilised. On immersion in water the jelly-like coat surrounding the small eggs swells to form a protective covering and a float. This jelly (albumen) is distasteful to some predators, and is too slippery for others to be able to grasp the eggs. The eggs remain stuck together for some time in clumps called *spawn*.

In practically all amphibians there is a larval stage, the tadpole, which develops in water (see fig 10.16). The adult form arises by a process called *metamorphosis* which is controlled by *hormones*. Mortality is high, and only one mature toad may survive from one batch of several hundred eggs. There is no parental care.

(a)

left forelimb close in to trunk

extended forelimb slows down water flow

(b)

fig 17.6 (a) Shape of a frog when swimming (b) Changing direction when swimming

17.30 Reptiles

This group includes turtles, crocodiles, snakes and lizards. Lizards are entirely terrestrial and they bask in the sun during most of the day, except at the hottest time of the day when they seek the shade. Their activity is controlled by the surrounding temperatures and they are only active during the daytime. This is because they are poikilothermic (cold-blooded) like fishes and amphibians. More information on this topic was discussed in Chapter 6 (Excretion and Body Temperature Control). None of these animals is able to maintain a constant body temperature, their temperature is that of the environment.

The common lizard (Lacerta vivipara)

This lizard has a distinct head, separated from its cylindrical trunk by a neck. Extending from the posterior end of the body is a long, tapering tail about the same length as the head and body. The horny overlapping scales that cover the body are largest on the head.

Neither the colours nor the patterns vary much compared with other European species of *Lacerta*. The dorsal surface varies between grey-brown and red-brown, broken up by pale or dark spots arranged in

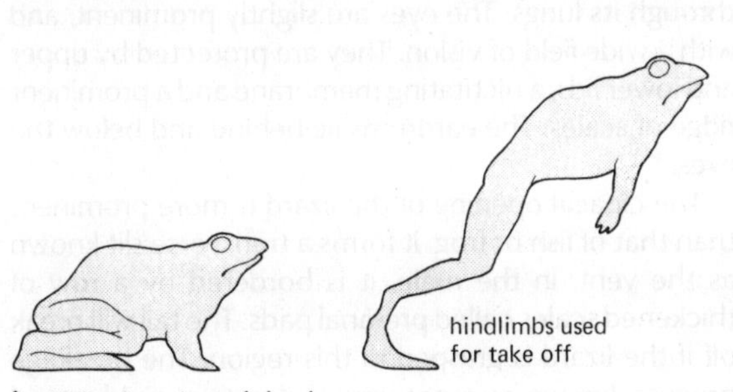

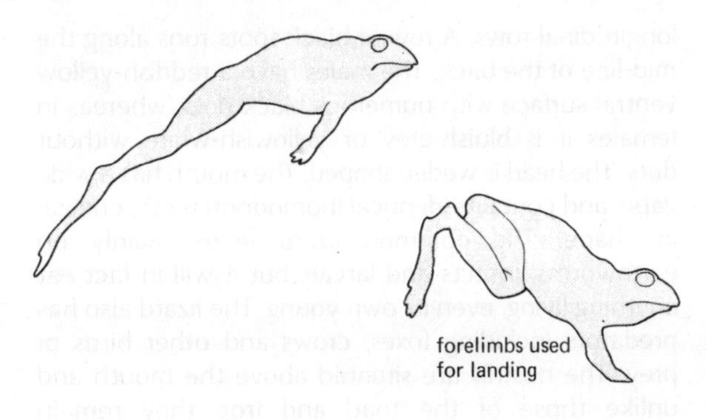

fig 17.7 Leaping of the frog

Table 17.1 Stages in the development of the frog tadpole

Tadpole	Mouth	Gaseous exchange	Belly, gut	Forelimbs	Hindlimbs	Tail
Newly hatched	unopened	2 pairs of external branched gills	swollen with yolk store	—	—	small but enlarging
One week after hatching	open – horny jaws; sucker posterior to mouth	3 pairs of external gills	not swollen yolk consumed; anus appears	—	—	tail enlarging
two weeks after hatching	horny jaws – sucker disappearing	3 pairs of external gills operculum covers right side gills – left side partly covered	gut long and coiled for plant digestion	—	—	long powerful fish-like tail
three weeks after hatching	mouth still rasping off plant material also takes in water	4 pairs of internal gills spiracle on left side	gut shortening	—	limb buds appear	powerful tail
five weeks after hatching	jaws changing to normal adult type	internal gills losing importance	gut shortens – now carnivorous, eating small water animals	buds appear beneath operculum	grow out with five webbed digits	powerful tail
seven weeks after hatching	normal jaws	lungs appear; replace gills; air taken in through mouth	carnivorous shortened gut	left forelimb grows out of spiracle – right limb bursts through operculum	long and powerful	tail as yet unchanged
ten weeks after hatching	mouth moves to anterior part of head	breathes by lungs	short gut	both limbs functional	both limbs functional	tail disappearing due to internal consumption by white blood cells
	METAMORPHOSIS ALMOST COMPLETE					
twelve weeks after hatching	Tiny frog now finally loses its tail – takes three more years to mature					

longitudinal rows. A row of black spots runs along the mid-line of the back. The males have a reddish-yellow ventral surface with numerous black dots, whereas in females it is bluish-grey or yellowish-white without dots. The head is wedge shaped. The mouth has a wide gape, and contains identical (homodont) teeth, conical in shape. The common lizard feeds mainly on earthworms, insects and larvae, but it will in fact eat anything living, even its own young. The lizard also has predators including foxes, crows and other birds of prey. The nostrils are situated above the mouth and unlike those of the toad and frog they remain permanently open since the lizard breathes exclusively through its lungs. The eyes are slightly prominent, and with a wide field of vision. They are protected by upper and lower lids, a nictitating membrane and a prominent ridge of scales. The eardrums lie behind and below the eyes.

The cloacal opening of the lizard is more prominent than that of fish or frog. It forms a transverse slit known as the vent. In the male, it is bordered by a row of thickened scales called pre-anal pads. The tail will break off if the lizard is grasped in this region. The breakage process, known as autotomy, is due to a sudden and

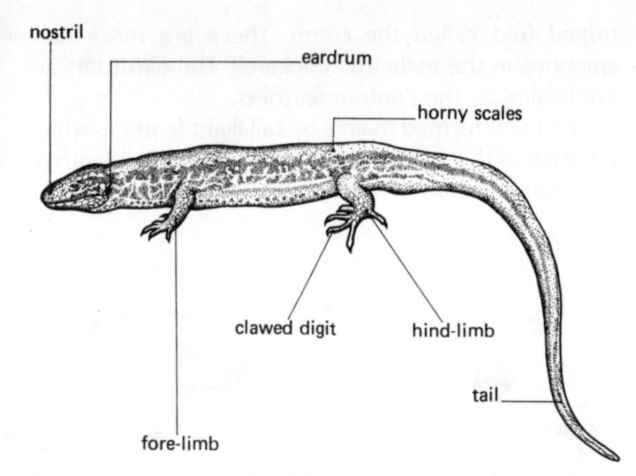

fig 17.8 A common lizard (Lacerta vivipara ♂)

powerful muscular contraction which is independent of whether or not the lizard is in the hands of an enemy. It does not take place at any fixed point along the tail. A new tail grows from the stump, but it never reaches the same length as the previous one. Furthermore the new tail has a central core of cartilage and not of vertebrae. The common lizard lives in damp places in meadows and moorland, at the edges of woods and on banks supporting hedgerows. It often lies on a tree stump or a rock especially on a sunny morning and in the afternoon. It is widespread in much of Europe but not in the Mediterranean regions, since it cannot tolerate tropical or sub-tropical climates. It hibernates during October, going into holes underground or under tree stumps. It emerges in March and mating takes place in May. The gestation period lasts about three months and the young are born alive at the end of July (hence the specific name *vivipara*). It is more correct to say that this lizard is ovoviviparous, for the young do not receive nourishment from a placenta as in the truly viviparous mammals. The female produces five or six young, each one being covered by a whiteish egg membrane. The mother shows no parental care for the young, which are about 11 mm long and can run about immediately after birth.

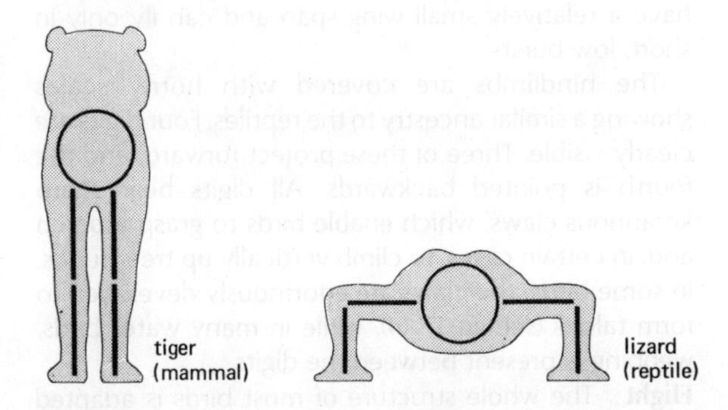

fig 17.9 The relative positions of limbs to trunk in mammals and reptiles

17.40 Birds

The domestic fowl is a bird commonly seen and is thus a convenient member of the group to study. Apart from its inability to fly, owing to its heavy body (which is the result of artificial selection by Man for meat) it is a fairly typical bird in structure and behaviour. The following description could apply to a number of breeds of common fowl e.g. Plymouth Rock, Light Sussex and New Hampshire.

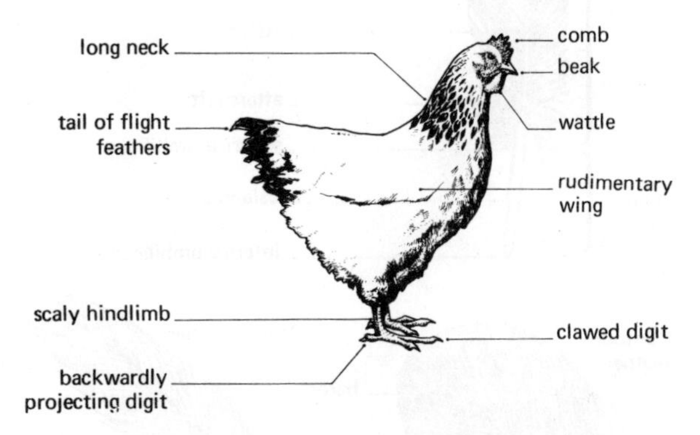

fig 17.10 An example of a Common Fowl (Light Sussex)

External features There is a significant difference in the external appearance of male and female birds. This is a result of sexual selection, since courtship involves visual display.

The outline of a fowl is smooth and relatively streamlined. Wild birds are more marked in this respect. This streamlining is produced by the close-fitting feathers which cover the bird. They are variously coloured and in many birds afford camouflage. They also serve for visual signals in courtship.

Feathers are formed in pockets or follicles in the skin, and consist of a horny protein called *keratin*. Their structure is best studied in the large flight or *quill* feathers. This type of feather has a stalk or *shaft* from which arises the *vane*. (See fig 17.11(a)). The stalk is hollow at the base, or *calamus*, for lightness. The terminal section of the shaft, the *rachis*, is partly solid to support the branches or *barbs* of the vane. At the base of the vane are a group of loose barbs called the *aftershaft*. The barbs forming the vane are held together by interlocking projections called *barbules* which bear short hooks for grasping the barbules of neighbouring barbs (see fig 17.11(b)). The result is a structure which will offer resistance to the air and thus make flight possible.

Other feathers possess smaller vanes than quill feathers. *Contour* feathers have a curved shaft so that their tips lie parallel to the body. They streamline the body and insulate it against heat loss. *Down* feathers are small, soft feathers found between the contour feathers. They provide extra insulation: young chickens

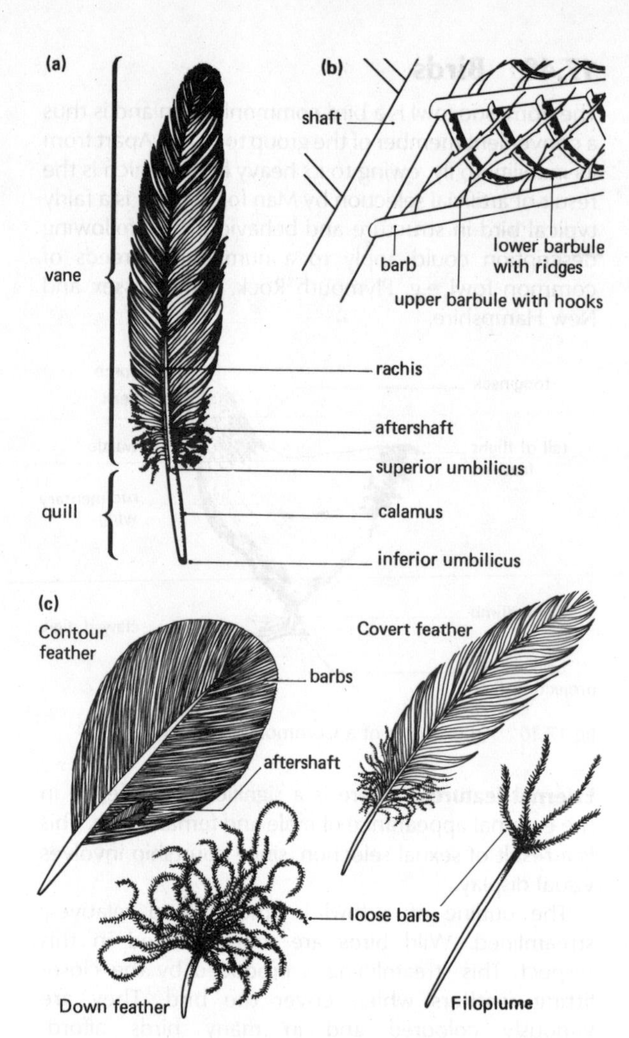

fig 17.11 Feathers (a) Flight or quill feather (b) Arrangement of barbs and barbules in a flight feather vane (c) Contour, covert, down feathers and a filoplume

are covered exclusively with down feathers. *Filoplumes* (see fig 17.11(c)) consist of a tiny shaft bearing a few loose barbs. Their function is not fully understood.

Common fowl have relatively small heads and long necks. The mouth has developed into a *beak*. This is a horny, hinged, toothless projection, which is pointed in domestic fowls for the grasping of grain. There is a wide variety of beak structure in birds, depending upon their diet. Some examples of beaks which have been adapted to deal with different foods are shown in fig 17.12. The external nostrils are positioned on the beak.

Fowl breathe using lungs and a system of air sacs inside the body. The small eyes are protected by eyelids. Actively flying birds have larger and keener eyes than fowl. Below the beak are red folds of skin called *wattles*, and on top of the head is a similar but ridged fold called the *comb*. These are more conspicuous in the male (the cockerel). The eardrums are concealed by the contour feathers.

The tail is formed mainly by tail flight feathers which emphasise the streamlining of the body. Tail feathers are more luxuriant in the cockerel than in the hen.

fig 17.12 A variety of bird beaks

The forelimbs are considerably modified to form the wings (see fig 17.13). The bones present can be identified as basic bones in a pentadactyl limb (Chapter 11), but they have been elongated, fused and reduced in number. The longest flight feathers are the *primaries* which are attached to the limb bones at the outer edge of the wing. The inner feathers are slightly smaller and are called *secondaries*. In most birds, the outermost layer of wing flight feathers are the even shorter *tertiaries* and *wing coverts*. Usually the surface area of the wing is very large in relation to the body, but fowl have a relatively small wing span and can fly only in short, low bursts.

The hindlimbs are covered with horny scales showing a similar ancestry to the reptiles. Four digits are clearly visible. Three of these project forward, and the fourth is pointed backwards. All digits bear sharp keratinous claws, which enable birds to grasp a perch and, in certain cases, to climb vertically up tree trunks. In some birds, the claws are enormously developed to form talons (see fig 17.14), while in many water birds, webbing is present between the digits.

Flight The whole structure of most birds is adapted for flight. Internally, the bones are hollow to reduce weight and a system of air sacs adds to the relative

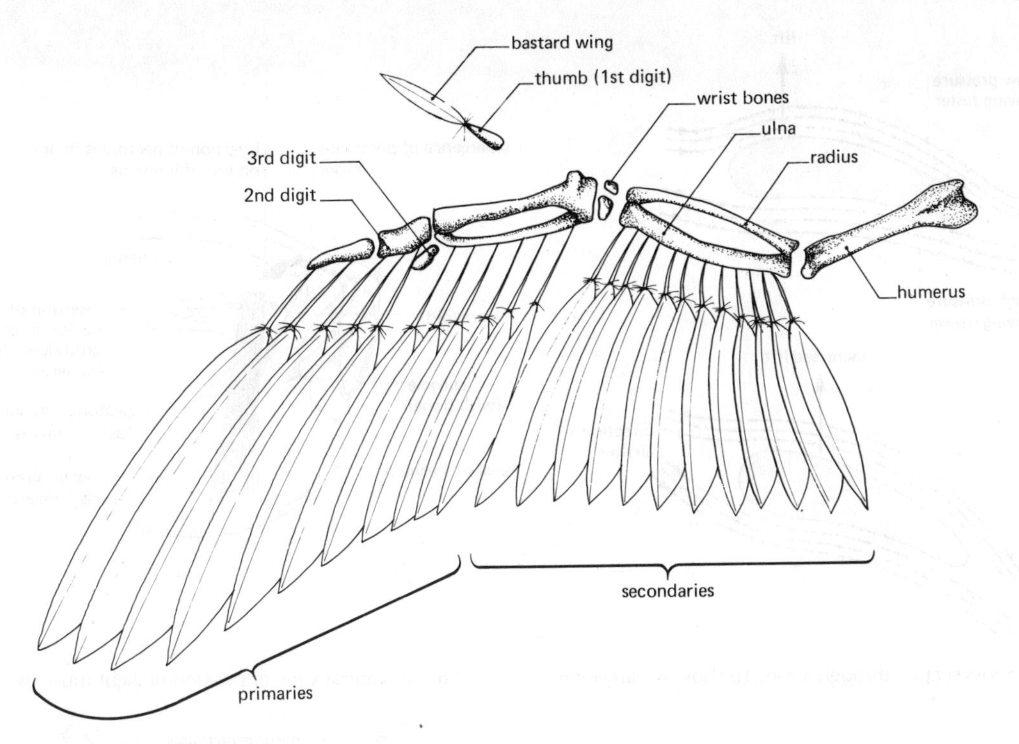

fig 17.13 Parts of a bird's wing

lightness. The air sacs are also involved in an efficient method of breathing. Flight is very demanding in terms of energy. Birds are homoiothermic (warm blooded) and small birds must feed continuously to obtain sufficient energy to conserve their body heat and to fly (See Chapter 5). Oxygen is extracted from inspired air twice in one breath. The first time is when the air inflates the lungs, the second time is when the air returns via the lungs from the air sacs during expiration. The double circulation of blood in birds ensures that the maximum amount of this oxygen reaches the respiring cells.

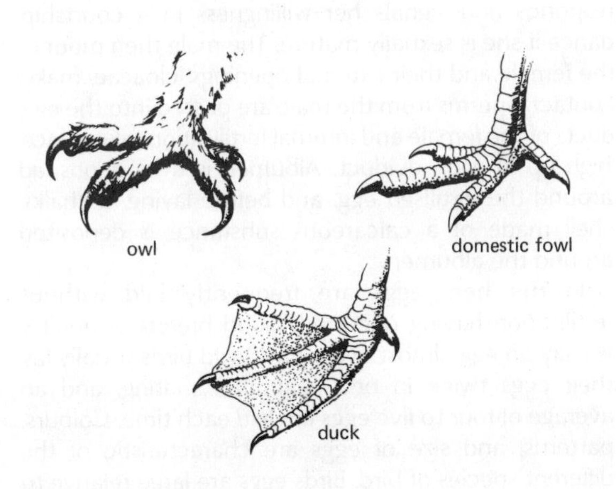

fig 17.14 A variety of bird's feet

The wing is the principal organ of flight. In cross-section, it presents a streamlined form against oncoming air (see fig 17.15). When taking off, a bird first raises its wings at an oblique angle and then brings the wings downwards and backwards with force. (The enormous flight or pectoral muscles in the breast region make up one-fifth of the body weight (see fig 17.16)). This latter movement forces air down and behind the bird to give upward and forward thrust. When the wings are raised, the barbules are not engaged and air can pass through the wing. During the downbeat of the wing, the barbules are engaged to offer maximum resistance to the air and to force it aside (see fig 17.17). The total movement of the wings is like a 'figure of eight'.

Steering is partly controlled by the tail, but the tips of the wings can be tilted to steer. Turning is effected by the more vigorous beat of one wing, comparable to the action of the paired fins in fishes. Balance is controlled by the tail feathers, which also act as a brake by spreading out to ease landing.

Life history Birds have a specialised behaviour in preparing for reproduction. Wild birds will first claim a territory by warning off all rivals principally by uttering a warning song common to the particular species. Nestbuilding follows. Nests may be constructed of many different materials such as sticks, animal hair, feathers and leaves. Birds are brought together for mating when they send out mating calls. Then follows a period of courtship, when the more brightly coloured male bird displays himself to the female. The female

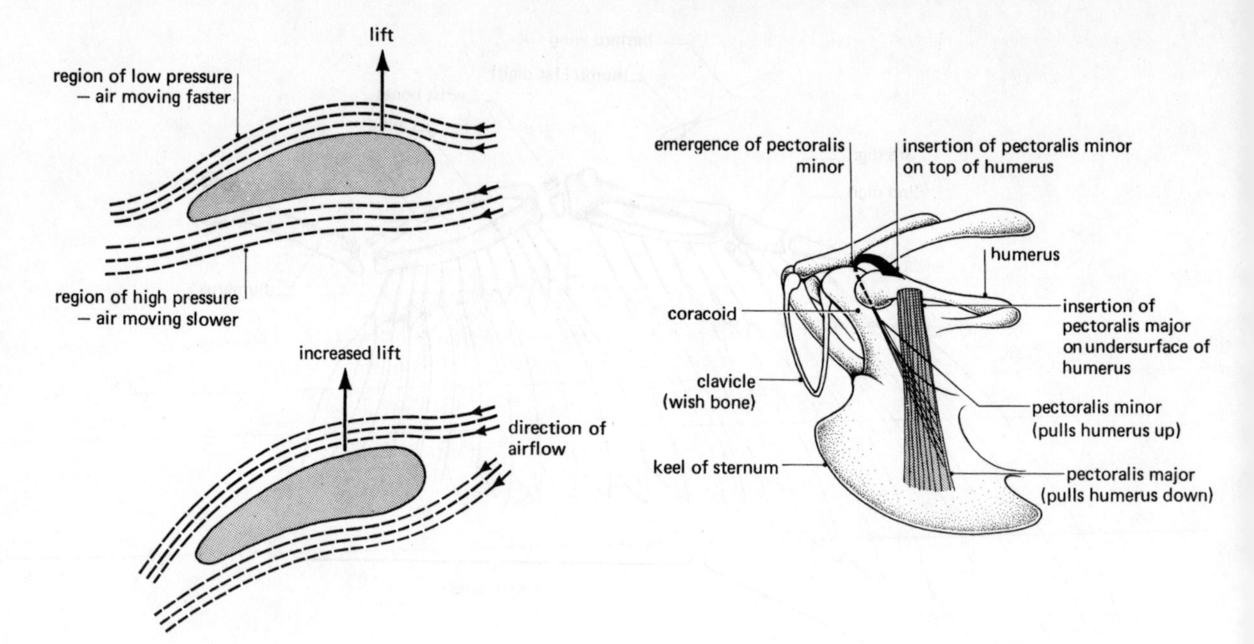

fig 17.15 Cross section through a wing to show streamlining

fig 17.16 (a) Lateral view of position of flight muscles

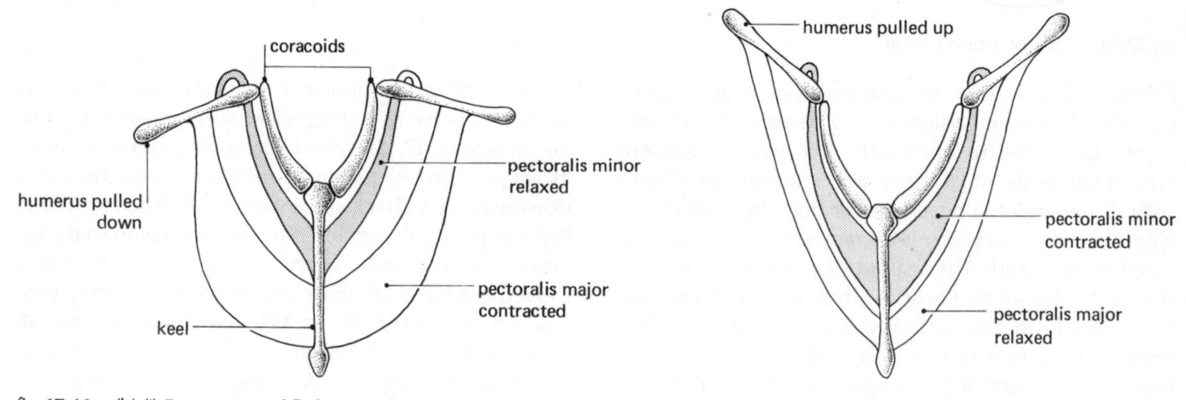

fig 17.16 (b) (i) Front view of flight muscles depressing the humerus (ii) Front view of flight muscles elevating the humerus

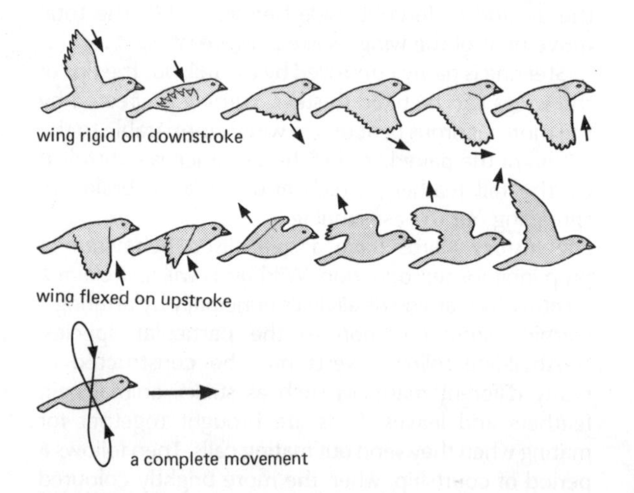

fig 17.17 Flight movements

responds and signals her willingness in a courtship dance if she is sexually mature. The male then mounts the female and their external openings (cloacae) make contact. Sperms from the male are passed into the egg ducts of the female and internal fertilisation takes place high up in each oviduct. Albumen is then deposited around the fertilised egg, and before laying, a chalky shell made of a calcareous substance is deposited around the albumen.

In the hen, eggs are frequently laid without fertilisation having occurred. Good breeds of poultry will lay an egg almost every day. Wild birds usually lay their eggs twice in one year after mating, and an average of four to five eggs are laid each time. Colours, patterns, and size of eggs are characteristic of the different species of bird. Birds eggs are large relative to the size of the adult, and well protected by the shell. Air is present between two internal membranes and there is an abundant supply of yolk (see fig 17.18).

Parental care is highly developed. During incubation of the embryos, which may last a varying length of time in different species of wild bird, but three weeks in chickens, the eggs must be kept at a constant temperature. In many cases the female bird sits on the eggs. Feathers are moulted from the underside of her body to form the 'brood pouch' which brings her warm blood in closer contact with the eggs. When wild birds are hatched, they are very vulnerable, and lack feathers. The parents continue to keep them warm by sitting on them, and feed them continuously. They are not left to fend for themselves until they can fly and obtain food independently. Despite this, the rate of mortality of young birds is high, probably around 50%.

Chickens are hatched with a covering of down feathers, and are able to feed themselves almost at once. This is because they are hatched at a more mature stage than in most wild birds. Ground-breeding birds are able to run and feed in a similar fashion.

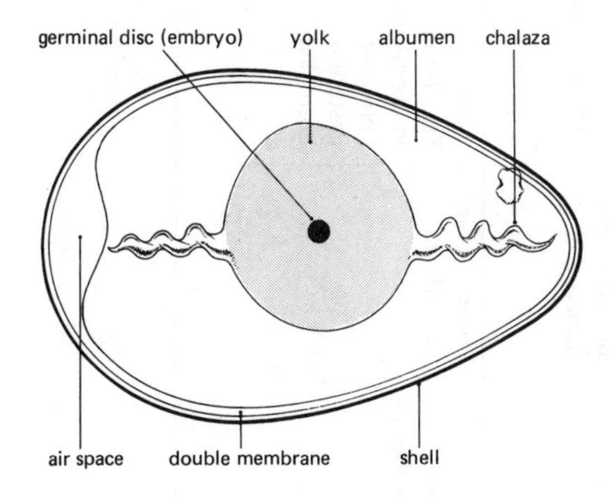

fig 17.18 The structure of a bird's egg

17.50 Mammals

The brown rat is a common rodent found around buildings and gardens. It digs long, deep burrows with several entrances, and stores there food which it carries in its cheek pouches. It eats grain and other stored foods.

External features The brown rat has a pointed head separated by a distinct neck from a long body which widens towards the posterior. It has a long, thin scaly tail. Apart from the tail, the animal is covered with dark grey and brown hairs. These tend to be darker on the back and lighter or even white on the lower surface – another example of camouflage. Hairs formed from the protein keratin occur in pockets or follicles in the skin.

The mouth of the rat has a divided upper lip which displays large incisor teeth for gnawing vegetables. These teeth have enamel which grows continuously to replace that which has been worn away. At the back of the mouth are the chewing teeth, – premolars and molars. Such a *heterodont* dentition, containing a

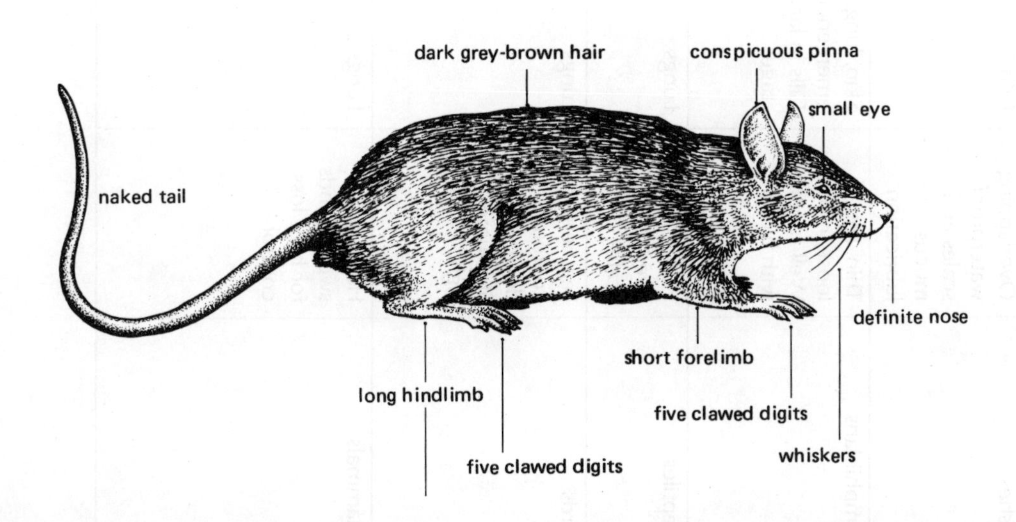

fig 17.19 The brown rat (*Rattus norvegicus*)

Table 17.2 Summary Table of Vertebrate Characteristics

Class	Skin	Breathing	Warm or cold blooded	Fertilisation	Eggs/batch	Nourishment of embryo	Parental care
Fishes	Overlapping waterproof scales and mucus lubrication	Gills	Cold blooded – poikilothermic	External in water	Thousands to millions 95% mortality	Yolk in egg	Usually nil
Amphibians	Thin, soft, loose, moist Well supplied with blood	Skin, lungs in emergencies Gills in larval stages	Cold blooded – poikilothermic	External in water	Hundreds Enclosed in jelly in batches or strings (spawn) 85% mortality	Yolk in egg Albumen	Usually nil
Reptiles	Horny, dry, overlapping scales	Lungs	Cold blooded – poikilothermic	Internal in oviducts	1–50 Leathery shell 30% mortality	Yolk in egg	Usually nil
Birds	Feathers:— contour down filoplume flight	Lungs	Warm blooded – homoiothermic	Internal in oviducts – high up, before shell is added	1–10 Chalky shell Protected in nest 50% mortality	Yolk in egg Water from albumen	Highly developed Territory, Nest building Incubating Feeding and Basic teaching
Mammals	Hairy skin, sweat glands for heat loss control	Lungs	Warm blooded – homoiothermic	Internal – high in oviducts	1–10 Eggs retained in uterus for period of gestation. Born alive	Yolk in egg Nutrients from mother via placenta	Highly developed, Nest building Feeding on milk from mammary glands. Shelter, protection, Teaching.

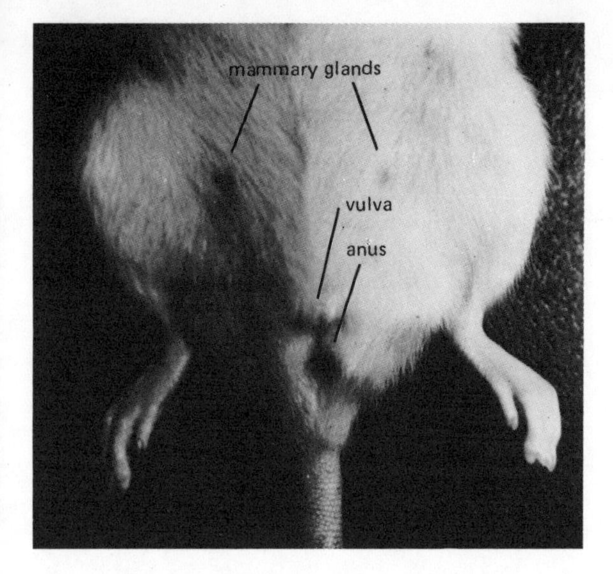

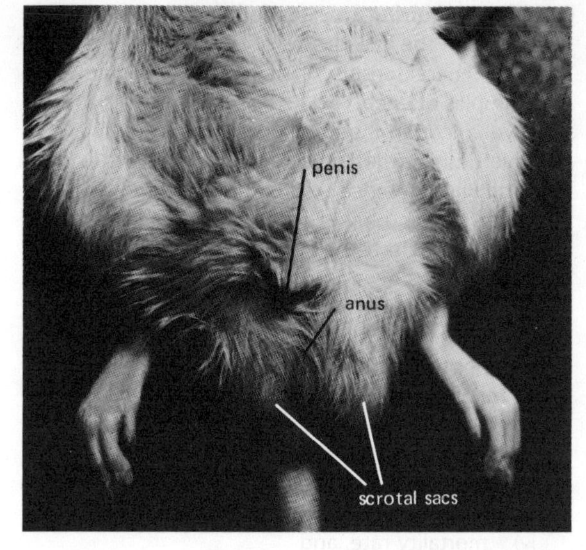

fig 17.20 The external genitalia of female and male rats

variety of teeth is characteristic of mammals. In a young rat, the teeth are incisors and premolars only and are later shed. This is the *deciduous* dentition (milk teeth) found in all mammals and it is later replaced by the *permanent* dentition (see Chapter 2).

External nostrils are found above the mouth and there is a distinct nose. Breathing is mainly through the nose. On either side of the 'muzzle' formed by the projecting nose and mouth are specialised long hairs called whiskers. These are sensitive to touch and enable the rat to judge the width of holes and crevices.

The small eyes are protected by upper and lower eyelids and a nictitating membrane. Behind the eyes are a pair of conspicuous protruding external ear flaps called 'pinnae'. The eardrum lies at the bottom of a short canal within the pinna. Pinnae are exclusive to mammals and in many species they are mobile and are used to locate the direction of sound vibrations. (See Chapter 6).

On the underside of the trunk of the female are the milk-producing mammary glands used to suckle the young. These are characteristic of all mammals. The male possesses external genital organs, consisting of a penis and scrotal sacs. These are situated towards the posterior of the underside of the trunk (see fig 17.20).

The naked tail is dark along the first half of its length and white along the terminal half.

The forelimbs are shorter than the hindlimbs. Both pairs of limbs have five clawed digits which enable the rat to dig burrows.

Movement The limbs are slender in relation to the body, and the characteristic movement of rats is 'scurrying', (they take short, quick steps). The long, powerful hindlimbs enable them to leap when necessary (see Chapter 11).

Life history The females and young are communal and live in burrows. The female makes a nest in the burrow from pieces of paper, straw, string and similar material. She carries the nest material to the burrow and heaps it up using pushing movements of paws and snout. She hollows out the heap by turning around in the centre, pushing the bedding to the side and pressing it down with her paws. Courtship is less ritualised than in most birds. When mating, the male rat mounts the female rat from behind and sperms are injected into the female through the copulatory organ (penis) of the male which is inserted into the vagina of the female. Internal fertilisation occurs in the fallopian tubes (oviducts) of the female. The fertilised eggs are retained in a specialised organ, the *uterus* (womb). After about 21–23 days, 6–8 young rats are born *alive*. The details of this are discussed in Chapter 8. This type of reproduction is called *viviparous* and is found in the majority of mammals. The young are 'nursed' by the mother, who feeds them on milk from the mammary glands. Milk contains all the basic nourishment required for growth. The young are also protected in the burrow and kept at a constant temperature until they are large enough to maintain their own temperature. The rate of mortality of the young is very low.

Humans display the highest form of parental care. Their young are not only fed, sheltered and kept warm by their parents, but they are also taught how to cope with life and society. In this way, human experience is handed down from one generation to the next. The young have the longest period of dependency of any mammal. Apes and monkeys are also dependent on their parents for a long period of their lives.

Questions requiring an extended essay-type answer

1 What are the five classes of vertebrates? State the major characteristics of each group and name **two** examples of each.

2 Give a comprehensive account of the ways in which fish are adapted for life in the water, with particular reference to breathing and movement.

3 Write an account of the life history of a *named* amphibian. Draw the egg stage, and no more than **three** larval stages.

4 a) In what ways is a bird adapted to flight?

b) Draw a flight feather, and explain how this is superior to a down feather for flying.

5 Compare and contrast reproduction in the five classes of vertebrates. Include in your answer

i) fertilisation,

ii) the nature of the eggs,

iii) nourishment of the embryo,

iv) mortality rate, and

v) parental care.

18 Answers and discussion

Introductory chapter

Section I 2.23

1 Fibres of cellulose can be seen at the higher magnification compared with the smoothness of the paper seen by the naked eye alone.
2 Yes.
3 It is now a letter quite different from letter 'p'. It appears to be the letter 'd'.
4 It is upside down and turned from right to left.

Section I 2.24

1 All of the wing can be seen.
2 It is much larger (more magnified) and more detail can be seen.
3 It moves to the left.
4 It moves towards the observer.

Section I 5.40

1 Hydrochloric acid 0.01 M has a pH of 2. Sodium hydroxide 0.0001 M has a pH of 11.
2 It has a pH of 7 which is between hydrochloric acid (pH 2) and sodium hydroxide (pH 11).
3 It is the same as the given solution of hydrochloric acid.
4 The pH value of the acid increases and the pH value of the alkali decreases. If the hydrochloric acid is diluted exactly 10 times, the pH increases by 1, and each time this is repeated the pH increases similarly.

Thus:	0.001 M	pH 3
	0.000 1 M	pH 4
	0.000 01 M	pH 5

If the sodium hydroxide is diluted 10 times the pH decreases by 1, and each time this is repeated the pH decreases similarly.

Thus:	0.000 1M	pH 10
	0.000 01 M	pH 9
	0.000 001 M	pH 8

1 Nutrition

Section 1.11

1 The colour changes from the brown colour of iodine solution to an intense blue-black. In testing foods for the presence of starch many different shades of blue may be observed, but all indicate a positive result.
2 The copper sulphate in the Benedict's or Fehling's solution is reduced to copper oxide, which forms a precipitate. This goes through a series of colour changes finalising at red or brown. If the tube is left to stand, the precipitate will sink to the bottom of the tube.
3 The colour changes in the following sequence: green; yellow; orange; red and brown. In the test for reducing sugars, the colour green indicates a slight amount of the sugar and brown means a substantial amount.
4 No change. The blue colour of the copper sulphate in the test reagents remains.
5 No change in the experimental or the control tube.
6 Positive reaction through green to red or brown.
7 The sucrose has been hydrolysed to monosaccharide molecules which can now reduce the test reagents and give a positive result. The test is thus positive for non-reducing sugar, but **not** reducing sugar.
8 The oil breaks up and dissolves in the ethanol.
9 A milky white emulsion is formed. This is the positive indication of the presence of oil or fat. When foods are being tested, particles of food should be allowed to settle before the ethanol and its dissolved fat is poured into the water.
10 The water spot has dried out, but the oil spot remains and the light shines through. When foods are being tested they can be rubbed on the paper to see if they make an oily mark.
11 The protein coagulates on heating and becomes red-brown in colour. There is no precipitate.
12 The test is specific for the amino acid, tyrosine, and there is no colour change with sugar or fat.
13 A violet colouration is a positive indication of peptides and hence all proteins. This is a much more general test than Millon's reagent.
14 Suppose that 10 drops of orange juice just decolourised 1 cm^3 of DCPIP solution, and that 8 drops of the ascorbic acid solution did the same.

Then the ascorbic acid solution contains 1 mg of ascorbic acid per cm^3, and the orange juice must have contained $1 \times \frac{8}{10} = .8$ mg of ascorbic acid per cm^3.

Repeat for all juices tested.
15 Exposure to the oxygen of the air destroys vitamin C in fruit and vegetables, therefore the values of

vitamin C (ascorbic acid) should be less after exposure for several days. Cooking also destroys vitamin C.

Section 1.21

1 It is green in colour. The chlorophyll has been extracted by the ethanol.

2 The leaf or leaf discs show a blue-black colour *between* the veins of the leaf. The colour of the veins is not blue-black. Starch must be present in the leaf, but only in the cells of the blade, not in the veins.

3 The leaf is the colour of the iodine solution, brown or orange-brown. No starch is formed in the absence of light.

4 Depending on the size of the leaf and the amount of sugar present, the colour could vary from green to brown. The test is positive for the presence of reducing sugar.

5 The green areas of the leaf show a blue-black colour indicating the presence of starch. The white areas of the leaf show the brown colour of the iodine solution indicating that no starch has been formed.

6 Starch is only formed in those areas of the leaf which are green in colour, that is where chlorophyll is present.

Section 1.22

1 No. The concentration did not drop below 0.032%.

2 After twelve midnight the carbon dioxide concentration rose to a peak at 7 a.m. and then fell steadily to reach its lowest concentration during the later afternoon. It then began to rise again through the hours of darkness.

3(a) 7 a.m. 0.054% carbon dioxide

(b) 4 p.m. 0.032% carbon dioxide

4 The carbon dioxide concentrations in the forest rose during the hours of darkness and fell during the hours of daylight. It could be that the vegetation is giving out carbon dioxide at night and taking it in during daylight.

5 *Tube A* changes to yellow

Tube B stays unchanged (or slightly yellow or slightly purple, depending on the density of the muslin screen).

Tube C changes to purple.

Tube D stays unchanged (orange-red).

6 *Tube C* The leaf in light has withdrawn carbon dioxide from the air above the indicator, and carbon dioxide from the indicator has diffused out decreasing the acidity thus changing the colour to purple.

Tube A This colour change to yellow represents an increase in acidity, indicating that carbon dioxide has been added to the indicator. This has occurred in the dark and is in accordance with the graph of carbon dioxide in the forest.

Tube B This tube has the leaf in half light or twilight and the uptake and output of carbon dioxide could be completely balanced. It depends on the thickness of the muslin screen; a thinner screen could tip the indicator towards a purple change, a thicker one could tip it towards a yellow change.

Tube D Has no leaf and the indicator is unchanged, showing that any change in the other tubes must be due to the enclosed leaf.

7 Control.

8 Yes. The indicator shows that carbon dioxide is given out in the dark and taken into the leaf in the light. In half light or twilight either there is a balance or it could be that there is none given out and none taken in. This can only be decided by further experimentation.

9 Emission of light is always accompanied by heat. The changes that take place in the indicator could be due to the action of heat emitted from the light bulb or the sun. To eliminate this variable the tubes are placed in water, so that the glass and the water act as a heat shield.

10 *Leaf A* (deprived of carbon dioxide) This shows no starch in the leaf, which remains the orange-brown colour of iodine solution after testing.

Leaf B (in the flask but carbon dioxide available) This gives a positive result for starch, but the blue-black colour is not strong.

Leaf C (in the air with more carbon dioxide available) This gives a much stronger positive result for starch with the colour a deep blue-black.

11 The carbon dioxide contains two of the elements needed for the formation of carbohydrate. The third element could come from water which is always present in living organisms. It would appear that the carbon dioxide contributes towards the manufacture of carbohydrate in the leaf.

12 Since the experiment investigates the supply of carbon dioxide to the leaf, it would probably be better to have flowing over the plant a supply of air from which the carbon dioxide has been removed. Enclosing the leaf in a flask limits the supply of other gases present in the air, and therefore a better experimental set-up

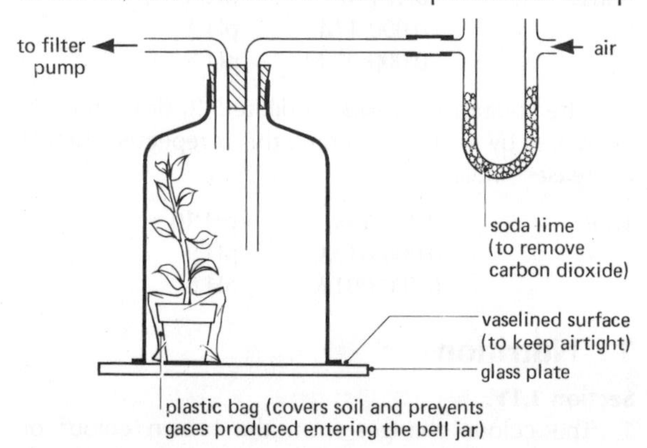

fig A Alternative apparatus to investigate the use of CO_2 by green leaves

would be as shown in fig A. What would be the control apparatus for this experiment?

13 *Leaf A* (deprived of light) This shows no starch in the leaf, that is orange-brown colour of iodine solution after testing.

Leaf B (in the flask but light available) This gives a positive result for starch with a blue-black colour.

Leaf C (in the air with light available) This gives a positive result for starch with a blue-black colour.

14 Light is essential for starch formation in the leaf.

15 Carbon dioxide might be used up in the flask and become a limiting factor. Another method is to shut off light from part of the leaf only, so that the remaining parts receive all conditions and therefore give a positive result for the starch test. The black cardboard must be raised slightly from the leaf in order to allow a free flow of air containing carbon dioxide.

Section 1.23

1 In order to stabilise the temperature of the gas bubble. Gas expands considerably with a small rise in temperature, thus when the apparatus is handled the body heat may cause expansion. The temperature and therefore the volume, is restored by returning the J-tube to the water trough.

2 In order to present the surface of the bubble with the potassium hydroxide and the potassium pyrogallate. The chemicals absorb the carbon dioxide molecules and the oxygen molecules respectively.

3 The results will vary according to the light intensity and the length of time the weed is exposed to light during each twenty-four hour period. The oxygen content can increase by up to 50% of the gas, and carbon dioxide by up to 10% compared with normal air.

4 Light intensity will alter the rate of photosynthesis: below are some figures which indicate the relationship:

	% of CO_2	**% of O_2**	**% of N_2 by subtraction** (not tested)
i) very dull weather	6	14	80
ii) dull	0	24	76
iii) fine	0	39	61
iv) very bright	0	49	51

Testing for the presence of oxygen with a glowing splint is often used in this experiment. It will clearly only relight the splint in the third and fourth examples (oxygen present substantially greater than 21%).

5 Greatest concentration at 17 00 h (5 p.m.)
Lowest concentration at 07 15 h (7 15 a.m.).

6 The period of time between 07 00 h to 14 00 h shows the most rapid increase, reaching a maximum at 17 00 h, broadly speaking from early morning to late afternoon.

7 8.05 – this value does agree with 'less carbon dioxide' since this pH is more alkaline or less acid indicating that there is less carbonic acid present. Carbonic acid is formed from carbon dioxide.

8 7.5 – this value does agree with 'more carbon dioxide' since this pH is more acid or less alkaline than the previous value for pH. Thus there is more carbonic acid and pH falls in value, but notice that it does not become acid since the pH 7 is neutral (see Introductory Chapter section 5.00).

9 Fig 1.13, graph of the concentration of carbon dioxide in a forest.

Section 1.24

1 The gas is produced in the leaves during photosynthesis. Assuming that there is a connection through the stem to the leaves then the gas will pass through the stem and bubble from the cut shoot.

2 The red filter between the lamp and the beaker causes the most rapid rate of bubbling.

3 The green light is reflected or passed through the leaf, whereas the red light is absorbed by the chloroplasts and used in photosynthesis.

Section 1.26

1 Cotton wool is largely composed of cellulose and if wet could be a medium for growth of fungi and bacteria which would damage the seedling.

2 No air is available. Roots in soil must have access to soil air containing oxygen. Bubble air through the solution each day by means of a glass tube.

3a) *Tube 1* containing complete solution

b) *Tube 4* containing distilled water.

4 *Tube 3* Calcium ions are required by the cell for middle lamella development between the cell walls. Deficiency of the element results in poor root growth.

5 Because poor root growth would limit the passage of water and salts to the leaves and hence there would be poor leaf growth. Conversely poor leaf growth would provide little food from photosynthesis for root growth.

6 The seedlings extract water and salts at different rates. Water is removed more quickly and therefore the addition of more culture solution would increase the concentration of salts. This could reach too high a level and result in water passing out of the roots.

Section 1.30

1 No.

2 No starch was present, but the glucose test was positive.

3 Glucose passed through the tubing. The glucose molecules are much smaller than the starch molecules and thus able to pass through the minute pores in the visking tubing. The pores are not large enough for the starch molecules to pass through.

4 Two controls would be appropriate:
i) the same experimental set-up but with starch only in the visking tubing.
ii) The same experimental set-up but with glucose only in the visking tubing.
5 The starch had disappeared from tube A.
6 Something in the saliva had caused it to disappear, because it had not disappeared from tube B where saliva is absent.
7 The starch had been hydrolysed to reducing sugar because the Benedict's test on tube A was positive. The Benedict's test on tubes B and C was negative.
8a) It shows that something in the saliva breaks down the large, insoluble molecules of starch to smaller, soluble reducing sugar molecules. This is the essential function of digestion.
b) Test saliva only with Benedict's solution. The result should be negative.
9 Tube C. The liquid in the tube was clear, indicating that the globules of albumen must have disappeared.
10 Enzyme activity must have been stopped by boiling. The enzyme is denatured above a temperature of about 45°C. Note that it is **not** killed since enzymes are not living material.
11 It could be that hydrochloric acid changes the egg albumen **only** in the presence of pepsin. What other control would be needed to test this hypothesis?
12 Since the pepsin works best in acid conditions, the pH must be less that pH 7 (neutral). Pepsin in the stomach works at about pH 1.5 to 2.0.

1.31

Investigation 1

1 Manure. This material is laden with bacteria and these can be conveyed to human food by the proboscis and thus cause disease in Man.
2 Bacteria can also be transferred to Man's food in the following ways:
i) on the bristles of the legs and body,
ii) the fly may defaecate on the food (the 'fly spot').

Investigation 2

3 The mandibles have a serrated edge for cutting plant material and a flattened ridged surface for grinding the plant food.
4 To direct the food to the mandibles.
5 From side to side. It feeds along the edge of the plant leaf.

Investigation 3

4 i) About 1 year of age ± three months.
ii) 4 in the upper jaw, 4 in the lower jaw.
iii) Incisors.
iv) Third molar in each half jaw, often called the 'wisdom teeth' since they appear at about the age of 19–21 years, when 'wisdom' has arrived!

Investigation 4

1 Incisors 0, canines 0, cheek teeth 6 (premolars 3, molars 3) in half jaw.
2 Incisors 3, canines 1, cheek teeth 6 (premolars 3 molars 3) in half jaw.
3 Front teeth (chisel shaped incisors and canines) on the lower jaw act against a horny pad at the front of the upper jaw. Sometimes food is chopped off and at other times it is pulled off the plant or out of the ground.
4 A large gap in the lower jaw, between the canines and the premolars, enables newly taken vegetable matter to be stored. The food is kept apart from that being chewed by the cheek tooth.
5 $i\frac{0}{3}\,c\frac{0}{1}\,p\frac{3}{3}\,m\frac{3}{3}$
6 The lower jaw performs a circular type of motion from outside to centre, on one side and then the other.
7 *Incisors* – chisel shaped, flattened surface not very sharp; chopping and pulling.
Canines – similar to incisors.
Premolars – ridged across the jaw so that the W-shaped ridges on the upper jaw fit into the M-shaped ridges of the lower jaw; chewing and grinding.
Molars – similar to premolars.
8 The root is open. Growth is continuous so that as material is worn away from the surface it is replaced by new material transported by the blood vessels through the root and into the pulp cavity.

Investigation 5

9 Incisors 3, canines 1, cheek teeth 6, (premolars 4, molars 2) in half jaw.
10 Incisors 3, canines 1, cheek teeth 7 (premolars 4, molars 3) in half jaw.
11 The wild dog uses its canines for killing its prey. Premolars and molars are used for cutting up and crushing the corpse. The domestic dog does not have to catch its prey, but still uses its teeth for cutting and crushing its food.
12 The hinge joint has no lateral movement so that the lower jaw moves up and down in the same plane. No circular movement is possible as in the goat, therefore chewing or grinding does not take place.
13 $i\frac{3}{3}\,c\frac{1}{1}\,p\frac{4}{4}\,m\frac{2}{3}$
Notice that in this animal the upper and lower jaws are not exactly similar, but the two halves of any one jaw are always symmetrical.
14 *Incisors* – chisel shaped, blunt; scraping food off the bone, the carrying of young by the female.
Canines – large, conical, curved, very pointed; killing prey, ripping food.
Premolars and molars – ridged along the jaw with sharp points, lower jaw fits inside upper jaw. Teeth work like a pair of scissors, particularly the *carnassial* teeth (i.e. upper premolar 4 and lower molar 1). Cutting and

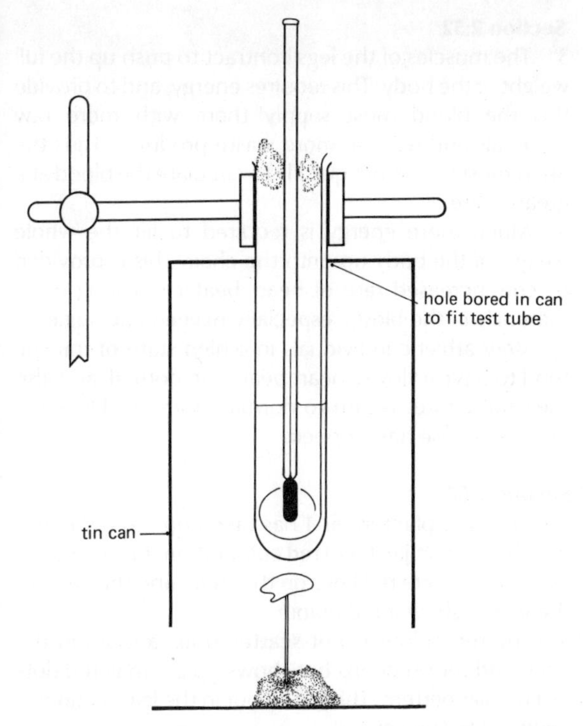

fig B Improved method for experiment and apparatus shown in Chapter 1, fig 1.48

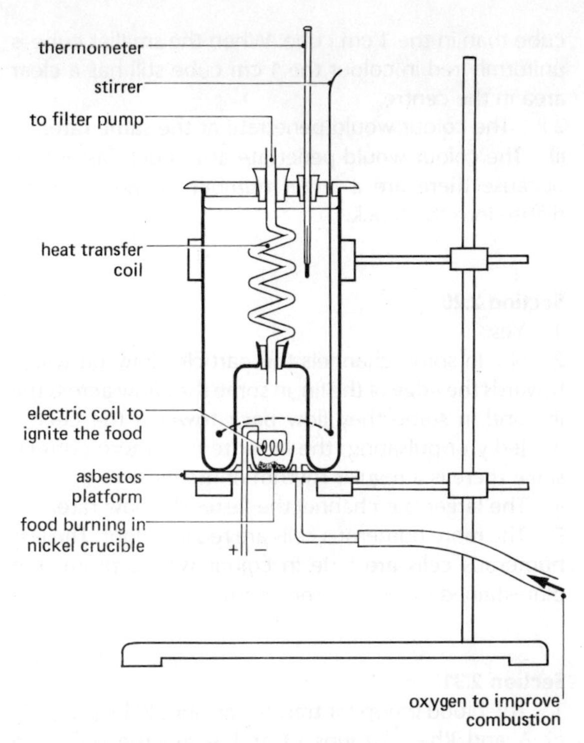

fig C A calorimeter for estimating the calorific value of foodstuffs

crushing (**not** chewing).
15 The tooth has been dissolved away on its surface. It has a pitted appearance.
16 The tooth is much softer.
17 The dentine.
18 Yes. The increase of the fluoride up to 2.0 ppm gives a rapid decrease in the incidence of cavities per child.
19 10.
20 3.
21 1.5 ppm (from 10 per child to 2.5 per child).

Section 1.40
1 .210 kJ.
2 1.050 kJ.
3 a) Some heat is lost around the edge of the tube.
b) Some heat is used up in warming up the glass of the tube.
c) The cashew nut or ground nut does not burn completely.
d) Heat could be lost from the water through the glass, or through the cotton wool.
e) The cashew nut is not a pure food, it contains oil, protein, carbohydrate and water.
Improvements:
i) Use a flat bottomed container so that heat is kept below the container and the water.
ii) Insulate the vessel in some way to stop heat loss.
iii) Use a jet of oxygen from an oxygen cylinder to make the nut burn freely and completely.
iv) Use a pure food such as sugar or oil.

Considerable heat loss can be prevented by simply surrounding the test-tube with a tin can (see fig B). A much more complex apparatus, satisfying many of the above criticisms, is shown in fig C.

4 Most of the shortcomings of the apparatus will tend to reduce the amount of heat entering the water in the test-tube and this will make the resultant energy value appear artificially small. Therefore, if we assume complete combustion externally is equivalent to the processes of oxidative decomposition of food in the body, more energy will be produced by the food in the body than would be indicated by this experiment because of the external losses in combustion.

Section 1.41
1 The abdomen is distended. Skin is peeling and blotchy.
2 There has been an acute lack of protein in the child's diet.

2 Transport

Section 2.10
1 The colour penetration is much faster in the 0.5 cm

cube than in the 1 cm cube. When the smaller cube is uniformly red in colour the 1 cm cube still has a clear area in the centre.
2 i) The colour would penetrate at the same rate.
ii) The colour would penetrate at a much faster rate because there are a larger number of molecules to diffuse into the block.

Section 2.20
1 Yes.
2 No. In some channels the particles flow outwards towards the edge of the fin, in some they flow across the fin, and in some they flow back towards the centre.
3 Jerky or pulsating; the latter term is more correct, since there is a regular rhythm of movement.
4 The larger the channel the faster the flow rate.
5 The more numerous cells are red in colour. The less numerous cells are blue in colour with a prominent blue-stained nucleus in the centre.

Section 2.31
1 The blood group for transfusion should be group O or A and Rh−. Groups O and A are the only two groups compatible with her blood (O is the Universal Donor). It is also advisable to have Rh− blood, since Rh+ antigens would produce antibodies in her blood. These could affect the unborn child in any future pregnancy, since the foetus could be Rh+. This situation is common when the father is Rh+.

Section 2.41 The mammalian heart – external examination
Experimental procedure 2. The auricles have wrinkled walls. They are smaller and darker in appearance than the ventricles. The ventricles are larger, smoother and pink.
Procedure 4. Arteries have semilunar valves at their base.
Procedure 5. To supply oxygen and dissolved food materials to the heart muscles. To remove the waste products produced by the working of the muscle.
Questions
1 No. There is no connection between right and left sides of the heart.
2 The left ventricle has the larger cavity and the thicker muscular walls.
3 The right auricle has the larger cavity.
4 A and E.
5 To stop the backflow of blood into the ventricles when the heart rests between beats. Also, the elastic walls of the arteries press on the blood.
6 To prevent the valve turning inside out when the muscles of the ventricle walls contract.

Section 2.52
3 The muscles of the legs contract to push up the full weight of the body. This requires energy, and to provide this the blood must supply them with more raw materials and remove more waste products. Thus the heart must beat more quickly to circulate the blood at a greater rate.
4 Much more energy is required to lift the whole weight of the body up onto the chair. This is provided by the increased rate of heart beat circulating more materials in the blood, especially oxygen and sugar.
5 Very athletic individuals in a high state of training tend to have a slower heart beat than normal, and also their pulse rates return to normal more quickly when heavy exercise has stopped.

Section 2.60
1 Only the plant in jar 1 has taken up the red dye.
2 The plant in jar 1 had red colouration throughout its length. There are red lines on the stem, and the veins of the leaves show a red colour.
3 The red colour is not scattered at random in the stem and root sections but shows up as small red dots in a regular pattern. The red colour in the leaf sections is confined to the veins.
4 From the parts named in fig 2.15, the red colour is carried up through the xylem.
5 These are radioactive particles which come from outer space, (i.e. cosmic radiation). They must be taken into account and deducted from the radiation of the plant.
6 A slight rise at 12 30 hours; a definite increase by 13 45 hours.
7 At 13 45 hours about the same time as the second leaf node.
8 The phosphorus salt is required for protein, ADP and ATP manufacture at the growing point of the shoot.

Section 2.61 Measure the difference between the new height of the liquid in the capillary tube and its initial level. This measurement will vary according to the diameter of the capillary tube and the diameter of the visking tubing.
2 Water molecules must have entered the visking tubing from the water in the beaker.
3 The liquid in the capillary tube must be exerting a hydrostatic pressure on the liquid in the visking tubing. As the liquid will continue to rise in the capillary tube, the inflow of water must be overcoming this pressure.
4 It looks distended or swollen. It feels firm; it is no longer as flabby as it was at the beginning of the experiment.
5 Water has passed into the visking tubing, and since the screw clip is closed the bag is distended.
6 The liquid inside the bag and tube shoots out in a fine stream from the glass nozzle. The water that has

entered the visking tubing has filled it and built up a pressure, possibly stretching it. When the clip is opened the pressure is released and the liquid spurts out.
7 It is still blue in colour, but a much paler blue.
8 The copper sulphate solution has diffused throughout the water and is therefore diluted. The dilution (mixing of water with the solution) has produced a paler coloured liquid. The same amount of copper sulphate is present but it is dispersed over a greater volume.
9 The cotton wool pieces soaked in phenolphthalein change colour to pink in succession along the glass rod. The cotton wool nearest to the ammonium hydroxide changes first.
10 The ammonia fumes cannot be transmitted in any way to the phenolphthalein except by diffusion through the air in the tube.
11 It is much faster in gas than in a liquid.
12 Piece number (iii).
13 Piece number (i) increases slightly in length. Pieces number (ii) and (iii) decrease slightly in length. This must be due to the movement of water into and out of the cylinders.
14 This is the process of osmosis.
15 More cylinders could be used, say five in each liquid. The length of each cylinder could be measured before and after the experiment, and the mean increase or decrease in length calculated. These multiple data ensure that sampling errors are reduced.
16 In (i) a liquid appears in the cavity and rises slowly upwards, in (ii) no liquid appears, and in (iii) no liquid appears.

In (i) the sugar crystals, moistened in the cell sap, form a strong solution and extract water from the adjacent cells by osmosis. The outer cells (nearest the water) take up water. Water molecules flow by osmosis across the tissue from the surrounding water to the inner cavity.

In (ii) no sugar is present. Although water will pass into the outer cells, none can emerge into the cavity.

In (iii) although sugar is present, dead cells are no longer capable of osmosis.
17 Numbers (ii) and (iii) are controls to indicate that observed results in (i) are due to the presence of sugar crystals in the cavity and to the action of living cells.
18 The pith cells in the centre of the stem, when released from the restrictions of the epidermis, expand and cause an increase in the length of the pith as a whole. Since the epidermal cells have thicker cell walls they are less elastic and the epidermis does not increase in length. This results in a curvature, with the pith cells on the outside and the epidermal cells on the inside.
19 The strip now curls inwards with the pith cells on the inside of the curve. These cells have lost water to the stronger solution by osmosis and have tended to shrink. Again the epidermis does not change in length, and remains on the outside of the curve.
20 Water evaporates from the strip of cells. The inner pith cells become less firm and the strip now gradually bends inwards with the strip of epidermal cells curved in the opposite direction from that at the beginning. (See fig 2.23.)
21 See fig D for sketches of the cells.

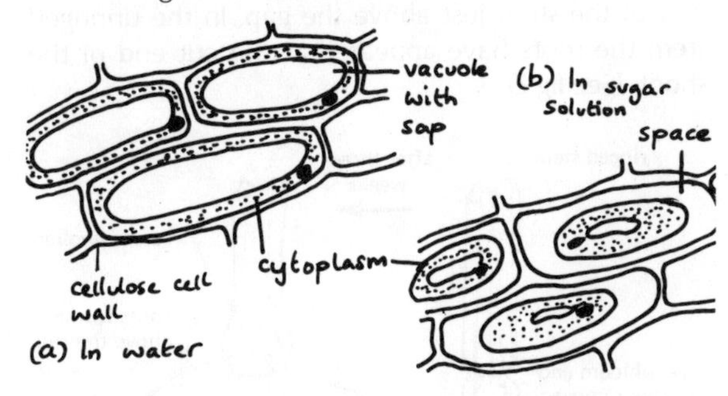

fig D A sketch of onion epidermal cells in different solutions

The contents of the cells in the sucrose solution have contracted from the cellulose cell wall to form a ball of cytoplasm. Cells in water are distended and the nuclei are flattened against the cell walls.
22 When the cells were in the sugar solution, water was withdrawn by osmosis from the cell contents and this caused them to fall away from the cell walls. In water the relatively concentrated cell sap caused water to move into the cell by osmosis; so that the contents pressed against the cell wall.
23 The level in the capillary tube would rise slowly.
24 Samples of water taken from the beaker would show an increase in density (weight per unit volume).
25 Blood cells could be seen under low and high power in tubes (ii) and (iii). No trace of cells could be seen in tube (i).
26 In tube (ii) the blood cells appeared as biconcave discs with a smooth outline. In tube (iii) the blood cells had a wrinkled appearance and the outline was no longer smooth.
27 In tube (ii) the blood cells were unchanged because their surroundings have the same salt concentration as normal blood. In tube (iii) the cells had lost water to the strong salt solution.
28 In tube (i) the liquid was a clear red colour. (ii) the liquid was red in colour but slightly opaque. (iii) the liquid was red in colour but slightly opaque.
29 i) The blood cells in this tube took up water from their surroundings by osmosis. This resulted in a stretching of the cell membranes until they burst, releasing the red pigment haemoglobin into the water.
ii) The liquid appears red because of the presence of blood cells, but it is not a clear red colour as in (i).
iii) The blood cells in this tube lose water to the surroundings by osmosis, so that the cell membrane

shrinks. The cells are still present and give the liquid its red colour, but again it is not as clear a red as in (i).

In both (ii) and (iii) the cells will sink to the bottom and in time the liquid will become clear.

Section 2.63

1 In the ringed stems the roots have appeared on the part of the stem just above the gap. In the unringed stem the roots have appeared at the cut end of the shoot. (See fig E).

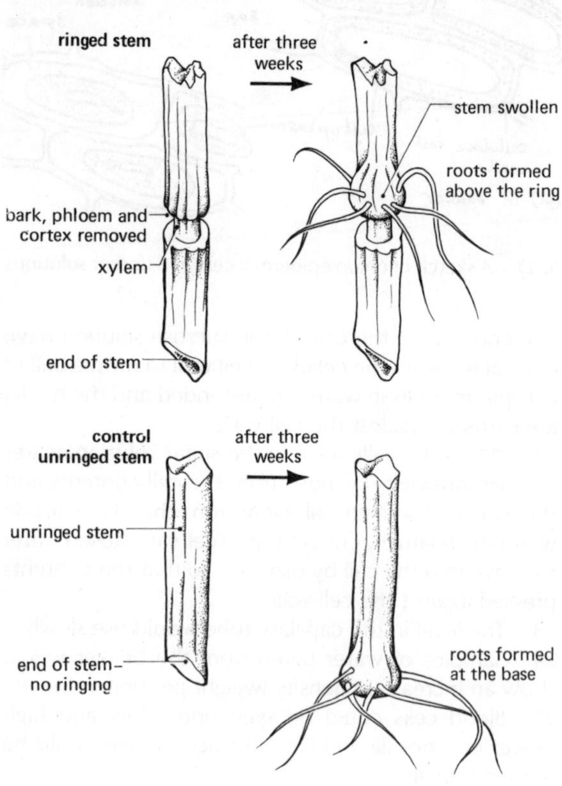

fig E The effect of ringing bark

2 The stem is swollen at the region where the roots have appeared.

3 A possible explanation is that the phloem transports food materials required for the growth of the roots. In the ringed stem the food materials accumulate above the gap, since they can pass no further. Thus these food materials are available for the growth of new roots at this region.

In the unringed stem the phloem transports food from the leaves down to the cut end of the stem, and it is here that they accumulate and are available for root growth.

4 The presence of radioactive sucrose, giving blackening of the radiograph, both in the stem and the leaves, above the treated leaf in the control shoot A.

5 The radioactive sucrose is present in the stem, both above and below the fed leaf, in shoot A.

6 In shoot B which has been ringed (i.e. the phloem removed) no radioactivity appears above the gap in the stem. This proves that the sucrose is transported in the phloem, because the xylem is still present and, if concerned in the movement of sucrose, would have produced radioactivity above the gap.

7 The sucrose is required for the growth of the stem and the leaves at the tip of the shoot.

Section 2.64

1 The rate of uptake of ions rises very rapidly from 0 to about 15% oxygen concentration, and after that it remains constant up to 55% oxygen concentration. In other words at oxygen concentrations below that of the atmosphere, the rate of ion uptake decreases very rapidly as the oxygen concentration decreases.

2 Living things require oxygen for energy production. Some of this energy is for the transport of ions into the cell. Thus when oxygen concentration decreases, energy production decreases and thus ion uptake falls rapidly.

Section 2.65

1 It prevents water evaporating directly from the water surface in the flask.

2 There was no fall in the water level, because there was no connection between the air and the water, and thus no evaporation.

3 The plant provided a pathway through the oil film. Water was absorbed by the plant and passed up to its stem and leaves where it evaporates.

4 If the initial weight is X g and the final weight is Y g; then the difference in weight is (X – Y)g.

5 If the volume added to the flask is A cm^3, then the mass of water added must be A g.

6 (From no. 3 above). Mass of water absorbed by the plant = A g
(From no. 5 above) Mass of water loss by the apparatus = (X – Y)g
Therefore mass of water retained by the plant = A – (X – Y)g.

The mass of water lost by the apparatus as a whole must also be that lost by the plant, since any water retained in the plant will still be included in the final weight.

Thus mass of water lost by the plant = (X – Y) g.

7 For photosynthesis, transport, turgidity and to act as a solvent.

8 Percentage of water lost by the plant

$$\frac{\text{Water loss}}{\text{Water uptake}} \times 100$$
$$= \frac{(X - Y)}{A} \times 100$$

The results of different class groups will depend upon

such variables as plant size, damage to roots, number of leaves present and so on.

Typically, about 95% of the water taken up by the roots is lost and less than 5% used in photosynthesis. A single maize plant will lose about 200×10^3 cm^3 of water in a growing season, about 2 000 cm^3 a day when mature.

9 Traces of sweat on the skin would turn the paper pink in colour.

10 To prevent the moisture of the air affecting the paper. Adequate controls should check two possible variables:

i) the sensitivity of the paper between glass slides in the absence of a leaf;

ii) the moisture content of the air, with the paper exposed (not between slides).

11 The papers all turn pink when placed next to the leaves.

In many plants this takes place more rapidly on the lower surface of the leaf than on the upper surface. Plants vary considerably in the speed at which they lose water and this variation is reflected in the speed at which changes take place.

12 Water vapour.

13 The stem and flowers could be involved in water loss. Fix cobalt chloride paper around the stem, and cover with transparent adhesive tape. Fix cobalt chloride paper on the floral parts and cover with glass slides, fixed with elastic bands.

14 Measure the loss in weight, but care must be taken to prevent loss of water from the pot and the surface of the soil by enclosing the pot and the surface of the soil in a plastic bag or other waterproof cover tied around the base of the plant stem.

This would be the loss of water from the plant since any water passed from the root to the stem and leaves and retained there is included in the final weight.

15 Placing the epidermis in water results in the largest opening between the guard cells.

16 Water passes by osmosis into the cells at first. When the cells are surrounded by the sucrose solution, water passes out of the cells and they become plasmolysed. When the guard cells are turgid they cause the pores to be larger, and when flaccid they collapse and close the pore.

17 A very useful model to consider is an uninflated inner tube of a bicycle tyre hanging on a nail. It hangs down limply with virtually no opening between its parts. If some air is pumped in, however, it opens out and an opening becomes apparent in its parts. The air is comparable to the water moving in and out of the guard cells. (See fig 2.34.)

18 We are investigating loss of water by the shoot and we have clearly shown by the eosin experiment that water passes up through the xylem. We must assume that if the shoot is removed from water, air will enter the xylem as the water gradually passes up and is lost from the plant.

19 The whole apparatus is airtight and liquid tight. If the plant sucks up water, pressure is reduced, and air pressure forces water into the only opening available at the end of the capillary tube.

20 Change in rate of water loss from the plant must be related to the rate of uptake, and therefore the rate of movement of the bubble in the capillary tube will speed up or slow down.

21 When external conditions are (i) warmer, (ii) less humid, (iii) more windy, (iv) or there is more light.

22 When external conditions are (i) colder, (ii) more humid, (iii) less windy, (iv) or there is no light.

3 Other Nutrition

Section 3.11

1 The bacteria obtain the energy to carry out their life processes from the chemical breakdown involved in decay.

Section 3.12

1 It is covered with fine filamentous or thread-like structures. The colour of the threads may be white or grey, green or even blue. In addition, small black dots may also be seen in large numbers.

2 A fungus.

3 It consists of a mass of very small filaments growing and branching horizontally on the surface of the bread. The tiny black dots are reproductive bodies, and these are carried on erect filaments above the substratum.

4 It grows on the substratum by extracting nutrients from the bread or fruit, and thereby obtaining its energy.

5 The atmosphere is full of microscopic fungal spores. These settle on the moist surface of the bread and start to grow into fungal plants.

6 The filaments are shaped like long thin cylindrical threads.

7 Each filament is surrounded by a cell wall.

Section 3.13

1 Each unit of cytoplasm contains a large number of nuclei. In some cases the individual hyphae are not divided into separate cells by cross walls. The presence of a central vacuole with a cytoplasmic lining also suggests that this is part of a plant structure. There are no chloroplasts present which are characteristic of the cells of green plants.

Section 3.30

1 The flea and louse are external parasites, and the trypanosome and tapeworm are internal parasites.

2 Small hooks, claws and bristles

3 A thin whip-like structure called a flagellum enables the organism to swim.

Section 3.41

1 Small swellings
2 Root nodules
3 Populations of the bacterium *Rhizobium*
4 The roots of the soya bean plant provide shelter and a place to live for the bacteria. The bacteria 'fix' atmospheric nitrogen, thus providing the roots of the soya bean with a steady supply of nitrogen compounds

Section 3.42

1 Wood
2 Cellulose
3 It helps in the digestion of cellulose
4 The alimentary canal of the termite provides shelter and a plentiful supply of food for the populations of *Trichonympha*. *Trichonympha* provides cellulase for the digestion of wood eaten by the termite, and turns it into soluble food which can be used by the termite.

Section 3.50

1 It is provided with food source by the external parasites of the rhinoceros.
2 The irritating external parasites are constantly removed by the birds. The birds also provide the advanced warning system similar to that provided by the egrets in their association with the buffaloes.

4 Storage

Section 4.11

1 Yellow. The colour is not evenly dispersed, but is restricted to the vascular bundles
2 Most of the sections between the vascular bundles are stained blue-black, showing that starch is present.

5 Respiration

Section 5.00

1 The candle burns longer in inhaled air (atmospheric air). The average burning time is 9.5 seconds for inhaled and 5.9 seconds for exhaled air in a class of 14 year old pupils.
2 Exhaled air contains (i) less oxygen, (ii) more carbon dioxide, (iii) more water vapour.
3 Condensation of vapour to form liquid drops. It is a colourless liquid. (You must not conclude that it is water yet – there is no evidence.)
4 The blue cobalt chloride paper turns pink or the anhydrous copper sulphate turns from white to blue. This test indicates that the colourless liquid is water.
5 The exhaled breath contains a great deal of water vapour which condenses on the glass.
6 When you are breathing in, bubbles stream out of the long glass tube in boiling tube B into the indicator. When you are breathing out, bubbles stream out from the long glass tube A.
7 When you are breathing in, the pressure decreases in boiling tube B, owing to the withdrawal of air, with the result that external air pressure forces air through the liquid. When you are breathing out, the increased pressure in the long tube in boiling tube A, forces exhaled air out through the liquid.
8

Tube A	Tube B
Bicarbonate indicator turns yellow. Lime water turns milky (white).	Indicator remains unchanged. Lime water remains unchanged.

9 The exhaled air passing through tube A contains much more carbon dioxide than the inhaled air passing through B.
10 This tube is the control.
11 To stabilise the temperature to a constant value, so that the volume of the gas bubble is always measured at the same temperature.
12 Mean values are of the following order:

Inhaled air		Exhaled air	
% CO	% O	% CO	% O
1	20	4	16

These results generally show clearly that the carbon dioxide concentration increases to 4–5% in exhaled air. This provides the cause of the change in the indicator in the previous experiment. It also indicates that about the same amount of oxygen is absorbed, so that its percentage in air reduces from 21–20% to 16%.

The proportion of carbon dioxide in atmosphere air is very small (0.03%) and the apparatus is not accurate enough to show this value. It generally indicates around 1–2%.

13 Exercise will increase the rate of breathing and also its depth, such that the air will be rapidly exchanged. The analysis of air from deep in the lungs generally gives the following, even for a resting subject:

Carbon dioxide 5.55%
Oxygen 14.08%

Arrange for a fellow student (or yourself) to perform vigorous exercises such as running up and down stairs, or stepping on and off a chair. Immediately the exercise is completed and the subject is breathing heavily, take a sample of exhaled air and analyse it by means of a J-tube.

Section 5.02

1 To ensure that no acid or alkaline deposit will be present on the apparatus to change the pH of the indicator.

2 Green plant material contains chlorophyll and under normal laboratory conditions of light and temperature, it would be photosynthesising in the daytime. This process results in a gaseous exchange opposite to that under investigation. For this reason the plant material used must not contain chlorophyll. There is of course another alternative, that all plant material could be used but it must be kept in the dark. The tubes could be surrounded by black paper or black polythene. In this way the gas exchange of photosynthesis is prevented.

3 The tubes containing animals turn yellow quite quickly, dependent on the activity of the animal. The tubes containing plant material do turn yellow eventually, but it takes longer.

4 The process which uses up oxygen and gives out carbon dioxide must proceed much more quickly in animals than in plants. It is clear from the rates of change of the indicator in the animal tubes that this depends to some extent on the activity of the animal concerned. For example, a cockroach will cause a more rapid change than a snail or slug.

5 To absorb carbon dioxide.

6 The coloured water moves up on the left side and down on the right side.

7 No. The oxygen is removed from the air by the gas exchange organs of the animal within the flask and an equal quantity of carbon dioxide produced. The carbon dioxide is absorbed by the potassium hydroxide and thus the volume of gas in the flask decreases by a volume equal to that of the oxygen produced.

Section 5.03

1 The following are examples of results with pupils of 15 years of age. Vital capacities shown are mean values for the groups indicated.

Age 15 years and

1 to 4 months	average vital capacity 3.25 l
5 to 8 months	average vital capacity 3.40 l
9 to 12 months	average vital capacity 3.90 l

Height

150 to 157 cm	average V. C. 2.40 litres
158 to 165 cm	average V. C. 2.75 l
166 to 173 cm	average V. C. 3.80 l
174 to 181 cm	average V. C. 3.90 l

Weight

45.0 to 51.3 kg	average V. C. 2.75 litres
51.4 to 57.7 kg	average V. C. 2.90 l
57.8 to 64.1 kg	average V. C. 3.30 l
64.2 to 70.5 kg	average V. C. 3.80 l
Over 70.5 kg	average V. C. 3.90 l

Activities

Very active	average V. C. 3.60 litres
Active	average V. C. 3.20 l
Some activity	average V. C. 2.75 l
No activity	average V. C. 2.70 l

If time permits the vital capacity of students over the whole age range of a school can be measured. The following results were obtained measuring about 100 pupils:

Age

12 and 13 years	average V. C. 2.6 litres
14 years	average V. C. 3.3 l
15 years	average V. C. 3.8 l
16 years	average V. C. 4.6 l
17 years	average V. C. 4.7 l
18 years	average V. C. 5.0 l

Weight

45 kg	average V.C. 2.6 litres
51 kg	average V.C. 3.7 l
57 kg	average V.C. 4.2 l
63 kg	average V.C. 4.9 l
69 kg	average V.C. 4.8 l
69 kg and over	average V.C. 4.9 l

Activity

Very active	average V.C. 4.9 l
Active	average V.C. 4.8 l
Some activity	average V.C. 4.5 l
No activity	average V.C. 3.9 l

2 From the figures quoted above there is a clear positive correlation of vital capacity with age, height, weight and degree of physical activity.

Section 5.10

1 Shiny and moist. This is an indication of the presence of fluid produced by the pleural membranes, which line the thoracic cavity and cover the lungs, so that they slide smoothly against each other.

2 Large blood vessels can be seen. The pulmonary artery from the right ventricle of the heart divides into two main vessels, one for each lung. Pulmonary veins discharge from each lung into the left auricle.
3 It is domed, with a tendinous centre portion attached to muscles at the edge.

Section 5.13
1 The ink particles flow into the mouth as it opens, and then flow out from under the operculum or gill flap on each side of the body of the fish.
2 When the mouth opens, the gill flaps are shut, but when the mouth closes, the gill flaps open.
3 Five arches on each side.
4 The rakers are on the inner side of each gill arch and they filter the water flowing over the gills. Suspended particles above a certain size are held back to protect the delicate gills. In some fish this is purely a cleaning function, and in others (plankton feeders) it is a means of capturing food.

Section 5.31
1 The energy comes from the heat of the flame which is released by the burning of the fuel (gas or paraffin).
2 The *sun's energy* was harnessed in the fuel. The burning fuel produced *heat energy* and *light energy*. The heat energy provided the latent heat of vaporisation of water. The steam expanded and passed out with *kinetic energy*. The steam moves the vanes of the wheel producing *mechanical energy*.
3 The level of liquid in the arm of the manometer nearest the flask falls, while the other level rises.
4 The air in the test-tube and the connection to the manometer must have increased in volume in order to depress the liquid in the manometer. Air expands considerably for a small rise in temperature. This rise in temperature must have been due to heat generated by the living organisms in the flask.
5 Use exactly similar apparatus, but with no living organisms in the flask. An equal mass of non-living material (e.g. small stones) could be substituted.
6 Flask B.
7 To prevent the growth of bacteria and fungi on the dead seeds. These organisms could produce heat and so affect the temperature in the flask.
8 The flask is not completely impermeable to heat flow. At night heat flows out of the flask and in the daytime some heat flows inwards. A graph of the temperatures in the flasks will show a fluctuation due to environmental temperatures.
9 Draw a graph of the temperature difference between the flasks which will indicate a steady rise in temperature due to the living germinating peas, or other seeds.

Section 5.32
1 To keep out light and prevent photosynthesis.
2 It absorbs carbon dioxide from the incoming air.
3 In the first flask the indicator remains unchanged, while in the second it turns yellow (bicarbonate indicator) or milky white (lime water) showing the presence of carbon dioxide.
4 Yes. The carbon dioxide is produced as a result of aerobic respiration proceeding in the dark.
5 Use the apparatus with the opaque covering removed from the bell jar. The mouse should be inside the bell jar.

Section 5.40
1 The amount of heat produced is very small and the vacuum flask prevents its loss to the air.
2 Yes. The temperature rises showing heat production in the yeast-glucose solution.
3 To prevent atmospheric oxygen from diffusing into the yeast-glucose solution.
4 Bubbles of gas appear, rising slowly through the oil layer and bubbling out of the end of the delivery tube.
5 The indicator has turned yellow, showing that the gas is carbon dioxide.
6 Alcohol (ethanol).
7 To prevent growth of bacteria (their respiration could destroy the effectiveness of the control).
8 The living, soaked peas have produced a gas which has depressed the mercury level in the test-tube.
9 The seeds could not be photosynthesising, so that the gas should be the carbon dioxide produced by respiration. Introduce some small pellets of sodium hydroxide into the open end of the test-tube. The pellets will rise up to the top of the mercury, absorb the gas and the mercury level will rise. Thus the gas must be carbon dioxide.
10 The soaked peas had speeded up their rate of metabolism, particularly respiration. Since all oxygen was excluded by the presence of the mercury, the respiratory activity must have been anaerobic.

Section 5.41
1 58 mg/100 cm^3 of blood.
2 40 mg/100 cm^3 of blood.
3 Assuming that the rate of loss of lactic acid continues to be the same as that shown by the last line on the graph, extend the graph until it cuts the 20 mg/100 cm^3 blood line. It cuts the line at about 73 minutes. Thus from the end of the exercise it takes: $73-9=64$ minutes.

6 Excretion

Section 6.10
1 The body temperature of the frog stays at the same level as the external temperature.

2 The body temperature of the students stays at a constant level, independent of the external temperature.
3 The students can remain active at night when temperatures fall, whereas the frog becomes inactive as the body temperature drops with the external temperature. Thus the students can study throughout the 24 hour period if necessary!
4 The constant body temperature enables Man to be independent of the external temperature and so live anywhere in the world, from the Poles to the Tropics.
5 The body temperatures of the insect larvae increase as the external temperature increases.

Section 6.12
1 To remove natural oils from the surface of the skin, and thus to prevent the liquid from running off.
2 The skin feels cool at this point.
3 The movement of air from the mouth, across the water drop, causes the skin to feel even cooler.
4 The ether feels much colder than the water.
5 The ether evaporates more quickly and thus draws more heat from the surface.
6 The movement of air from the fan cools the body because the sweat produced by the body at high temperatures is evaporated.
7 Figures for this experiment will depend upon the environmental temperature. For an environmental temperature of 20°C the following were calculated:
a) 0.8°C per minute
b) 1.26°C per minute
c) 0.89°C per minute (table A and see fig F).

Time (mins)	Fall in temp. first 10 minutes (°C)	Fall in temp. when wiped with ethanol	Fall in temp. second 10 minutes (°C)
1	94.0	84.5	77.4
2	93.2	82.2	76.5
3	92.4	80.9	75.5
4	91.5	79.5	74.6
5	90.6	78.2	73.6
6	89.8		72.6
7	89.0		71.5
8	88.0		70.5
9	87.0		69.4
10	86.0		68.5

Table A Figures showing effect of ethanol on the rate of cooling
(refers to 6.12 no. 7)

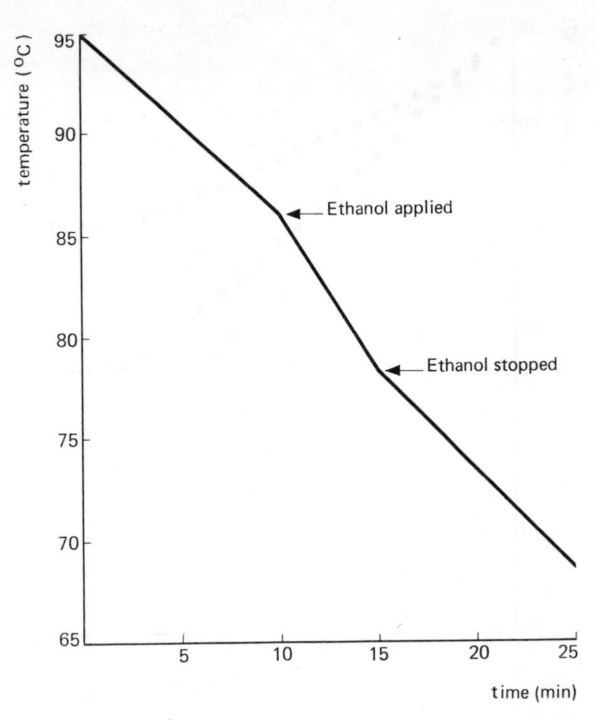

fig F A graph to show the cooling effect of ethanol on the skin

8 The latent heat of evaporation of the ethanol draws heat more quickly from the flask, and the rate of fall of temperature increases.
9 Prevent heat loss to the air by surrounding the flask with insulating material. This slows down convection and radiation of heat.

Section 6.13
1 As in section 6.12 the figures for this experiment will depend upon the environmental temperature. For an environmental temperature of 20°C the following were calculated: Flask A 0.79°C per min., flask B 0.44°C per min. (See table B and fig G).
2 The slower rate of fall of temperature for flask B shows that the insulating material reduces heat loss.

Section 6.14
1 The ears of the foxes show adaptation to different environmental temperatures. The African fox has large ears, as does the elephant. They enable a quicker heat flow from the blood to the surrounding air.
2 The Arctic fox has the greatest amount of fur, and this slows down heat flow to the cold air surrounding the animal.
3 The African fox has the smallest head and the Arctic fox the largest. Thus the head of the African fox has a higher S.A./V. ratio to enable greater heat loss to take place.

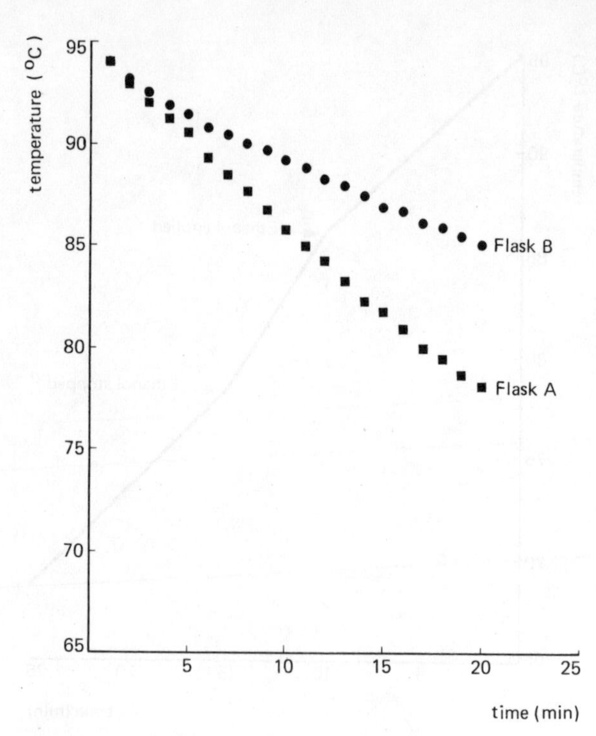

fig G A graph to compare cooling curves in insulated and uninsulated flasks

Time (mins.)	Flask A (°C)	Flask B (°C)
1	94.0	94.0
2	93.0	93.2
3	92.0	92.5
4	91.4	91.8
5	90.5	91.5
6	89.2	90.8
7	88.4	90.5
8	87.6	90.0
9	86.7	89.7
10	88.8	89.3
11	85.0	88.8
12	84.2	88.3
13	83.2	88.0
14	82.3	87.6
15	81.8	87.0
16	81.0	86.8
17	80.0	86.2
18	79.5	86.0
19	78.8	85.6
20	78.2	85.2

Table B Information for the plotting of cooling curves
(refers to 6.13 no. 1)

Section 6.20

1 We have to drink more water, although under these conditions loss of water in urine is reduced.

2 If the external temperature is low, the rate of sweating is reduced and urine output increases.

3 Diarrhoea is a condition in which water is not reabsorbed from the gut, and so more than 8 litres a day could be discharged. In conditions other than intensive hospital care this water cannot be replaced.

Section 6.21

1 Renal artery; renal vein; ureter.

2 Blood vessels are distinguishable from the ureter since they contain blood. The ureter is much paler in colour. The artery has thicker walls than the vein.

3 It is divided into an inner and outer portion (the medulla and cortex).

4 It receives the urine from the pelvis.

Section 6.22

1 No protein is filtered through the Bowman's capsule. The molecules are too large.

2 It must be reabsorbed during its passage down the nephron.

3 They become very concentrated in the urine. Urea is 60 times more concentrated in the urine, ammonia 40 times and creatinine 75 times.

4 There are two possible reasons:

i) these three substances have been actively secreted into the tubules along their length;

ii) water has been removed from the fluid of the tubules thus concentrating the solution.

The latter is the correct explanation. From the text previous to the table, it can be seen that the production of filtrate is of the order of 100 litres a day whereas urine production is 1 000 to 1 500 cm³ per day.

Section 6.23

1 Two hours.

2 This strength of salt solution is near to that of the blood and thus it does not alter the osmotic pressure of the blood. Urine output is not increased, but body weight increases by the weight of 1 litre of salt solution.

7 Response

Section 7.10

1 Typical percentage results are as follows:

Larvae in the dark 85%	Larvae in the light 15%
Larvae in moist air 88%	Larvae in dry air 12%
Larvae on dry meat 22%	Larvae on raw meat 78%

The figures indicate that the strongest stimulus causing tactic movement is probably moisture but also that larvae move away from light and towards the raw meat.

2 In the second and third experiment you were told to

keep the larvae in the dark since from the first experiment it is clear that they move away from the light stimulus. Thus the only choice is between the dry air or the moist air, raw meat or dry meat.

In the first experiment, therefore, the environment in which the larvae are making their choice between light and dark would be more suitable if the air was moist. Thus water should be placed in each of the bottom dishes.

3 Larvae of this genus are found on decaying carcasses while the meat is still raw and moist, but as soon as the dead body dries up in the sun the larvae pupate. Alternatively as the carcase dries out the larvae move down underneath where the flesh is still moist and they are away from the light.

4 There may be a touch stimulus operating in that the larvae need to be in a habitat where their bodies are surrounded on all sides. Another factor could be that there is a temperature gradient which is part of the light-dark stimulus. How could you investigate this factor?

5 See table C.

6 The projecting eyebrow ridges above the eye socket together with the bridge of the nose protect the eyes from damage if the head hits a blunt object. The eyebrows and eyelashes also form a protective screen. The eyelids form some protection but their principal purpose is to wash dust from the front of the eyeball when blinking. The eyeball is lubricated by the watery fluid discharged from the tear gland.

Experiment

How do the eyes react to light and dark? Responses to questions posed in procedures 1 and 2.

1 Owing to the lack of light in the left eye the right eye adjusts to allow in more light. The pupil of the right eye can be seen to enlarge.

2 When the left eye is covered the pupil enlarges to its maximum in the absence of light. Immediately the eye is uncovered the light causes the eye to react, and the pupil closes due to the contraction of the iris. Thus the eye is now adjusted to the external light intensity.

7 The dot disappears when the left eye reaches the region of numbers 5, 6 and 7. Examine fig 7.8 in Chapter 7. Can you give a possible explanation why this should occur?

8 The discs have been cut to different sizes, but one eye is unable to distinguish this fact, or which disc is nearer to the eye.

9 The discs appear to be their normal size, that is when the right eye is opened the two eyes can see the discs at their correct distance and correct size.

10 Two eyes set in front of the head can distinguish clearly the different distances of objects from the head, whereas only one eye finds difficulty in interpretation of distance.

Question	Answer and conclusions
a) Are the eyes fixed in their sockets? If not how do they move?	No. They move up and down and from side to side, but always together in the same direction.
b) Can we judge how far an object is in front of us?	Yes. This is possible through the possession of two eyes in the front of the head.
c) Can we see the shape of objects?	Yes. The sense cells detecting light must be spread over a surface enabling us to detect shapes.
d) Can we see the colour of objects?	Yes. We do not see the world in black and white only, thus the sense cells of the eyes must detect colour.
e) Can we see in light only, in twilight, or even in the dark?	Yes. We can see clearly in light, but we also have vision in poor light conditions. Even at night the eyes are able to see after some adjustment. The sense cells of the eyes must be capable of twilight vision.

Table C Versatility of the human eye
(refers to 7.10 no. 5)

11 An animal which depends for its life on judging distance correctly is the monkey. It locomotes by swinging from branch to branch. This animal, together with others like the marmoset and hawk has eyes in the front of the head, thus providing overlapping fields of vision.

12 This is muscle tissue and the strips are concerned with the movement of the eyeball.

13 There are six strips.

14 The other end of each strip is attached to the socket of the skull so that the contraction of the muscles moves the eyeball within the socket.

15 This is the nerve connecting the receptor cells of the eye with the brain. It conveys nerve impulses.

16 The black choroid prevents the reflection of light rays inside the eyeball. In a similar fashion, the camera is black inside to prevent internal reflection.

17 The humour helps to keep the eyeball spherical. It carries nutrients to the non-vascular portions of the eye. For example, it contains large amounts of vitamin C.

18 The pupil shape is elongated compared to our own eye when seen in the mirror. This is typical of the ungulate eye. See the discussion in the early part of section 7.20 (p. 149).

Section 7.21

1 The comparison is very close except for the focusing mechanism. In the camera, the lens moves forward or backwards, whereas in the eye the lens stays in the same position, but alters its shape to increase or decrease its power.

2 At a certain position of the flask an image of the window and the outside scene appears on the white paper screen. It is obvious from inspection of the image that it is upside down (inverted).

3 No. Although one can see the pencil because an image must be present on the retina, this must be formed by rods only. No cones are present which permit colour perception. The image must have just come on to the edge of the retina so that in this region the sense cells are rods only.

Section 7.30

1 To equalise air pressure on either side of the tympanum.

Section 7.31

1 Perforation of the tympanic membrane would stop it reacting to the small changes in pressure caused by sound waves, so these could not be transferred to the ear ossicles and hence to the inner ear.

2 The vibrations of the tympanum are transferred, by the ear ossicles, to the oval window which also vibrates at the same frequency. Since the liquid of the inner ear is incompressible there must be some mechanisms by which pressure can be transferred. The round window is a membrane which acts in unison with the oval window, except that as one moves inwards, the other moves outwards and vice versa.

Section 7.40

1 Yes. The four taste areas are shown in fig H.

2 There is overlap of each area.

3 The basic pattern is similar but the overlapping areas may vary considerably. Some people are unable to separate sourness and bitterness, while others are able to differentiate types of sweetness.

Section 7.50

1 Yes, the two fingers feel at the same temperature. To ensure that they were initially at the same temperature before commencing the experiment.

2 The right index finger coming from the ice water

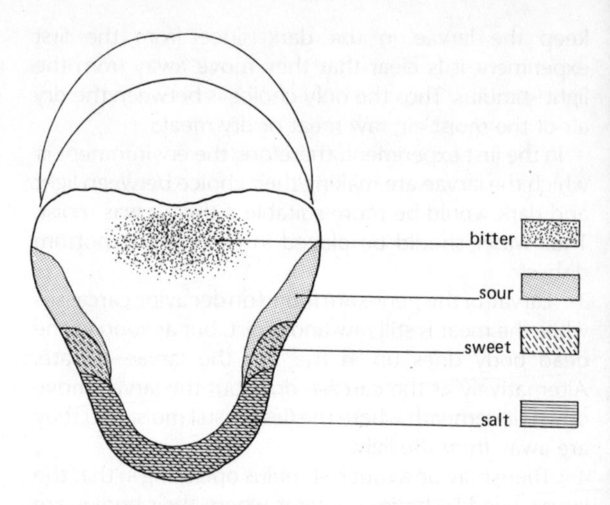

fig H Taste areas of the tongue

indicates a much warmer liquid than does the left index finger which has come from the hot water. The water feels warm to the right finger, and feels cold to the left.

3 The temperature receptors cannot indicate absolute temperatures although there are detectors of cold and heat. It can detect changes in temperature and in fact is sensitive to quite small temperature differences, such as 0.2°C on the arms and 0.5°C to 1.0°C by the fingers.

4 In some parts of the skin the receptors must be further apart than 2 mm since the pins cannot be detected as two separate points.

5 The areas of the skin vary considerably. The most sensitive are tongue, nose, lips and fingertips. The least sensitive are middle of the back, shin and thick parts of the sole.

Section 7.63

1 The lower part of the leg and the foot jerk forwards.

2 The upper muscle of the thigh can be felt to bunch up as it contracts.

3 The tap below the knee is a stimulus which initiates a nerve impulse that travels by way of sensory neurons to the spinal cord. The return impulse travelling along the motor neurons causes the leg muscle to contract.

4 The vapour from the cut onion causes the eyes to discharge a liquid (tears) which can spill over onto the cheek and run down into nasal ducts.

Section 7.70

1 Less than 1, say 11/13 or 0.85.

2 Greater than 1, say 5/3 or 1.7.

3 The word 'a'.

4 Making a tick.

5 Tapping the pencil.

6 'Simon says'. Children are told to obey a command *only* if it is preceded by the phrase 'Simon says'. For

example, they should sit down if told 'Simon says sit down', but not if told simply 'Sit down'. They must react *immediately*. Inevitably some will obey the simple command (equivalent to the bell) when it is *not* prefixed by 'Simon says' (equivalent to the food).
7 The mouse tends to find the food more and more quickly with successive repetitions of the experiment. Thus he may be said to have 'learnt' the quickest way through the maze.
8 Early morning, before the mice have been fed. Remember, you should not deprive mice of food for long periods.
9 A hot or cold object, or gentle prodding with a pencil. Do *not* use electric shock.
10 To remove the possibility of a mouse following the scent of its predecessor.
11 To reduce the possibility of a mouse locating it by smell.
12 So that the mouse has time to become hungry again.

Section 7.81

1 Yes.
2 In *Experiment 2* the new blood supply developed but no nervous connection appeared. In *Experiment 3* the transferred testes developed a new blood supply, but again no nervous connection appeared. In both these experiments the cockerel appeared normal and so the maturation must have been controlled by a chemical discharged into the blood and not nervous control.
3 Kamimura showed that even with the pancreatic duct tied off no sugar was discharged into the urine, but if a pancreas was removed (Mering and Mindowski) the animal died. Thus the hormone controlling blood sugar must have been discharged into the bloodstream.
4 The extract only lowers the glucose circulating in the blood, and therefore has to be repeated again when the glucose builds up. It is not a cure for the disease.

8 Reproduction

Section 8.10

1 It contains both root and shoot systems.

Section 8.31

1 It bears an artery and a vein.
2 It is the vas deferens, which carries the sperm from the testis to the urethra.
3 Six or possibly seven: the bladder; the two vasa deferentia; the two seminal vesicles (which are very difficult to separate from the coagulating glands; which lie anterior to them), and the prostate gland (which may or may not appear paired).

Section 8.32

1 At the base of the bladder, near the urethra.
2 No. The right kidney is more anterior than the left.
3 Conduction of urine only to the exterior. It is separate from the reproductive tract.
4 The following structures are absent: penis; testes; vasa deferentia; prostate; Cowper's glands and seminal vesicles.

The following is present in the female but not in the male: a large Y-shaped structure with 'knobs' of tightly-coiled tubules at the tips of the arms.

Section 8.41

1 Five fused sepals.
2 Five petals. The uppermost one (the standard) is single and separate. The sides of the corolla are formed by two partly-fused petals (the wings). Within the wings is a boat-shaped keel, formed of two more partly-fused petals.
3 There are ten stamens; five with long, thin anthers and five with rounded anthers. The bases of the filaments are fused forming a sheath around the ovary.
4 One.
5 The stigma is covered with fine hairs.
6 A, B, C and F.
7 A.
8 To ensure that pollination has not already taken place before the start of the experiment.
9 The stigma. Group F flowers show that cross-pollination can result in fertilisation.
10 To show that self-pollination can result in fertilisation.
11 To show that only pollination by pollen of the same species is likely to result in fertilisation. In fact, this is impossible to prove conclusively without an infinite number of experiments, but a fairly safe generalisation can be made by selecting your species carefully.
12 Attraction of insects for pollination.
13 To prevent self-pollination.
14 Having the stigma higher than the anthers.
15 The female parts might develop before the male parts in the same flower (or vice versa).
16 Five sepals: petals fused to form a tube with two lips and a bulge at the base.
17 Four stamens in two pairs.
18 Commonly, more than twenty ovules, inside a two–chambered capsule.
19 Brightly coloured petals; frequent visits of bees to the plant in its natural state; male and female parts enclosed by corolla tube.
20 There are no large, brightly-coloured petals or strong scents to attract insects. The stamens hang outside the flower, exposed to the wind. The anthers are attached loosely to the filaments, so that the slightest air current will shake them, thus releasing pollen. The

stigmata also hang loosely outside the flower, exposing a large surface area, covered with fine hairs for trapping pollen.

9 Genetics

Section 9.00

1 i) Both contain a nucleus.
ii) Both contain cytoplasm.
iii) Both are surrounded by a cell membrane.

2 i) It has chloroplasts
ii) It has a large vacuole.
iii) It has a cell wall outside the membrane.

Section 9.02

1 i) No asters are present.
ii) The parent cell does not constrict at telophase: instead a new cell wall grows inwards at the equator.

2 They are a part of the plant that is growing rapidly, and therefore, the cells are continually dividing.

Section 9.04

1 Theoretically the answer is four. However, the details of meiosis vary according to whether the organism is an animal or a plant, and whether the reproductive organs are male or female. For example, in mammals one spermatogonium gives rise to four spermatozoa, but one oogonium will produce one ovum plus three polar bodies (see section 8.30).

2 The gametes are said to contain the *haploid* number of chromosomes, which is half the (*diploid*) number found in somatic cells.

3 This would result in less variation in the offspring.

4 This allows *variation* of gametes to occur.

Section 9.11

1 Bell-shaped, with the greatest numbers occurring at the middle of the range of heights. Since there is a complete range of heights, from the shortest to the tallest, this type of variation is known as *continuous variation*.

2 No. The fact that no two sets of fingerprints are exactly the same has made fingerprinting an extremely useful technique in crime detection. Since there is a complete range of the different types of fingerprint, this is another example of continuous variation.

3 The approximate proportions in a large population are: loops 70%, whorls 13%, arches and other types (including combinations) 17%.

4 No. Either the tongue can be rolled or it cannot be rolled. This is an example of *discontinuous variation*.

5 This is clearly different from the type of inheritance for height and fingerprints, which show a wide range of variation because a number of different factors are involved. Inheritance of tongue rolling is controlled in a relatively simple way: the following is a full explanation, which you will find easier to understand after you have worked through sections 9.12 to 9.16.

The ability to roll your tongue is inherited through a single dominant gene. Thus the homozygote for this gene (RR) and the heterozygote (Rr) both have the ability to roll their tongues. Only the homozygote for the recessive gene (rr) cannot roll his tongue.

6 i) Small numbers of offspring.
ii) Long generation time.
iii) Unsuitable for experimental crosses.
iv) Unsuitable for breeding of close relatives.

Section 9.12

1 25% (approximately).

2 i) The time between generations is very short (10–14 days).
ii) It has large numbers of offspring.
iii) It is easily bred in the laboratory.
iv) It has a number of characters which show discontinuous variation.
v) These characters are easily visible.
vi) Flies can be crossed with their parents or *siblings*.

Section 9.15

1 The greater the numbers used, the nearer the ratio obtained approaches the ideal ratio of 3:1.

2 Long; long; vestigial.

3 Approximately 3:1.

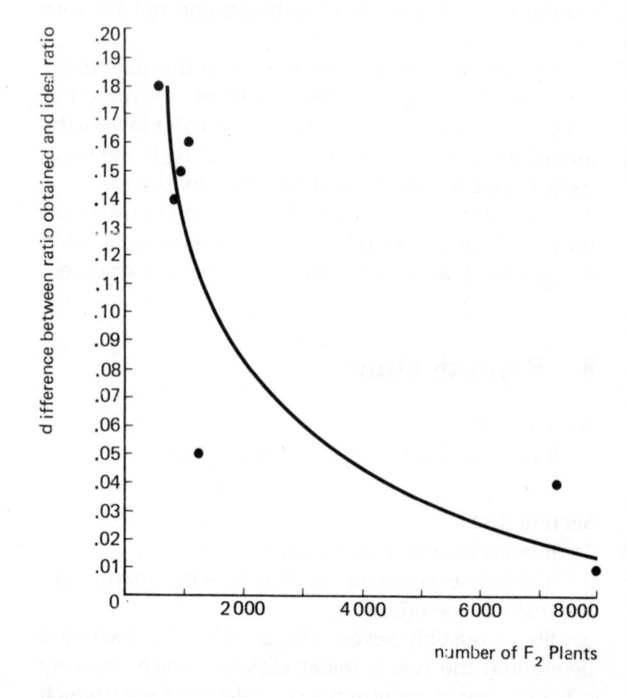

fig I The difference between the ratio obtained and the ideal ratio

Section 9.16

1 Ll and ll.
2 L and l.
3 l only.
4 P_2 Ll × ll
ll

	gametes	(l)	(l)	
Ll	(L)	Ll	Ll	2 long wings
	(l)	ll	ll	2 vestigal wings

F_2 phenotypes

The ratio of longed winged offspring to vestigial winged offspring is 1:1.

Section 9.17

1 P_1 TT × tt
tall short
F_1 Tt × TT
TT

	gametes	(T)	(T)
P_2 Tt	(T)	TT	TT
	(t)	Tt	Tt

F_2 All tall
or:
P_2 TT × Tt
gametes

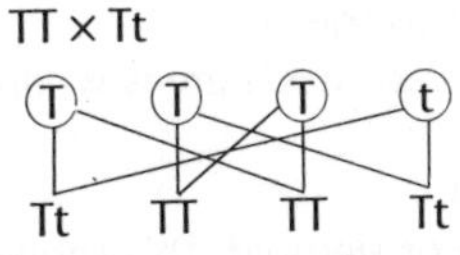

F_2 Tt TT TT Tt

2 50%. The other 50% are heterozygous: that is to say they (like the F1 plants) possess genes for tallness and shortness.
3 i) They can be self-pollinated.
ii) Since they are naturally self-pollinating, pure-breeding strains are easy to obtain.
iii) The numbers of offspring can be even greater.
4 50% smooth seeds and 50% shrunken seeds.
5 The time between pollination and appearance of the characters is very short. Experimenters do not have to wait until a plant has grown and matured.

Section 9.20

1 P $G^AG^O \times G^BG^O$
AO × BO
BO

	gametes	(B)	(O)
AO	(A)	AB	AO
	(O)	BO	OO

Genotypes: G^AG^B, G^AG^O
G^BG^O, G^OG^O
Phenotypes: AB A
B O

Section 9.52

1 Malaria
2 People possessing the sickle cell trait also have a high resistance to a certain type of malaria. Thus, people living in regions where malaria is common are less likely to die before producing children if they possess a single S gene than if they possess none at all: the S genes in these areas are therefore selectively advantageous.

Section 9.54

1 Light moths are more vulnerable in Birmingham and dark moths in Dorset, because they stand out against the tree backgrounds.
2 With the advent of the industrial revolution, smoke and soot from factory chimneys blackened the bark of trees in industrial areas. The prevailing winds carried this smoke and soot to the east.
3 The overall colour of tree bark would lighten and dark moths would become more obvious to predators than light moths. More light moths than dark moths would survive to reproductive age and would pass on the genetic character for light wings to their offspring. Thus the next generation would contain a greater proportion of light moths. This process would continue until almost all peppered moths in the area would have light wings. This is an example of evolution taking place by natural selection responding to changes in the environment.

10 Growth

Section 10.00

1 i) Accumulation of alcohol (ethanol) from anaerobic respiration will produce a toxic effect.
ii) Alternately, the substrate may be exhausted.

Section 10.10

1 The radicle. The food reserves of the seed must be made soluble, so the absorption of water is most important.

Section 10.11

1 In the case of the seed in box A, the plumule penetrates the surface of the soil with its tip slightly bent, protecting the delicate leaves at its apex. The cotyledons remain underground. This is known as hypogeal germination.

In the seeds in box B, there is a marked difference. The first structure to appear at the surface of the soil is a loop or arch of stem (the hypocotyl) from below the cotyledons. Eventually, the loop straightens out, pulling the cotyledons and the tip of the plumule clear of the soil. The cotyledons have lost most of their original food reserves by this time, they turn green and they begin to act like a pair of leaves, manufacturing food by photosynthesis. This type of germination is called epigeal.

2 B – germination has reached a stage similar to jar A.
C – little or no germination.
D – at 5°C there will be appearance of a radicle and plumule i.e. some growth, but little compared with A.
E – no germination.

3 i) Air (in particular oxygen present in air).
ii) A minimum temperature.
iii) Water.

Light and dark have no effect on pea seed germination. Certain plants have seeds which need conditions other than the three listed above. Some seeds require light, some need to be partly digested in an animal's gut, while others need dormancy and sometimes a period of low temperature (vernalisation) before germination.

4 There is a decrease in the dry weight of the pea seedlings at the start of germination.

5 The food stores in the cotyledon have been used up for cell growth and for respiration. The latter process accounts for the weight reduction.

6 The endosperm, which stains blue-black with iodine. Maize belongs to a large group of plants, including other cereals, onions, lilies, and palms, in which the seeds have only a single cotyledon. Thus they are referred to as monocotyledons. Food reserves are not stored in this single cotyledon, but in the endosperm tissue (see section 8.42).

7 Hypogeal.

Section 10.20

1 A few mm behind the root tip.

2 An organism grows by increasing the number of its cells (see Chapter 9, section 9.00), not, as a rule, by increasing the size of its existing cells. We have seen, just behind the protective root cap, that there is a region of cell division.

Immediately after division of a cell, the two daughter cells have only the same combined volume before undergoing a further division of differentiating into a specialised root tissue. This increase in volume occurs mainly by elongation of the cells in a region just behind the region of cell division.

Section 10.31

1 4 years.

Section 10.40

1 The radicles have grown downwards and the plumule has grown upwards, except the left one which is orientated correctly. This seedling is the control. A useful instrument which can be used as a control apparatus in experiments on the influence of gravity and light is a *klinostat*. It consists of a cork disc to which the germinating seeds are pinned through their cotyledons. The seedlings must be kept moist and a cover is fixed over the cork disc, lined with damp blotting paper. The disc is rotated by either clockwork or an electric motor, with one turn every 15 minutes. If a seed is positioned horizontally on the cork, there is no change in the direction of growth of the radicle or plumule. Can you suggest why this should be?

2 The shoots would be exposed to light and air for photosynthesis, since only by this process can growth continue. The roots would grow into the ground for anchorage and for water and mineral nutrition to further aid the growth by synthesis of organic compounds. Finally the reproductive structures must be borne up by the shoot.

3 They remain unchanged: there has been no growth.

4 They have grown taller in a vertical direction.

5 They have grown at an angle towards the hole in the cardboard box: i.e. towards the light.

6 It would enable the shoot to grow towards a region where maximum sunlight would be available for photosynthesis.

7 Many of them have grown straight downwards and become shrivelled through lack of water.

8 Many of these have grown along the lower surface of the netting, and some have grown back through the netting and into the soil.

9 The root can reach areas where water is plentiful.

Section 10.51

1 S-shaped or sigmoid. This shows a slow increase in weight at first, followed by a greatly increased rate of weight increase, followed in turn by a gradual levelling off of the rate of weight increase at maturity. Compare this curve with that of the population of yeast in fig 10.2. You will note that colonies of micro-organisms, where the individuals are separated from each other, conform to the general growth pattern of a multicellular organism when treated as one mass of cells.

2 Sigmoid.

3 The curve for males. In the curve for females, the increase in rate of growth starts earlier and flattens out both earlier and at a lower weight. This reflects the facts that girls tend to mature earlier than boys and that men tend to be heavier than women (see fig. J).

4 The 'height spurt' appears earlier in girls (12–13 years) than in boys (13–15 years). This indicates earlier maturity. Boys on average are 12 cm taller than girls when the growing period has been completed (see fig. K).

Section 10.53

1 Growth occurs in a series of 'steps'. There are rapid bursts of growth after each moult (edysis), followed by a period of no growth (see fig. L).

2 Table D (see p. 350.)

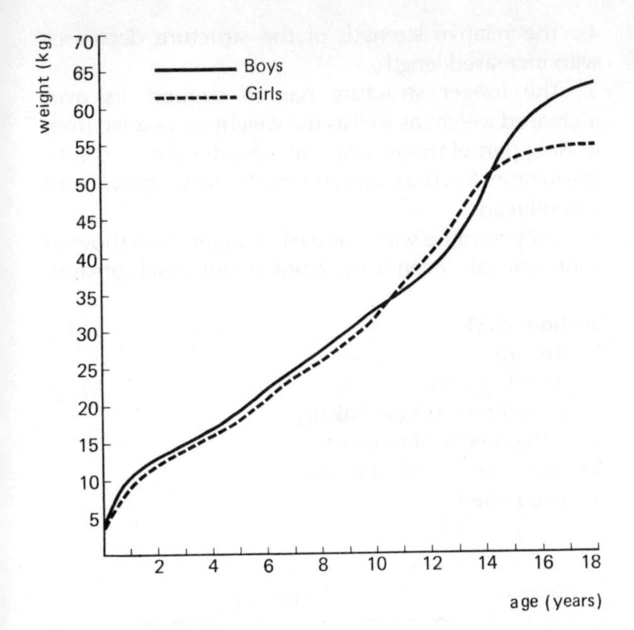

fig J A comparison of weight gains in boys and girls

fig K A comparison of height increases in boys and girls

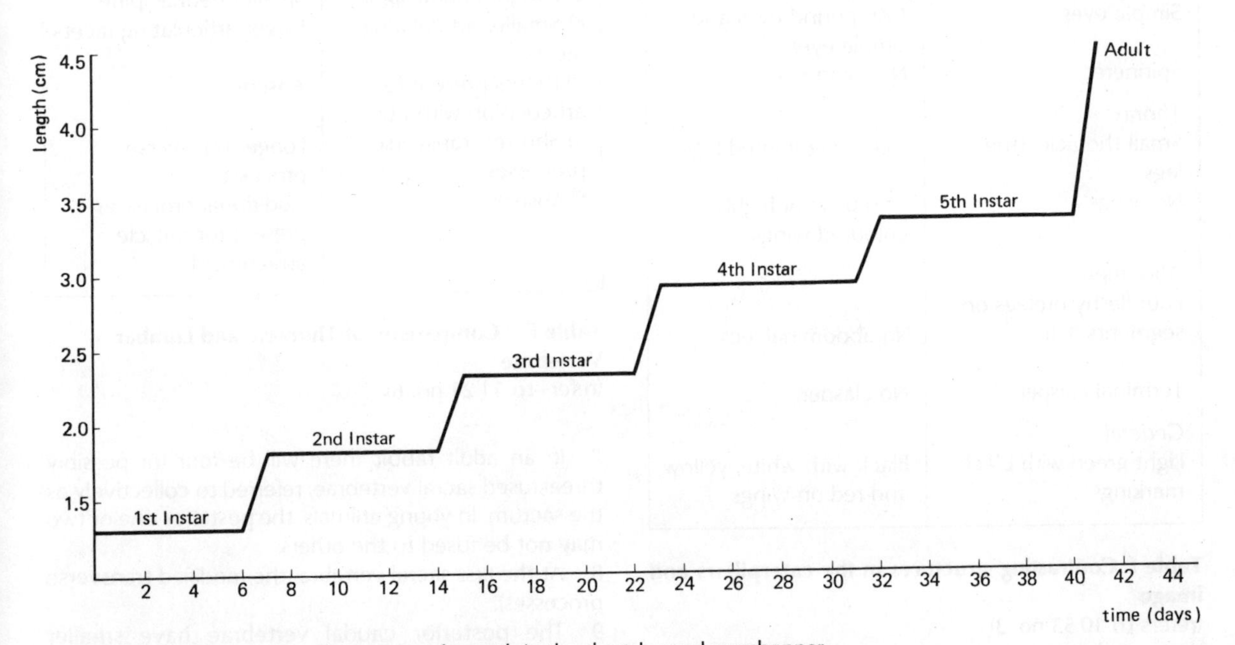

fig L Graph to demonstrate the pattern of growth in the short-horned grasshopper

3 Table E (see p. 350.)
4 None, apart from occasional twitching as internal changes occurred. The pupa neither feeds nor moves.
5 Feeding for growth. The caterpillar feeds constantly on the leaves of cabbage plants and grows rapidly. It moults about four times during its short existence (one week on average).
6 The caterpillar feeds on solid food, the imago sucks nectar and plant juices through the proboscis.

11 The skeleton and locomotion

Section 11.00
1 It has become shorter and thicker as it pulls on the forearm to bend it forward. In other words, it has contracted.
2 It has become longer and thinner as it allows the elbow to bend. In other words, it has relaxed. These two muscles act against each other antagonistically to

Instar	Changes
2	Leg spines developing
3	Abdomen larger in proportion, and upturned. Hind legs larger in proportion
4	Wing buds appear. Pronotum well developed
5	Further increase in proportional size of hind legs and wings.
Adult	Wings enormously developed.

Table D Changes in Nymphal Instars
(refers to 10.53 no. 2)

Caterpillar	Imago
Head	
No antennae	Long antennae
Biting mouthparts	Long proboscis
Simple eyes	Compound eyes and simple eyes
Spinneret	No spinneret
Thorax	
Small thoracic (true) legs	Very long jointed legs
No wings	Two pairs of highly coloured wings
Abdomen	
Four fleshy prolegs on segments 3–6	No abdominal legs
Terminal clasper	No clasper
General	
Light green with black markings	Black with white, yellow and red on wings

Table E Contrasting structures in the caterpillars and imago
(refers to 10.53 no. 3)

produce bending and straightening of the arm.

Section 11.12
1 The tube.
2 A tubular structure is stronger than a solid structure of the same mass.
3 Having the heavier compact bone tissue arranged in the form of a tube around the lighter marrow enables the limb to be relatively light and strong.
4 The relative strength of the structure decreases with increased length.
5 The longer structure has to support its own increased weight as well as the weight suspended from it. Also, part of this weight is at a greater distance from the fulcrum (i.e. the chair) and thus a greater movement is produced.
6 They must be wider, as well as larger, than those of short animals, even if the trunk is not much greater.

Section 11.21
1 The atlas.
2 Nodding only.
3 Partial rotation or shaking.
4 Attachment of muscles.
5 Articulation with the ribs.
6 See table F.

Thoracic	Lumbar
a) Small centrum	Large centrum
b) Longer neural spine	Smaller neural spine
c) Smaller articulating facets	Larger articulating facets
d) Facets present for articulation with ribs	Absent
e) Shorter transverse processes	Longer transverse processes
f) Absent	Additional processes present for muscle attachment

Table F Comparison of Thoracic and Lumbar Vertebrae
(refers to 11.21 no. 6)

7 In an adult rabbit there will be four (or possibly three) fused sacral vertebrae, referred to collectively as the sacrum. In young animals, the posterior one or two may not be fused to the others.
8 At the first sacral vertebra (the modified transverse processes).
9 The posterior caudal vertebrae have smaller projections. The very last ones consist only of centra.
10 Twelve.
11 Seven pairs.
12 Two pairs.
13 Three pairs.

Section 11.31
1 Lateral (or dorso-lateral).
2 The breast-bone or sternum (see fig 11.12).
3 It is cup-shaped and smooth.
4 Attachment of muscles.

5 During birth it 'gives' to allow the passage of the foetus through the vagina.
6 Articulation with the sacrum.

Section 11.32
1 Attachment of muscles.
2 Lateral.
3 Facing downwards.
4 No.
5 Prone.
6 Because Man walks on two legs instead of four, the forelimbs can be used for a variety of functions, such as feeding, carrying and manipulation of tools. Thus, during the course of evolution, specialisation and strength has been sacrificed for versatility. The act of turning the hand from a prone to a supine position is an essential feature in such complex activities as using a screwdriver.
7 There is a ridge (the *deltoid tuberosity*) on the anterior face, for the attachment of the deltoid muscle.
8 Seven: arranged as one row of two and a more distal row of five.
9 Seven: arranged as one row of three and one row of four.
10 Its weight is carried on the bases of the digits of its hind feet: i.e. it walks on the balls of its feet. This means that the elongated metatarsals are free to increase the effective length of the leg and thus increase the potential speed. The same is true of humans when they run on the balls of their feet. Almost all fast-running mammals have elongated metatarsals that function in this way. The next time that you see a dog or a cat, note that the backward-pointing joint halfway up the animal's back leg is in fact the heel.

In ungulates, such as horses or antelope, the walking or running surface is the tip of the toe: this allows an even greater effective length of leg.

Section 11.33
1 i) There is a segment corresponding to the femur.
ii) There is a segment corresponding to the tibio-fibula.
iii) There are small segments corresponding to the tarsi, metatarsi and phalanges.
iv) The limb ends in claws.
v) The joints permit movement in approximately the same planes.
2 i) The skeleton is external.
ii) The distal end of the limb is not divided into digits.
3 Growth cannot take place without the shedding of the skeleton (moulting).

Section 11.40
1 Greyish-pink with faint lines running along the main axis of the limb.
2 Near the heel bone, the tissue takes on an opaque, white appearance. This is in fact the tendon that joins muscle to bone.

Section 11.50
1 1st The atlas vertebra and the skull
2nd Ankle joints operated by the calf muscle
3rd The elbow operated by the biceps

12 Disease

Section 12.10
1 *Tubes* 1, 2 and 3 have changed in appearance.
2 The broth has become cloudy (turbid).
3 The broth has gone bad. It has an unpleasant smell, and it is probable that this has been caused by micro-organisms.
4 *Tube 1* goes cloudy even though heated, but it is open to the air.
Tube 2 although protected from the air is not heated and goes cloudy, but rather more slowly.
Tube 3 is open to the air through the glass tube and although heated turns cloudy.
Tube 4 is the only tube to remain clear. It has been heat treated but is open to the air. Thus air itself does not cause change in the broth but something present in the air. The micro-organisms present in the air must be trapped in the s-shaped glass tube.
5 Water condenses inside the Petri dish and if it were the correct way up the drops on the lid could fall onto the culture. Therefore the cultures are placed upside down and marking them on the base enables the markings to be seen clearly. Furthermore since the lid can be twisted around markings on the lid could be moved out of their correct position.
6 To ensure that they are not taken off accidentally or fall off as the dishes are picked up.
7 Growth of micro-organisms show up as patches on the jelly.
8 This could be regarded as a control to make sure that the broth is sterile, but it could be contaminated by the hairs when placed on the jelly. It is better to have a separate dish with jelly only as a control.
9 There are fewer colonies on the jelly in 1B than in 1A, showing that soap and water can remove many bacteria, but not all.
10 i) Hands and other parts of the skin should be kept very clean.
ii) Hair should be contained under a hat in order that stray hairs should not fall on the food.
iii) Mouth masks would prevent the transfer of micro-organisms by coughing or sneezing.
11 At a temperature of 37°C which is the mammalian body temperature.
12 Dish 3 showed fewest or no colonies of bacteria, and so food stored at the freezing point of water (or slightly below) would be prevented from decaying.

Heat treatment of food as shown in dish 5 is also effective in that it destroys micro-organisms, and no decay will occur if micro-organisms can be stopped from reinfecting the food.

Section 12.35

1 The time is temperature-related. At 10°C eggs may take 4 weeks to develop into larvae, at 30°C, only 3 days.
2 At 10°C larvae may take 8 weeks to pupate, at 30°C, 3 days.
3 Larvae are actively feeding on the decaying meat. They moult and increase in size. When fully grown, they seek a sheltered, moist place to pupate. The pupa is inactive but inside there are considerable changes as the larval form changes to the adult.
4 It crawls out of the pupa. Its wings are limp and it waits for these to dry and harden before flying away.

13 Soil and cultivated crops

Section 13.10

1 Recent rainfall can make a considerable difference, so that soil examined in the rainy season will show a much greater amount of water than that examined in the dry season. Nevertheless at any one season, the amount of water depends on the drainage characteristics of the soil.
2 Forest soils will have the greater amount of humus compared with garden soil. Swampy ground, where humus cannot decay, will show considerable amounts of humus. Sandy soils lack humus (see table 13.3).
3 Humus is missing since plants and animals cannot live easily in the hard-packed soil. Water also shows a smaller percentage.
4 Topsoil has more humus and less mineral matter than subsoil. If the soil contains much chalk (calcium carbonate) some of the weight loss could be due to the decomposition of chalk on heating.
5 An identical sample of soil could be dried to eliminate water content and the amount should be subtracted from the calculated air.
6 The soil could be compressed into the tin beyond its natural state. The soil in the cylinder was not stirred properly and thus all the air was not released.
7 It provides oxygen for the respiration of soil organisms and plant roots.
8 Limewater turns milky or bicarbonate indicator turns yellow in tube A but not tube B.
9 Carbon dioxide has been produced in tube A, indicating the presence of living organisms. The organisms in the soil of tube B were killed by heating.
10 The light, heat and dryness produced by the bulb are all factors causing organisms to move away from the upper parts of the soil.
11 To reflect the light and heat into the soil.
12 There are more different types of organism and they are more numerous in the topsoil. This is due to the presence of decaying plant and animal matter on which they feed, while some of the carnivorous forms feed on the other living organisms.

Section 13.11

1 Sand.
2 The larger air spaces enable the water to rise more rapidly in the first hours (see fig 13.5).
3 Clay.
4 The smaller air spaces, although slowing the early rise of water, enable the water finally to rise higher than in the sand. (See fig M).

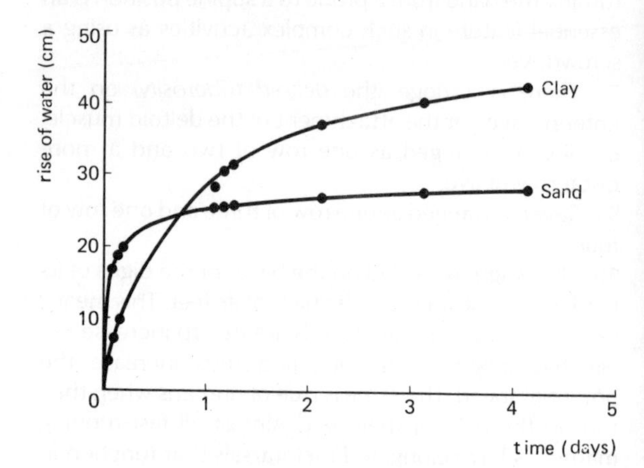

fig M A graph to compare capillarity in clay and sand soils

5 Sand
6 The larger air spaces enable the water to drain through more rapidly.
7 Clay.
8 Clay can hold more water around its small particles.
9 In beaker A more clay has settled out. There is a greater depth of clear (or slightly milky) water at the top of the beaker and a deeper layer of clay at the bottom. The clay particles in A have clumped together, whereas those in B have not. The clumping together (flocculation) of clay particles into larger lumps has caused them to settle out.
10 Lime causes flocculation of clay into larger particles. This improves the porosity of the soil and allows greater penetration of air and water to root systems.

Section 13.20

1 *The box of soil inclined at 40°*. No plant material was present to bind the soil. The 40° slope allowed the water to travel down quickly taking soil particles with it into the container.
2 *The box containing the cut turf*. The leaves deflected

Dried soil	Colour before burning	% loss of weight on burning	Colour of remaining mineral matter
Forest (topsoil)	Black	30	grey
Garden (topsoil)	Dark brown	15	grey
Soil from a school field	Light brown	5	red
Subsoil	Red	3	red

Table G Soil data

the water drops thus protecting the soil, the roots bound together the mineral fraction and the humus absorbed the water.
3 The process is called erosion.

14 Ecology

Section 14.20
1 Three.
2 No, they are distributed unevenly throughout the world.
3 They are distributed, generally, within the tropics, although *Neoceratodus* (the Australian lungfish) is found in subtropical areas.
4 Each type of lung fish is confined to a particular continent. *Lepidosiren* is found only in the rivers of South America, *Protopterus* only in the rivers of Africa and *Neoceratodus* only in the rivers of Australia.
5 Yes. The two distribution patterns are identical.
6 The Golden Eagle is found only in mountainous regions of Europe.
7 Yes. The distribution of another organism or an environmental factor (such as mountains) may help to explain the reasons for the distribution of an animal or plant species.
8 The distribution patterns of sickle celled individuals and Anopheles mosquitoes. These two distribution patterns showed great similarities. (See Chapter 9).

Section 14.54
1 Plant species D.
2 A, 5%; B, 9.5%; C, 1%; D, 15%; E, 3%.

15 Lower plants and animals

Section 15.00
1 Segmented body; body divided into two or three parts; jointed legs; head with antennae; large compound eyes;
2 Nos 1,8,7,2 and 13 body divided into three parts (head, thorax and abdomen); three pairs of jointed legs;

Section 14.83
Results of experiment: numbers and condition of fronds

Experiment beakers	Number of days after beginning experiment					
	2	4	7	14	21	28
A	10	10	10	13	18	24
B	10	losing green colour	decolourised	–	–	–
C	10	10	10	10	10	dead – fungal attack
D	10	10	10	10	10	dead
E	10	10	12	13	19	36
F	6 sunk	9 sunk	all sunk	–	–	–
G	10 yellow	brown	–	–	–	–
H	10	pale	pale	decolourised	–	–
I	10	10	13	20	29	48
J	10	10	11	12	16	19

Nos 2, 3, 5 and 10	body divided into two parts (prosoma, abdomen); four pairs of jointed legs;	
Nos 4 and 9	body divided into two parts; many legs;	
Nos 6 and 11	body divided into two parts; numerous legs including large pincers;	

Investigation based on figs 15.3 to 15.6:

A hyphae forming mycelium; PLANT, THALLOPHYTA, FUNGI (Mushroom)

B unsegmented body with a shell; large muscular foot; ANIMAL, INVERTEBRATA, MOLLUSCA (Snail)

C leaf-like structures; rhizoids; spore capsule; PLANT, BRYOPHYTA, MUSCI (Moss)

D body divided into two parts; four pairs of legs; ANIMAL, INVERTEBRATA, ARTHROPODA, ARACHNIDA (Spider)

E segmented body; no limbs; bristles or chaetae; ANIMAL, INVERTEBRATA, ANNELIDA (Earthworm)

F frond with leaflets; fibrous roots; PLANT, PTERIDOPHYTA (Fern)

G head, thorax and abdomen; three pairs of legs; one pair of antennae; ANIMAL, INVERTEBRATA, ARTHROPODA, INSECTA (Beetle)

Index

Acetobacter bacteria, 230
Acetic acid, 244
Acetylcholine, 155–6
ADP (adenosine diphosphate), 125, 129, 334
Adrenalin, 161, 162
Air: inhaled and exhaled, 111–16; content, 114; volumes moved in breathing, 115–16; pollution, 121–2
Airborne diseases, 235
Albumen, 19, 43, 319, 324, 325, 332
Alcohol: fermentation, 127; production, 127
Algae, 286–8, 304
Alimentary canal, 40, 43, 51–5, 96, 97, 203, 235, 241, 296, 338; structure of, 51–2
Alleles, 194–5, 201
Alveoli, 117–20
Amino acids, 18–20, 39, 53, 54, 101, 109, 142, 198, 329
Ammonia, 36, 141, 142, 335, 342
Amoeba, 164, 167–8, 291–2
Amphibians, 212, 318–9
Ampulla, 153
Amylase, 52, 53, 54
Anabolism, 125, 141, 145
Anaemia, 58; pernicious, 109; sickle-cell, 198–9
Anaphase 185–6, 188–9
Androecium, 176
Ankle bones, 222, 351
Annelids, 294–6
Annuals, 101, 305
Anopheles, **see** Mosquito
Antennae, 299, 301
Anther, 177, 194, 345
Anthrax, 229
Antibodies, 66–8, 241, 334
Antigens, 67, 241, 334
Antiseptics, 241
Anus, 51, 55, 296
Aorta, 69–70, 73
Aphids, 84
Appendix, 51, 52
Arachnids (spiders), 297, 303
Arteries,71–3, 74, 334; hepatic, 73, 109; pulmonary, 73, 120; renal, 140, 141, 342
Arterioles, 72, 74, 134, 141
Arthropoda, 282
Arthropod limbs 223, 226
Arthropods, 282, 297–303
Ascorbic acid, 25, 329
Asexual reproduction: in lower organisms, 167–9; in plants, 164–7; runners and suckers, 165–6; perennation, 166; use by farmers, 167
ATP (adenosine triphosphate), 125, 129, 142, 170, 225, 334
Auditory canal, 152–3
Auxins, 210
Axillary glands, 74
Axons, 155, 156

Bacilli, 229
Bacteria, 36, 50, 53, 55, 66, 98, 108, 117, 119, 204, 235, 240–42, 286, 331, 337, 351; aglutinated, 66; saprophytic, 93, 230; sterile techniques, 227–9; growth, 227–9; and disease, 229–31; aerobic, 229; anaerobic, 229; useful, 230
Bacteriophages, 231
Bamboo, 307
Bananas, 107
Banting, Sir Frederick, 161
Bar charts, 11, 262
Basal metabolism and Basal Metabolic Rate (BMR), 55
Beaks, 322
Bees, 160, 300–301
Bernard, Claud, 130
Berries, 181, 182
Berthold, Adolf, 161
Best, C. H., 161
Bicarbonate indicators, 14, 28
Biceps, 224, 351
Biennials, 101, 105–6, 305
Bile and bile duct, 52, 53
Binary fission, 167–8
Birds, 321–5; flight and steering, 324; nests, 323
Bivalents, 187, 201
Bladder, 140, 345
Blights (of plants), 234
Blood; contents, 64–5; structure, 65–8; red cells, 65–8, 72, 197, 217; white cells, 65, 66, 72, 241; groups, 67–8, 334; vessels, 71–2; sugar, 109, 144, 161, 345; **see also** Circulation of blood
Blowflies, 236
Body functions, regulation of, 130–31
Boils, 229, 241
Bone and bones, 215–17, 219, 220; marrow 216, 217
Bowman's capsule, 141, 142, 342
Bracts, 179
Brain, 155–7, 215, 216, 220, 231
Bread production, 127
Breastbone, **see** Sternum
Breasts, 176
Breathing, 111–16; mechanism, 118–19
Bristleworms, 295
Bromide ions, 86
Bronchioles, 117
Bronchitis, 122
Bronchus, 117
Bryophyta, 289–90
Budding, 167
Bulbs, 104, 166, 167
Butterflies, 214, 299–301

Caecum, 51, 52, 98
Cabbage white butterfly, 299
Calamus, 321
Calcium, 37, 215; hydroxide, 14; chloride, 14; phosphate, 217; hypochlorite, 241; oxalate, 145, 289
Calories, 55
Calyx, 176–7
Cambium, 208, 310
Camels: fat store, 108; metabolism, 138
Cancer of the lung, 122
Canines (teeth), 46, 332
Capillaries, 71–3, 117, 120, 134, 141–2, 241
Capillarity, 84, 249
Carbohydrates, 17–20, 23–4, 35, 36, 43, 53, 55, 57, 104, 106–10, 144–5, 225, 330; formation in plants, 29–32; storage, 106
Carbon, 17, 18, 123; cycle, 37
Carbon dioxide, 14, 19, 20, 27–9, 32, 34–5, 37, 62, 66, 72, 73, 98, 112, 113, 123, 125–6, 141, 145, 225, 330, 331, 338, 352; gaseous exchange, 114–21
Carbon monoxide, 121
Carbonic acid, 14, 32, 331
Caries, dental, 50
Carnivores, 26, 48–9, 109, 267–8
Carpals, 222
Carpels, 177, 179, 180, 181
Cartilage, 216–18
Caryopsis, 181
Castration, 161
Catabolism, 125, 141, 145
Caterpillars, 214, 299, 350
Caucasoid race, 189, 190
Cell division, 168, 184–9
Cells, plant and animal, 80–84, 85–7, 203
Cellulase, 53, 98
Cellulose, 18, 53, 55, 77, 78, 84, 94, 98, 108, 233, 287, 329, 331, 335, 338
Centimetre-gramme-second (c.g.s.) system, 1
Centipedes, 298–9
Centromere, 185, 201
Cereal crops, 20, 106–7, 306
Cerebellum, 156, 157
Cerebrum, 156–7
Chaetae, 296
Chemistry, elementary, 13–14
Chicken pox, 241
Chilopods, 298–9, 303
Chitin, 18, 297, 298
Chlorination, 240
Chlorophyll, 20, 26–7, 29, 30, 33, 89, 95, 129, 287, 288, 305, 330, 339
Chloroplasts, 33, 89, 287, 304, 331, 337
Chloroquine, 238
Chloroxylenol, 241
Cholera, 140, 229, 235, 239, 242
Chordata, 283
Chromatids, 185, 201
Chromosomes, 173, 185, 187, 189, 192, 194–5, 201, 346; homologous, 187, 192, 195
Chyme, 51–3
Cilia, 117, 122, 241, 292
Circulation of blood, 68–9, 72–3, 109, 334; pulmonary, 72; body, 72–3

Classification, 278
Clavicles (collar bones), 221
Clay, 249, 250, 352
Climate, 255
Clitellum, 169, 295, 296
Cocci, 229
Cochlea, 153
Cockroaches, 236, 299
Coelenterata, 283, 293
Colon, 52
Commensalism, 99–100
Compost, 253
Conidiophores, 169
Conservation, 275
Contagious diseases, 236
Co-ordination, 155–63; without nerves, 161–3
Coprophagy, 53
Copulation, 174
Corms, 103–4, 166, 167
Corolla, 176–7, 345
Corti, organ of, 153
Cotyledons, 181, 347
Cowper's gland, 171, 345
Crabs, 298
Cranium, 220
Crayfish, 298
Crop rotation, 252–3
Crustaceans, 298, 303
Cuttings, 84, 167
Cytoplasm, 81, 82, 84, 86, 90, 125, 155, 170, 174, 203, 233, 287, 291, 335, 337
Cytostome, 292

Daraprim, 238
Decapods, 298
Decussate leaves, 305
Deficiency diseases, 20, 23, 37, 57–8
Dehydration, 138, 143, 144, 243
Dendrites, 155
Density, estimate of, 261
Dermis, 135, 154
Desmids, 286, 292
Diabetes, insipidus (water), 144; mellitus (sugar), 144, 161
Diaphragm, 51, 117–19
Diarrhoea, 58, 140, 240, 342
Diatoms, 286, 291, 292
Dicotyledons, 305–9
Diet, balanced, 57–60
Diffusion, 78–80, 84
Digestion, 40, 46, 51–4; mechanical, 46; chemical, 52, 53; extracellular, 93, 94
Digits and digit bones, 222, 322
Diphtheria, 241
Diplococci, 229
Diplopods, 298–9, 303
Disaccharides, 17–18, 42
Disease: and bacteria, 229–31, 236; and viruses, 231–2, 236; and protozoa, 232–3; and plants, 234; and fungi, 233, 234, 236; and worms, 234–6; transmission of, 235–40
Disinfectants, 241
Dogs: jaw movements, 49; cross-breeding, 189, 190
Dominance, 192, 201; incomplete, 194, 196
Drupes, 181, 182
Duodenum, 52
Dysentery, 235, 239, 242

Ears, 152–4, 344; middle ear, 152; inner ear, 153, 220, 344; of rats, 327
Earthworms, 295–6; worm casts, 296
Ecosystem, 254
Eggs, 164, 169–74, 192, 212, 293–4, 297, 299, 301, 302, 318–21, 324–5; **see also** Ova
Elbow joint, 224, 351
Elm, 312, 313
Embryo, 174, 175, 181, 318, 325
Endocarp, 177, 182
Endocrine glands and endocrine system, 67, 74, 157, 161–3, 169, 176
Endosperm, 107, 181, 348; tissue, 177
Energy, 19, 122–9, 268; chemical, 33; light, 33; food value, 55–60; obtaining, 122–6, 340; types, 123–4; without oxygen, 126–8
Enzymes, 19, 23, 37, 42, 43, 53, 94, 97, 98, 170, 174, 206, 230, 241, 332; co-enzymes, 23; anti-enzymes, 97
Epididymis, 171, 174
Epiglottis, 117
Epiphytes, 290, 307
Equator (in genetics), 185, 201, 346
Erosion,251–2, 253; gully erosion, 251; sheet erosion, 251
Erythrocytes, 65, 109
Ethyl alcohol (ethanol), 126, 128, 329, 330, 340, 341, 347
Euglena viridis, 304
Eustachian tube, 152–3
Evolution, evidences, 196
Excretion, 141; in plants, 145
Exponential curves, 11
Eyes, 147–52, 215, 220, 343–4; visual fields, 149; focusing, 150–52; defects, 152; in fish, 315; in fowl, 322; in rats, 327

Faeces, 54, 55, 235
Fallopian tubes, 172–4, 328
Fallow land, 253
Fats, 18–20, 24–5, 36, 53, 55, 108–9, 125
Fatty acid, 18, 53, 54, 144
Feathers, 321–2, 325
Feeding, types of, 43–6
Femur, 223, 351
Fermentation, alcoholic, 126
Ferns, 286, 290–91
Fertilisation, 164, 169, 170, 174, 175, 177–8, 187, 194, 212, 297, 317, 318, 319, 324, 327, 345
Fibrinogen, 66, 241
Fibula, 223
Filoplumes, 322
Filtration: pressure, 141; beds, 239
Fins, 315
Fishes, 315–17; gas exchange, 120; swimming, 316–17
Flagellum, 170, 340
Flatworms, 234, 294
Fleas, 96, 337
Flowers, 176–80, 306
Flukes, 234, 294
Fluoridation, 50–51
Foetus, 175, 217, 236, 351
Follicle cells, 170, 173
Fontanelle, 220
Food: classes of, 19–21; energy value, 55–61; preparation, 238; preservation, 242–4; poisoning, 242–4; frozen, 243
Food chains, 266–8; webs, 267–8; and energy, 268
Foodborne diseases, 235–6
Fovea, 152
Foxes, 137, 341
Frequency, estimate of, 261
Frog, 131, 318–19, 340; life history, 212–13, 318–19; swimming, 319
Fruit fly (*Drosophila*), 192, 193
Fruits, 181–2; food storage in, 106–8; classification, 181–3
Fungi, 95, 204, 242, 286, 288–9, 331, 337, 340; saprophytic, 93–5; and diseases of man, 233–4, 241; reproduction, 288–9

Galactose, 17
Gametangia, 289
Gametes, 164, 170, 177, 192, 287–90, 346
Gamma radiation, 241
Gaseous exchange, 114–21, 339–40
Gastric juice, 53
Gastrocnemius, 223, 224
Genes, 192, 194–5, 201, 346, 347
Genotypes, 192, 194, 201
Geotropism, 210
Germination, 204–7, 347; types, 206; conditions for, 206–7
Gestation, 175
Gibberellic acid, 210
Gills, 117, 120, 121, 212, 298, 303, 315, 340
Glomeruli, 141
Glottis, 117
Glucose, 17, 18, 20, 41, 42, 53, 54, 109, 123, 125–9, 142, 144, 225, 331, 340, 345
Glutamic acid, 36, 198
Glycerol, 18, 53, 54, 108
Glycogen, 18, 95, 108, 109, 127, 144, 225
Gonads, 176
Gonorrhoea, 236
Graafian follicle, 173
Grafting, 167
Graphs, 11, 262
Gravity and plant growth, 208–10
Growth, 207–14; definition of, 203; in roots, 207–10; in shoots, 207–10; secondary thickening, 208; plant growth and external stimuli, 208–10; responses, 210; in mammals, 211–12; in lower vertebrates, 212–13; growth in insects, 213–4
Gut, 51–5
Guttation, 145
Gynaecium, 176, 177, 180

Habitat, 254, 255, 261; abiotic factors, 256–9, 264; biotic factores, 259
Haemoglobin, 19, 65, 66, 197, 198, 335
Haemolytic disease, 67–8
Haemophilia, 195, 196
Halteres, 302
Harvey, William, 68, 69, 72
Haversian canals, 217
Hearing, 153
Heart, 69–71, 216, 334; auricles and ventricles, 69–73; coronary artery, 69; valves, 70, 71, 334; heart beat, 70–71, 74; muscles, 71; diseases, 122
Heat: production of, 124, 340; loss and gain, 134–7
Heel-bone, 223, 351
Height, measurement of, 258–9
Herbivores, 26, 49, 98, 99, 108, 267; and birds, 98, 99, 338
Herbs, 305, 310
Heredity, 189–94
Heterozygote, 192, 194, 201, 346
Hibernation, 132
Hilum, 181
Histograms, 11
Holophytic nutrition, 93
Homeostatic mechanism, 119, 143, 161
Homoiotherms, 131–8; heat loss and gain, 134–8; heat retention, 134–5; body size, 135–7

Homozygote, 192, 194, 198, 201, 346
Hopkins, Gowland, 23
Hormones, 143, 155, 161–3, 169, 176, 319, 345
Houseflies, 236
Humerus, 222–3
Humidity, 257
Humus, 245, 246, 253, 296, 352
Hydrocarbons, 121–2
Hydrochloric acid, 53, 329, 332
Hydrogen, 17, 29, 122; ions, 15, 32; oxidation of, 22
Hydrolysis, 42
Hydrotropism, 210
Hygiene, 238–41; personal, 238; social, 238–9; community, 239–40
Hypermetropia (long sight), 152
Hyphae, 94, 95, 169, 233, 288
Hypochlorites, 241
Hypothalamus, 132, 157

Identification of specimens, 262
Ileum, 52
Ilium, 221, 226
Imago, 213, 214, 302, 350
Immunity, 241–2
Incisors, 46, 327
Indoleacetic acid (IAA), 210
Indusium 291
Infection, prevention of, 238–44
Inflammation, 241
Inflorescences, 179–80
Influenza, 232, 241
Inguinal glands, 74
Innominate bone, 221, 226
Insectivorous plants, 210
Insects, 299–302, 303; gas exchange, 121; growth, 213–14; skeleton, 215; insect borne diseases, 236–8; herbivorous, 267
Insemination, 169
Instar, 213
Insulin, 19, 144
Intervertebral discs, 218
Intestines, 52–4
Inulin, 18, 20
Invertebrate animals, 291, 315
Invertebrata, classification, 283
Iodine, 241, 329, 330
Ionising radiations, 244
Ions, 143, 336
Iron, 37, 109
Irritability, 146
Ischium, 221, 226

Jaws, 48–9, 220, 332
Jejunum, 52
Jenner, Edward, 242
Joints, 225–6
Joule, James, 55
Joules, 55

Kamimura, experiment on pancreas, 161, 345
Keratin, 321, 327
Keys, 278–82
Kidneys, 82, 140–44, 345; nephron, 141–2; artificial, 144
Knee jerk reflex, 157–8, 344
Koch, Robert, 227
Kwashiorkor, 20, 57, 58

Labium, 44, 302
Lactic acid, 109, 127–9, 225, 244
Lactose, 17–18
Landsteiner, Karl, 67
Lamina, 305
Larvae, 214, 320; and stimuli, 146–7, 342–3
Larynx, 117, 119
Leaf structure, 88–9, 305
Learning, 159
Leaves, 305, 306–8
Leeches, 295
Leeuwenhoek, Anton von, 227
Legumes, 181
Leguminous crops, 20, 36, 106
Lens of eye, 150–52
Leucocytes, 65, 241
Lice, 96, 337
Lichens, 97
Ligaments, 217, 221
Light, 257; and plants, 29, 31–3, 209, 210
Limb girdles, 221–2
Limbs, 221–3; pentadactyl, 222–3; arthropod, 223, 226
Lipids, 18, 108, 109
Lister, Joseph (Lord), 227, 240
Liver, 109; storage functions, 109; cells, 109; as regulator, 144
Liverworts, 282
Living things, characteristics, 15
Lizards, 319–21
Lobsters, 298
Locomotion and the skeleton, 215–26
Locust, 213, 254
Lungfish, 256, 353
Lungs,117–18, 128, 216, 339; gas exchange, 120; cancer, 122
Lymph and lymphatic system, 73–4; nodes, 74
Lymphatic tissue, 65
Lymphocytes, 65, 66, 242

Maggots, 214
Magnesium, 37
Maize and maize plant, 107, 194, 309, 348
Malaria, 96, 197, 236–8, 302, 347; control of, 238
Malpighi, Marcello, 69
Maltose, 17, 18, 42
Mammary glands, 327
Manson, Sir Patrick, 237
Marasmus, 20, 58
Measles, 58, 241
Medulla oblongata, 119, 157
Meiosis, 170, 171, 177, 187–9, 346; stages of, 187–9
Mendel, Gregor, 193
Menopause, 170
Menstrual flow and menstruation, 171, 176
Mepacrine, 238
Mering and Minkowski, experiment on pancreas, 161, 345
Mesentery, 52
Mesocarp, 177
Mesophyll cells, 88, 92
Metacarpals, 222
Metamorphosis, 212, 213, 302, 319; incomplete, 213, 299; complete, 214, 299
Metaphase, 185–6, 188–9
Metatarsals, 223, 224, 351
Metazoa, 293–303
Metre-kilogramme-second (MKS) system, 1
Micro-organisms, 227–8, 238, 241–4, 351, 352; discovery of, 227; destruction of, 240, 243–4
Microscopes, 3–6; and magnification, 5–6
Microvilli, 54, 142
Milk, 327
Millipedes, 298–9
Mineral salts, 20–21, 38–40, 54, 57, 306; uptake in plants, 85–6
Mistletoes, 96
Mitochondria, 170
Mitosis, 184–7; in animal cells, 185; in plant cells, 185, 187; stages, 185
Molars, 46, 327, 332
Molluscs, 297
Monocotyledons, 305, 306–10
Monocytes, 241
Monohybrid crosses, 193
Monosaccharides, 17, 329
Mosquito, 198, 327–8, 301–3, 353; control of, 237–8
Mosses, 289–90
Moulds, 94–5, 288–9
Mucor, 288–9
Mucous membranes, 241
Mucus, 53, 55, 97, 174
Mumps, 241
Muscles, 109, 127, 128, 215, 217, 219, 223–6, 349; intercostal, 118–19; skeletal, 134, 225; ciliary, 151, 152; thigh, 157, 344; calf, 223, 224; movement of skeleton, 223–4; structure, 224–5; cardiac, 225; smooth, 225; striated, 225; fatigue, 225
Mutation, 196–8
Mycelium(a), 94, 98, 169, 233, 289
Myonemes, 304
Myopia, 152
Myriapods, 298, 303

Nasal cavity, 117, 220
Nastic responses, 210
Navel, 175
Negroid race, 189, 190
Nematodes, 234, 235
Nephron, 141–2, 342
Nerves: optic, 152, 215; cranial, 157; spinal 157; autonomic, 157
Nervous system: central, 155–7, 163, 215; peripheral, 156
Neuronemes, 292
Neurons, 155–7, 215, 344
Nitrates, 36
Nitric acid, 36
Nitrogen, 18, 19, 36–7, 114, 144, 253; cycle, 36–7; atmospheric, 36; oxides, 121
Nitrogenous wastes, 145
Nitrous acid, 36
Nutrition: definition, 19; in green plants, 26–40; in animals, 40–55; holophytic, 93; saprophytic, 93–5; parasitic, 95; symbiotic, 97–8
Nymph, 213

Oak, 312
Occupations and energy, 56–7
Oesophagus, 51, 117
Oestrogen, 176
Oestrus and oestrus cycle, 169
Oil seeds, 106
Oils, 18, 20, 24–5, 106, 341
Olecranon process, 222
Oleic acid, 18
Olfactory epithelial cells, 154
Olfactory lobes, 156
Omnivores, 48
Oocyte, 173
Oogonia (egg cells), 172, 346
Operculum, 120, 121, 315, 340
Optic lobes, 157
Osmoregulators, 82

Osmosis,76–7, 79–84, 335; osmotic pressure, 77, 79, 84, 88, 90, 143, 342; osmotic gradient, 88; passive osmosis, 143
Ossicles, 152, 344
Ova, 164, 170, 297, 346 **see also** Eggs
Ovaries, 169, 170, 173, 176; in plants, 177, 181, 182, 345
Oviparity, 169
Ovulation, 169, 170, 172–4, 176
Ovules (in plants), 177, 345
Oxygen, 18, 31–2, 62, 66, 67, 72, 73, 86, 112, 137, 241, 257, 302, 323, 331, 333, 336, 338, 339, 340, 348, 352; gaseous exchange, 114–21; and energy, 122, 125
Oxyhaemoglobin, 66

Palate, hard, 220
Palm bones, 222
Palmitic acid, 18
Paludrine, 238
Pancreas and pancreatic duct, 52, 144, 161, 345
Pancreatic juice, 53
Paramecium caudatum, 292–3
Paramylum granules, 304
Parasites, 95–7, 254; ectoparasites, 95, 96, 97; endoparasites, 95, 96; fungal, 95; higher plant, 96; animal, 96–7; and disease, 97
Paratyphoid, 235, 242
Parental care, 175–6, 328; in birds, 325
Parthenogenesis, 169
Parturition, 175
Pasteur, Louis, 23, 126, 227, 240, 242
Pavlov, Ivan, 158
Pectoral (shoulder) girdle, 221–2
Pellicle, 292, 304
Pelvic (hip) girdle, 221–2
Penicillium and penicillin, 94, 169, 288
Penis, 169, 171, 174, 327, 345
Pepsin, 42–3, 53, 332
Peptide bond, 18
Perennials and perennation, 101, 102, 105, 166, 305; herbaceous, 101, 104; woody, 101, 305
Pericarp, 177, 181
Periodontal disease, 50
Peristalsis, 51, 157
Peristomium, 295
Petals 177, 179, 210, 345
Petioles, 305
pH, 14, 53, 329, 331; scale, 15, 250, 257; indicators, 15
Phalanges, 222, 223, 351
Pharynx, 117, 152, 296
Phenol, 241
Phenotypes, 192, 194, 199, 202
Phloem and phloem tubes, 75, 76, 84–5, 310–11, 336
Phosphorus, 18, 37, 76, 215, 334
Photoreceptors, 296
Photosynthesis, 15, 28–35, 83, 85, 86–90, 93, 98, 102, 125, 166, 208, 254, 304, 305, 331, 336, 339, 348
Phototropism, 210
Phytoplankton, 286
Phytophthora, potato blight, 95
Pie charts, 12, 262
Pine tree, 313
Pinna (ear trumpet), 152, 153, 327
Pituitary gland, 143, 169, 176
Placenta, 174
Plant kingdom, classification, 284
Plasma, 65–7, 86, 241; hormones, 67
Plasmagel, 291–2
Plasmasol, 291–2
Plasmodium parasite, 96, 237; control of, 238
Plasmolysis, 82
Platyhelminthes, 234, 294
Plumule, 181, 207, 347
Pneumonia, 58, 229, 240
Poikilotherms, 131–2
Poliomyelitis, 232, 241
Pollen, 177, 194, 345
Pollination, 177–8, 345, 347
Pollution, 273–5; of air, and health, 121–2
Polymorphs, 65, 66
Polypeptides, 18, 53
Polysaccharides, 18, 144
Pomes, 181
Population, 254, 268–73; fluctuation 254; distribution, 254–7; checks, 268; human, 270; increase, 270–2
Potassium, 37; hydroxide, 14, 16, 112, 331; pyrogallate, 14, 112, 331; sulphate, 253
Potatoes, 102–3, 165–6
Prawns, 298
Pregnancy, 174–5
Premolars, 46, 47, 327, 332
Progesterone, 176
Propagation, artificial, 166–7
Prophase, 185–6, 188–9
Prostate gland, 171, 174, 345
Prostomium, 295
Proteins, 18–20, 35–7, 43, 53, 55, 57, 106, 107, 109, 125, 145, 197, 321, 329, 333, 334, 342; lack of, 20, 58
Prothallus, 291
Prothrombin, 66
Protoplasm, 18, 203
Protozoa, 291–3, 304; and disease, 232–3, 237
Pseudopodium, 292
Pteridophyta, 290–91
Ptyalin, 42, 53
Puberty, 170, 176, 217
Pubis and pubic bones, 221, 226
Pulse and pulse rate, 74–5
Pulvini, 210
Pupa, 214, 302, 349
Pus, 241
Pylorus, 52–3
Pyorrhoea, 50
Pyrenoids, 287, 304

Quinine, 238

Radicle, 181, 207, 210, 348
Radius (bone), 222
Rainfall, 255, 256–7; measurement, 256–7
Ramenta, 290
Rats, 325–8
Reafforestation, 253
Receptacle (of plant), 176, 181
Receptors, 146; light, 147–52; sound and gravity, 152–4; chemical, 154; skin, 154–5; stretch, 157
Rectum, 52, 55
Reflex action, 157–9; conditioned reflex, 158–9
Rennin, 53
Reptiles, 319–21
Respiration, 17, 19, 206; cellular, 125–6; anaerobic, 126–8; energy flow, 128–9; and photosynthesis, 129
Retina, 151, 152, 344
Rhesus factor, 67–8, 334
Rhizomes, 102, 166, 290
Rhizopus and rhizoids, 94, 95, 288, 289
Ribose, 17
Ribs, 215, 216, 218, 350
Ringworm, 233, 236
Root system, 83–4, 85, 251, 306, 308, 331, 336, 337, 348; tap roots, 105; root tubers, 105; adventitious roots. 103, 165, 290; growth, 207
Ross, Sir Ronald, 237
Roughage, 55
Rusts (of plants), 234

Sacrum, 221, 351
Salinity, 257
Saliva, 52–3, 241, 302, 332
Sampling, 12–13, 261
Sand, 249, 250, 352
Saprophytic nutrition, 93–5, 98, 230, 288, 304
Scapula (shoulder blade), 221, 223
Schizocarps, 181
Seaweeds, 286
Seeds, 181–2, 347; food storage in, 106–8; classification, 181; germination, 204–7
Selection: natural, 198
Semen, 171, 172, 174
Sepals, 177–80, 345
Septic tanks, 239
Septicaemia, 74
Sex determination, 194–5
Sex linkage, 195
Sexual cycles, 169–71
Sharks, 317
Shrubs, 305, 310
S.I. units, 1–3, 55
Skeleton: mammalian, 215–18; tissues, 216–17; axial, 215, 218–21; appendicular, 215, 221–3; movement by muscles, 223–6
Skin, 154–5, 241, 344; functions of, 134–8
Skull, 215, 216, 218, 220, 226, 351
Slope of the land, 258
Smallpox, 236, 241
Smell, 154
Smog, photochemical, 122
Smoking of meat, 244
Smuts (of plants), 234
Snails, 297
Snakes, 319
Sodium: chloride, 21; hydroxide, 14; hypochlorite, 241
Soil, 245–50; organisms, 247; types, 249; acidity and alkalinity, 250; fertility, 250–3
Soot, as pollutant, 121
Sori, 291
Spawn, 318, 319
Spallanzani, Lorenzo, 227
Sperm(s) (spermatozoa), 164, 169–74, 176, 192, 195, 296, 317, 319, 324, 327, 345, 346
Spermatids, 174
Spermatocytes, 174
Spermatogonia, 173–4
Spermatophore, 169
Spiders, 297
Spinal cord, 155, 156, 216; reflex, 157–8, 344
Spiracle, 121
Spirilla, 229
Spirochaetes, 229, 236
Spirogyra, 168, 287–8
Sporangiophores, 169
Sporangium, 169, 289, 290
Spores, 168–9, 288-90
Stamens, 177–9, 312, 345
Staphylococci, 229, 241

Starch, 18, 20, 23, 26–7, 29, 35, 41–2, 105, 181, 287, 304, 330–32; digestion, 52–3; storage, 106
Stearic acid, 18
Stem tubers, 102–3
Stems, 307–9
Sterilisation, 241
Sternum (breastbone), 215, 218, 350
Stevin, Simon, 1
Stickleback, 315–17
Stigma, 177, 180, 194, 345
Stimuli, 146–7, 154, 158, 215; and plant growth 208–10; directional, 208; non-directional, 210
Stomach, 51–2
Stomata, 88–92
Storage, 101–8, 108–9
Streptococci, 229, 241
Strip cropping, 252
Style (in plants), 177
Subsoil, 245
Succus entericus, 53
Suckling, 176, 327
Sucrose, 17, 18, 20, 84–5, 329, 335–6
Sugar, 101, 104, 244; non-reducing, 18, 24, 329, 335; reducing, 18, 23, 26–7, 42, 329; breakdown, 123, 125
Sugar cane, 104
Sulphur, 18, 37; dioxide, 121
Sunflowers, 309, 310
Surface area to volume (SA/V) ratio, 62–5, 135, 176, 341
Sweat and sweat glands, 134, 135, 138, 341
Swimmerets, 298
Symbionts, 98
Symbiotic nutrition, 97–8
Synapses, 155
Synthesis, 30, 348
Syphilis, 236

Tadpoles, 212, 320
Tapeworms, 97, 234, 294, 337
Tarsals, 222–3, 351
Taste, 154, 344; buds, 154
Taxis (tactic response), 146–7
Tears, 158, 241, 344
Teeth, 46–8; types, 46; internal structure, 47; decay, 49, 50; dental care, 49–50; diseases, 50
Telophase, 185–6, 188–9, 346
Temperature, 256; control, 131
Tendons, 217, 351
Termites, 98
Terracing, 252
Testa, 180
Testes, 161, 171–2, 176, 345
Testosterone, 176
Thorax and thoracic cavity, 117–19
Thrombocytes, 66
Tibia, 222
Ticks and tick fever, 97
Tissue fluid, 73
Tongue, 154
Topsoils, 245, 248, 251, 352
Trachea(e), 117, 119, 121
Tracheoles, 121
Translocation, 35, 101
Transpiration, 88–92
Trees, 262–6, 310–14; associated community, 264–6
Trichonympha, 98
Trichocysts, 292
Tristearin, 18
Tropisms, 210
Trypanosoma, 96, 337
Tubers, 102–3, 165–6; root tubers, 105
Turbellaria, 294
Turgor in plant cells, 82–3, 210
Tympanum (eardrum), 152–3, 344
Typhoid, 229, 235, 239, 242

Ulna, 220, 222–3
Ultrafiltration, 141
Umbilical cord, 174–5
Universal Indicator, 15
Urea, 67, 109, 134, 142, 342
Ureter, 140
Urethra, 140, 171–2, 174
Uric acid, 67
Urine and urinary system, 82, 138, 140–2, 161, 171–2, 342
Uterus, 169–70, 172–5, 327
Utriculus, 154

Vaccines and vaccination, 242
Vagina, 170, 172, 174, 175, 327
Valine, 198
Variation, 189–92, 346; continuous and discontinuous, 190–2, 346
Vas deferens, 171, 345
Vasopressin, 143
Veins, 71–3; pulmonary, 73, 120, 342; innominate, 73; renal, 140, 141; hepatic portal, 73, 144
Venae cavae, 73
Venereal disease, 236
Vents, in lizards, 320
Venules, 72, 120
Vertebral column and vertebrae, 218–20, 221, 350
Vertebrate animals, 315–28
Vesicles, seminal, 171, 174, 345
Vibrios, 229
Villi, 54
Viruses, and disease, 229, 231–2
Vitamins, 22–3, 54, 57; vitamin A, 22, 109; vitamin B, 22, 109; vitamin C, 22 57, 329; vitamin D, 22, 57, 109; deficiency, 57–8
Viviparity, 169, 327
Vocal cords, 117
Vulva, 172

Water, 21–2, 82–4, 139–40; loss in plants, 86–92, 337; balance, 139–44; and root growth, 209–10; supply, 239–40; drinking, 240; properties, 13–14
Waterborne disease, 235
Wattles, 321, 322
Waxes, 18
Wheat, 106
Winds and wind measurements, 257–8
Wings and vestigial wings, 192, 322–3, 324
Wood, 208
World Health Organisation, 240
Worms, 266, 294–6; skeleton, 215; and disease, 234–5, 240; larvae, 235; segmented, 294–6
Wright, Sir Almroth, 242
Wrist bones, 222

Xylem and xylem tubes, 76, 84–5, 88, 91, 208, 310–11, 337

Yeast, 126–7, 204, 288, 340, 348

Zooplankton, 291
Zygospore, 289
Zygote, 174, 291

Acknowledgements

The authors and publishers wish to acknowledge the following photograph sources:

Aerofilms Ltd, figs 13.9, 13.10 A. C. Allison & G. N. C. Crawford, fig 9.14(a) Heather Angel, figs 3.5, 3.11, 5.11(b), 8.1, 15.24(a), (b), (c), 15.26(b), (c), 17.5 Peter Baker, fig 9.21 Dr. Alan Beaumont, fig 9.23 J. Brownbill, fig 7.3 Cambridge Scientific Instruments, fig 1.54 Camera Press, figs 9.8(b), 9.8(c) J. Allan Cash, figs 6.12, 9.8(a) Bruce Coleman, figs 2.16, 2.34, 8.9, 9.1(a), 9.2(a), 9.5, 10.11(a), 11.17(a), 15.23(c), 15.26(a), 15.26(d), 16.6(a), 16.8 Gene Cox, figs 6.10, 6.18, 8.18 Peter A. Dean, figs 9.22(a), 10.11(b), 11.5 *Farmer & Stockbreeder*, figs 9.20, 9.22(b) *Farmers Weekly*, fig 9.25 F. A. O., fig 8.27 T. Gerrard & Co, fig 1.51 Glass Crops Research Institute, figs 1.24, 1.25, 1.26 Dr. Hannay, fig 2.28 Philip Harris Biological Ltd, fig 11.10(b) Horniman Museum, London, figs 1.44(a), 1.45, 1.46 Eric J. Hosking, figs 16.6(b), 16.18, 16.22 Imperial Chemical Industries Ltd, fig 3.4 Dr. Kettlewell, Department of Zoology, Oxford, figs 9.19(a), 9.19(b) Dr. Morley, figs 1.56, 1.57 L. Hugh Newman, Natural History Photo Agency, fig 14.2(b) Dr. Knut Schmidt-Nielsen, figs 10.1(a) & (b) M. Nimmo, figs 16.12, 16.13, 16.17, 16.21 N.N.I.S., fig 4.11 Ron Oulds, fig 8.4 Oxfam, fig 1.7 Queen Charlottes & Chelsea Hospital, fig 8.21 R.T.H.P.L., fig 6.11 Rothampstead Experimental Station, fig 1.30 Shell Photographic Service, figs 1.36, 12.7 Topham/Coleman, fig 3.7 C. James Webb, figs 2.12, 15.23(b), 16.7(a), 16.7(b), 16.9 Thomas A. Wilkie, fig 14.2(a) Zoological Society of London, fig 9.24 Taken from the publication McLeish & Snoad *Looking at Chromosomes*, figs 9.4(a), (b), (c), (d), (e), (f), 9.6(a), (b), (c), (d), (e), (f)

Cover Photograph Heather Angel

The authors and publishers wish to thank the following who have kindly given permission for the use of copyright material:

Academic Press Inc. (London) Limited, for a Key to the Order Hymenoptera from *Introduction to Experimental Ecology* by Lewis and Taylor. The Biochemical Journal for fig. 7 from the paper by Butcher, Pentelow & Woolley in *Biochem. J.* 21: 1423–1435, 1927 British Ornithologists' Union for a figure in an article by May published in *Ibis*, 91:24, 1949. Cambridge University Press for the Key to 'Species of the Genus Ranunculus' from *Flora of the British Isles* by Clapham, Tutin and Warburg. Churchill Livingstone for 'Graph of drinking salt water from *Textbook of Physiology and Biochemistry* 3rd Edition, 1956, by Bell, Davidson and Scarborough. The Conseil de L'Europe for two figures from *Naturopa*, No.27, 1977. Faber and Faber Limited for illustrations from *The World of a Tree* by Arnold Darlington. W. H. Freeman & Company for adaptation of graph from p. 144 of *The Human Thermostat* by T. H. Benzinger. Copyright © 1961 by Scientific American, Inc. All rights reserved Heinemann Educational Books Limited for figs 7.2 and 7.8 from *Man Against Disease* by Clegg and Clegg. Longman Group Limited for fig 1.48 adapted from 'Peristalsis: how food is moved along the gut'; fig 1.52 'How the surface area of the gut wall is increased'; fig 1.47 'Fluoride concentration of water supply (ppm)'; and fig C 'Bomb Calorimeter' all from *Nuffield Biology Text III*. 'Graph of uptake of Bromide Ions' from *Nuffield Biology Text IV*. Two figures from *Nuffield Biology Text IV, Pupils Book*, New Revised Edition, and one figure from the previous edition. Fig 10.4 adapted from *Biology for Tropical Schools* by Stone and Cozens. D. G. MacKean for fig 11.15 'Movement of the forearm muscles' from *Introduction to Biology* published by John Murray (Publishers) Limited. W. B. Saunders Company for extract from *Introduction to Embryology* by Professor B. I. Balinsky, 2nd Edition, 1965, reprinted by permission of the authors and publishers.

The publishers also wish to thank Illustra Design for the line illustrations.

The authors wish to acknowledge the stimulation given to the teaching of Biology by the work of the Nuffield Biology Teaching Project and the Schools Science Project. Some of the ideas and experiments in this book are based on those projects

The publishers have made every effort to trace copyright holders, but if they have inadvertantly overlooked any they will be pleased to make the necessary arrangement at the first opportunity.